D1270243

Nuclear Wastelands

Nuclear Wastelands

A Global Guide to Nuclear Weapons Production and Its Health and Environmental Effects

By a Special Commission of International Physicians for the Prevention of Nuclear War and The Institute for Energy and Environmental Research

edited by Arjun Makhijani, Howard Hu, and Katherine Yih

The MIT Press
Cambridge, Massachusetts
London, England

This book was set in Palatino by Asco Trade Typesetting Ltd., Hong Kong and was printed and bound in the United States of America.

Library of Congress Cataloging-in-Publication Data

Nuclear wastelands: a global guide to nuclear weapons production and its health and environmental effects / by a special commission of International Physicians for the Prevention of Nuclear War and the Institute for Energy and Environmental Research; edited by Arjun Makhijani, Howard Hu, and Katherine Yih.
p. cm.
Includes bibliographical references and index.
ISBN 0-262-13307-5 (alk. paper)
1. Nuclear weapons plants—Environmental aspects. 2. Nuclear weapons—Testing—Environmental aspects. 3. Nuclear weapons plants—Health aspects. 4. Nuclear weapons—Testing—Health aspects. I. Makhijani, Arjun. II. Hu, Howard. III. Yih, Katherine. IV. International Physicians for the Prevention of Nuclear War. V. Institute for Energy and Environmental Research (Takoma Park, Md.)
TD195.N85N83 1995
363.17′99—dc20 95-945
CIP

Contents

Illustrations and Tables

ILLUSTRATIONS

TABLES

Foreword

Much of the public perceives the nuclear nightmare to be over with the end of the Cold War. Many have laid this issue on the back burner of their social concerns. Yet the reality is starkly different. Military budgets remain high, and secret weapons laboratories are continuing the modernization of nuclear weapons.

The threat of proliferation, a widely acknowledged problem, is a catalytic, self-begetting process. Acquisition of nuclear know-how by some countries promotes intense urgency among their neighbors to do likewise. This is abetted by past superpower propaganda that nuclear deterrence kept peace in Europe during more than four decades of cold war. The same logic is applicable to the Middle East as well as to South Asia. The ongoing spread of nuclear technology fosters regional instability and increases the likelihood of the use of these weapons during local conflicts.

Proliferation is facilitated by the growing global stockpiles of nuclear material. The ongoing dismantling of weapons leaves behind thousands of plutonium pits (the core nuclear component of the warhead) and other fissile materials. A typical pit consists of about 3 to 5 kilograms of plutonium. The disposed of plutonium presents enormous challenges.

The collapse of the Soviet Union adds an awesome dimension to proliferation. The once mighty empire has splintered into contending nations and into numerous fractious ethnic groups. Economic chaos now sweeps this huge land mass, corruption is widespread, and frontiers are increasingly porous. Nuclear widgets and expertise—once carefully guarded secrets of national security—may soon become lucrative commodities for sale to the highest bidder. There are ominous signs that a black market in nuclear materials may already be forming.

NUCLEAR LEGACIES

While no nuclear weapons have been detonated in war since Hiroshima and Nagasaki, a kind of secret low-intensity radioactive warfare has been waged against unsuspecting populations. In the name of protecting national security, health and environmental safety have been sacrificed. The Physicians for Social Responsibility (PSR) designated radiation pollution as a "national public health and safety emergency," a kind of "creeping Chernobyl."[1] This is happening not in one place but at many sites, slowly and steadily affecting the health and lives of people in nuclear weapons states, in the countries where uranium is mined, and, indeed, throughout the world due to fallout from nuclear weapons testing in the atmosphere.

In the last decade a deluge of disclosures has documented egregious mismanagement, safety violations, environmental contamination, nuclear accidents, unsafe working conditions, and equipment failures at weapons plants in the United States. The same has happened since 1989 in the former Soviet Union. The Hanford Site, a 570-square-mile U.S. government reservation along the Columbia River, is perhaps the most polluted place in the Western world. Russia's plant in the Urals near Chelyabinsk may be the most radioactively contaminated area on Earth.

The U.S. government is facing enormous technical challenges in disassembling the nation's nuclear arsenal and coping with the growing accumulation of garbage of the atomic age. Millions of pounds of highly radioactive reactor fuel are sitting in Energy Department storage pools that are rusting and spreading radioactivity. Until the late 1980s no one seriously considered the possibility that plutonium would not be needed for military purposes and would be too expensive to use as an energy source. Today both prospects have become a reality, and plutonium is now a terrible security and environmental liability. Yet there is currently no accepted plan for long-term disposal of plutonium. Notwithstanding the expenditure of $4 billion on the search for a permanent high-level radioactive waste repository, no solution is presently in sight. The Department of Energy indicates that the earliest date for a permanent underground waste storage repository will be sometime around the year 2013.

Nuclear waste is accumulating and has nowhere to go. In large measure this is a disease without a visible cure. High-level nuclear wastes can be dangerous for many thousands of years. The U.S. Environmental Protection Agency insists on guarantees that the material stay isolated for 10,000 years. Even 10,000 years may not be enough, because many radionuclides have half-lives longer than that. This is totally outside the range of human vision and perhaps beyond the capacity of human imagination.[2]

If the situation is grim in the United States, it is a mounting catastrophe in Russia. In order to maintain nuclear parity, the Soviet

1. PSR 1988.

2. Erickson 1994.

government callously disregarded environmental and human safety, especially in the early years of its nuclear program. It has dumped hundreds of tons of nuclear garbage into the Sea of Japan, despite an agreed 30-year worldwide ban against such ocean dumping of radioactive waste. And Russia is not the only country committing nuclear mayhem. The action of others is shrouded in secrecy.

SECRECY AND DECEPTION

The public has tolerated nuclearism in no small measure because it has been kept in abysmal ignorance and deliberately deceived. These practices were not unique to any one country. Every government with a nuclear establishment pursued similar policies, justified by the dictates of national security. Archival material was gratuitously classified as secret, environmental and dosimetric data were kept poorly, journalists were fed slanted intelligence, and public relations pros spewed propaganda to soothe an anxious citizenry.

In the 1950s intense nuclear testing with multimegaton devices at the Marshall Islands in the Pacific roused public anxieties. The U.S. government issued assurances that the radiation hazard was no worse than that involved in a chest X ray. These words lost credibility when a Japanese fishing trawler encountered a radioactive cloud, causing acute radiation sickness of all on board and the death of one sailor by the time they reached their home port two weeks later.[3]

U.S. authorities argued that strontium-90—a component of fallout from atmospheric tests—posed no dangers, until scientists and physicians demonstrated the presence of elevated concentrations of this isotope in the deciduous teeth of infants. Strontium-90 accumulates in bones in the same way that calcium does, increasing the risk of certain cancers.

The Atomic Energy Commission (AEC), the agency in charge of testing, issued pamphlets to reassure ranchers living downwind from the Nevada Test Site "not to worry if their Geiger counters went crazy."[4] Believing the government, many residents in Utah and Nevada stayed outdoors to observe the atomic flash and the radioactive cloud as it drifted by, thereby subjecting themselves unsuspectingly to high doses of radioactive fallout. When thousands of sheep died from radiation near the Nevada-Utah line, the government falsified the evidence and denied any relation to the ongoing nuclear tests nearby.

The U.S. government even deceived the public about the very detectability of underground nuclear explosions. In September 1957, after the first underground nuclear test in the Nevada flats, the AEC claimed that the explosion could not be detected beyond 250 miles. The

3. Lapp 1958.

4. Shusterman 1983.

objective was to make it seem that detection required seismic ground stations so numerous and intrusive that the Soviets would balk at a test ban agreement. This deception was rapidly laid to rest by an investigative reporter who found that seismic stations throughout the United States detected this explosion—including one station in remote Fairbanks, Alaska, more than 3,680 kilometers away.[5]

A spate of recent revelations has dismayed an already cynical public. In fact, the U.S. government conducted experiments on people who were completely unaware that they were being used as guinea pigs.[6] These experiments reminded Hazel O'Leary, the present U.S. Secretary of Energy, and millions of others of Nazi experiments during World War II. According to officials, over a thousand people were exposed to radiation, often without informed consent, for medical and military research. Plutonium was injected into nineteen patients in one of several experiments with unknowing human subjects. In another, researchers gave radioactive pills to about 800 pregnant women.[7] In an effort to develop a radiation weapon—one that would kill or incapacitate enemy soldiers by fallout—the Army and what was then the Atomic Energy Commission tested weapons charged with large quantities of radioactivity in the atmosphere. These deliberate releases of large amounts of radiation were conducted as part of a huge program involving 250 experiments in the years from 1944 to 1973 conducted under both Democratic and Republican administrations.

Secretary Hazel O'Leary, who publicly acknowledged the government's role in these experiments in December 1993, has initiated an investigation into how widespread the experiments were, how often the tests violated standards for human research, and to what extent people suffered harm. At the time of this writing, more than thirty million pages of documents remain sequestered out of reach of public scrutiny. The work of fully uncovering this history will be long indeed.

IPPNW

The International Physicians for the Prevention of Nuclear War (IPPNW) emerged in the early 1980s to educate the public on the consequences of nuclear war and nuclearism. Recipient of a Nobel Peace Prize in 1985, it now encompasses a constituency of 200,000 health workers in 83 countries. The mission of IPPNW has been to inform and rouse people to the threat of the atomic age. To overcome a dearth of information, in December 1988 IPPNW created a commission to study the health and environmental effects of nuclear weapons

5. *I. F. Stone's Weekly* 1958.

6. *Lancet* 1994.

7. *The New York Times*, 17 December 1993.

production, testing, and deployment. It retained the Institute for Energy and Environmental Research (IEER) as its principal scientific consultant. IEER is well known for the excellence of its scientific and technical work in assessing the environmental problems arising from nuclear weapons production. It has done many analyses of plants in the U.S. nuclear weapons complex.

IPPNW and IEER have issued two previous books, *Radioactive Heaven and Earth*[8] and *Plutonium: Deadly Gold of the Nuclear Age.*[9] These have been scrupulously exacting and scientific. At the same time they eschewed academic jargon. The aim was to produce reports that could be grasped by an intelligent layperson.

The current book, *Nuclear Wastelands: A Global Guide to Nuclear Weapons Production and Its Health and Environmental Effects,* comes at a most propitious time, when the public is no longer ready to trust government propaganda on nuclear issues. The book, with seventeen distinguished authors, demonstrates the vastness of the nuclear weapons enterprise and documents the hazards at every step along the way. The danger to health begins with the substantial risks of lung cancer and silicosis faced by miners of uranium ore. It is not widely known that in terms of damage to workers, this has probably been the deadliest part of nuclear weapons production. Many of these highly exploited workers live in Third World countries having poor working conditions. The production of weapons-grade fissile material exposes workers to noxious chemical hazards. Every step of the weapons production process creates environmental hazards, the culmination of which is the accumulation of highly radioactive waste, for which no satisfactory disposal method has been found.

Our knowledge of the consequences of nuclear weapons production in the United States is now beginning to take some shape due to the Freedom of Information Act, congressional inquiries, and the declassification of documents. Our knowledge of the situation in Russia, while far from comparable to that for the United States, is improving. But secrecy still holds sway in the United Kingdom, supported by the Official Secrets Act. Secrecy is worse in France and worst in China, where our knowledge of the health and environmental effects of weapons production is very sketchy indeed.

This third and final volume in the collaborative effort of IPPNW and IEER is one contribution toward educating the public about the true consequences of nuclear weapons production. The deadly nuclear shadow will not vanish without public education, arousal, and involvement. Politicians rise to the challenge only when a powerful, informed public opinion perseveringly clamors for change.

8. IPPNW and IEER 1991.

9. IPPNW and IEER 1992.

We in the health profession have a profound historic responsibility to generations yet unborn to stop the madness of nuclear weapons production and testing for good and to make rapid progress toward complete nuclear disarmament. In the atomic age our life gains meaning when we join with others to heal a sick planet.

Bernard Lown
Cofounder and Emeritus Co-President,
International Physicians for the Prevention of Nuclear War
June 1994

Preface

Even after the end of the Cold War, the news brings regular reminders that the world has yet to learn the dangerous lessons of 50 years of nuclear weapons production and testing. Despite billions of dollars of projected cleanup costs looming over the nuclear weapons powers and a terrible legacy of environmental contamination and damage to human health, the twin threats of nuclear war and the health hazards of nuclear weapons production persist. Whatever the outcome of the Nuclear Non-Proliferation Treaty review conference in April and May 1995, proliferation dangers will persist as well. North Korea is only the most publicized recent example. There has been considerable reluctance in Ukraine to give up the strategic nuclear weapons on its territory at the time of the breakup of the Soviet Union. Other warning signs in this period include:

• In September 1993 the *New York Times* reported that high on the agenda of India's Hindu nationalist Bharatiya Janata Party is to declare the country a nuclear weapons state.[1] At the same time, Pakistan's nuclear weapons development continues.

• In October 1993, Dr. Bruce Blair, senior fellow at the Brookings Institution, noted that an automated system had been designed by the Russians that would retaliate against a Western nuclear attack with a massive nuclear weapons strike, even if top military commanders had been killed and the capital incinerated.[2]

• Also in October of that year, while the authors and many contributors to this book attended the World Congress of the International Physicians for the Prevention of Nuclear War, China exploded a nuclear weapon, despite a global moratorium on such tests by the other nuclear powers.[3]

1. *New York Times*, 17 September 1993.

2. As quoted in the *New York Times*, 8 October 1993.

3. *New York Times*, 6 October 1993.

• A November 1993 report in the *Journal of the American Medical Association* tied an excess of thyroid tumors to exposure to radioactive isotopes generated from U.S. tests of nuclear weapons. The article, by R. A. Kerber and his colleagues, was based on a study of 4,818 people potentially exposed as schoolchildren to fallout from nuclear weapons tests conducted at the Nevada Test Site in the 1950s.[4]

• In December 1993, Secretary of Energy Hazel O'Leary announced that the U.S. government had conducted radiation experiments on its own people, sometimes without informed consent.

There have been occasional encouraging signs that the lessons of the past 50 years were in fact finally being appreciated by some. To cite two examples:

• In May 1993, the World Health Assembly, the governing body of the World Health Organization, passed an IPPNW-supported petition calling on the International Court of Justice in the Hague to issue an advisory opinion on the following question:

> In view of the health and environmental effects, would the use of nuclear weapons by a state in war or other armed conflict be a breach of its obligations under international law including the WHO Constitution?

By June 1994, at least eight nations, including Kazakhstan, Mexico, Moldova, North Korea, Papua New Guinea, Solomon Islands, Sweden, and Ukraine, had submitted briefs to the World Court in support of IPPNW's position that the use of nuclear weapons should be declared illegal.

• On 15 July 1994, General Charles A. Horner, who as leader of the North American Aerospace Defense Command is responsible for defending the United States and Canada from nuclear attack, publicly called for the elimination of all nuclear weapons, stating that "the nuclear weapon is obsolete. . . . I want to get rid of them all. . . . Think of the high moral ground we secure by having none."

Unfortunately, the "high moral ground" of nuclear abolition remains far from an achieved reality. Despite the waning of superpower confrontation, the world is still burdened with tens of thousands of nuclear weapons, mounting surpluses of weapons-usable plutonium and highly enriched uranium, and lands, aquifers, rivers, lakes, and seas polluted by a multitude of poisons, even apart from the potential devastation of nuclear war. What price has the world paid for ushering in the age of nuclear weapons? And what risks will future generations inherit? These two questions led International Physicians for the Prevention of Nuclear War (IPPNW) to create a research commission in

4. Kerber et al. 1993.

December of 1988. In early 1989, IPPNW chose the Institute for Energy and Environmental Research (IEER), a leader in independent research on the effects of U.S. nuclear weapons production, as its principal consultant and partner in the enterprise that has led to this book.

The objective of the commission has been to describe the health and environmental effects of nuclear weapons production and testing in terms that are both scientifically accurate and accessible, in order to help the public understand the true costs of building and testing nuclear weapons. Although the Berlin Wall has fallen and the Soviet Union and the United States have begun reducing their arsenals, other events have dramatically demonstrated the continuing freshness and importance of these lessons. Indeed, the list of states considering developing their own nuclear weapons is growing, making the work of this commission ever more relevant. The commission has striven both to testify to the destruction and contamination wrought by the superpowers' nuclear arms race and to warn of the threats that loom as other countries aspire to nuclear weapons status. As during the Cold War, the damage will be done even if not a single additional weapon is exploded in conflict. Destruction before detonation is the hallmark of nuclear weapons production.

This book represents the third and most encompassing publication of the commission. It reviews the entire process of nuclear weapons production, from uranium mining and milling through plutonium reprocessing and weapons assembly. It identifies the major pollutants resulting from these processes, presents a country-by-country review of the major nuclear weapons production sites, and reviews all available information on their emissions and health and safety records.

In confronting such a massive subject, we deliberately chose to exclude topics that were covered in our two earlier books. Our first publication, *Radioactive Heaven and Earth*, explores the health and environmental effects of atmospheric and underground nuclear weapons testing. Thus, except for new information regarding fallout from Chinese nuclear tests, the present book does not address this topic. Similarly, our second publication, *Plutonium: Deadly Gold of the Nuclear Age*, looks closely at the hazards surrounding the production of plutonium and the storage of high-level radioactive wastes associated with nuclear weapons production, which therefore are not considered in such detail here.

In spite of increasing revelations in recent years about the historical origins and motivations behind the development of nuclear weapons, far too much information remains hidden. Indeed, secrecy, deception, and outright lies characterize the nuclear weapons industry. This secrecy conceals not only information on the financial costs and technological processes that underlie the production of nuclear weapons, but also information relevant to gauging the impact of these industries on the environment and human health—information that its keepers no doubt recognize would threaten the public image of their enterprises. Trying to assess the situation from publicly available information is a difficult, costly, and uncertain undertaking. The quality and quantity of information available is uneven, rendering coverage of various countries similarly uneven.

The difficulties arising from secrecy are compounded by barriers of language and geography. Nonetheless, awareness of the health and environmental problems due to nuclear weapons production and testing has grown and spread dramatically in recent years, and many local movements committed to obtaining accurate information have sprung up in the nuclear weapons states. IPPNW and IEER have used information from such sources, including materials translated for this project from Russian into English. We have made considerable efforts to review this data critically, but the fast-moving nature of the field and the far-flung nuclear weapons enterprise mean that we have not been able to tap many potential sources. To mitigate the obstacles to extensively examining regional news and information sources in languages other than English, this book has used U.S. State Department compilations of news from around the world, especially in countries where secrecy is still great. (Names of nuclear facilities are translated in different ways in various sources; for the sake of clarity we have tried to use one name consistently for each facility.)

As experience with U.S. data shows, official documents are crucial for investigating the issues raised here, but even a total release of government reports would not reveal the full impact of producing nuclear weapons. Even in the United States, by far the most open of the nuclear powers, official data are generally incomplete, often of poor quality, and sometimes simply wrong. Cover-ups and deception have been common. Thus, this book should be viewed only as a starting point to understanding the vast global nature of the health and environmental legacy of the nuclear weapons industry. Most of the information in the book is updated to that available in sources published through the first half of 1994. Some data are more recent, updated to about November 1994, while some information is as of the early 1990s, in cases where no more recent information is available.

One of the great challenges the world over is to document with ever greater certainty the extent of contamination and potential health threats. The principal purpose of making this initial effort is to empower all those who would take up that challenge.

Howard Hu
Arjun Makhijani
Katherine Yih

Acknowledgments

Production of such a large book with so many authors spread all over the world requires immense coordination and logistical effort, especially in the editing phase. The greatest effort in this regard was made by David Kershner. He gave of himself totally in keeping track of a huge list of references, various draft versions in process with authors around the world, figures, photos, and tables. Coordination of the production of the first review draft as well as a preliminary edit were done by Marc Miller. This book was conceived and a first outline for it was prepared in 1989, when Anthony Robbins of IPPNW was the director of this project to investigate the health and environmental effects of nuclear weapons production worldwide.

Checking, copyediting, proofreading, and word processing of the manuscript were also done by Bret Leslie, Julie Barnet, Freda Hur, Diana Kohn, Anne Bryant, Tessie Topol, Noah Sachs, and Annie Makhijani. Lois Chalmers provided research assistance.

As with most projects of this complexity, scope, and length, many individuals and organizations around the world helped by providing research materials, reviewing drafts, and making suggestions. We are grateful to all of them. In particular, we would like to thank Philip Crouch, who provided invaluable information for both chapters 4 and 5; Lydia Popova and David Rush for their comments on chapter 7; Bruno Barrillot and Mary Davis for their comments on chapter 9; Dingli Shen for his comments on chapter 10; and Masa Takubo for his helpful comments on the entire manuscript. Nick Thorkelson prepared the maps and some other figures.

We also gratefully acknowledge the following individuals for their assistance: Robert Alvarez, Lenore Azaroff, Till Bastian, Abraham Béhar, Albina Beichminova, Alex Belenki, Alexander Bolsunovsky, Marc Chao, Tom Cochran, Marina Degteva, Robert Del Tredici, Alexander Emelyanenkov, Murray Feshbach, Richard Fieldhouse, Lachlan Forrow, Ellen Ginzbursky, Eloi Glorieux, Leanne Grossman, Xanthe Hall, Joshua Handler, Frank von Hippel, Vladimir Iakimets, Gulsim Kakimjanova, Todd Karl, Victor Khokhryakov, Daryl Kimball, Eliza

Klose, Mira Kossenko, Randy Kritkausky, Terry R. Lash, Linda Lehman, Richard M. Levine, John Lewis, Annie Makhijani, Khanyiso Mbulawa, Larissa McMahan, Vladimir Miheev, Natalia Mironova, Scott Monroe, Raj Mutalik, Robert S. Norris, Diane Noserale, Marlene Odell, Alexander Penyagin, D. J. Peterson, Ilya Popova, Lydia Popova, William M. Potter, Steve Raymer, Michael Renner, Lynne Richards, Jordan Richie, Paul Robinson, Lyubov Rubinchik, David Rush, Randy J. Rydell, Mycle Schneider, Katherine Schultz, Vyacheslav Sharov, Dingli Shen, Paul Soler-Sala, John Spalding, John Sturino, Valery Thomas, David Wright, Alexei Yablokov, Elena Zhukovskaya, and Elvina Zlobinskaya.

We are especially grateful to the activists of the Military Production Network, who have intimate knowledge of U.S. nuclear weapons sites, for their assistance: Beverly Gattis, Trish Neusch, and Doris Smith (Pantex Plant); Brian Costner (Savannah River Site); Beatrice Brailsford (Idaho National Engineering Laboratory); Ralph Hutchison (Oak Ridge); Chris Brown (Nevada Test Site and Yucca Mountain); LeRoy Moore and Tom Rauch (Rocky Flats Plant); Lance Hughes (Sequoyah Fuels); Bill Mitchell (Hanford Site); and Lisa Crawford (Fernald). Other members of the Military Production Network, too numerous to list here, were also of tremendous help.

It has been a real pleasure to work with the editors responsible for this book at the MIT Press: Madeline Sunley, the acquisitions editor, and Janet Abbate and Matthew Abbate, the manuscript editors. Their professionalism, patience, courtesy, and consideration have helped sustain our energy through the long process leading from contract negotiations to putting the finishing touches on the final manuscript. We are very thankful to all three of them.

Finally, this book would not have been possible without the generous financial support of the following: The W. Alton Jones Foundation, The John D. and Catherine T. MacArthur Foundation, The Ruth Mott Foundation, The John Merck Fund, The Educational Foundation of America, The New-Land Foundation, The Simons Foundation, and Svenska Lakare Mot Karnvapen (Swedish Physicians Against Nuclear Weapons).

Nuclear Wastelands

1 A Readiness to Harm

Arjun Makhijani

This is the greatest thing in history.
—President Harry Truman, 6 August 1945[1]

When the United States dropped an atomic bomb on Hiroshima on 6 August 1945, inaugurating an era, many political leaders viewed this new weapon as a vehicle that could give those who possessed it a dominant role in world affairs. U.S. Secretary of War Henry Stimson, who oversaw the Manhattan Project to build the first atomic bombs during World War II, put it thus in April 1945: "If the problem of the proper use of this weapon can be solved, we would have the opportunity to bring the world into a pattern in which the peace of the world and our civilization can be saved."[2]

However, Stimson also saw the perils of a world with nuclear weapons. In notes he took during a May 1945 meeting, Stimson wrote: "May *destroy* or *perfect* International *Civilization*/May [be] *Frankenstein or* means for World Peace."[3]

Since 1945, nuclear arsenals have been viewed as prime symbols of international power and prestige. Within the nuclear weapons states, discussions about producing, testing, and planning to use nuclear weapons have consistently fostered the impression that the principal deleterious effects are reserved for external enemies during wartime. Yet as early as the 1950s, strontium-90 fallout from atmospheric nuclear tests had been discovered in children's teeth. Today, it is evident that nuclear weapons have profoundly damaged the very people and lands they were supposed to protect through the adverse environmental and health consequences of production and testing. Moreover, even if all nuclear weapons production and testing ceases, nuclear weapons are likely to endanger future generations well into the twenty-first century.

1. As quoted in Sherwin 1987, p. 221.

2. Stimson 1945.

3. As quoted in Rhodes 1988, p. 642.

Figure 1.1 Duplicates of the Nagasaki plutonium bomb ("Fat Man") and the Hiroshima uranium bomb ("Little Boy"), and a glovebox for the handling of plutonium, at the Bradbury Science Museum in Los Alamos, New Mexico. Photo by Robert del Tredici.

The damage and risks fall into three categories:

• *damage to health and environment*, with workers in the nuclear weapons industry, people living near production and testing centers, experiment victims, and some armed forces personnel suffering disproportionately;

• *security risks* to all of humanity arising from the existence of vast arsenals of weapons and related materials; and

• *subversion of democratic principles* by the governments and academic and corporate institutions that design, make, test, and plan for the use of nuclear weapons.

The first threat is the immediate subject of this book—an investigation into the harm to human health and the global environment that has arisen, and may be projected to arise, as a result of producing nuclear weapons. For instance, a weapons plant in the Soviet Union discharged highly radioactive waste directly into a river and into a lake. Today, the waste discharge point near Lake Karachay in the Ural Mountains is so radioactive that a person would get a lethal dose of radiation in less than one hour. As another example, plutonium production at Hanford, Washington, in the United States resulted in the release of hundreds of thousands of curies (tens of thousands of terabecquerels) of radioactive iodine and high doses to the thyroids of children who drank milk produced downwind from the plant. In both countries, the risk of cancer among those exposed was substantially increased. In neither case were the potential victims warned of the dangers. Through such damage, the bombs have come home to the people of the weapons states.

Although indirect, the other two types of damage are equally important. For example, the potential for illicit diversion of even a small quantity of the large stockpile of plutonium stored at sites across the former Soviet Union threatens the security not only of its successor states but of the whole world. The plutonium could be used to make nuclear weapons, or even radioactivity-dispersion weapons that require little technical expertise. Moreover, secrecy and undemocratic practices have enabled nuclear weapons establishments to endanger human health in ways that would have been hardly imaginable under open and accountable decision-making systems. This has included the conduct of human experiments without informed consent.

A SECRET OPERATION

A large part of the environmental and health damage inflicted by nuclear weapons production has arisen out of a disregard for democratic norms, even in states whose legal systems incorporate such principles. In 1989, Deputy Secretary of Energy W. Henson Moore described the prevailing attitude during the Cold War. Nuclear weapons production, he said, had been "a secret operation not subject to laws ... no one was to know what was going on." He added that "the way we've [the government and its contractors] operated these plants in the past ... was: This is our business, it's national security, everybody else butt out."[4]

That view persists in Russia, Great Britain, France, and China, as well as in undeclared and aspiring nuclear states, from India and Pakistan to Israel and North Korea. In every case, nuclear weapons establishments have been set up and run as laws unto themselves. Even in the United States, the most open of the weapons powers, the Atomic Energy Commission, created in 1946, had jurisdiction over all aspects of atomic energy and was, in effect, a government within a government. David Shea Teeple, who worked on nuclear issues both in the AEC and in Congress, wrote in 1955 that, as a governmental commission, the agency had broad authority. While most commissions were limited to one or two special powers, Congress gave the AEC:[5]

4. As quoted in *The Washington Post*, 17 June 1989.

5. Teeple 1955, pp. 29–30.

- the *quasi-legislative* power to create regulations with the force of law;
- the *quasi-executive* power to enforce regulations;
- the *managerial* power to oversee nuclear weapons production in plants it owned;
- the *quasi-judicial* power to direct enforcement of penal statutes; and
- the *investigative* power to maintain security within the nuclear weapons complex.

Even the *policy/planning* function of the AEC to carry out studies and make recommendations for legislation turned into a power. The Joint Committee on Atomic Energy, which Congress created to oversee the AEC, functioned instead essentially as its booster.

Once Winston Churchill, with Truman at his side, had formally declared the Cold War in 1946 at Fulton, Missouri, the extreme secrecy that had characterized the wartime nuclear weapons establishment spilled over into public political assaults on U.S. citizens who challenged the weapons establishment. Perhaps nothing so clearly symbolized that attitude as the AEC's public assault on J. Robert Oppenheimer, the "father of the atomic bomb" and director of the Los Alamos Scientific Laboratory where the first bombs were designed. Even though no one accused Oppenheimer of giving away any secrets, the AEC refused to renew his security clearance in 1954 based on the former left-wing affiliations of some of his family members, even though these were known at the time of his appointment.[6]

Congress, the executive branch, and the AEC itself defined the agency's mission as producing nuclear weapons. If that meant that no exposure standards would protect uranium miners from radioactive radon gas for almost two decades or that pilots would be asked to fly through the mushroom clouds of nuclear weapons tests, that became part of the price.

Cover-ups and fabrications targeted at the very citizens the weapons were supposed to protect have been a hallmark of nuclear weapons establishments across ideological boundaries. The situation was even more extreme in the Soviet Union than in the United States. Faced with a hostile power already in possession of nuclear weapons, and operating under Stalin's dictatorial authority, Soviet nuclear scientists had first call on all national resources after World War II. They used these amply and operated with wanton disregard for the environment and the inhabitants of the country.

A 1957 accident near Chelyabinsk in the Ural Mountains is a good illustration. On 29 September, a tank containing highly radioactive wastes from plutonium production exploded at a nuclear weapons plant, but until 1989 the Soviet government tried to cover up the

6. Philip M. Stern in foreword to U.S. AEC 1971 and Rhodes 1988, p. 448.

accident; much about it is still unknown. Soviet authorities evacuated over 10,000 people from the zone of heaviest fallout but never told them why they had to move. The U.S. Central Intelligence Agency found out about the accident by 1959, but it chose not to criticize its Cold War adversary.[7] Apparently, the CIA feared that any possible propaganda advantage did not outweigh the fact that Americans might awaken to the dangers posed by their own country's weapons production.

Consider also France. To date, the French government has released no significant data on its weapons-testing program. On the contrary, it has gone to extreme lengths to maintain that secrecy. In July 1985, two French agents blew up a Greenpeace vessel, *Rainbow Warrior*, at Auckland harbor in New Zealand to prevent activists from symbolically trying to block tests. The agents killed Fernando Pereira, a Greenpeace photographer, and, although these agents were convicted of the crime, the French government secured their early release from New Zealand jails. It even promoted one of them after her return to France.

Or take Great Britain. On 10 October 1957, a fire ignited at a reactor at Sellafield. As in the much larger 1986 fire at Chernobyl, a graphite moderator caught fire, although no explosion ensued. The U.K. government covered up the consequences, denying at the time that any harm had resulted. Yet the fire exposed both workers and people living nearby to radiation: in 1983, the British National Radiological Protection Board estimated that the doses received by the public could cause hundreds of thyroid cancers. The government released its report on the accident only in 1988. Some of the health data are still secret.[8]

China is the most secretive of the declared nuclear weapons powers. The government has released essentially no data on the health and environmental effects of its nuclear weapons program. One rather cryptic 1986 statement by Qian Xuesen, a senior military official, appears to represent the limit of China's public admission regarding the health and environmental effects of its nuclear weapons testing program: "Facts are facts. A few deaths have occurred, but generally China has paid great attention to possible accidents. No large disasters have happened."[9]

In general, the undeclared nuclear weapons states have been even more secretive. Of these, Israel probably has the largest arsenal, comprising 60 to 100 nuclear warheads, plus materials for making many more. But Israel has revealed no details of its nuclear weapons

7. U.S. CIA 1959. For further discussion, *see* IPPNW and IEER 1992, chapter 4.

8. McSorley 1990, pp. 13–14.

9. As quoted in May 1989, p. 145.

program. On the contrary, in 1986 Israeli agents kidnapped whistle-blower Mordecai Vanunu from Italy and forced him to stand trial in Israel, where he remains imprisoned. Vanunu gave the first inside account of the large-scale production of nuclear weapons materials at the Dimona complex in Israel. India and Pakistan also continue to keep their nuclear weapons programs secret, going so far as to deny their very existence from time to time.

A DISREGARD FOR PUBLIC HEALTH

Nuclear establishments have attempted to keep the citizens of their own countries in the dark partly because people were already wary of exposure to radiation. Official denials of harm from radiation have been routine, and public relations have often taken the place of public education when it comes to weapons production and testing. In the United States, "It appeared that the idea of making the public feel at home with neutrons trotting around is the most important angle to get across," noted Major William Sturges, who participated in safety discussions of nuclear testing.[10]

One result of this attitude is that even precautions that could have been taken often were not. No one warned people living downwind from Hanford not to drink locally produced milk, which was tainted with iodine-131. Downwind from the Nevada Test Site, the AEC routinely reassured people that there was little or no danger, despite knowledge of heavy fallout after many tests.

The disregard for health has also been prevalent in other countries. For instance, a journalist investigating problems in India's nuclear complex concluded that the regulatory "limit for radiation workers exposure means very little in practice: it has been breached so frequently as to make one wonder why it exists at all."[11] As another example, the British government has routinely discharged large quantities of radioactive materials into the Irish Sea. In 1983, the radiation rate from seaweed and flotsam in the area of the discharges was so high that the government closed a 12-mile stretch of beach for several months.

The nuclear establishments of every single nuclear weapons power have displayed a shocking readiness to harm people. In the Soviet Union, some of the work producing nuclear weapons was performed under conditions resembling slave labor (see chapter 7). There are also unconfirmed allegations of experiments in the Soviet Union on human beings; in the United States, where access to documents is far greater, similar allegations have been confirmed. For instance, Dr. Eugene

10. As quoted in Fradkin 1989, p. 97.

11. *Business India 1978.*

Saenger of the University of Cincinnati conducted whole-body radiation experiments on terminally ill patients under contract to the Department of Defense.[12]

The effects of radiological warfare on human reproductive organs appears to have been of particular interest. For example, researchers irradiated the testicles of prisoners in the states of Washington and Oregon, sometimes to very high levels. In all, 131 prisoners were irradiated, including 67 in Oregon State Prison from 1963 to 1971 and 64 in Washington State Prison from 1963 to 1970. Because the level of radiation for many prisoners was so high, each subject had to agree to have a vasectomy at the end of the test. Catholics were not allowed to take part in this experiment, presumably in deference to papal edicts against birth control. The prisoners were paid $5 to $10 for each irradiation and $100 at the time of the vasectomy.[13]

The formal object of this experiment "was to obtain data on the effects of ionizing radiation on human fertility and the function of testicular cells."[14] But it seems connected to an issue raised in a Manhattan Project review of radiological warfare. During World War II, it was feared that Germany might carry out radiological warfare on a British city or against British and U.S. troops preparing to invade Europe. In 1944, officials of the Manhattan Project estimated the effects of radiation and their effect on soldiers in the battlefield from animal data:

> It is well known that localized radiation to the extent of 500 to 600 roentgens will produce complete sterility, both in the male and female. There is experimental evidence on animals to indicate that lesser amounts may not produce sterility but may definitely affect the progeny. The question of permanent mutations has not been established fully. There is some evidence to indicate that this will occur with radiation dosage well below the recognized amount for sterilization. It is important to recognize that sterility produced by radiation usually has no effect on the libido.[15]

Other experiments were done on pregnant women, supposedly retarded children, and military personnel.[16]

Nuclear weapons powers have caused damage to the health of peoples outside their countries as well. Nuclear testing has been carried out in the Pacific by the United States, France, and Britain, causing widespread damage and contamination.[17] The United States, the Soviet

12. Rapoport 1971, pp. 90–96.

13. U.S. Congress, House 1986, p. 15.

14. U.S. Congress, House 1986, p. 15.

15. Eyster 1944.

16. Kershner 1994.

17. IPPNW and IEER 1991.

Union, Britain, France, and China have all obtained uranium from other countries for their weapons programs. From South Africa, Namibia, and Australia in the south to (former) East Germany, North Korea, and Canada in the north, nuclear weapons powers have created radioactive and nonradioactive toxic residues from mining ore and processing it chemically into uranium oxide (''yellowcake'').

All too often such damage has been done to ethnic minorities or on colonial lands or both. The main sites for testing nuclear weapons for every declared nuclear weapons power are on tribal or minority lands. Uranium mining has also largely taken place on—and defaced—tribal lands or colonial countries.

The willingness to take liberties with the lives of people outside their borders in order to acquire nuclear weapons is illustrated by a 1960 editorial that justified inflicting birth defects around the world for the sake of a U.S. nuclear deterrent. It appeared in the engineering school alumni magazine of the University of California, which was (and continues to be) a contractor for designing and testing nuclear weapons:

> The increase in radiation one receives from fallout is about equal to the increase one receives from cosmic rays when moving from sea level to the top of a hill several hundred feet high.... It means, though, your babies' chances of having a major birth defect are increased by one part in 5,000 approximately. Percentage wise, this is insignificant. *When applied to the population of the world,* it means that nuclear testing so far has produced about an additional 6,000 babies born with major birth defects [emphasis added].
>
> Whether you choose to look at ''one part in 5,000'' or ''6,000 babies,'' you must weigh this acknowledged risk with the demonstrated need of the United States for a nuclear arsenal.[18]

Health and environmental problems have also arisen because of the large quantities of nonradioactive hazardous materials, such as hydrofluoric acid, beryllium, and carbon tetrachloride, that have been used in the manufacture of nuclear weapons. There are indications that many of these toxic chemicals may have been particularly damaging to workers exposed to them. Yet, the record keeping of occupational exposure has been so poor that, despite considerable effort, we have been able to find little quantitative information on exposure either of workers or of off-site populations to nonradioactive toxic materials.

Ironically, one of the greatest threats nuclear weapons pose to the environment and human health is growing even as these weapons become less important as instruments of state policy. The end of the Cold War has turned thousands of nuclear weapons and huge quantities of weapons-usable materials into surplus. At the same time, the chaos occasioned by the disintegration of the Soviet Union has aggravated the problem of proliferation and increased the chances that some

18. April 1960 editorial in *California Engineer,* reprinted in *California Engineer* 1990.

countries and even groups could acquire enough materials to build nuclear or radiological weapons.

This possibility has intensified and greatly complicated the problem of nuclear proliferation. Already, German authorities have arrested smugglers carrying small quantities of radioactive materials, including plutonium, from the former Soviet Union; there are unconfirmed reports of sales of weapons, which Russian authorities have denied. There is still no comprehensive international system for verifying stockpiles of either weapons or weapons-usable materials, which continue to accumulate. The quantities of materials needed to make nuclear warheads are small compared to the uncertainties arising from the lack of a comprehensive verification system. Because the present system of international controls and inspections does not fully cover the weapons states, it is an inadequate instrument for nonproliferation policy.

Since World War II, the principal focus of the nuclear weapons states on power and deterrence has created immense direct liabilities in the form of health and environmental costs from emissions of pollutants and poor waste management practices. In the United States, where the situation is not as bad as in the former Soviet Union, estimates of the costs of cleanup range from one hundred to several hundred billion dollars. Adding to the burden, the technologies to manage some of the wastes and to clean up much of the pollution do not yet exist.

For the former Soviet Union sufficient information regarding the extent of the damage does not yet exist and it is impossible to assess even tentatively the costs to clean up the sites. However, it is clear that damage in the former Soviet Union is far worse, and hence the cleanup costs will be far higher than in the United States. Considerable damage requiring cleanup has also been done in the countries where the nuclear weapons states have obtained uranium or conducted tests.

Even when "cleanup" activities have been carried out, hundreds of millions of metric tons of radioactive waste, much of it mixed with other toxic materials, will remain. These wastes will need to be disposed. Some are highly radioactive and contain long-lived radio nuclides, but there are at present no accepted methods or sites for disposal of such wastes. The immense problems of cleanup and radioactive waste management, as well as the increased proliferation threats at the end of the Cold War, are among the principal legacies of half a century of producing nuclear weapons, and must be addressed in any serious discussion about the risks of allowing nuclear weapons production and testing.

2 Methodology

Arjun Makhijani and Howard Hu

This book details the environmental and health problems arising from nuclear weapons production throughout the world.

Studies covering U.S. nuclear weapons production alone (such as *Complex Cleanup*, a study by the congressional Office of Technology Assessment)[1] have helped reveal the scope of actual and potential health and environmental problems. Our focus on production-related impacts is intended to assist similar inquiries in other countries, so that the costs of limiting future health and environmental damage can be assessed. Moreover, the global approach here makes it possible to compare, to the extent that data allow, the health and environmental effects of nuclear weapons production in different countries.

We have limited our scope in several ways to create a coherent project and one of manageable proportions. Thus, we omit issues arising from transporting, storing, and deploying nuclear weapons. Nor do we consider other aspects of the military use of nuclear energy, such as the use of nuclear reactors for naval propulsion, with their associated environmental effects. We also exclude questions associated with the use of depleted uranium in "conventional" munitions, tanks, and armored personnel carriers. Finally, as mentioned earlier, we do not cover nuclear weapons testing here because IPPNW and IEER have produced a book specifically on that subject, *Radioactive Heaven and Earth.*[2]

ANALYTICAL APPROACH

In this book we analyze generic health and environmental impacts according to the processes that are associated with the various materials production and fabrication steps required to make nuclear weapons. We begin with a description of various types of weapons, including the main materials and the major processes involved in their

1. U.S. Congress, OTA 1991.

2. IPPNW and IEER 1991.

production. In many cases, there is more than one approach to manufacturing a particular component or material; this book describes all major processes to some extent. We focus on the production and processing of the two main radioactive materials in nuclear weapons, highly enriched uranium and plutonium-239. Then we provide information on the health effects of many radioactive and nonradioactive materials used in or generated as a result of nuclear weapons production.

Uranium mining and milling, which constitute the common starting point for the production of nuclear weapons materials, take place in many countries that are not nuclear weapons states. Therefore, we gather the information and analysis on the environmental and health effects associated with these processes globally in a single chapter.

The bulk of this book, however, analyzes the situation in each nuclear weapons country, beginning with the historical context for producing such armaments. Because the policies of the United States and the Soviet Union have played a large role in influencing decisions made in other countries, these historical overviews clarify the connection between foreign policy and proliferation considerations and internal production activities. The country-by-country chapters list the production facilities for the processes that are generically described earlier and analyze the overall impact of each country's nuclear weapons complex. Major facilities are described in greater detail, along with available health and environmental data.

In many cases, little or no information is available on production facilities, not to mention data on environmental and health effects. This is especially true for aspects of nuclear weapons production that are relatively unresearched from an environmental point of view. To cite one example, the manufacture of lithium has involved the use of mercury, the fate of which at most plants is not documented.

The question of where civilian research, development, and production stop and a nuclear weapons complex begins is a difficult one that can involve many judgments. In fact, there are many actual and potential overlaps between civilian and military applications of nuclear energy. The problem begins with scientific and technical expertise, much of which is common to both realms. Physicists, engineers, chemists, and technicians who can build and operate facilities associated with nuclear energy can also build and operate most facilities associated with weapons.

The mining, milling, chemical conversion, and enrichment of uranium form a second large overlap between nuclear weapons and civilian nuclear energy. However, to determine the extent of uranium use in weapons would require detailed official data on production. In the absence of such data, we separate military from civilian uranium production by making assumptions about the fissile material content of each weapon and combining this with published figures on the size of a country's nuclear arsenals.

We consider a country's capacity to produce low-enriched uranium to be part of a weapons complex only when such plants provide feed materials for producing highly enriched uranium elsewhere. For instance, the Paducah, Kentucky, plant produces low-enriched uranium and supplies it for making highly enriched uranium at Portsmouth, Ohio. Thus, Paducah is included in the description of U.S. nuclear weapons plants.

A third overlap involves plutonium production. This presents a difficult and controversial line to draw. All civilian nuclear power reactors have considerable amounts of uranium-238 in their fuel rods. Some of this uranium-238 is converted to plutonium-239 when irradiated with neutrons from fission reactions. In this way, all civilian nuclear power reactors produce plutonium; it even contributes to energy production, since some plutonium fissions to yield energy before the fuel rods are withdrawn and replaced. Irradiated fuel rods withdrawn from a reactor are also called "spent fuel" even though a quarter to a third of the original fissile uranium-235 remains. After fission products build up to a certain level in the fuel rods, consuming the rest of the fuel without first chemically processing it is no longer economically or technically efficient.

For several reasons, this book does not include civilian nuclear reactors or the plutonium in unreprocessed spent fuel rods in its definition of weapons-usable materials. Plutonium in civilian reactors is mixed with far larger quantities of uranium and highly radioactive fission products. Only about 1 percent or less of the content of a typical spent fuel rod is plutonium; a few percent consists of fission products; the rest is uranium. Moreover, spent fuel rods are very bulky, thermally hot, and highly radioactive. Therefore, a great deal of physical handling and chemical processing is necessary before spent fuel can be converted into weapons.

On the other hand, plutonium is weapons-usable after it is separated from the fission products and residual uranium, and this book's coverage does include facilities for performing this separation. Plutonium produced in most civilian reactors consists not only of plutonium-239 but typically also 19 percent or more of plutonium-240 as well as other isotopes (plutonium-241 and -242); yet this mixture can be used to make nuclear weapons. Since plutonium is a highly carcinogenic, radioactive material, it can also be fashioned into deadly radiation weapons.[3]

It might be argued that this distinction is somewhat artificial, since plutonium from a civilian nuclear power plant can be separated and made into weapons. That is why many people consider the spread of nuclear power itself to be a proliferation threat, the Nuclear Non-

3. IPPNW and IEER 1992, pp. 141–145.

Proliferation Treaty notwithstanding. However, the time, money, and effort needed to convert civilian spent fuel into weapons-usable plutonium are considerable. In contrast, a country with nuclear expertise and stocks of separated plutonium could rapidly build nuclear weapons.

Governments of countries with civilian plutonium-separation plants (such as Japan) often deny any intention to use plutonium for nuclear weapons. In many cases (including Japan), the plutonium is under international safeguards of the International Atomic Energy Agency. However, this book does not take the intentions of governments to be a criterion for determining potential nuclear weapons status. The intentions of government are changeable; indeed, governments and even systems of government are far less durable than the half-life of plutonium-239—over 24,000 years. Therefore, we rely on the characteristics of the materials and the existence of facilities and technological capability as the primary guides in discussing actual and potential nuclear weapons powers. That this is a sound and practical approach is illustrated by recent expressions of interest in Japan in producing nuclear weapons as a response to North Koreas suspected plutonium production.[4]

SELECTION OF COUNTRIES

Nuclear weapons powers, actual and potential, fall into seven categories:

1. *Declared nuclear weapons states:* United States, Russia, United Kingdom, France, China.

2. *De facto nuclear weapons states:* Israel, India, Pakistan.

3. *Potential nuclear weapons states with some present production of weapons-usable materials:* Japan, North Korea.

4. *Former nuclear weapons states:* South Africa, which claims to have dismantled all its nuclear weapons.

5. *Potential nuclear weapons states with some past production* and *some technical capability for producing weapons-grade materials, including states that own plutonium reprocessed in other countries:* Argentina, Belgium, Brazil, Germany, Iraq, Italy, Netherlands, Sweden, Switzerland, perhaps Iran.[5]

6. *States holding nuclear weapons but currently of uncertain intentions as to whether they will continue to do so:* Ukraine, Kazakhstan, Belarus. All of them have issued declarations that they will sign the Non-Proliferation Treaty as non-nuclear weapons states.

4. *The Washington Post*, 31 October 1993.

5. Albright, Berkhout, and Walker 1993, chapters 6, 10, 11, and 12; *The Washington Post*, 25 November 1994.

7. *States or groups that may become nuclear weapons powers or near-nuclear weapons powers by purchasing weapons or weapons-usable materials:* Libya is thought to belong to this category.

This book addresses in greatest detail the first four categories of actual and potential nuclear weapons powers because the health and environmental effects of production occur principally in these countries. It also briefly discusses some countries in the fifth and sixth categories. In the case of non-nuclear weapons countries that have provided uranium for use in nuclear weapons, it discusses only the relevant mining and milling facilities. This category includes Namibia, Canada, and several other countries.

SOURCES OF DATA AND EVALUATION

This book uses country-specific data to the extent the project's resources have allowed. Besides the vast scope of the book, however, these efforts have been limited by:

• government secrecy surrounding nuclear weapons, including health issues and environmental data and analyses;

• the difficulty of accessing and translating data from diverse sources in many countries and many languages;

• the dearth of health and environmental data, even in the best of circumstances; and

• the poor quality of some of the data.

In many cases, information exists but its quality is hard to verify. Official statements sometimes contradict those made by environmental activists. Statements in the media can conflict with one another and with officially published data. Because it is generally impossible to get access to the assumptions and methods used to produce official estimates, verification is a difficult, often impossible, exercise.

These issues are less problematic regarding the United States. First, the amount of official data that is public is far greater than in any other country, due to the Freedom of Information Act, congressional investigations, public hearings, lawsuits, and the like. While the record is far from complete, a great deal of it is reasonably well established. Thus, more evaluations of official data are available for the United States than for any other nuclear weapons state. These evaluations indicate that systematic biases may distort official estimates of environmental releases from nuclear weapons plants. Two examples discussed in chapter 6 are uranium production at Fernald, Ohio and iodine-131 releases from the Hanford, Washington facility.

While scattered information and in-depth analyses are available in some other countries—for instance, Russian publications about a 1957 tank explosion at the Chelyabinsk-65 nuclear weapons complex (formerly called Chelyabinsk-40)—it is not yet possible to make overall critical assessments of the global data or of the extent of the pollutant discharges into the environment from nuclear weapons production. So far as is possible with countries besides the United States, this book reflects a critical use of official sources of information. We have checked the information for internal consistency as well as its consistency with what is known about various generic processes and their characteristics, with information available from the United States, and with other data that is relatively more reliable.

Due to government secrecy and the lack of data collection at the time of weapons production, official data are all too often not available at all. This is notably the case for France, China, and Russia, as well as for undeclared de facto and near-nuclear weapons states.

We also review a considerable number of nonofficial sources, including press reports, peer-reviewed journals, and publication of nongovernmental organizations. But we have tried to avoid material that contains obvious problems. When such information is cited, a comment is included on its quality. Since even elementary data are often lacking, we have provided some default estimates of waste production and environmental contamination at the end of this book based on generic analyses or inferences from U.S. data.

ANALYSIS OF ENVIRONMENTAL ISSUES

This book describes the generic environmental impacts of the main processes involved in nuclear weapons production. It also presents as much country-specific data on such impacts as possible.

However, even when a good deal of information is available about production facilities and their locations, data on environmental and health effects are far less complete, partly because of secrecy and partly because the data do not exist. Still, it is reasonable to make preliminary, order-of-magnitude assessments of the scope of problems by starting from a knowledge of the approximate size of a country's nuclear arsenal and the size of its weapons-usable stocks of fissile materials.

The problem of adverse health and environmental consequences of nuclear weapons production extends well beyond the question of radioactive waste and emissions of radioactive materials. Nuclear weapons production involves large quantities of inorganic and organic toxic materials that have been stored or discharged as wastes or emitted as pollutants. There are many issues of data inadequacy concerning these chemicals. First, there is little or no data on the quantities of emissions of many of the chemicals. Second, the historical coverage of the data is typically poor. For instance, in the United States, with the best records, data on these materials have only been gathered mainly in the past decade or at best since the early 1970s. As a result, environmental and exposure data do not exist for the early, crucial years, when pollution per unit of production tended to be far higher than in recent years.

Indirect methods of calculation, such as studying purchase records for materials and the engineering processes in which they were used, can yield estimated releases to the environment and potential exposures. Such studies are arduous and expensive and far beyond the scope of this book. A number of dose-reconstruction efforts are underway in the United States. These are shedding new light on such questions, and a better picture is likely to emerge in a few years.

ANALYSIS OF HEALTH ISSUES

This book uses available information to describe the generic health impacts of the main processes involved in nuclear weapons production and also presents as much country-specific data as possible. It pays attention to effects on both the workers involved in nuclear weapons production and the residents who may have been affected by exposure to the products of the industry. It does not cover health effects related to testing, since these are covered in *Radioactive Heaven and Earth*.

As much information as possible has been gleaned from environmental and occupational epidemiology studies that are in the public domain, published either in peer-reviewed journals or as technical reports. However, the paucity of such studies leaves gaping holes in knowledge regarding health impacts, particularly with respect to the states that remain the most secretive, such as France and China. Moreover, the relatively few studies that exist, including those from the United States, often suffer from methodological problems. (For more on this point, see chapter 4). And some studies, such as recent published works on the health effects of the Russian nuclear weapons program, are based on raw data of doubtful validity.[6]

On many occasions, studies that are crucial to understanding the generic health effects of hazardous exposures also provide some of the main information available on the extent of health effects experienced within particular countries. Thus, discussions of some studies appear both in chapter 4, where generic health hazards are summarized, and in individual country chapters.

Unfortunately, it was beyond the scope of this project to undertake independent epidemiological investigations. Wherever possible and warranted, we interpret the information already available and relate it to what might be expected from the levels of environmental pollution sustained and our knowledge of the generic risks posed by the toxins involved.

6. Rush 1992.

Some use of technical terms and concepts is unavoidable in a work such as this. Explanations for terms that are used in the text, as well as concepts from medicine, radiation biology, toxicology, and epidemiology, can be found in chapters 3 and 4, with brief definitions also provided in the glossary.

TERMS AND UNITS

This book uses metric and standard international units unless otherwise noted. In the case of radioactivity units, both the standard international unit, becquerel (one disintegration of a nucleus per second), and the traditional unit, curie (which is equal to 37 billion becquerels), are in common use. Therefore, we have given figures in both units. For instance, if a waste tank contains 100 curies, as cited in a source publication, we present the data as "100 curies (3,700 gigabecquerels)," using the standard prefix "giga-" for billion (10^9). The other common standard prefix in radioactivity units is "tera-" for trillion (10^{12}).

The standard international unit of radiation dose is a gray, which measures energy deposition. One gray equals the deposition of 1 joule of energy per kilogram of tissue. The older unit is a rad. One rad equals 0.01 grays—that is, one rad equals one centigray. Due to this fortunate equivalence between the traditional and standard international unit, we use centigrays to express doses whenever it is not cumbersome to do so.

Gamma radiation measurements are often cited in the traditional unit, roentgen. This is an amount of radiation that produces 2.58×10^{-4} coulombs of electrical charge due to ionizations in 1 kilogram of air. It is approximately equivalent to 0.93 rads (0.93 centigrays). Since radiation measurements are generally associated with considerable uncertainty, using a precise conversion factor of 0.93 to go from a measurement expressed in roengtens to one in centigrays would create a misleading impression of accuracy in many cases. It is therefore customary to convert measurements in roengtens to centigrays (or rads) approximately, simply by using a multiplication factor of 1. We indicate conversion by an approximate factor by using the words "about" and "approximate."

The biological effects of alpha particles and neutrons are generally much greater than those of beta particles and gamma rays per unit of energy deposited in tissue. This is because alpha particles and neutrons transfer most of their energy to a single cell or an even smaller volume, greatly increasing the chance of damage to DNA.

Relative biological effectiveness (RBE) is the concept used to compare the ability of different kinds of radiation to cause different amounts of damage per unit of energy. It can vary from one cell type to another and with energy even for a single type of radiation. Since relative biological effectiveness is quite complex, a simple empirical radiation weighting factor, w_R (also referred to as the quality factor, QF), is used in dosimetry to account for the differing biological impacts per unit of energy deposited. Using a reference value of 1 for gamma rays, the current recommended values in dose calculations are 1 for most beta particles, 5 to 20 for neutrons (depending on energy), and 20 for alpha particles. The equivalent dose so far as biological damage is concerned is defined as the absorbed dose times the weighting factor. The equivalent dose is measured in sieverts, Sv (standard international unit). (Another unit of equivalent dose was the rem, which is also still in use: 1 rem = 0.01 sievert.)

For uranium mining doses, the most convenient and widespread unit in use is the working-level-month (WLM). This corresponds to exposure to radon-222 and its decay products in equilibrium (see chapter 3) present in a concentration in air such that there is a release of 130,000 MeV of alpha radiation energy in 1 liter of air for one working month assumed at 170 hours. This corresponds approximately to air with about 100 picocuries (3.7 becquerels) per liter of radon-222 with its decay products in equilibrium.

CANCER RISK COEFFICIENTS

We have used the most recent study of the Committee on the Biological Effects Of Ionizing Radiations (BEIR) of the U.S. National Academy of Sciences as the starting point for making cancer risk estimates. This study, called "BEIR V," since it was the fifth report in the series done by the committee, analyzes risks posed by low levels of radiation, that is, radiation whose health effects are not immediately manifest (see chapter 4 for further discussion).

The BEIR V report analyzes data from studies of the survivors of the bombings of Hiroshima and Nagasaki and from studies of people subjected to radiation in the course of medical treatment or examinations. These are the two principal sources of human data for estimating effects of low levels of radiation. In addition, there are large numbers of animal studies that provide supplementary and complementary information.

There are considerable uncertainties in the effects of low-level radiation below the levels estimated to have been received by the group of Hiroshima-Nagasaki survivors who exhibit excess cancers. BEIR V provides several different estimates for cancer risk of low-level radiation. One set of estimates relates to the risk of fatal leukemias. This risk is generally acknowledged to be greater for a given amount of dose if the dose is delivered suddenly rather than in slow increments. Thus, the fatal cancer risk coefficients for repeated exposures at low dose rates are lower than those for single exposure to a relatively high dose. Based on BEIR V, we use a leukemia risk factor of 75.5 fatal leukemia cases for a population dose of 10,000 person-sieverts, delivered in increments of 0.01 sievert per year to individuals between the ages of 18 and 65.[7] Person-sieverts measures population dose since it is the total radiation dose received by all members of an exposed population group.

Another set of risk factors relates to solid tumors. As discussed in chapter 4, it is also customary to use a reduction in risk by about a factor of 2 for low dose rates. The evidence for this comes largely from animal data and hence this conclusion is less firm than the use of a dose rate reduction factor for leukemia.

According to BEIR V, for most cancers other than leukemia the estimated factor for dose rate is approximately 1, as indicated by models fitting data from relatively high doses. However, animal data indicate lower effectiveness for low dose rates, with risk reductions per unit of dose ranging from 2 to 10 compared to high dose rates. However, since there are great uncertainties in these risk reduction factors, BEIR V simply provides cancer risk for solid tumors assuming a risk reduction factor of 1 (that is, no risk reduction), thereby leaving it to its readers to apply any risk reduction factors that they deem prudent.[8] In this book, we present fatal cancer estimates derived directly from unadjusted BEIR V coefficients (that is, using a risk reduction factor of 1 for solid tumors), and also present fatal cancer estimates using a risk reduction factor of two for solid tumors, following common regulatory practice in the United States.

BEIR V cites a risk of 695 fatal solid tumors per 10,000 person-sieverts for high dose rates. When leukemias are included, the total risk for fatal cancers is 790 per 10,000 person-sieverts.[9] When a dose rate reduction factor of two is applied for low dose rates for solid tumors, the resultant risk coefficients are about 350 to 450 fatal cancers per 10,000 person-sieverts, depending on the dose rate and ages over which irradiation takes place.

In this book, we have estimated cancer risk only corresponding to data from the former Soviet Union. Since Soviet dose and health data are questionable, we do not find them suitable for estimating precise risk coefficients. We have used a risk coefficient of 400 fatal cancers per 10,000 person-sieverts (or 0.04 fatal cancers per person-sievert) as the basis for estimation. This includes a dose rate reduction factor of approximately 2 for solid tumors. As an approximate indication of risk for a dose rate reduction factor of 1 for solid tumors, we have used a risk coefficient of 0.08 fatal cancers per person-sievert, which is the rounded risk factor for single doses cited above.

7. National Research Council 1990, pp. 172–173. This risk is an average for males and females.

8. National Research Council 1990, pp. 22–23, 171–174.

9. National Research Council 1990, pp. 23, 172–173.

In sum we use a range of 0.04 to 0.08 fatal cancers per person-sievert to estimate fatal cancer risk for all cancers (including leukemia), and about 0.0075 fatal cases per person-sievert for leukemia alone.

There are uncertainties around such estimates. These also vary according to dose rate. One approximate indication is provided by the BEIR V range of 585 to 1,200 fatal cancers as the 90 percent confidence interval for fatal cancers (central estimate = 790) in the case of a single dose of 0.1 sievert per person delivered to a population of 100,000 people.

3 The Production of Nuclear Weapons and Environmental Hazards

Arjun Makhijani and Scott Saleska

In order to understand what nuclear energy is and how nuclear weapons function, it is necessary to understand some basic concepts about chemical elements and the atoms of which they are made. Chemical elements like gold, oxygen, or lead are different from one another precisely because they are each made up of distinct kinds of atoms. Atoms, in turn, differ primarily as a result of differences in their nuclei. Every atom has a nucleus composed of elementary particles called protons (with a positive electrical charge) and neutrons (with no electrical charge). The nucleus is surrounded by negatively charged particles called electrons that whirl around it. An electron's weight is just 1/1836 that of a proton.

The number of protons in the nucleus of an atom (known as that atom's "atomic number") determines what element that atom is. For example, an atom with atomic number 1 (one proton in its nucleus) is the chemical element hydrogen. Atomic number 2 is helium. Uranium, with atomic number 92, has 92 protons in its nucleus.

In addition to atomic number, elements are characterized by their mass number, which is the sum of the numbers of protons and neutrons in the nucleus. There are often several different mass numbers for a given element, each containing the same number of protons but different numbers of neutrons. These are called the isotopes of that element. For instance, uranium-238 has 92 protons and 146 neutrons, and uranium-235 has 92 protons and 143 neutrons.

During nuclear reactions called fission (splitting of the nucleus) and fusion (fusing the nuclei of two elements), a portion of the mass of the nucleus (or nuclei, in the case of fusion) is converted into energy. The conversion follows Albert Einstein's famous formula, $E = mc^2$, where E is energy, m is the amount of mass being converted into energy, and c^2 is the speed of light multiplied by itself. Since the speed of light is a very large number (300 million meters per second), a small amount of mass can be converted into a large amount of energy.

Nuclear energy can be derived from certain heavy elements by splitting them into two fragments. Because the binding energy per nuclear particle, which is the energy that is needed to break apart one of the particles in a nucleus from the rest of it, is lower for elements of heavy weight than for intermediate elements, energy is released when a heavy element is split into two intermediate elements. Similarly, binding energy is also lower for lighter elements than for intermediate ones, so that energy can be obtained by fusing the nuclei of light elements.

While these principles are theoretically valid for these classes of elements, other conditions must be fulfilled to actually break apart or fuse nuclei. Fission is the easier of the two to accomplish because the nuclei of heavy elements can be bombarded with particles, notably neutrons, to cause them to split. Some nuclei can be split simply by adding a neutron of zero energy (ideally) or (practically) of very low energy. Elements that possess nuclei that can be split in this way are called fissile elements; they are central to the production of nuclear energy. Uranium-235 and plutonium-239 are the best-known examples. Other nuclei, such as uranium-238, will not split unless the bombarding neutron possesses a considerable amount of energy, which is absorbed by the nucleus and is essential to the fission process for these nuclei. Such nuclei are called fissionable, but not fissile. Some nuclei release neutrons when they are split. This raises the possibility of a self-sustaining chain reaction. The energy released during fission and chain reactions are the principal physical phenomena that made the first nuclear weapons possible.

Fusion is far more difficult to accomplish because it requires very high temperatures, comparable to those in the sun (where fusion reactions are the main mechanism for energy release). Fusion reactions in nuclear weapons depend on fission reactions first creating the high temperatures needed for them.

Many heavy elements, as well as the fragments resulting from fission, are radioactive—that is, their nuclei are unstable. Unstable nuclei tend toward stability in various ways. Most do so by releasing a particle and thereby becoming spontaneously transmuted into another element. The spontaneous transmutation of unstable nuclei into other elements is called radioactivity.

Many heavy nuclei, such as uranium-238 or plutonium-239, decay by emitting alpha particles, which are themselves helium nuclei containing two protons and two neutrons. Other nuclei decay by emitting beta particles, which can be electrons or positrons (which are identical to electrons except that they are positively charged). Sometimes an unstable heavy nucleus will fission spontaneously, emitting neutrons in the process. Some nuclei transmute by capturing an electron from outside the nucleus.

Frequently, there is still some excess energy in the nucleus after radioactive decay (or spontaneous transmutation). This energy can be released by emission of high frequency electromagnetic radiation

Table 3.1 Properties of Uranium-235 and Uranium-238

Property	Uranium-235	Uranium-238
Nominal atomic weight	235	238
Atomic number	92	92
Fissionable	Yes	Yes
Fissile	Yes	No
Percent in natural uranium	0.711	99.284
Half-life	4.46 billion years	704 million years

called gamma rays (which are identical in nature to X-rays, but often of higher energy). Some elements that result from transmutation remain unstable enough to undergo further radioactive decay, giving rise to a series of elements known as a decay chain.

Uranium and all other heavy elements with atomic numbers greater than 83 are radioactive. Natural uranium consists of three isotopes: uranium-238 (the raw material for producing plutonium-239), uranium-235 (the fissile isotope used in nuclear weapons and to generate power), and uranium-234 (a trace isotope that constitutes almost half the radioactivity of natural uranium). Uranium-235 and uranium-238 have somewhat different properties and are used in different ways in making nuclear weapons (see table 3.1).

Both uranium-235 and uranium-238 are fissionable—that is, their nuclei can be split into two fragments, called fission products. While every such fission releases energy, there is a crucial difference between these two isotopes. Uranium-235 is "fissile." This means that it can be assembled into a *critical mass* that can sustain a chain reaction without an external source of neutrons. The amount of a fissile material required to make a critical mass depends on the properties of its nucleus, the arrangement of the material, the density of the material, and other factors.

OVERVIEW OF WEAPONS TYPES

Over the last half century, many countries have fabricated and tested three kinds of nuclear weapons, broadly speaking:

- *pure fission weapons* of either the implosion type or "gun" type;
- *boosted fission weapons* in which tritium-deuterium fusion reactions increase the quantity of neutrons available for a fission reaction but essentially all the explosive power derives from fission reactions;
- *thermonuclear weapons* in which a primary fission reaction triggers a secondary thermonuclear reaction (see figure 3.1). Such weapons may also have boosting of neutrons and may incorporate a third

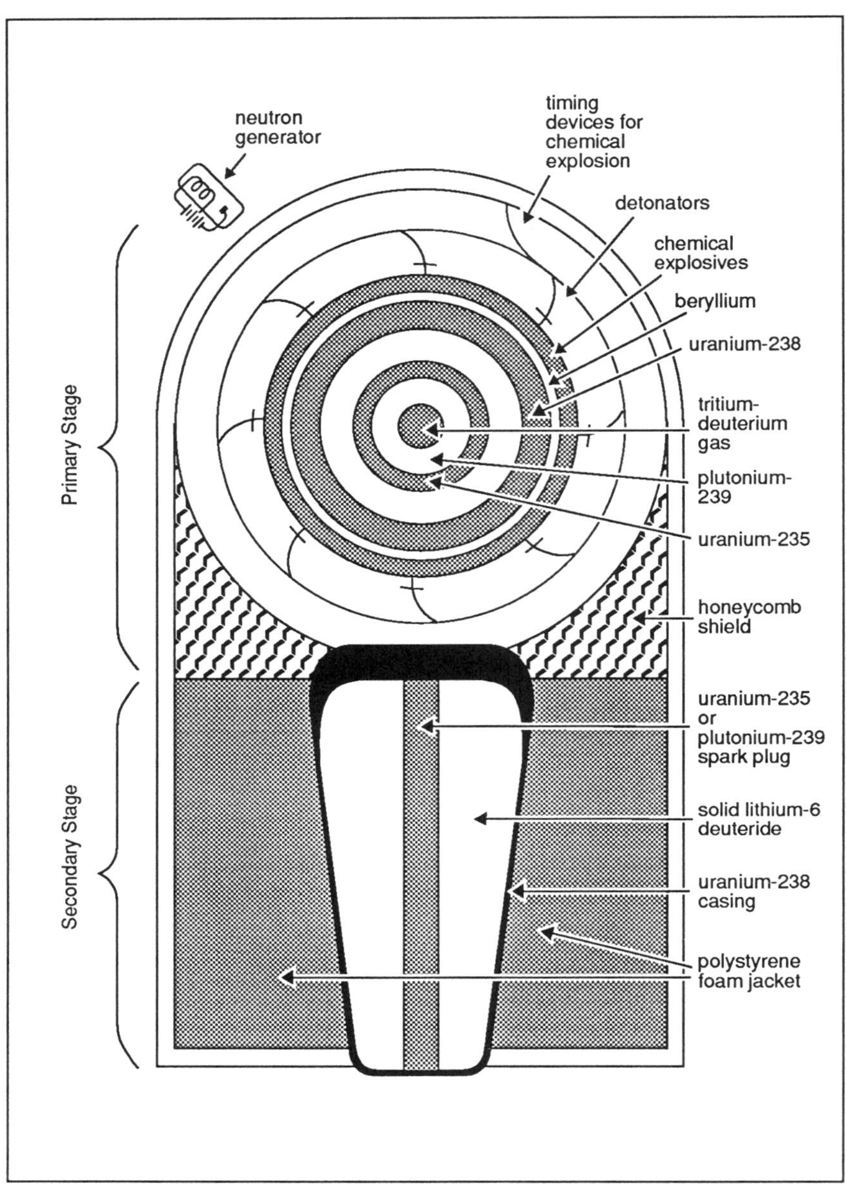

Figure 3.1 Schematic cross section of hybrid weapon. Source: Coyle et al. 1988, p. 20, and Morland 1980, as cited in del Tredici 1987, pp. 130, 131.

fission stage, including explosive energy derived from a uranium-238 "blanket."[1]

The primary fissile material in a nuclear weapon won't explode until various parts come together or are imploded to produce a *supercritical mass* (see below). Conventional explosives provide this initiating event. In the case of gun-type weapons, such as the Hiroshima bomb, a conventional explosion hurtles a piece of uranium metal toward another piece at the far end of a barrel. In implosion weapons, such as the Nagasaki bomb, precisely timed conventional explosives compress a central sphere of fissile material. This difficult engineering feat puts a premium on the timing devices and the exact shape and placement of explosive charges.

Thermonuclear weapons are more complex. Conventional explosives compress a spherical nuclear "trigger." This trigger, consisting of plutonium, highly enriched uranium or both, is the primary stage of the explosion. A neutron initiator or source initiates the chain reaction. Various electronic and mechanical components provide for the external triggering command and for safeguards against accidental or unauthorized detonation. This primary stage of the nuclear explosion compresses a mixture of tritium (a radioactive isotope of hydrogen) and deuterium (a stable isotope of hydrogen) inside the hollow fissile sphere. Fusion of tritium and deuterium provides an additional source of neutrons:

deuterium + tritium → helium-4 + neutron + energy

These neutrons produce additional fission reactions in the primary stage in the short period before the assembly blows apart. By doing this, they increase the efficiency of use of the uranium and/or plutonium in the weapon. The energy from fusion at this stage does not contribute much to the overall explosive power.

The fusion (or thermonuclear) stage of the bomb comes next; the explosive energy is derived from the same deuterium-tritium reaction stated above. But in the secondary stage, tritium is first produced from lithium-6 when the latter is bombarded with neutrons generated by the detonation of the primary stage. The energy from X-rays from the first, mainly fission stage is used to compress the secondary stage. In this way, the primary stage creates extremely high temperatures for a very short time (on the order of a microsecond), but long enough for the tritium and deuterium nuclei to fuse and yield a large explosive burst. The uranium-238 blanket fissions from bombardment with fast neutrons, providing a substantial portion of the explosive energy.[2] The secondary stage may also contain uranium-235 or plutonium-239. In

1. Cochran, Arkin, and Hoenig 1984, pp. 27–28.

2. Coyle et al. 1988, pp. 19–21; Cochran, Arkin, and Hoenig 1984, pp. 27–28.

sum, thermonuclear weapons combine fission and fusion energy in complex ways to yield large explosions.

The environmental and health effects of nuclear weapons production depend on the nature of the specific materials and manufacturing processes that are used. For example, for weapons containing plutonium, the effects of nuclear reactors and plutonium-separation plants, called reprocessing plants, must be factored in, along with all the steps for uranium processing. For weapons containing only uranium, plants that "enrich" natural uranium (see below) use a great deal of energy and must be considered. Boosted fission weapons require tritium production. Thus, while some steps, such as uranium mining, are common to all nuclear weapons, others are not. Moreover, the amounts of uranium ore and the magnitude of subsequent processing needed for a weapon depend on its yield and design.

The nature of health and environmental problems also depends on the specific pollution controls, standard operating procedures, maintenance and waste-handling procedures, and standards for worker and off-site populations protection that were in place in the various countries at various times. Equally important is the question of whether these procedures and standards were actually followed. Topics that are specific to countries will be considered in later chapters. Here we consider the generic environmental effects of the various steps necessary to produce nuclear weapons.

URANIUM MINING AND REFINING

Uranium, a radioactive element, occurs in nature as an oxide; it is metallic in its elemental form. Although discovered in 1789 by the German chemist H. M. Klaproth, it had only minor commercial uses before the invention of nuclear explosives in the 1940s.[3] Since the mid-1950s, uranium has also fueled nuclear reactors that generate commercial electricity.

It is not always possible to distinguish uranium mining and processing activities on the basis of the uranium's end uses or to separate uranium facilities into those "for nuclear weapons" and those "for nuclear power." In some cases such distinctions are possible, and it is also sometimes possible to estimate how much of the work of a particular nuclear facility is devoted to military or to civilian purposes. However, once uranium finds its way to the world market, tracing its origins and use is not always possible.

In nuclear weapons, the nuclear reaction must not only sustain itself but very rapidly increase in scale to yield an explosion. A slower chain reaction would cause a bomb to fizzle. Therefore, the fissile material

3. Cochran et al. 1987a, p. 78.

must be compressed into a "supercritical" mass so that the number of neutrons—and hence the number of fissions—escalates very rapidly, yielding a sudden burst of energy. Surrounding the fissile material with a heavy material, called a tamper, helps prevent a premature blowing apart of the bomb before it can yield its design explosion. Finally, a neutron reflector helps prevent the loss of neutrons outward by reflecting them back into the fissile material. The tamper and reflector may be made of the same material.[4]

A supercritical mass is a prerequisite for the nuclear chain reaction that results in a nuclear explosion. Since uranium-235 can sustain a chain reaction and can be compressed into a supercritical mass, it can be the heart of a fission bomb, such as the one dropped on Hiroshima. One does not need pure uranium-235 to accomplish this. Typically, highly enriched uranium, containing more than 90 percent uranium-235, is used in nuclear weapons. It can also trigger thermonuclear fusion bombs. In the military sector, uranium-235, fuels reactors that propel nuclear-powered naval vessels and also is used in "materials production reactors" that yield other key nuclear weapons materials, notably plutonium and tritium.

One of the most important uses of uranium-238 in weapons production is its conversion via neutron bombardment in a reactor into plutonium-239. The uranium-238 absorbs a neutron, then converts to neptunium-239 by a beta decay, which in turn becomes plutonium-239 by another beta decay:

uranium-238 + neutron → uranium-239

uranium-239 → neptunium-239 + beta particle

neptunium-239 → plutonium-239 + beta particle

Uranium-238 has a half-life of about 4.5 billion years. It gives off radiation in the form of alpha particles. Since alpha particles have very little penetrating power relative to other forms of ionizing radiation, pure uranium-238 poses little health hazard as long as it remains outside the human body and contact is occasional (see chapter 4). However, uranium-238 does emit some X-rays, and prolonged exposure can cause significant external doses. If inhaled or ingested, uranium increases the risk of lung cancer and bone cancer. As a heavy metal, it is also a chemical toxin. The heavy metal toxicity of uranium is relatively more important than its radioactive toxicity when it is present in soluble form in the body (see chapter 4).

In natural uranium ores, uranium-238 is always found with its radioactive decay products (also called "daughter products" or "progeny"), including uranium-234, thorium-230, radium-226, and the gas radon-222. These decay products form a series or decay chain

4. Cochran, Arkin, and Hoenig 1984, pp. 25–26.

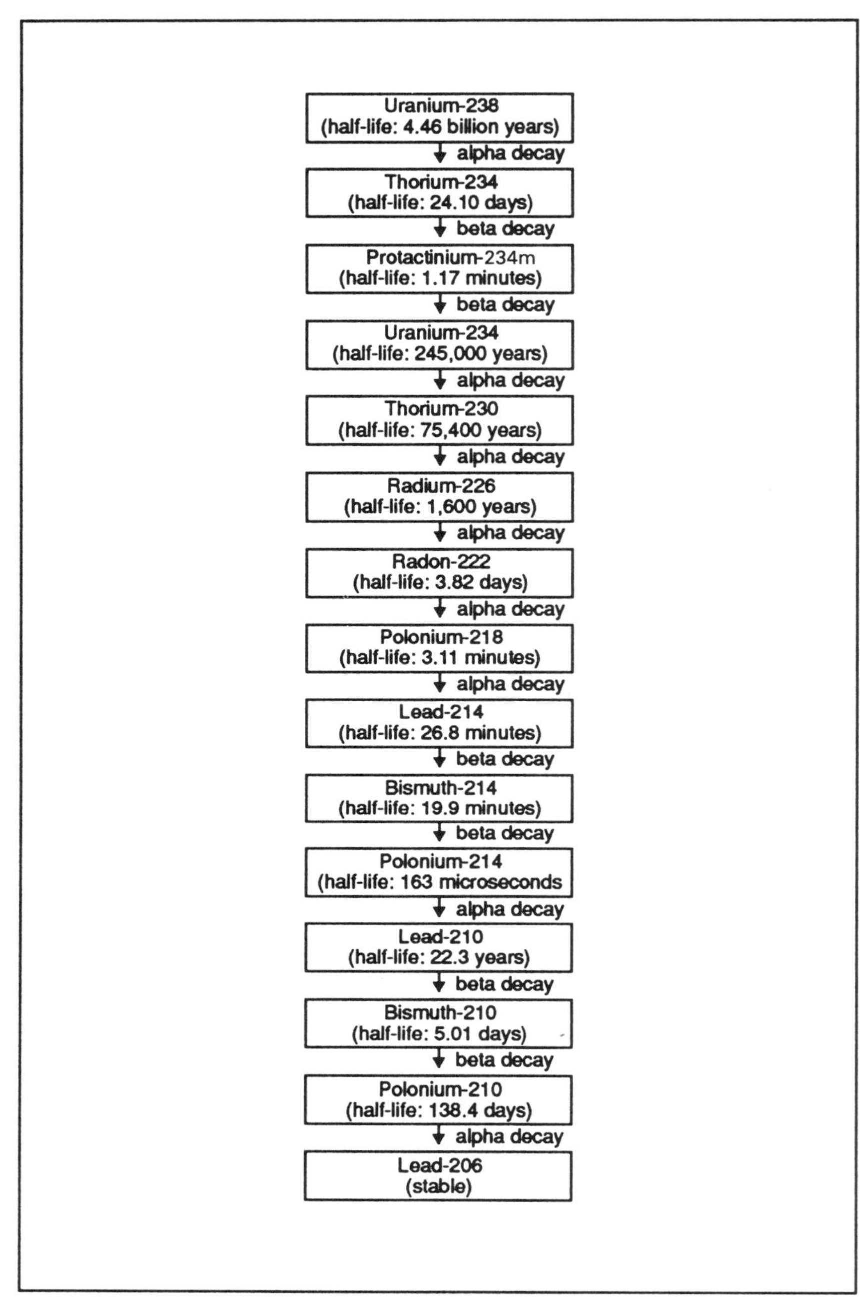

Figure 3.2 Main branches of the uranium-238 decay series. Source: Weast 1988.

Table 3.2 World Uranium Production and Resources in 1992

Country	Production (metric tons)	Uranium Resources (metric tons)[1]
Canada	9,385	275,000
Niger	2,965	166,000
Kazakhstan/Kyrghystan[2]	2,800	95,000
Uzbekistan/Tadjikistan	2,700	171,000
Russia	2,600	127,000
Australia	2,346	474,000
France	2,127	22,000
United States	1,808	117,000
South Africa	1,769	317,000
Namibia	1,692	115,000
Other	5,340	267,000
Total	35,532	2,146,000

1. Recoverable U_3O_8 @ $80 per kilogram. The estimate of recoverable reserves is a function of price.
2. Official estimates did not identify production and resources for these republics.

Source: Uranium Institute 1993.

of radioactive elements, starting with uranium-238 (see figure 3.2). Each element in a decay chain is called a decay or daughter product. Uranium-235 also has a decay chain, but it and its daughter products contribute less than 5 percent of the total radioactivity. Hence, hazards from the uranium-235 decay chain are small compared to the uranium-238 decay chain.

Uranium Mining

Uranium occurs in mineral ores in nature in various chemical compositions. A common form is called uraninite (more commonly known as pitchblende), which is an oxide with the chemical formula U_3O_8.[5] The main reserves of uranium are in the United States, Canada, Australia, South Africa, and Russia and several other republics of the former Soviet Union (see table 3.2). Significant reserves exist in the former East Germany and Czechoslovakia in a region known as the middle European "Uranium Province." These deposits have been heavily mined.

Most uranium mining takes place at open-pit and underground mines, and the ore is trucked to nearby mills for processing to separate

5. Benedict, Pigford, and Levi 1981, p. 232.

the uranium from the rock. Uranium mining produces large quantities of waste. This consists of unusable material that is below ore grade but may contain enough uranium (and its decay products) to pose a substantial health hazard. In the case of surface mining, the soil overburden (covering the extractable ore) also sometimes constitutes a waste, although this can be used to partially restore the original landscape.

A growing fraction of uranium mining occurs via solution mining (also known as *in situ* mining). Solution mining involves injecting chemicals (such as hydrochloric acid, alkaline carbonate solution, hydrogen peroxide, sodium hydroxide, or brine) into underground ores to dissolve the uranium. The solution, with the uranium, is pumped to the surface and processed to extract the uranium. The resulting liquid waste is injected into deep wells or sprayed over the surface,[6] while solid "low-level" radioactive wastes[7] are disposed of at conventional low-level radioactive waste dumps. Although not without problems, this type of mining presents relatively less environmental damage than does conventional mining because it produces no tailings.[8]

Uranium can be recovered as a by-product of other materials production from extremely low-grade ores (0.01 to 0.03 percent uranium) via a process known as "heap leaching" or "lixiviation." A leaching solution (typically sulfuric acid or ammonium carbonate) is percolated down through an ore pile. The leachate that accumulates at the bottom is collected and recirculated until the uranium concentration in the leachate solution is high enough to make uranium extraction using ion-exchange resins economical.[9] Uranium is also recovered as a by-product of gold and phosphate mining.

Uranium Refining: Milling

The amount of uranium in conventionally mined ore is typically only 0.1 to 0.2 percent (although in rare cases it can reach 10 percent or even higher). Thus, uranium ore is typically refined at nearby uranium mills.

In the first refining step, ore is crushed and mixed with water in a wet-grinding process to produce a fine-grained mixture. This mixture is put through either an acid-leaching or alkaline-leaching process

6. Texas Department of Agriculture 1986, p. 2.

7. There is no consistent international usage of terms to describe radioactive wastes. In the United States the term "low-level" wastes encompasses materials ranging from those that truly have small amounts of radioactivity (and hence fit the term "low-level") to wastes that are more radioactive than some officially classified as "high-level" waste. See Makhijani and Saleska 1992, p. 26.

8. National Research Council 1986, p. 151.

9. See Cochran et al. 1987a, pp. 124–125.

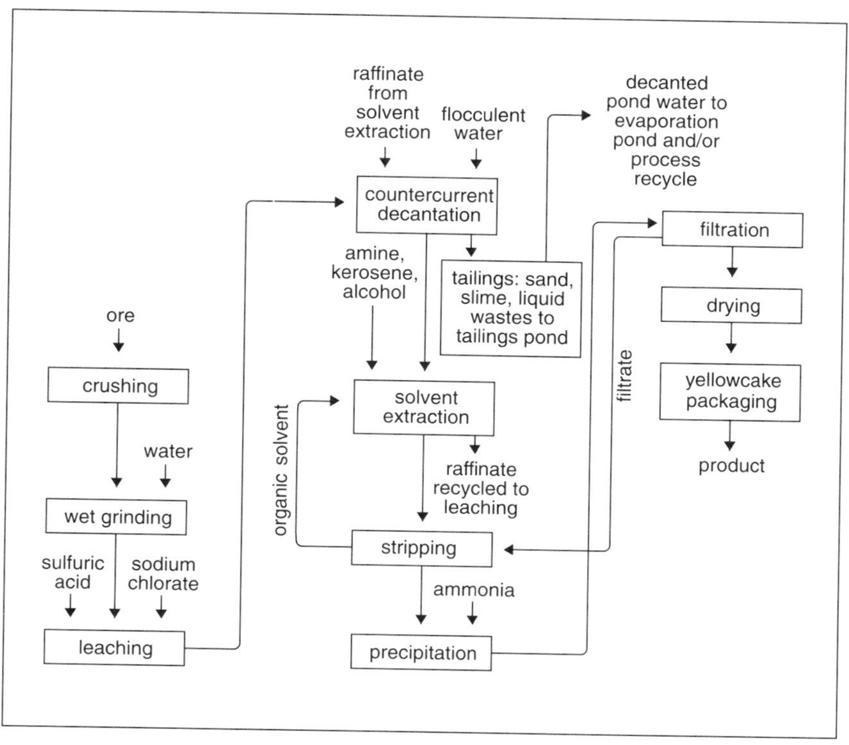

Figure 3.3 Yellowcake production: acid-leach process. Source: U.S. NRC 1980, p. 5-3.

(depending on the chemical composition of the ore) to convert the uranium to a soluble form. The resulting solution is passed through ion exchangers to maximize the uranium content. Subsequent steps to strip, dry, wash, remove impurities, and roast the product result in what is known as yellowcake (see figures 3.3 and 3.4).[10] Yellowcake is mainly U_3O_8; its yellow color derives from another chemical form, ammonium diuranate, created during refining and present to varying extents in the final product. Pure U_3O_8 itself is black.

ACID-LEACH AND ALKALINE-LEACH PROCESSES

The waste remaining after the uranium is extracted (for low-grade ores, virtually all the weight of the originally mined material) is discharged to settling ponds. This waste, called uranium mill tailings, often remains at the mill site. At the time of discharge, the tailings consist of a slurry that is about 40 percent solids and 60 percent liquid. Some of the liquid is recycled for further use in milling; the water content of the remaining liquids evaporates or percolates into the soil;

10. National Research Council 1986, p. 25.

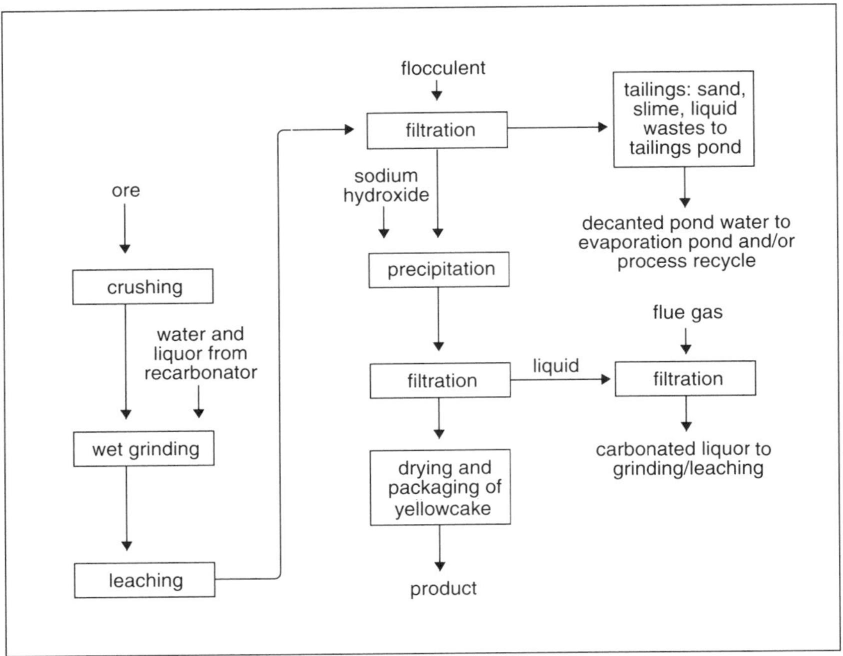

Figure 3.4 Yellowcake production: alkaline-leach process. Source: U.S. NRC 1980, p. B-8.

the solid portion of tailings continues to accumulate. A typical conventional mill has the capacity to process 2,000 metric tons of uranium ore per day, most of which ends up as mill tailings.[11] Such a mill would produce 2 to 4 metric tons of uranium per day, assuming a typical feed concentration of 0.1 to 0.2 percent U_3O_8.

The primary wastes from uranium refining are these mill tailings. The tailings contain over 85 percent of the radioactivity of the original ore, primarily in the form of radium-226 (half-life, 1,600 years) and thorium-230 (half-life, 75,400 years). The average radioactivity of mill tailings is fairly low—usually on the order of a nanocurie (37 becquerels) per gram or less[12]—but mill tailings are produced in huge quantities, and their radioactivity is extremely long-lived. Moreover, their radioactivity is typically hundreds of times higher than that naturally present in surrounding soils near mill sites. Not only are the tailings radioactive, but they also contain a variety of nonradioactive but nonetheless toxic chemicals, including chlorides, sulfates, and heavy metals.[13]

11. National Research Council 1986, p. 29.

12. National Research Council 1986, pp. 36–37. Mill tailings sometimes can have concentrations of radium-226 and thorium-230 greater than this.

13. National Research Council 1986, p. 1.

Mill tailings account for over 95 percent of the total volume of radioactive waste from all stages of the nuclear fuel cycle.[14] If not isolated, they will release radioactivity into their surroundings for hundreds of thousands of years. Mill tailings have all too often been neglected. Many tailings piles remain unremediated.

To prevent air contamination, tailings can be kept underwater. This essentially eliminates radon-222, radium-226, and thorium-230 air emissions. However, a further major problem at mill tailings sites is groundwater pollution that could result from unlined tailings ponds. Surface water pollution and soil contamination can also result via runoff from an overflow of the ponds. Radioactive materials and nonradioactive heavy metals in milling waste could contaminate the food chain via this contaminated water.

Protecting groundwater from future contamination involves constructing lined tailings areas to reduce the amount of contaminants seeping out. Extensive groundwater monitoring is generally necessary to check on the continued efficacy of the liner. Existing groundwater contamination can be addressed by pumping contaminated water back into the lined tailings ponds if the evaporation capacity of the ponds and the amount of pumping required are compatible.

The principal difficulties with such remedial measures result from improper installation of the liners—they tend to tear—and a failure to keep all the tailings under enough water to prevent air emissions. The long-term problem is that the half-life of thorium-230, the parent radionuclide for radium-226, is over 75,000 years. It is highly doubtful that the above measures can be sustained even a small fraction of that time. (U.S. regulations, for example, cover only 1,000 years.)

URANIUM CONVERSION AND ENRICHMENT

The various steps in processing uranium generate a variety of wastes and pollutants. Solid, sludge, and liquid wastes are generated and discharged, and radioactive and nonradioactive pollutants are emitted to the air. The uranium processing plant near Fernald, Ohio, processed about 500,000 metric tons of uranium for conversion to metal over about 35 years. The air emissions are estimated to have been between 370 and 600 metric tons.[15] In addition, large quantities of radioactive and nonradioactive materials were dumped into pits on the site.

Naturally occurring uranium contains about 0.711 percent uranium-235 by weight. Another 99.284 percent is uranium-238, with trace

14. Based on the total volume of all radioactive waste (spent fuel, high-level waste, transuranic waste, low-level waste, and uranium mill tailings) from all sources (commercial and military) produced in the United States since the 1940s. Based on U.S. Department of Energy records as compiled in Saleska et al. 1989, appendix C.

15. Voillequé et al. 1993, p. 39.

amounts of uranium-234 (about 0.005 percent). This concentration of uranium-235 is too low to produce the supercritical mass needed for nuclear weapons. Therefore, natural uranium is "enriched" to increase the proportion of uranium-235 relative to the other isotopes.

The extent of enrichment determines the potential uses of the uranium. Enriched to about 1 percent U-235, uranium fuels some plutonium production reactors, although most military production reactors have used natural uranium. Enriched to about 3 to 5 percent, it fuels commercial light-water power reactors. At 20 percent or greater, it fuels many research and test reactors, and at 97.3 percent, it fuels U.S. naval reactors. As the primary component in a nuclear bomb, uranium is enriched to about 93.5 percent U-235.[16] Theoretically, an enrichment of 5 to 6 percent uranium-235 is enough to make a nuclear explosion, but the quantities of bulk uranium required would be too large for practical explosives.[17]

The Uranium Conversion Process

In preparation for enrichment, uranium is typically converted from yellowcake into uranium hexafluoride. The conventional "wet process" first purifies yellowcake by dissolving it in nitric acid. This produces an impure uranyl nitrate solution from which the uranyl nitrate is extracted in tributylphosphate, an organic solvent (see figure 3.5). The organic stream is brought into contact with a dilute acidic solution to strip out the uranium. The aqueous uranyl nitrate is concentrated and sent into denitrators that convert it to uranium trioxide powder. The powder is reduced with hydrogen to produce uranium dioxide. The uranium dioxide is then reacted with hydrogen fluoride gas, producing uranium tetrafluoride, and finally with fluorine gas to produce uranium hexafluoride.[18]

The "dry process" bypasses the initial solvent-extraction step (see figure 3.6). Uranium ore concentrates are reduced with hydrogen to impure uranium dioxide, which is reacted with hydrogen fluoride to produce impure uranium tetrafluoride, which is fluorinated to give impure uranium hexafluoride. Purification is done by fractional distillation after the yellowcake is completely converted to "dirty" uranium hexafluoride. A plant at Metropolis, Illinois, uses the dry conversion process.[19]

16. Cochran et al. 1987a, p. 125.

17. Cochran, Arkin, and Hoenig 1984, p. 23.

18. Benedict, Pigford, and Levi 1981, pp. 269–271.

19. Benedict, Pigford, and Levi 1981, pp. 272–273.

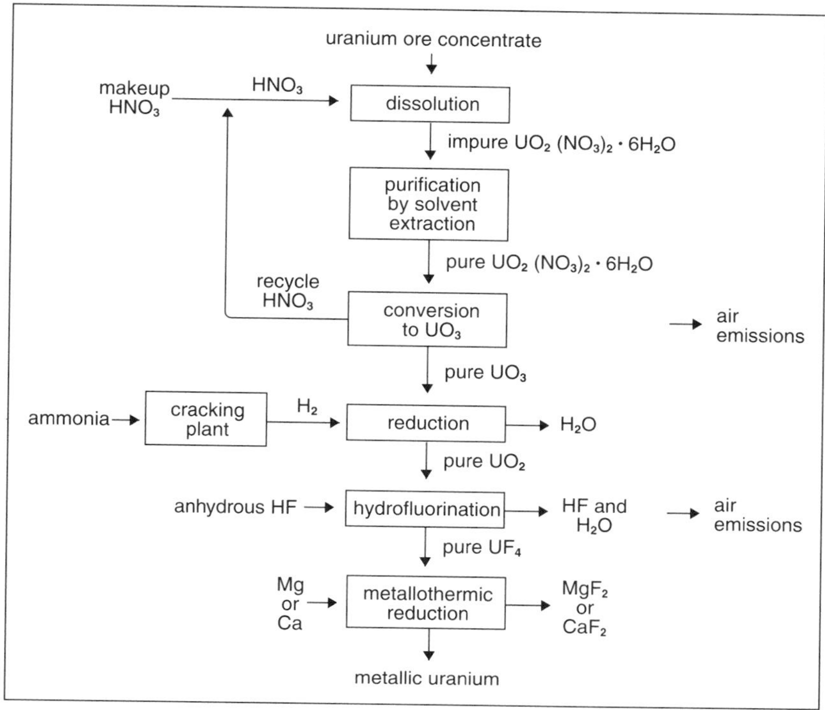

Figure 3.5 Steps in uranium metal production. Adapted from Benedict, Pigford, and Levi 1981, p. 269, with permission of McGraw-Hill, Inc.

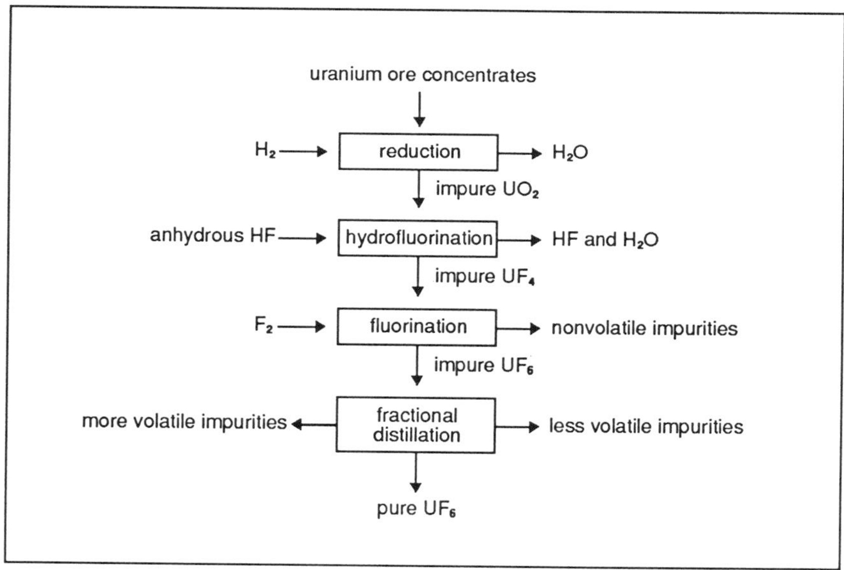

Figure 3.6 Dry-process uranium hexafluoride production. Source: Benedict, Pigford, and Levi 1981, p. 273; reproduced with permission of McGraw-Hill, Inc.

The Uranium Enrichment Process

After conversion, uranium hexafluoride goes to a plant that produces enriched uranium hexafluoride. Such plants produce low-enriched uranium for use in various power and military reactors, and highly enriched uranium containing about 93.5 percent uranium-235 for use in military weapons and 97.3 percent uranium-235 for U.S. naval reactors.

The enrichment process neither creates nor destroys uranium. The amount of both uranium-235 and uranium-238 that enters an enrichment plant equals the amount that leaves, but the uranium that leaves is divided into two streams: the enriched product (with a higher-than-natural concentration of uranium-235) and the depleted uranium tails (with a lower-than-natural concentration of uranium-235). The concept that the total amount of material involved remains constant throughout enrichment is called "materials balance."

The effort expended in enrichment is expressed in kilogram separative work units, often referred to as SWUs (pronounced "swooze"). SWUs are a measure of the work required to separate uranium-235 from uranium-238. The greater the amount of enriched uranium output and the higher the enrichment required, the greater the amount of separative work required. Separative work units are generally measured in kilograms or metric tons.

It takes about 4.3 kilogram SWUs of separative work to produce 3 percent enriched uranium and about 236 kilogram SWUs to produce 93 percent enriched uranium, if one begins with natural uranium and the tails content of uranium-235 is 0.2 percent.[20] (We will use kilogram units of SWU in this book—hence the SWU below means kilogram SWU.)

Enrichment services are priced in dollars per SWU. In general, capital investment is proportional to an enrichment plant's capacity in SWUs. Similarly, annual operating costs are approximately proportional to the amount of separative work done.

Enrichment Technologies

Since uranium-235 and uranium-238 are isotopes of the same element, they have very similar chemical properties. Separation of the isotopes into one stream relatively rich in uranium-235 ("enriched uranium") and another relatively depleted in uranium-235 ("depleted uranium") is generally based on the slightly different weights of these isotopes.

The two principal commercial enrichment methods, gaseous diffusion and gas centrifuge, use uranium hexafluoride in gaseous form. Detailed descriptions and theoretical bases for these methods appear

20. Cochran et al. 1987a, p. 127.

Figure 3.7 The K-25 gaseous diffusion plant at Oak Ridge. Photo by Robert del Tredici.

in many textbooks.[21] Uranium hexafluoride is a solid at room temperature. It is converted into a gas by heating to at least 64°C.

In the gaseous diffusion method, uranium hexafluoride gas is pressurized and fed into a chamber, called a converter, with a diffusion barrier. Uranium-235 is slightly lighter than uranium-238, so it diffuses more easily through the barrier. Because uranium-235 and uranium-238 weigh almost the same, only a little separation occurs at each stage. Typically, thousands of separation stages are required to produce highly enriched uranium.

The gas centrifuge method also exploits the differences in weight but uses a rapidly spinning centrifuge to separate the isotopes (see figure 3.8). The heavier uranium-238 concentrates at the outer radius and is made to flow in one direction; the uranium-235, which is enriched near the centerline of the rotating cylinder, is made to flow in the opposite direction. As with gaseous diffusion, only a small amount of separation occurs at each stage. The details of arranging the large numbers of converters or centrifuges and the flow of uranium hexafluoride through them are quite complex. The gaseous diffusion process typically takes 5 to 20 times the energy of a gas centrifuge plant.

Electromagnetic separation was used to enrich uranium at Oak Ridge during World War II for the Hiroshima bomb, and it was at-

21. For example, see Benedict, Pigford, and Levi 1981, chapter 14.

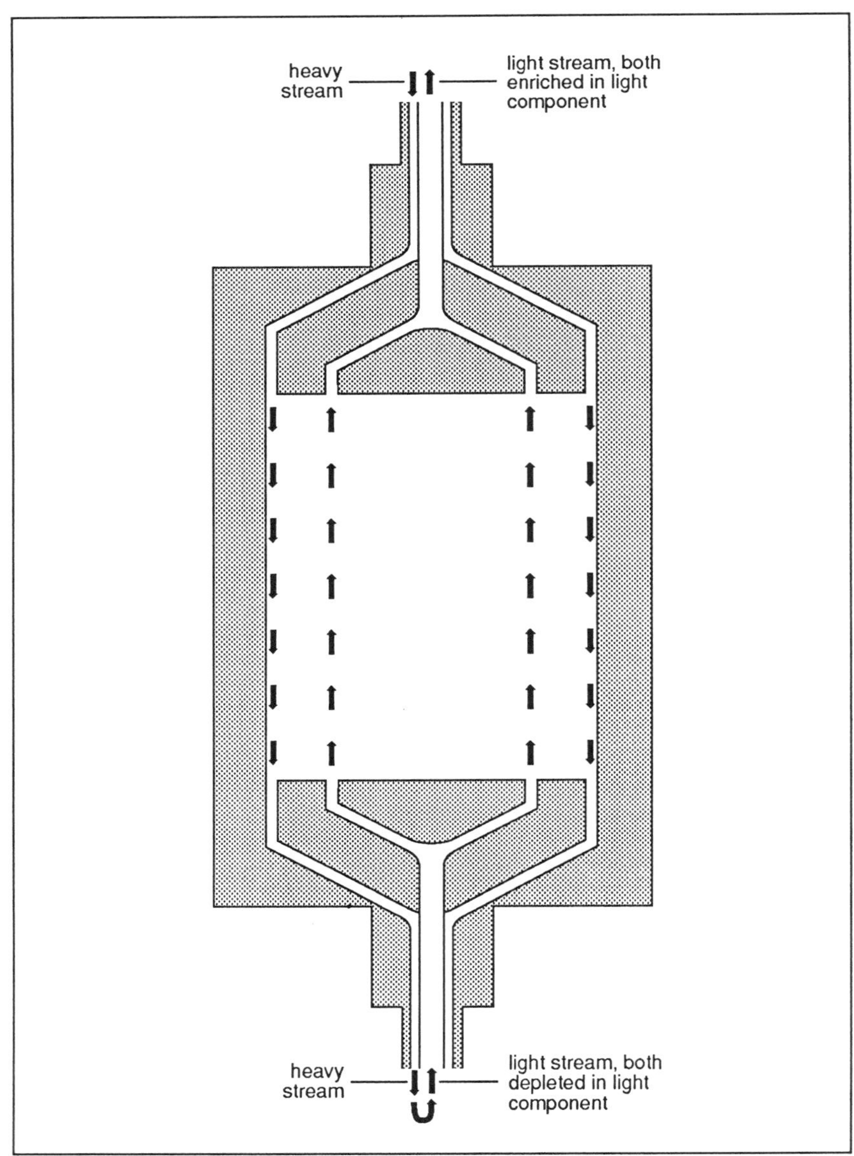

Figure 3.8 Gas centrifuge. Source: Benedict, Pigford, and Levi 1981, p. 848; reproduced with permission of McGraw-Hill, Inc.

tempted in Iraq more recently. In this technique, uranium is ionized —that is, one or more electrons in its outer shell are knocked out— accelerated, and injected into a magnetic field. The particles bend while they travel in the magnetic field, but the lighter particles bend more than the heavier ones, thus achieving a separation. Due to its relative expense, the United States stopped using this method in 1946.[22]

22. Benedict, Pigford, and Levi 1981, p. 815.

The more advanced atomic vapor laser isotope separation (AVLIS) process uses uranium vapor and intense radiation in a narrow frequency band to make atoms containing uranium-235 undergo a different physical or chemical process than those containing uranium-238.[23] At a demonstration plant at the Lawrence Livermore National Laboratory in California, electric fields separate the ionized isotopes. (The plant has not been opened, however.)

Other, less common enrichment technologies have been researched and in some instances developed as well. These include a chemical enrichment process (under development primarily in France, which uses a process called "Chemex") and a plasma separation process that was under development in the United States. In several aerodynamic processes—once used in West Germany, South Africa, and Brazil—a mixture including uranium hexafluoride gas flows rapidly along a sharply curved path, causing a centrifugal action that partially separates the isotopes.[24]

To complete the conversion of uranium hexafluoride enriched product or depleted tailings into uranium dioxide or metal, the usual approach takes an intermediate step, reducing the uranium hexafluoride to uranium tetrafluoride by reaction with hydrogen. By-product hydrogen fluoride is recovered for recycling in the yellowcake-to-uranium-hexafluoride conversion. Uranium dioxide may be obtained by reacting the uranium tetrafluoride with steam and reducing it to uranium dioxide, producing more hydrogen fluoride.[25] Reducing the uranium tetrafluoride with magnesium yields uranium metal and magnesium fluoride waste.[26]

Enriching uranium produces a variety of wastes. Notable among these is depleted uranium, which contains 0.2 percent to 0.4 percent uranium-235, the rest being uranium-238 (with a trace of uranium-234). Tens of thousands of metric tons of depleted uranium tails are discharged each year from uranium enrichment plants around the world. In the United States, most of the waste is stored as uranium hexafluoride in cylinders stacked outdoors at enrichment plants. As of the end of 1990, Department of Energy enrichment plants had about 320,000 metric tons of depleted uranium on site in uranium hexafluoride tailings that were stored in some 40,000 cylinders; the plants continued to produce 20,000 metric tons annually.[27] In Western

23. Benedict, Pigford, and Levi 1981, p. 817.

24. Cochran et al. 1987a, p. 134; Albright, Berkhout, and Walker 1993, p. 12.

25. Benedict, Pigford, and Levi 1981, p. 274.

26. Benedict, Pigford, and Levi 1981, p. 276; Makhijani 1988, p. 3.

27. Martin Marietta 1990. The weight of depleted uranium is on an elemental basis.

Europe, the URENCO enrichment enterprise, which operates several gas centrifuge enrichment plants, stores over 30,000 metric tons of depleted uranium in uranium hexafluoride tails.[28] Some of the depleted uranium tails are converted into uranium oxide or metal and have a variety of military and civilian applications, such as artillery shells, tank armor, and gyroscopes. However, so much depleted uranium is generated that most of it should be considered waste. Further, the use of this radioactive material in tank armor and artillery shells has given rise to considerable controversy since the 1991 Persian Gulf War because of its possible link to health effects.

Each kilogram of 93 percent uranium-235 (weapons-grade uranium) requires about 182 kilograms of natural uranium to produce, assuming 0.2 percent uranium-235 in the tailings.[29] With an average of 15 to 20 kilograms of highly enriched uranium per nuclear weapon, each weapon would require the mining of about 1,500 metric tons of ore, and would generate about 3,000 kilograms of depleted uranium waste containing about 1 curie (37 billion becquerels) or more of radioactivity. The mill tailings would contain about 1 curie (37 billion becquerels) each of radium-226 and thorium-230. Additional wastes are generated from further uranium processing and from the plutonium components.

Potential Hazards from Uranium Conversion and Enrichment

The handling of uranium hexafluoride, which is highly corrosive and chemically toxic as well as radioactive, presents a major occupational and health hazard during both uranium conversion and uranium enrichment. When uranium hexafluoride touches water—including moisture in air—it breaks down into hydrofluoric acid, a highly corrosive toxin, and uranyl fluoride, a heavy-metal compound.

Each of these compounds poses health hazards. Hydrofluoric acid causes pulmonary edema (excess body fluids in the lungs) and other respiratory damage, severe burns, and irritation. Uranyl fluoride poses a radiological hazard. It also has toxic effects—principally temporary or permanent kidney damage—because uranium is a heavy metal.[30]

Waste from conversion and enrichment plants can pose environmental problems if not adequately managed and disposed of. Waste from the dry conversion process includes low-level radioactive solids containing long-lived alpha emitters as well as chemical wastes. The wet conversion process generates a liquid stream (called raffinate) con-

28. Lippard and Davis 1991, p. 5.

29. Cochran et al. 1987a, p. 127.

30. U.S. Congress, House 1987, pp. 4,6.

Table 3.3 Annual Waste Generation from a Typical 10,000-Metric-Ton-Capacity Uranium Conversion Plant (cubic meters)

Conversion Technology	Solid Low-Level Radioactive Wastes	Chemical Wastes	Fluoride Settling Pond
Direct fluorination (dry process)	457	63	617
Solvent extraction fluorination (wet process)	595	375	—

Source: U.S. DOE 1992, pp. 268–269.

taining dissolved radioactive solids; this stream generally is dumped into a settling pond. Pond evaporation creates a sludge—containing radium, thorium, and uranium—that is handled as low-level radioactive waste (see table 3.3).[31]

Enrichment produces small quantities of radioactive effluents, including uranium (which is sometimes discharged to the atmosphere) and low-level radioactive liquids (which are released to holding ponds).[32] For instance, a 10-million-SWU gaseous diffusion enrichment plant produces about 230 cubic meters of low-level radioactive waste per year.[33] A typical gas-centrifuge plant of the same size produces many times more waste, about 1,700 cubic meters of low-level radioactive waste. Waste from enrichment facilities also includes nonradioactive toxic chemical waste, such as polychlorinated biphenyls (PCBs), chlorine, ammonia, nitrates, zinc, and arsenic.[34] Many enrichment plants routinely release fluorine gas and hexavalent chromium to the atmosphere.[35]

The low-level radioactive waste from conversion and enrichment plants poses radiological hazards and must be isolated from the environment for long periods of time. It is typically buried in landfills; in some cases, it has been dumped into the ocean. In the past, many radioactive landfills have leached radionuclides (radioactive elements) into the surrounding environment. For example, radioactive contamination has been measured outside four of the six commercial low-

31. Lipschutz 1980, p. 36.

32. Lipschutz 1980, p. 36.

33. Based on source-term in U.S. DOE 1989a, p. 213, of 2.29 × 10–5 cubic meters of low-level waste per SWU.

34. U.S. GAO 1985, p. 31.

35. U.S. enrichment plants routinely release various fluorides. For instance, see U.S. GAO 1985 for Portsmouth plant emissions. The U.S. Department of Energy's Portsmouth enrichment plant releases 30 to 40 pounds of hexavalent chromium to the atmosphere daily, U.S. DOE, as cited in the *New York Times*, 7 December 1988.

level waste-disposal sites in the United States.[36] To varying degrees, groundwater is contaminated at all the principal military nuclear production sites in the United States, most of which dispose of low-level radioactive waste on-site.[37]

Accidents at Conversion Facilities

One of the dangers posed by enrichment facilities is the possibility that an accident could allow enough enriched uranium hexafluoride to concentrate to cause a self-sustaining nuclear reaction—a "criticality" accident. A number of incidents have occurred that might have led to such an accident had they not been discovered in time. At least one such incident occurred in 1982, in 1983, and in 1984 at the Portsmouth, Ohio, enrichment plant.[38] In 1981, a near-critical event occurred at the Oak Ridge, Tennessee, enrichment facility: a compressor failed, and friction from metal rubbing against metal caused a uranium hexafluoride-metal reaction.[39]

Since many uranium conversion plants are far from enrichment facilities, uranium hexafluoride must often be transported long distances—in many cases, from continent to continent. This poses some hazard as well, and at least one serious transport accident has taken place. In August 1984, the French freighter *Mont Louis* sank in 14 meters of water in the English Channel near Belgium after colliding with a passenger ferry. No one was hurt in the collision, but the *Mont Louis* was carrying 375 tons of uranium hexafluoride in 30 containers, sent by the French nuclear agency Cogéma and destined for enrichment facilities in the Soviet Union. Because of uranium hexafluoride's violent reactivity with water, there was concern that some of the containers would be punctured if leaks developed in them. Fortunately, after a difficult recovery operation, the last container was recovered unbreached about six weeks after the accident.[40]

PLUTONIUM PRODUCTION REACTORS

Two key facilities are needed to produce plutonium. First, a nuclear production reactor transmutes uranium-238 into plutonium-239. In the

36. Three of the six U.S. commercial sites have been closed and radionuclides have migrated into the surrounding environment (Jordan 1984, pp. 4–6). At one still-operating site (Barnwell in South Carolina), radioactive tritium has been detected in monitoring wells.

37. U.S. DOE, as cited in *Wall Street Journal*, 5 December 1988.

38. Coyle et al. 1988, p. 86.

39. Coyle et al. 1988, p. 75.

40. See May 1989, p. 275.

Figure 3.9 The plutonium-producing L-reactor at the Savannah River Site. Photo by Robert del Tredici.

reactor the plutonium is mixed with some remaining uranium and fission products. Second, a chemical separation or reprocessing plant separates the plutonium from the other materials.

A nuclear reactor produces plutonium-239 by irradiating uranium-238 with neutrons. These neutrons generally come from fission of uranium-235 fuel. Since uranium-235 can sustain a fission chain reaction driven by neutrons, any nuclear reactor can produce plutonium.

There are two approaches to producing plutonium-239 in a reactor. In one, all the fuel rods inserted into the reactor drive the fission reaction. These same rods contain uranium-238 that captures some of the neutrons to get converted to plutonium-239. This scheme produces plutonium in some military and all civilian power reactors. Some military reactors use the other approach, in which special "target rods" containing depleted uranium (approximately 99.8 percent uranium-238) are inserted into the reactor to make plutonium. "Driver" rods (another name for fuel rods), which contain enriched uranium, supply the neutrons required for plutonium production.

Plutonium-239 in a reactor can absorb a neutron to yield plutonium-240; additional absorption of neutrons yields the isotopes plutonium-241 and -242 successively. Long, intense irradiation in a nuclear reactor, typical of civilian power reactors, results in higher proportions of these three isotopes,[41] which are relatively undesirable for weapons.

Various types of reactors produce plutonium. Some are dedicated military reactors that produce no usable power or energy. Except for the N-reactor at Hanford (which produced steam sold for commercial electricity production, in addition to plutonium), this is the case in the United States. Some production reactors in the former Soviet Union also produce steam for district heating and/or electricity.

Nuclear reactors can sustain fission reactions for long periods of time, limited essentially by the need to remove the accumulations of waste products in the fuel. Fission reactors derive their energy from chain reactions in fissile materials assembled into critical masses, most commonly uranium-235. The main difference between fission reactions in a nuclear reactor and those in a nuclear weapon is that in the former, the reactions are controlled and sustained at an even and controllable level; in a weapon, the number of reactions grows very rapidly until the device explodes. The control of nuclear fission in a reactor is maintained by keeping overall neutron production at a level that results in each fission reaction yielding exactly one more fission reaction. Control rods, made of materials such as boron, absorb neutrons, and their positions are adjusted to change the level of power production.

Almost all nuclear reactors today are based on chain reactions sustained by "slow" or "thermal" neutrons (see figure 3.10) obtained by slowing down the "fast" neutrons emitted from a fissioning uranium-235 nucleus. A "moderator," a light atom like hydrogen or carbon, slows down the neutrons without absorbing them. Frequently, the moderator also serves as a coolant, carrying away the heat generated by fission so the reactor core doesn't melt or burn. Regular (light) water is the coolant and moderator in most civilian reactors. In early U.S. production reactors, Soviet production reactors, and Chernobyl-type power reactors, graphite is the moderator and water the coolant. Gas-graphite reactors use a gas (such as air, helium, or carbon dioxide) as a coolant. Heavy-water reactors, such as those at the Savannah River Site (for military purposes) or in Canada (for generating power), use deuterium, a stable isotope of hydrogen, in place of the ordinary hydrogen in water as a coolant and moderator.

Older production reactors have "once-through" cooling systems: the cooling water that carries heat away is discharged directly to the environment. Such reactors can discharge considerable fission and activation products. Newer reactors, more typical of the ones built in the United States in the 1950s and after, filter the coolant to collect the radioactive materials in filter resins. These are disposed of as low-level

41. Lamarsh 1983, p. 105.

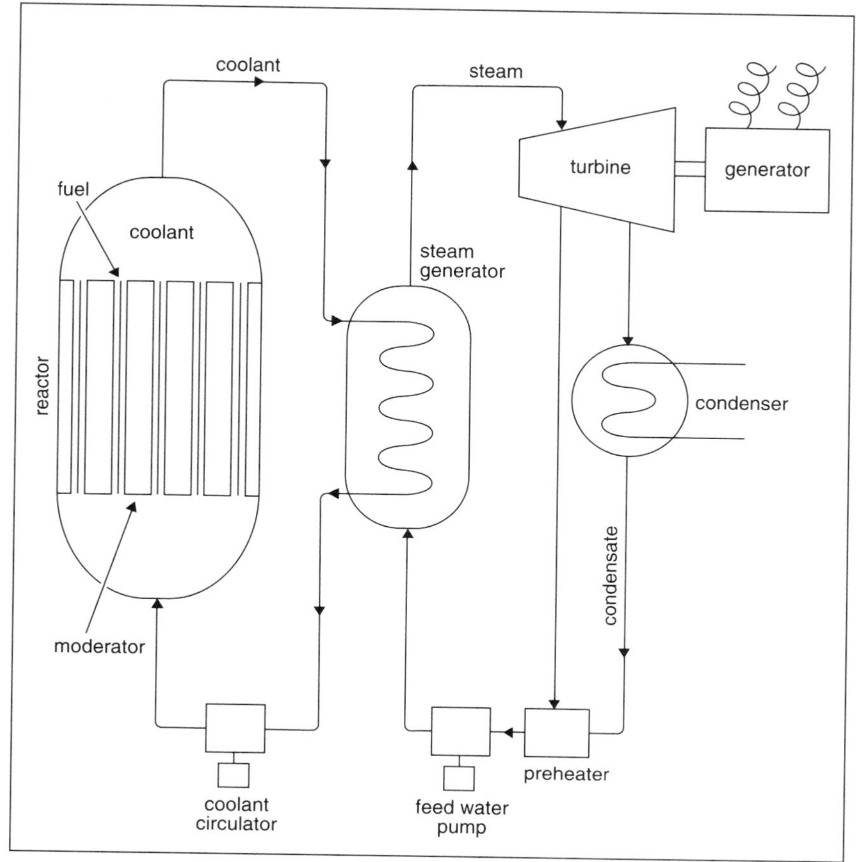

Figure 3.10 Schematic of an electric power production reactor. Source: Benedict, Pigford, and Levi 1981, p. 8; reproduced with permission of McGraw-Hill, Inc.

radioactive wastes in the United States and intermediate-level wastes in the United Kingdom and France.

Nuclear reactors are vulnerable to various kinds of accidents, the most dangerous of which are fires, explosions, or meltdowns from the loss of cooling fluids. The now-closed N-reactor at Hanford, for instance, is of a design similar to the Chernobyl reactor, which had the worst reactor accident by far in nuclear history in 1986. A similar accident, on a much smaller scale, occurred at a graphite-moderated British nuclear reactor at Sellafield in 1957. (The facility was then called Windscale.)

PLUTONIUM SEPARATION: REPROCESSING

Reprocessing chemically separates plutonium and uranium from fuel that has been irradiated in reactors. Generally, it is the plutonium that is of interest for weapons production; uranium is essentially a by-product that can be recycled as fuel. Reprocessing is also a part of the civilian nuclear fuel cycle in some countries, notably the United Kingdom, France, Russia, and Japan. It is a key link between civilian nuclear power and nuclear weapons production.

The operation of nuclear reactors, whether civilian or military, converts some uranium-238 in the fuel rods into plutonium-239. Plutonium-239, in turn, can be used to generate electrical power or to make nuclear weapons. Through the reprocessing of spent fuel, this plutonium can be separated chemically from uranium and other elements. Spent fuel in power plants is typically "high burn-up" spent fuel—that is, it has been irradiated for extended periods to generate a large amount of energy. Conversely, uranium irradiated for extracting plutonium for weapons is "low burn-up" fuel. Typically, weapons-grade fuel might be irradiated on the order of 1,000 megawatt days per ton of uranium or less,[42] while fuel in a light-water power reactor would be irradiated for about 30,000 megawatt days.

Fuel for producing plutonium for weapons is irradiated for a short time to assure a larger proportion of the fissile isotope plutonium-239 relative to the undesirable isotope plutonium-240. "Super-grade" plutonium contains 2 to 3 percent plutonium-240. "Weapons-grade" plutonium is less specific but is used to describe plutonium generally containing less than 7 percent plutonium-240. "Fuel-grade" plutonium contains 7 to 18 percent plutonium-240, while "reactor-grade" contains over 18 percent plutonium-240.[43] Because plutonium-239 and plutonium-240 are difficult to separate—they are close in atomic weight and both are highly toxic—the principal means of assuring adequate purity of plutonium-239 is to control reactor burn-up. Another means is to mix "super-grade" weapons plutonium, which contains less than 3 percent plutonium-240, with fuel grade plutonium to obtain weapons-grade material.[44]

Since civilian nuclear reactors can also produce low burn-up irradiated fuel and since plutonium from high burn-up fuel, although less desirable for weapons production, can still be used for this purpose, *it is the existence of a reprocessing plant that essentially gives a country the ability to produce plutonium for nuclear weapons*. All countries with operating reprocessing plants are therefore classified as actual or potential nuclear-weapons powers (see tables 3.4 and 3.5). Even in countries that have reprocessing plants but no declared nuclear weapons production, the potential materials availability and technical capabilities are prime indicators of the ability to make nuclear weapons.

42. Albright, Berkhout, and Walker 1993, p. 16.

43. Albright, Berkhout, and Walker 1993, p. 15.

44. Albright, Berkhout, and Walker 1993, p. 28.

Table 3.4 Commercial Reprocessing Plants

Country	Owner/ Location	Operator[1]	Facility[2]	Design Fuel[3]	Capacity[4]	Years of Operation
UK	Windscale/ Sellafield	BNFL	B205	metal	1,500	1964–2010 (?)
			B204/205	oxide	300	1969–1973
			THORP	oxide	700	1994–
	Thurso	UKAEA	DNPDE	oxide (MTR)	<1	1959–1966 (?)
				oxide (FBR)	7	1958–1995 (?)
France	Marcoule	Cogéma	UP1	metal	400	1958–2000 (?)
	La Hague	Cogéma	UP2	metal	400	1966–1987
				oxide	400	1976–1990
			UP3	oxide	800	1990–
			UP2-800	oxide	800	1994–
	Marcoule	CEA	APM	oxide (FBR)	6	1988–
USA	West Valley	NFS	West Valley	oxide	300	1966–1972
Russia[5]	Chelyabinsk-65	Minatom	Mayak	oxide	400	1978–
Japan	Tokai-mura	PNC	Tokai	oxide	90	1981–
	Rokkasho-mura	JNFS	Rokkasho	oxide	800	2002–
India	Tarapur	DAE	PREFRE	oxide	100	1982–
	Kalpakkam	DAE		oxide	100–200	1993/94–
FRG	Karlsruhe	KfK/DWK	WAK	oxide	35	1971–1990
Belgium	Mol	Eurochemic		oxide	30	1966–1974

1. BNFL = British Nuclear Fuels plc, UKAEA = U.K. Atomic Energy Authority, Cogéma = Compagnie Générale des Matières Nucléaires, CEA = Commissariat à l'Énergie Atomique, NFS = Nuclear Fuel Services Company, Minatom = Ministry of Atomic Energy, PNC = Power Reactor and Nuclear Fuel Development Corp., JNFL = Japan Nuclear Fuel Limited Company, DAE = Department of Atomic Energy, KfK = Kernforschungszentrum Karlsruhe, DWK = Deutsche Gesellschaft für Wiederaufarbeitung von Kernbrennstoffe.
2. THORP = Thermal Oxide Reprocessing Plant, DNPDE = Dounreay Nuclear Power Development Establishment, APM = Atelier Pilote Marcoule, PREFRE = Power Reactor Fuel Reprocessing, WAK = Wiederaufarbeitungsanlage Karlsruhe.
3. LWR fuels unless stated otherwise. MTR = materials test reactor, FBR = fast breeder reactor.
4. Capacity is in metric tons of heavy metal per year.
5. Construction began on another plant at Krasnoyarsk-26 (RT-2), but environmental opposition and financial problems led to its cancellation when only 30 percent complete. See Cochran and Norris 1993, pp. 101–104.

Source: adapted from Albright, Berkhout, and Walker 1993 (*World Inventory of Plutonium and Highly Enriched Uranium, 1992*), p. 90; courtesy of Stockholm International Peace Research Institute (SIPRI) and Oxford University Press. Dates for THORP and UP2-800 are from chapters 8 and 9 respectively.

Table 3.5 Military Plutonium Separation Sites

Country	Facility
United States	Hanford
	Savannah River
Soviet Union	Chelyabinsk-65
	Tomsk-7
	Krasnoyarsk-26
United Kingdom	Sellafield
France	Marcoule
China	Jiuquan (Subei County)
	Guangyuan (Sichuan)
India	Trombay
Israel	Dimona
Pakistan	New Labs

Source: IPPNW and IEER 1992, p. 40. Some of these plants are not formally designated as military reprocessing plants.

The reprocessing plants with the longest operating history are U.S. military plants. Large-scale civilian facilities are at La Hague in France and Sellafield in Britain. A smaller plant at Tokai-mura, Japan, has operated on and off since 1975. Reprocessing plants in Russia have separated plutonium in both civilian and military irradiated fuel rods.

Two kinds of uranium are in the fuel rods from which plutonium is extracted. The most common kind in civilian nuclear power plants have uranium in the form of uranium dioxide. However, many graphite-moderated reactors, such as the British Magnox reactors and the N-reactor at Hanford, have uranium in metal form. The heavy-water reactors at the Savannah River Plant also use uranium metal for both the drivers to sustain the nuclear reaction and the target rods.

The PUREX Process

The steps to recover plutonium depend on the type of fuel and reactor as well as the choice of process chemistry. The most common approach to chemical separation is the PUREX process—*P*lutonium *UR*anium *EX*traction (see figure 3.11).

In the first step in the PUREX process, the contents of the irradiated fuel rods are separated from the cladding (the sheath or tube containing the irradiated fuel). Various processes are used to remove cladding, depending in part on the material from which it is made. Broadly speaking, there are two types of decladding processes. In chemical decladding, a chemical is used to strip the cladding from the fuel. For instance, the aluminum-silicon bonding of the uranium metal slugs used at Hanford was removed by dissolution in a mixture of sodium

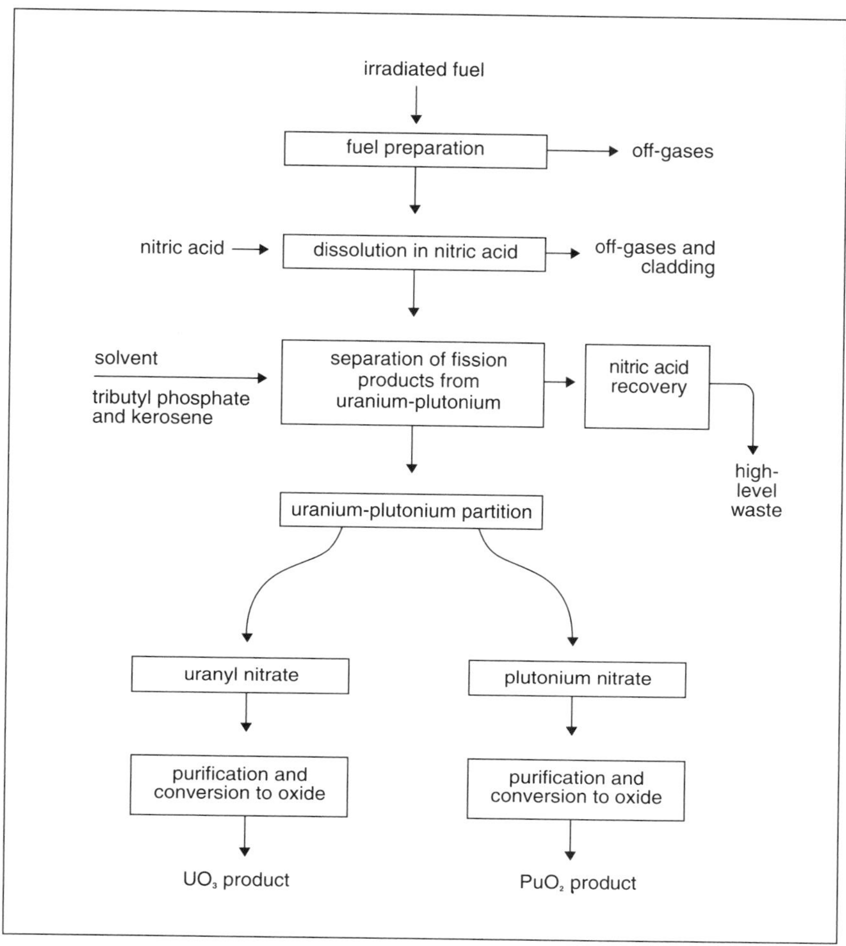

Figure 3.11 PUREX process. Adapted from Benedict, Pigford, and Levi 1981, p. 467, with permission of McGraw-Hill, Inc.

hydroxide and sodium nitrate.[45] The most recent PUREX plants use the mechanical "shear-leach" process. The fuel rods are chopped transversely by specially designed shears.

Next, the chopped-up fuel rods or the separated fuel-rod contents are dissolved in nitric acid to form a nitrate solution. In the case of chopped-up fuel rods, the cladding is separated and collected at this point, then stored or discarded as nuclear waste.[46] The nitrate solution is exposed to the solvent tributyl phosphate, which is mixed with kerosene to improve its physical properties. The tributyl phosphate selectively separates out the plutonium and uranium from the rest of the solution.

45. Benedict, Pigford, and Levi 1981, pp. 470–471.

46. Benedict, Pigford, and Levi 1981, pp. 475, 573.

Finally, the plutonium and uranium are separated from one another. The products are two forms of plutonium nitrate and uranyl nitrate, all liquids. Plutonium and uranyl nitrates may be further processed before shipping to reduce the consequences of transportation accidents. Generally, at least the plutonium is converted to solid oxide form before shipping. U.S. safety regulations now ban the shipping of plutonium in liquid forms.

Other Processes

A number of other processes have been used to extract plutonium from irradiated fuel:

• The *Butex process* was used at Windscale in England until the 1970s to chemically separate plutonium.[47] Like the PUREX process, the Butex process also uses nitric acid, but the solvent is dibutyl carbitol.

• An earlier chemical separation process used at Hanford was the *Redox (REDuction OXidation) process*. This involved a chromium compound and an organic solvent, hexone. The process was used from 1951 to 1967.

• The earliest process for separating plutonium was the *bismuth-phosphate process*, also used at Hanford.

• The first process used at the Chelyabinsk-65 reprocessing plant involved extracting uranium and some of the plutonium in an acetate form (sodium uranyl acetate and sodium plutonyl acetate) from a nitric acid solution.[48]

Processing Wastes for Plutonium Extraction

Various kinds of emissions and radioactive wastes result from extracting plutonium from irradiated nuclear fuel:

• *Highly radioactive wastes:* Known as high-level wastes, these are liquids and are stored in tanks.

• *Wastes from decladding the fuel rods:* If the decladding process is chemical, these wastes are liquid. In the ''shear-leach'' process, the wastes are solid. The cladding shells collect in a basket for storage or dumping in solid form. Solid decladding wastes are highly radioactive.

• *Low-level radioactive waste:* This is produced in large volumes from purifying uranium and plutonium.

47. Benedict, Pigford, and Levi 1981, p. 461. See also pp. 459–60, for a history of reprocessing.

48. Cochran and Norris 1993, p. 52.

• *Gaseous emissions from reprocessing:* The main radioactive components of this have been krypton-85 and iodine-131, which are fission products in the fuel rods. Iodine-131 emissions have been greatly reduced in the last three decades or so.

• *Degradation products of solvents:* For instance, radiation degrades (or breaks up) tributyl phosphate and yields new organic compounds.

• *Resins, sludges, and other wastes:* These come from the chemical processing of radioactive materials as well as from solvent recovery and purification.

THE HAZARDS OF REPROCESSING

Reprocessing poses a number of dangers to workers, the public, and the environment. Some dangers arise from the process itself, while others are associated with radioactive and hazardous nonradioactive wastes released into the environment.

A number of these problems are associated with both civilian and military reprocessing operations. Accidental criticalities, for example, can occur in reprocessing plants for either use. The likelihood of this is increased if the dissolver tank or plutonium separation columns are designed poorly.

Fires Associated with Reprocessing

Uranium metal can combust spontaneously, particularly at high temperatures. Zirconium fuel cladding can also spontaneously ignite in air. In the United States, there have been several fires during the reprocessing of metal fuel, as well as instances of cladding catching fire.[49]

Metal fires have started in the fuel rod–dissolving step. These result from the ability of metals (such as uranium or zirconium) to burn in air if the temperature is high enough. This is especially a threat if fuel has only recently been withdrawn from the reactor, and thus is hot, and is allowed to dry out in air. It may also be a hazard at lower temperatures for water-stored fuel with defective cladding that has formed uranium hydride after interacting with water. When exposed to air, uranium hydride can ignite even at close to room temperature.

High-Level Liquid Radioactive Wastes

Some of the most severe dangers arising from reprocessing are associated with the high-level radioactive wastes it generates. Most high-level radioactive wastes are stored in steel tanks. A typical tank in the U.S. holds about one million gallons (about 4 million liters). In other

49. Gydesen 1981.

countries tanks are usually far smaller than is typical in the U.S. since wastes in the latter are more dilute. In some cases, as at the Idaho National Engineering Laboratory, the wastes are dried ("calcined") and converted into a powder for storage. Fresh high-level liquid radioactive waste must be cooled to prevent boiling.

If highly radioactive wastes are stored in the acidic form in which reprocessing plants generate them, they require storage in stainless steel tanks. This method is used at La Hague and Idaho National Engineering Laboratory. Wastes can also be neutralized with lye (sodium hydroxide) before storage. This is the practice at Hanford and Savannah River. Neutralization allows storage in cheaper carbon-steel tanks. Neutralization also reduces the amount of radioactivity per unit volume, but it vastly increases the volume of wastes and makes them physically more difficult to handle because the treatment produces a highly radioactive sludge.

The tanks at Savannah River contain three different kinds of wastes. Sludge at the bottom of many tanks contains most of the radioactivity, except for cesium-137 (in most cases). Sludge also contains other chemicals. Supernatant liquid contains a substantial portion of the cesium-137, as well as some other radioactive materials and chemicals. Salts crystallize out of evaporated, concentrated wastes; these salts contain a large portion of the cesium-137. Hanford also has "capsules" of dry cesium-137 and strontium-90 that were separated from the liquid high-level waste.

The total volumes of wastes and the radioactivity per unit volume per unit of plutonium produced vary from country to country and from one plant to another because of differences in fuel irradiation levels, ages of the waste, and chemical treatment processes. A generic figure that can be used across countries when no direct information is available is about 3 curies (111 gigabecquerels) each of strontium-90 and cesium-137 per gram of plutonium produced. For a five-kilogram plutonium weapon, this yields a figure of about 15,000 curies (555 terabecquerels) each of strontium-90 and cesium-137. Newer weapons designs may contain as little as three kilograms of plutonium; waste generation would be proportionately lower.

The rising quantities of high-level wastes without a satisfactory long-term-disposal or management method have created a number of problems. The following is a list of some of the possible problems based principally on the U.S. experience and sketchy data from other countries:

• The need to evaporate wastes to reduce volume—which entails moving the wastes from one tank to another—and additional processing can lead to equipment problems, spills of radioactivity, and worker exposures.

• Various mechanisms can lead to catastrophic explosions in high-level waste tanks. These include the failure of cooling systems, radiolytic generation of hydrogen, buildup of organic vapors, and special mixtures of chemicals that have been put into the tanks. (See catastrophic risks, below.)

• Leaks from tanks and associated piping can contaminate soil and groundwater. For instance, about 750,000 gallons (2.8 million liters) of high-level waste have leaked into the soil from Hanford tanks, including a single leak of over 100,000 gallons (about 400,000 liters).[50]

• Soil and groundwater may be contaminated by the deliberate discharge of high-level wastes into them, as in the former Soviet Union.

Catastrophic Risks: Waste-Tank Explosions

Explosions or fires may occur in high-level radioactive waste tanks in a number of ways (not all of which are applicable to all forms of high-level waste):

• *Radiolytic hydrogen:* Radiolysis, or the disintegration of chemicals in waste tanks by the action of radioactivity, generates hydrogen. Ventilation systems must remove this hydrogen. Should the ventilators fail, hydrogen could build up to explosive levels. Tanks with fresh wastes and high concentrations of radioactivity are especially vulnerable. A spark is required to initiate a fire or explosion.

• *Cooling system failure:* This could result in wastes boiling and pressure building up in the tanks, with an accompanying increase in chemical activity. In 1957, such a failure and explosion occurred at the Chelyabinsk-65 complex in the Soviet Union. A failure of cooling due to power supply failure occurred at La Hague on 15 April 1980. (See chapter 9.) Tanks with non-neutralized and relatively new high-level wastes are most vulnerable.

• *Generation of explosive gases:* Organic vapors or hydrogen may evolve due to the specific chemicals a tank contains. These gases, too, can build up to explosive levels during a ventilation system failure. At least 20 tanks at Hanford and an unspecified number of tanks at Savannah are susceptible. (See chapter 6.)

• *Chemical explosions:* Chemicals may have been added to tanks that increase the risk of explosions. Ferrocyanide was added to some tanks at Hanford to precipitate cesium-137 and make room for more wastes.

Tank explosions are the most serious catastrophic environmental risk associated with reprocessing. The 1957 explosion at Chelyabinsk-65, the Chernobyl of nuclear weapons production, is instructive. The tank

50. U.S. GAO 1989, p. 27.

exploded with a force estimated at 70 to 100 metric tons of TNT (see chapter 7). The long-lived radioactivity in the fallout cloud consisted mainly of about 110,000 curies (4,070 terabecquerels) of strontium-90.[51]

In the United States, tanks at the Savannah River Site contain much more long-lived radioactivity than the Chelyabinsk-65 tank, since cesium-137 is not removed. Moreover, cesium-137 is not only a beta emitter; it also emits gamma radiation. Therefore, it poses dangers of both external and internal radiation, in contrast to strontium-90, a beta emitter, which is highly dangerous only if taken internally.

MANUFACTURE OF PLUTONIUM COMPONENTS

After PUREX reprocessing, plutonium is in a liquid nitrate form. It must be converted to metal and shaped in a foundry into the solid or hollow spherical form suitable for nuclear weapons. The first stages in this process are chemical.

After plutonium nitrate is converted by heating (calcining) into plutonium oxide, there are various routes to convert the oxide into metal. One common approach resembles that for producing uranium metal. Plutonium oxide is combined with hydrofluoric acid to produce plutonium tetrafluoride. Combining this with calcium produces plutonium metal.[52]

The chemical reactions are as follows:

$$PuO_2 + 4\ HF \rightarrow PuF_4 + 2\ H_2O$$

$$PuF_4 + 2\ Ca \rightarrow Pu\ (\text{metal}) + 2\ CaF_2$$

Another route is to produce plutonium oxalate first and react that with hydrofluoric acid to produce plutonium tetrafluoride.

Wastes heavily contaminated with transuranic elements (elements with atomic numbers greater than 92, the atomic number of uranium) are generated from this stage of plutonium processing. In the United States, which has a separate category for wastes with a content of more than 100 nanocuries (3,700 becquerels) per gram of transuranics, producing 60,000 weapons (most contained some plutonium) generated about 400,000 cubic meters or more of transuranic waste as of 1990. The waste contained more than 1.2 million curies (4.44×10^{16} becquerels) of radioactivity.[53] Weapons production has essentially stopped since 1990, but additional wastes continue to be generated as equipment is cleaned and remaining materials are processed. In addition, substantial plutonium residues, liquids, metal, and oxide stored

51. Nikipelov 1989.

52. Benedict, Pigford, and Levi 1981, pp. 435, 446, and 447.

53. Makhijani and Saleska 1992, p. 20.

at various plants contain more than 30 metric tons of plutonium, with an unclear designation and future.[54] (The U.S. Department of Energy relaxed its definition of transuranic waste in 1984, increasing the minimum amount of transuranic material in the waste from 10 nanocuries per gram to 100 nanocuries per gram.)[55]

Once the crucial materials are manufactured, they are fabricated into the components that make up a nuclear weapon. The central component is the "pit," which is the fission trigger of a thermonuclear weapon. An outer layer of chemical high explosives detonates the device.[56]

The principal component of a pit is a central sphere of fissile plutonium-239. In the case of boosted weapons a gaseous mixture of tritium and deuterium is bled into the hollow core of this sphere prior to firing. The outer layer of the pit is a tamper and/or neutron reflector made of uranium and beryllium. This is a fission warhead or the first stage, or "fission primary," of a thermonuclear weapon.

The principal industrial processes involved in making pits are the metallurgical and chemical processing of plutonium, uranium, and beryllium metals. Some conventional metals (such as stainless steel) are also used in making warhead pits.[57]

Required metallurgical operations include the reduction of plutonium dioxide to convert it to a metal. This is followed by rolling, blanketing, forming, and heat-treating the metal to produce ingots that can be formed, machined, and joined into the pit subcomponents. Similar metallurgical operations are conducted with uranium, beryllium, and other more conventional metals used in pit fabrication.[58]

Chemical processing is essential to recover and purify plutonium from scrap and from retired weapons. Chemical processing also may recover beryllium and extract americium-241 from plutonium.[59]

Hazards from Component Manufacturing

A number of potential hazards are associated with pit fabrication, due mostly to the nature of the principal materials. The primary hazards come from plutonium, but potential hazards also arise with beryllium, americium-241 (a plutonium-241 decay product), and organic compounds.

54. U.S. DOE 1993d.

55. Makhijani and Saleska 1992, p. 18.

56. Cochran, Arkin, and Hoenig 1984, pp. 25–28.

57. Cochran et al. 1987b, p. 83.

58. Cochran et al. 1987b, pp. 83–84.

59. Cochran et al. 1987b, p. 83.

• *Plutonium-induced fires:* Because plutonium metal gives off heat when it reacts with air, and because it conducts heat poorly, it can ignite spontaneously in air. Plutonium lathe turnings are particularly susceptible. There have been several plutonium fires in the U.S. nuclear weapons complex.[60]

Depending on the location of a plutonium pit fabrication facility, such fires could pose considerable risk to civilians. Large amounts of plutonium could be oxidized and released to the atmosphere if a fire breached a building structure.

• *Accidental criticality:* Nuclear criticality occurs when enough fissile material assembles in a small enough space to sustain a chain reaction. Criticality is required inside nuclear reactors to generate heat in a sustained fashion and inside a bomb to produce the nuclear blast. It is not desirable when it occurs by accident. Intense radiation is likely to be produced; damage and injury could be significant. The consequences are likely to be much less severe than from a full-scale nuclear explosion, however: during an accidental criticality, the energy from the nuclear reaction would blow the material apart, resulting in a loss of criticality far before it reached the level of energy released from a nuclear bomb.

About 11 kilograms of plutonium are required for bare criticality of a plutonium sphere,[61] but the amount can vary widely depending on various factors. The chemical form (metal or oxide) and shape of the sphere, as well as the presence of neutron reflectors or moderators (such as water) all affect the critical mass.[62] The critical mass of plutonium in water is only a few hundred grams.

At least eight accidental criticalities have occurred in the U.S. nuclear weapons complex, some with fatal consequences. Such events have typically produced radiation that is potentially lethal within about 10 meters. None of the events resulted in the explosive release of energy. The energy release in the first few seconds was on the order of one kilowatt-hour, and the greatest energy release is estimated at 100 kilowatt-hours over a period of hours.[63]

• *Americium:* Americium-241 (half-life, 432.2 years) is not deliberately used in making nuclear warheads. However, it is inevitably present as a result of the radioactive decay of plutonium-241, which makes up about 0.5 percent of weapons-grade plutonium. Since plutonium-241 has a half-life of 14.4 years, significant americium-241 builds up after several years.

60. Coyle et al. 1988, p. 91.

61. National Research Council 1989, p. 115.

62. National Research Council 1989, p. 115.

63. National Research Council 1989, p. 117.

Americium-241 presents an occupational radiological hazard due to its emission of penetrating gamma radiation. It is also an alpha emitter. In the United States, americium-241 is separated from the plutonium via chemical processes before the plutonium is processed into pit components.

FABRICATION OF OTHER RADIOACTIVE AND NON-RADIOACTIVE HAZARDOUS WEAPONS COMPONENTS

Lithium-6

Lithium-6 has two important applications in nuclear weapons. Military production reactors for making tritium use it as a target material, and it is a crucial ingredient for the second stage of thermonuclear warheads. Lithium-6 is combined with deuterium to yield lithium-6 deuteride, which constitutes the nuclear fuel for the second stage of thermonuclear weapons. The lithium-6 absorbs a neutron and yields tritium. The intense heat and pressure fuse the newly created tritium and deuterium, producing the thermonuclear blast. Supplemental fusion energy also results from deuterium-deuterium fusion reactions. These fusion reactions are:

deuterium + tritium → helium-4 + neutron + 17.6 MeV

deuterium + deuterium → tritium + hydrogen + 4.0 MeV

Lithium (atomic number 3) is a metal found in nature as a mixture of two stable isotopes: lithium-6 (7.5 percent), and lithium-7 (92.5 percent). Lithium constitutes about 0.006 percent of the earth's crust, making it more abundant than lead or tin.[64] The United States is the world's leading producer of lithium minerals and chemicals. Other major suppliers are Australia, Chile, China, the former Soviet Union, and Zimbabwe.[65]

Although lithium-6 is important for making nuclear warheads, it is not used commercially. However, natural lithium and lithium enriched in lithium-7 have many commercial and industrial applications. The main ones are in aluminum manufacturing (in the smelting process), the manufacture of ceramics and glass, and the production of lithium-based lubricants and greases. Lightweight lithium-aluminum alloys are part of the skins of some experimental aircraft. Lithium is a primary component of lithium/manganese dioxide batteries, as well as lithium carbonate pharmaceuticals for treating manic depression.[66]

64. Cochran et al. 1987a, p. 90.

65. U.S. Bureau of Mines 1988, pp. 619, 625.

66. U.S. Bureau of Mines 1988, pp. 620–621.

Weapons production requires separating lithium-6 from the more abundant lithium-7. The concept matches that for enriching uranium, although the processes may differ. Historically, the United States has used chemical separation to produce lithium-6 for military ends. The principal method, the Colex (Column Exchange) process, separates lithium-6 and lithium-7 by exchange between lithium amalgam and an aqueous solution of lithium hydroxide. The lithium-7 concentrates in the amalgam phase.[67] Other methods use a lithium amalgam in conjunction with lithium chloride or bromide in an organic solvent.[68] Laser techniques for enriching lithium, planned for producing lithium-7 for commercial uses,[69] could presumably be directed toward producing lithium-6 for military purposes.

After lithium-6 is separated, it is processed into target rods that can be inserted into a nuclear reactor. In the United States, this is done by alloying the lithium with aluminum. The alloy is extruded into target tubes.[70]

Lithium separation uses large amounts of mercury, a heavy metal noted for its severe neurotoxicity, particularly when in the form of methyl mercury. If mercury is not adequately managed and contained, severe and long-lasting contamination of water bodies can result.

Through neutron bombardment, lithium-6 is converted to tritium, a key bomb component. This occurs through the reaction:[71]

neutron + lithium-6 $\rightarrow$ tritium + helium-4

Deuterium

Deuterium occurs naturally as hydrogen deuterium oxide (HDO) mixed with regular hydrogen oxide (H_2O) in seawater in a concentration of about 0.0156 percent.[72] The principal application of deuterium in the form of heavy water (D_2O) is as a moderator and coolant in nuclear reactors. A very small portion of the production goes directly into nuclear weapons as lithium-6 deuteride. There are a number of ways to produce deuterium from sea water, including electrolysis and various types of distillation.[73]

67. Cochran et al. 1987b, p. 75.

68. Benedict, Pigford, and Levi 1981, pp. 800–801.

69. Cochran et al. 1987a, p. 91, note 157.

70. Cochran et al. 1987b, p. 114.

71. Friedlander et al. 1981, p. 546.

72. Benedict, Pigford, and Levi 1981, p. 708.

73. Benedict, Pigford, and Levi 1981, chapter 13, section 2.

Neutron Generators (Triggers)

Nuclear weapons need an initial burst of neutrons to initiate fission chain reactions. In some early weapons, a polonium-210 and beryllium trigger provided the neutrons. However, polonium-210 has a short half-life (about 138 days), necessitating frequent replenishment. Tritium-containing electronic neutron generators have replaced this approach in many modern weapons because of polonium's short half-life.[74]

Beryllium Neutron Reflectors and Amplifiers

Beryllium (atomic number 4) is a nonradioactive substance used in the warhead pit as a tamper to reflect and amplify neutrons. It was used along with radioactive polonium-210 in early warhead designs to start a chain reaction: alpha particles generated by the decay of polonium interact with beryllium to produce neutrons.

helium-4 + beryllium-9 → carbon-12 + neutron

Although not radioactive, beryllium is highly toxic, posing a high risk of chronic lung disease if inhaled.[75]

Organic Solvents

Trichloroethylene, tetrachloroethylene, carbon tetrachloride, chloroform, and other volatile organic compounds are used in chemical processes and metal-working operations needed for pit production. Many of these substances are toxic to humans and could contaminate groundwater if managed in unsound ways. Carbon tetrachloride is also a strong ozone-depleting compound. Gaseous diffusion plants use chlorofluorocarbons (CFCs) as coolants, some of which also deplete the ozone layer. At the Portsmouth, Ohio, plant, for instance, the coolant is CFC-114. Tens of metric tons of coolant can be vented each year, either directly to the air or after degradation to other chemical forms, such as carbon tetrafluoride and chlorine trifluoride.

ASSEMBLING THE WEAPONS

Final assembly begins with fabricating the chemical high-explosive components that detonate a nuclear weapon. These chemical high explosives create an imploding shock wave to compress the fissile plutonium and/or highly enriched uranium into a supercritical mass. While chemical high-explosive materials are usually available from

74. National Research Council 1989, p. 127.

75. See chapter 4; also see Meyer 1994.

commercial suppliers, they must be formed into the precise configuration required to produce a nuclear detonation. This may involve casting, pressing, or machining the explosive material to the desired density and shape. Molten casting is an older method. Fabricating high explosives for most modern nuclear weapons relies on precision machining.[76]

Subassembly operations for high explosives involve mating two or more high-explosive parts or joining high-explosive parts to a case or liner. Such operations can occur before introducing the radioactive components, but they involve exposed high explosives. For this reason, in the United States they are conducted in reinforced-concrete rooms that are covered with earth (called sub-assembly bays). These are designed to prevent an accidental explosion in one bay from causing an explosion in another. Subassembly bays at the Pantex Plant near Amarillo, Texas, the principal U.S. weapons-assembly plant, are designed to accommodate up to about 136 kilograms of high explosive each.[77]

In final assembly, the high-explosive parts are mated with the fissile components of the weapon pit. The entire high explosive/nuclear material unit is encased in a protective shell or liner that might consist of stainless steel, aluminum, or titanium. The encased unit is often referred to as a "physics package." In the United States, the physics package is put together in assembly cells called "Gravel Gerties" that are designed to absorb the shock and to prevent the spread of radioactivity in case of an accidental explosion. Pantex has 13 cells.[78]

Assembly Hazards

The most obvious occupational hazard in assembling chemical high-explosive components is an accidental detonation. This has occurred at least once in the United States, killing three workers at Pantex on 30 March 1977.[79] This is also a principal hazard to workers in disassembly of warheads.

The most significant radiological accident risks to the public and the environment are associated with the accidental dispersion of tritium or plutonium materials.[80] An accidental detonation of high explosives could also release radioactive materials to the environment. Moreover, as with any facility that processes dangerous materials, irresponsible handling can result in a release or in the contamination of groundwater

76. Cochran et al. 1987b, pp. 77–78.

77. Cochran et al. 1987b, p. 78.

78. Cochran et al. 1987b, p. 78.

79. Coyle et al. 1988, p. 82.

80. Ahearne 1989, p. 2.

by hazardous or radioactive materials. Finally, the presence of large amounts of fissile material poses some risk of accidental criticality.[81]

RESEARCH, DEVELOPMENT, AND DESIGN AND ASSOCIATED EXPERIMENTAL LABS

A variety of laboratories are involved with nuclear weapons, and there is considerable overlap with civilian nuclear energy work. Research, development, and design activities associated with uranium mining, milling, processing, and enrichment are common to both military and civilian applications. The overlap is even greater in countries with breeder reactor programs since these reactors are designed to produce more plutonium-239 (from uranium-238) than they consume. Reprocessing is an essential component of a large-scale breeder reactor program.

Research, development, and design of nuclear components involve discovering basic properties of fissile materials and light elements and ways in which these properties can be used to build nuclear weapons of specific characteristics. The physical size of the weapons is a major consideration: smaller weapons can be carried into battle, and a larger number of compact warheads can fit into a single missile.

Every nuclear weapons state has at least one laboratory to design and fabricate prototype weapons. These labs are, in effect, small-scale production facilities in addition to being research, development, and design centers. Production brings with it the entire array of pollutants, from transuranic wastes to toxic, nonradioactive chemicals. It is not a given that small-scale production means that contamination is proportionately smaller than in the larger factories. This is because environmental and other controls in a laboratory or small-scale production setting can be much more lax than in a large-scale facility.

INACTIVE WEAPONS PRODUCTION AND RADIOACTIVE WASTE DUMPING OR STORAGE SITES

One legacy of the long history of nuclear weapons production in the United States is a large number of "formerly used" sites that have been contaminated and abandoned or restored to civilian use without appropriate cleanup. There is evidence of a similar picture in Russia. The extent to which each site represents a health hazard varies greatly from case to case. We present some examples of such sites in this book, but this remains an area of even greater uncertainties than at major sites that have been used for some aspect of nuclear weapons production in the last decade or so.

81. Ahearne 1989, p. 1.

4 Health Hazards of Nuclear Weapons Production

David Sumner, Howard Hu, and Alistair Woodward

The manufacture of nuclear weapons involves many toxic substances, both radioactive and nonradioactive. Some of these materials pose significant health hazards, both to the workers directly involved and to the public.

This chapter summarizes known important aspects of the health effects of nuclear weapons production, including radiation and non-radiation hazards. In so doing, it draws on three disciplines. In general, *radiobiology* studies the effect of radiation on living systems, while *toxicology* studies the harmful effects of any toxic hazard on living systems. *Environmental and occupational epidemiology* studies the experience of populations or groups of workers to investigate the relationship between potentially hazardous exposures and disease.

These disciplines overlap considerably. For instance, much of radiobiology is part of toxicology, although the former also includes therapeutic uses of radiation. Classical toxicology focuses on the effects of individual toxins on humans and animals, as well as on the cellular and subcellular levels. In many cases epidemiologists who are investigating populations and particular toxins will look at biological indicators of exposure dose, or an early effect that toxicologists had first developed.

This book integrates information from these three disciplines as well as others to summarize the state of knowledge on individual hazards or processes that arise in the production of nuclear weapons.

There is some justification for the claim that radiation has received relatively more attention than nonradioactive toxins in the study of nuclear weapons plants. Certainly, an enormous amount of research has been conducted over the past half century on the health effects of low-level radiation. In part, this attention has arisen from the rapid exploitation of nuclear fission in weapons and nuclear power during and since World War II. Another imbalance in the study of health hazards has been the principal focus on cancer, sometimes to the exclusion of other health effects within the nuclear industry.

One way of considering toxic exposure is to distinguish between carcinogens—that is, cancer-causing agents, including radiation—and noncarcinogens. A basic idea behind the establishment of chemical-specific standards and guidelines is that noncarcinogens cause toxic effects only if exposure levels exceed a certain threshold. Experiments can determine the threshold for a specific chemical, and that level can form the basis for a regulation or guideline. In contrast, carcinogens are often treated as though they can carry a finite risk of causing cancer at any dose—as though there is no threshold below which exposure is considered to be safe.[1] However, recent environmental epidemiology has begun to challenge the "threshold" distinction between carcinogens and noncarcinogens. For example, sophisticated studies indicate that the toxic effect of lead on children's nervous systems appears to have no threshold.[2]

UNDERSTANDING RADIATION HAZARDS

Radioactive decay occurs when the nucleus of an atom spontaneously transmutes by emitting or capturing a particle. The term *ionizing radiation* usually refers to the high-energy electromagnetic radiation (X rays and gamma rays) and alpha and beta particles that are emitted when radioactive substances decay. All these can have enough energy to ionize atoms—that is, remove one or more of the electrons whirling around their nuclei. Lower energy electromagnetic radiation (visible light, ultraviolet, infrared, radio waves) is nonionizing and is not considered in this book.

An alpha particle is a helium nucleus and consists of two protons and two neutrons. A proton is a positively charged particle; a neutron has no electrical charge. Alpha decay occurs mostly in elements heavier than lead, such as uranium-238, thorium-232, and plutonium-239. Because of its relatively high mass and double positive charge, an alpha particle readily ionizes material it encounters and transfers energy to that material's electrons. As it transfers its energy, the alpha particle slows down. The energy transfer is faster in denser materials: an alpha particle travels several millimeters in air, but its range in a solid is extremely short. For example, alpha particles do not penetrate keratin, the outer dead layer of human skin.

Most common radionuclides lighter than lead undergo beta decay. A beta particle, which is an electron or positron, is over seven thousand times lighter than an alpha particle. As a result, beta particles generally lose energy over far larger distances than alpha particles. A medium-energy beta particle has a range of about one meter in air and a millimeter in body tissue.

1. Frumkin 1995, p. 294.

2. Goyer 1991; Mushak 1992.

Gamma rays are electromagnetic radiation similar to X rays, but usually of higher energy. They often accompany alpha and beta emissions. A radioactive element emits gamma rays in discrete quanta called photons in cases where the nucleus remaining after alpha or beta decay carries excess energy and is in an excited state. This is analogous to a vibration in a body that occurs after a piece breaks off from it. In general, gamma rays penetrate much more deeply than alpha or beta particles. A high-energy gamma ray photon may pass through a person without interacting with tissue at all. When gamma rays interact with tissue, they ionize atoms or molecules. A gamma-ray emission changes the energy state of a nucleus but not its identity.

As discussed in chapter 3, neutrons are produced from fission, for instance of uranium-235 and plutonium-239 nuclei; fusion of some light nuclei; and in other nuclear reactions. Neutrons have no electric charge, so they do not interact with electrons or cause ionization directly. Rather they ionize indirectly, in a number of ways. Low-energy neutrons (thermal and near-thermal neutrons) can be captured by nuclei, which become radioactive. The two cases of importance to tissue doses are neutron capture by nitrogen-14, which creates radioactive carbon-14, and neutron capture by ordinary hydrogen (hydrogen-1) which creates deuterium (hydrogen-2). While deuterium is not radioactive, neutron capture by hydrogen is accompanied by the emission of a 2.2 MeV gamma ray.[3]

For fission neutrons and other fast neutrons with energies less than 20 MeV, elastic collisions (similar, in principle, to collisions of hard balls) with hydrogen nuclei are the most important mechanisms of energy transfer. Other important mechanisms for fast neutron energy transfer are inelastic collisons, in which some kinetic energy is converted to gamma rays, and nonelastic scattering, in which neutrons are absorbed by a nucleus that then emits an ionizing charged particle, such as a proton or an alpha particle.[4]

Linear energy transfer (LET) is the energy transferred from alpha, beta, gamma, neutron, or other radiation to the surrounding medium per unit distance. It varies enormously, depending on a particle's speed and charge. In general, an alpha particle causes much more ionization per unit distance—and has a higher LET—than a beta particle. Alpha particles and neutrons are considered high-LET radiation; beta particles and gamma rays are low-LET radiation.[5] (See figure 4.1.)

Ionized molecules can be chemically more reactive, and some radiation damage results from chemical reactions that follow upon ionization.

3. National Research Council 1990, pp. 16–17.

4. National Research Council 1990, pp. 15–17.

5. National Research Council 1990, p. 11.

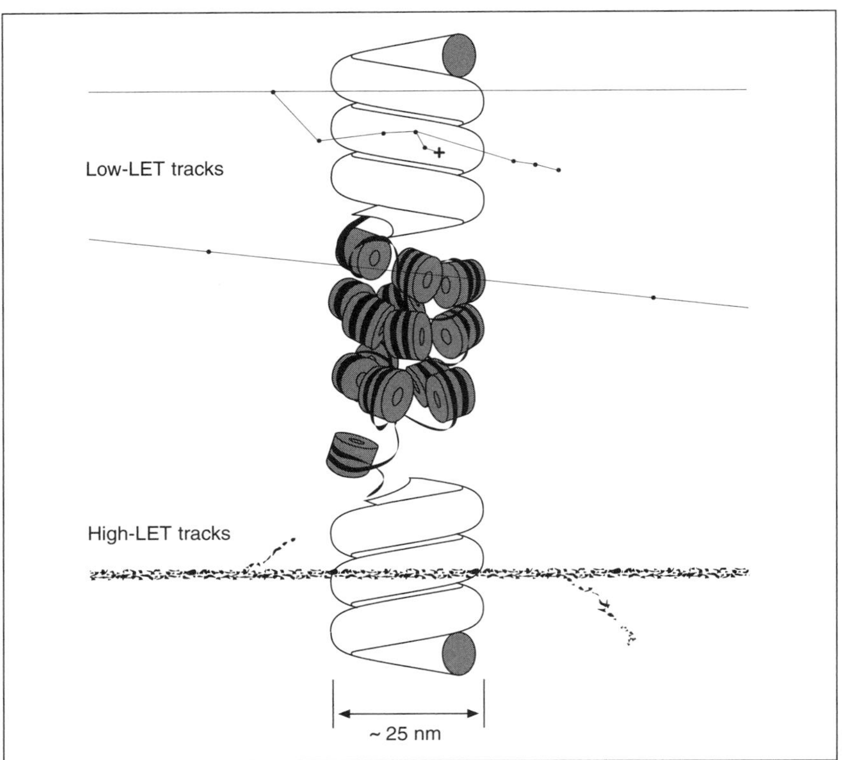

Figure 4.1 Diagram of high- and low-LET tracks passing through a section of chromatin (a mixture of DNA and protein). Source: ICRP 1991, p. 95.

Radioactive materials fall into three main classes:

• *Heavy elements:* uranium-235, polonium-210, radium-226, and plutonium-239, as well as other isotopes of plutonium (238, 240), americium-241, and neptunium-237, all belong to this category. Many are alpha emitters, although several, like plutonium-241 and thorium-234, emit beta particles. A small fraction of plutonium-240 undergoes spontaneous fission, yielding neutrons that increase the radiation hazard from plutonium. A particular sub-class of heavy elements, grouped together as "actinides" from their location in the periodic table of elements, encompasses all elements with atomic numbers from 90 to 103. The actinide group includes thorium and uranium as well as all transuranic elements of importance in nuclear weapons production.

• *Fission products:* when heavy nuclei undergo fission, they split into two fragments (usually radioactive). Among the very large number of these are cesium-137, strontium-90, iodine-131, iodine-129, and krypton-85. These are mainly beta emitters and many also emit gamma radiation as their nuclei decay.

• *Activation products:* formed when stable atoms capture neutrons. Among these are carbon-14, sulfur-35, iron-55, iron-59, cobalt-60, zinc-65, manganese-54, and argon-41. These isotopes are mainly beta emitters, accompanied by gamma ray emissions in many cases.

In addition, tritium, an isotope of hydrogen, is formed when lithium-6 captures a fast neutron and subsequently undergoes alpha decay. It is a beta emitter.

Radiation Doses

Radiation exposures to individuals are measured by the amount of energy deposited in their bodies; exposures to populations are measured by adding up the individual doses in that population. The unit of radiation dose is the gray. It is a measure of the amount of ionization caused by the radiation and is a strictly physical unit. Other factors such as the type of radiation involved (alpha, beta, etc.) and the parts of the body exposed affect the biological effect of the radiation. When corrections are made for these factors, the unit used is the sievert. One sievert of radiation should have about the same biological effect, whether it be one sievert of gamma radiation to the whole body or one sievert of alpha radiation to the lung. Where the total dose to groups or populations is being considered, units such as person-sieverts are used. Population doses are measured in person-grays and person-sieverts (person-Sv), depending on whether energy deposition or biological damage is being measured.

Two further units apply to uranium mines. The working level (WL) is the quantity of radon decay products (also called radon daughters or radon progeny) in one liter of air that will result in the emission of 130,000 million electron volts of alpha-particle energy. If the radon progeny are in equilibrium with radon in the air (that is, if the radon has remained in the air for some time), then about 100 picocuries (3.7 becquerels) of radon per liter of air equals one working level. The working level month (WLM) measures the total radiation dose a miner would receive by breathing air containing a concentration of 1 working level for one working month (170 hours).

Radiation doses may be due to sources outside the body or to substances that have entered the body in the course of eating, drinking, or breathing, or through a wound. It is relatively straightforward to estimate radiation doses due to gamma rays and beta particles from outside the body, provided a person wears appropriate measuring equipment, such as a film badge. However, it is generally much harder to estimate doses from substances inside the body. The size of the dose will depend on the chemical form of the material, its pathways and distribution in the body, and the rate of its elimination from the body, among other factors. The elimination of a radionuclide from the body is generally quite a complex phenomenon; it can be very approximately described by the concept of "biological half-life"—the time it takes for half the material to be eliminated from the body.

When estimating doses from environmental radioactivity, direct measurements are almost never available for the amounts of particular radionuclides in the body. Complex computer models have to be used, often with large numbers of parameters and associated uncertainties. This is especially true of dose estimation for off-site populations where there are no direct measurements for dose or for body-burdens of radioactive materials. However, radionuclides in food, water, and air can be measured. If done carefully, such measurements can provide a basis for estimating doses. If internal burdens are large, techniques such as whole-body counting (called *in vivo* measurements) and urine sampling can also be used.

Health Risks of Ionizing Radiation

Ionizing radiation can cause stochastic (random) and deterministic (or nonstochastic) effects. Deterministic effects appear if a minimum radiation dose is exceeded. Above that threshold, the effects are readily observed in most or all exposed people and the severity increases with dose. The occurrence and severity of a deterministic effect in any one

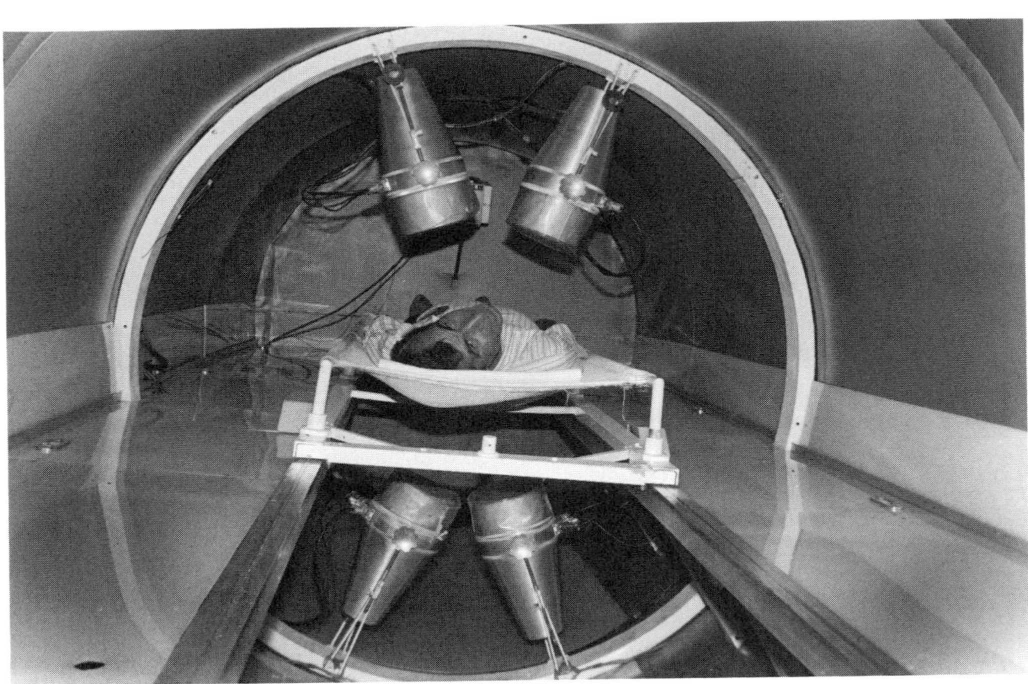

Figure 4.2 The strontium-90 whole-body counter, Chelyabinsk. Photo by Robert del Tredici.

individual are reasonably predictable. A radiation burn is an example of a deterministic effect.

In adults, nonstochastic effects dominate when the dose to the entire body is more than about one sievert. An exception is temporary sterility in the male, which can occur with a single absorbed dose to the testis of about 0.15 grays.[6] With respect to children, the threshold for congenital malformations and other developmental abnormalities has been estimated to be 0.25 grays of radiation exposure up to 28 days of gestation.

Single radiation doses over about 1 gray cause radiation sickness; acute effects include nausea, vomiting, and diarrhea, sometimes accompanied by malaise, fever, and hemorrhage. The victim may die in a few hours, days, or weeks. Other acute effects can include sterility and radiation burns, depending on the absorbed dose and the rate of the exposure. The dose at which half the exposed population would die in sixty days without medical treatment is called the LD50 dose (LD for lethal dose, and 50 for 50 percent). It is about 4 sieverts for adults. The sixty-day period is sometimes explicitly identified, and the dose is then called the LD50/60 dose. In general, a number of different LD50 doses can be specified, depending on the number of days, T, after which the observations of death are cut off.

For radiation doses less than about 1 sievert, stochastic effects have been the greatest concern. The most important stochastic effects, cancer and inheritable genetic damage, may appear many years or decades after exposure. It is thought that there is no minimum threshold for these effects; as dose decreases the effects are still expected to occur, but with lower frequency. However, the uncertainties at low doses (10 millisieverts or less) are very large. Estimates of the magnitude of low-dose radiation effects have tended to rise over the years, but remain the subject of controversy.

Because ionizing radiation can damage the genetic material of virtually any cell, cancer can occur in many sites or tissues of the body. The actual effect depends in part on the route of exposure. For example, external radiation, such as X rays or gamma radiation, can affect DNA in blood-forming cells or in many organs in ways that cause cancers of these organs decades later. It should be noted that tissues vary in their sensitivity to radiation damage. For instance, muscles are less sensitive than bone marrow.

There are many pathways by which the body can be exposed to internal irradiation. Decay products of radon, which are present in an underground uranium mine, may be inhaled by miners and end up in their lungs. Particles of plutonium-239 or other actinides, which emit mostly high-LET alpha particles, may be inhaled and deposited on the epithelial lining of bronchi in the lung. A radiation dose from such

6. ICRP 1991, p. 15.

exposure pathways increases the risk of lung cancer. In addition, soluble particles may be absorbed and distributed through the blood or lymph systems to other parts of the body. Some elements, such as radium, strontium, or iodine, tend to accumulate in certain organs. For example, iodine-131 delivers its principal ionizing radiation dose to the thyroid gland, making that the most likely site of a resultant cancer. Iodine-131 is also used to combat thyroid cancer, since the emitted radiation destroys the cancerous cells along with healthy ones. But when there is no disease in the thyroid, the radiation affects only healthy cells.

Estimating the Risk of Cancer from Ionizing Radiation

Various institutions have estimated the risk of cancer following exposures to ionizing radiation, particularly the United Nations Scientific Committee on the Effects of Atomic Radiation (UNSCEAR), the U.S. National Academy of Sciences Committee on the Biological Effects of Ionizing Radiation (BEIR), and the International Commission on Radiological Protection (ICRP). These estimates are derived mainly from studies of the survivors of the Hiroshima and Nagasaki bombings, and also from various groups of people given radiation for therapeutic and diagnostic purposes or who have been exposed at work, such as radium dial painters and uranium miners. (The latter are discussed in more detail below.)

Studies of survivors of the atomic bombings of Hiroshima and Nagaski indicate statistically significant excess cancers for doses greater than 0.2 grays. These doses were delivered suddenly, following explosions. A number of problems arise when using such data to estimate cancer risks for lower doses of ionizing radiation or doses delivered in gradual increments.

The first problem is how to extrapolate the dose-response relationship down to low doses. It is usually assumed that a "linear no-threshold" model applies—that is, the risk is directly proportional to dose, with no threshold. Because the main effect of low-dose radiation is the induction of cancer, and cancer is a common disease with many causes, it is not yet possible to verify the linear no-threshold model; nevertheless, there is considerable radiobiological evidence for this theory and it is generally used for public health protection purposes, such as setting standards.

The second problem is that some assumption has to be made about how calculations of cancer risk will change in the future. After all, more than half the Hiroshima and Nagasaki survivors are still alive. At present, the data best fit a relative-risk model—that is, the cancer risk is proportional to the "spontaneous" or "natural" cancer risk. If this is correct, there will be an increasing number of radiation-induced cancers later in life.

A third problem is that the relative biological effectiveness of radiation depends partly on the energy of the radiation. For instance, data indicate that low energy neutrons and alpha particles may be more effective in producing biological damage than high energy particles (per unit of absorbed energy).[7] Thus, assuming a constant quality factor, as is common practice, can sometimes yield an inaccurate estimate of the dose.[8]

Finally, there are uncertainties related to the effect of low doses and low dose rates of low-LET radiation. The conclusion of the BEIR Committee, ICRP, and others is that low doses and dose rates of low-LET radiation are less effective in producing cancer, particularly leukemia, than would be expected based on linear extrapolation of data for low-LET radiation at high doses and high dose rates (i.e., the effect is nonlinear at low doses and dose rates). Unfortunately, the epidemiological database for evaluating the validity of DREF adjustments is sparse.[9]

Despite these potential limitations, most cancer projections continue to utilize the cancer risk factors estimated by established radiological protection committees. Their current estimates are as follows:

• UNSCEAR, 1993:[10] 0.11 fatal cancers per person-sievert for high doses (comparable to those experienced by the survivors of the Hiroshima and Nagasaki bombings). For low doses, UNSCEAR states that "no single figure can be quoted" for the risk reduction factor, "but it is clear that the factor is small. The data from the Japanese studies suggest a factor not exceeding 2."[11] For a population between the ages of 18 and 64 (corresponding to the ages of people in a typical industrial

7. National Research Council 1990, pp. 27–30.

8. There is also some experimental evidence that alpha particles may be more effective than recognized heretofore in producing chromosomal abberations. Kadhim et al. exposed stem cells in the bone marrow of mice to a range of doses of alpha particles from plutonium-238. Many surviving cells were either not traversed by a particle or suffered ionization from only a small part of the track. When the chromosomes of these surviving cells were analyzed, a surprisingly high frequency of aberrations was evident many cell divisions after the original exposure. Comparable aberrations were not induced by similar doses of X rays. The implication of this research, if confirmed, for the biological effectiveness of alpha radiation needs to be evaluated (Kadhim et al. 1992).

9. One study that provides some evidence against application of DREF adjustments for cancer was a recent study of British workers that suggested an excess cancer mortality risk of 1 death per 10 sieverts of low-LET radiation exposure. However, the lower 95 percent confidence bound for the risk was below zero and the finding was not statistically significant. In view of the lack of statistical significance, the authors of the study did not feel that any changes in existing practice of risk estimation were warranted due to their findings (Kendall et al. 1992).

10. UNSCEAR 1993, pp. 16–17.

11. UNSCEAR 1993, p. 17.

work force), a factor of 2 yields a fatal cancer risk at low dose rates of 0.04 per person-sievert.

• BEIR Committee, 1990:[12] 0.08 fatal cancers per person-sievert for a single dose of 0.1 sievert, based on Hiroshima and Nagasaki survivor data. This figure is unadjusted for any reduction of risk at low dose rates.

• ICRP, 1991:[13] 0.05 fatal cancers per person-sievert for the entire population and 0.04 fatal cancers per person-sievert for adult workers, with both estimates being for low doses and incorporating a dose rate reduction factor of 2.

• The U.S. Environmental Protection Agency uses a cancer *incidence* risk factor of 0.06 per person-sievert.[14] Since the cancer incidence rate is about 50 percent greater than the cancer fatality rate, the implicit risk for fatal cancers is about 0.04 per person-sievert.

Estimates of the risk per unit dose may be revised substantially again (upward or downward). As the BEIR committee points out:

> Most of the A-bomb survivors are still alive, and their mortality experience must be followed if reliable estimates of lifetime risk are to be made. This is particularly important for those survivors irradiated as children or in utero who are now entering the years of maximum cancer risk.[15]

Radiation Protection

Radiation protection regulations are based on three basic recommendations originally made in 1977 by the ICRP and reaffirmed later:[16,17]

• *Justification:* No practice involving exposures to radiation should be adopted unless it produces enough benefit to the exposed individuals or to society to offset the radiation detriment it causes.

• *Optimization:* Exposures to radiation should be as low as reasonably achievable.

• *Individual dose and risk limitation:* No individual should receive radiation doses higher than the maximum allowable limits.

The most difficult of these principles, and certainly the one that is rarely adequately addressed, is justification. Assessing the likelihood

12. National Research Council 1990, pp. 5–6.

13. ICRP 1991, pp. 69–70.

14. U.S. EPA 1993, p. 7.

15. National Research Council 1990, p. 8.

16. ICRP 1977, p. 3

17. ICRP 1991, p. 28.

that any practice will produce a net benefit involves many value judgments that are difficult, if not impossible, to quantify. ICRP recognizes this:

> The Commission recommends that, when practices involving exposure, or potential exposure, to radiation are being considered, the radiation detriment should be explicitly included in the process of choice. The detriment to be considered is not confined to that associated with radiation—it includes other detriments and the costs of the practice. Often, the radiation detriment will be a small part of the total. The justification of a practice thus goes far beyond the scope of radiological protection.... To search for the best of all the available options is usually a task beyond the responsibility of radiological protection agencies.[18]

This point is expanded in a statement by the Committee on Radiation Protection and Public Health of the OECD Nuclear Energy Agency:

> Decisions about the justification of a practice or activity involving radiation exposure usually involve a broad range of social, economical and political issues in addition to those concerning radiological protection.... Justification is essentially a political decision-making process, in which the technical and purely radiation-related advantages or detriments play an important, but relatively limited role.[19]

In the early years of nuclear weapons development, the scientists and administrators involved implicitly assumed that national security justified the risks of the enterprise. According to J. Newell Stannard, "In 1947, the data for plutonium and the other actinides were used at a series of three-nation conferences on radiation exposure limits.... They required careful interpretation, for the most conservative interpretation could have closed Los Alamos."[20]

The principle of justification continues to be a cornerstone of ICRP philosophy, but the application of this principle to a particular situation in the nuclear industry, whether civil or military, is rarely discussed.[21]

Optimization implies that measures will be taken to reduce exposures until the benefits of further reductions do not justify their cost. It is not clear how this principle can be rigorously applied, particularly as it requires some quantitative estimate of the monetary value of a life saved. In practice, optimization is applied in two ways: as an exhortation to use "best available technology" and as a recognition that merely complying with dose limits is not enough. If further dose

18. ICRP 1991, para. 115.

19. Nuclear Energy Agency 1992, p. 15.

20. Stannard 1988.

21. QUEST Radiation Database (1992) gave just 5 references to "justification" but 91 to the principle of optimization.

reductions are practicable at reasonable cost, they should be made. Optimization generally refers to collective rather than individual radiation doses.

The principal dose limits recommended in ICRP Publication 26 (1977) were 50 millisieverts (5 rem) per year for radiation workers and 5 millisieverts (500 millirem) per year for members of the public. A subsidiary recommendation to keep doses to the public below 1 millisievert per year if possible has slowly become the primary long-term dose limit for the public, with short-term exposures of 5 millisieverts per year allowed.

The ICRP intended these limits to apply to the total exposure from all sources except natural background radiation. It has developed a methodology for combining the doses from different sources—such as combining exposures from inhaling ore dust with those from gamma exposure—and it is this total that should be compared with the appropriate limit.

In 1991, the ICRP revised its radiation protection standards, largely in response to reevaluation of dosimetry and cancer risk among atomic bomb survivors.[22] The most significant change lowered the worker's annual limit to 20 millisieverts. Regulations do not yet widely reflect this change.

UNDERSTANDING NONRADIATION HAZARDS

In addition to radiation hazards, nuclear weapons production requires and generates enormous quantities of a wide variety of toxic substances. Estimating the potential harm of each of these substances requires the application of basic principles of toxicology. Among the major principles are that toxins, or their metabolic products, must come into close contact with the target organ for which they have the potential to cause injury; that the observed toxicity should be quantitatively related to the degree of exposure to the toxin, i.e. there should be a "dose-response relationship;" and that toxicity varies according to a toxin's chemical and physical form, the route of exposure to humans, the level and duration of the exposure, the mechanism of toxicity (i.e., the fundamental chemical and biologic interactions and resultant aberrations that are responsible for the toxic response), and the presence of modifying factors (e.g., species, sex, and environmental conditions).

Since it is far beyond the scope of this book to provide a summary of basic toxicology, readers are referred to recent texts.[23] Nevertheless, it is possible to provide an overview of toxic outcomes that can be

22. ICRP 1991.

23. See for example Sullivan and Krieger 1992.

expected from exposure to general classes of nonradiation hazards that are commonly experienced in the nuclear weapons industry.

Mineral Dusts—Silica

Mineral dusts in the nuclear weapons industry are primarily encountered in uranium mines, where dusty conditions prevail if industrial hygiene controls are not applied. Of greatest concern is the inhalation of silica dust generated by drilling hard rock, which can lead to a variety of pulmonary diseases collectively known as silicosis. Most characteristic is the gradual development of nodular fibrosis (a type of scarring) in the lungs. This typically produces in an individual restricted lung function, a pattern of round opacities (multiple spots) on chest X ray, and the main clinical symptom of progressive shortness of breath. Common complications include respiratory insufficiency with diminished oxygenation and an increased risk of tuberculosis. Sufferers also are at increased risk of rheumatoid arthritis, scleroderma and other rheumatic disorders, kidney disease, and possibly lung cancer. Very high exposures to silica dust can lead to an acute form of silicosis in which the lung alveoli (air sacs) fill with a thick proteinaceous material, ultimately causing death by asphyxiation within a few years of exposure.

Heavy Metals

Heavy metals in the nuclear weapons industry include highly toxic substances that have wide commercial applications (such as chromium and mercury), metals that may be encountered in mining (such as molybdenum, vanadium, and arsenic), and other metals that are used for specialized purposes (such as barium, nickel, and chromium). Uranium itself is a heavy metal and has toxic effects in addition to its radiological effects.

Significant exposure to metals occurs mainly through inhalation of dust or fumes (tiny particles with the appearance of smoke, e.g., welding fumes), or by ingestion of contaminated food, water, or by smoking cigarettes. The degree to which a body will absorb a metal (i.e., the internal dose) depends largely on the metal's chemical form. For instance, trivalent chromium (Cr^{+3}) compounds are less toxic than hexavalent chromium (Cr^{+6}) compounds, due to differences in both absorption and oxidation. Once metals gain entry into the body, the circulatory system widely distributes them to hard and soft tissues.

In general, the most toxic metals, particularly lead, mercury, cadmium, and arsenic, are poisons that interfere with metabolism and enzymatic function on a cellular level, thereby leading to damage to many different organs and the possibility of fatal injury at high enough doses. Lower amounts of persistent exposure can lead to gradual accumulation and increase the risk of chronic disease. The most common targets of metal toxicity are the kidneys and the neurological system. Some also affect the lung, and others increase the risk for cancer.

Lead is by far the most studied metal toxin, due in part to the widespread nature of lead exposure in contemporary society from the inhalation of combusted leaded gasoline and air from polluting lead industries, and the ingestion of lead originating from the unfortunate use of lead historically in domestic products such as paint, ceramics, solder, and plumbing. Decades ago, children who ate lead paint chips were noted to develop constipation, abdominal pain, seizures, mental retardation, anemia, limb weakness, and other manifestations of systemic toxicity. Adults suffer similar effects, although their already developed nervous systems are less vulnerable to lead's neurotoxicity. Most recently, lead has received increased attention because of investigations indicating an adverse effect of lead on neurological development (affecting, for instance, learning and hearing acuity) in children and blood pressure in adults at levels of exposure far lower than previously thought.[24]

Due to its low vapor pressure, mercury exposure can occur readily through the inhalation of elemental mercury vapor. Alternatively, mercury contained in pollutant discharges, as has occurred in lithium-6 production for nuclear weapons, can be converted to organic mercury forms by microorganisms, leading to concentration of organic mercury in the food chain and the threat of human exposure through ingestion (particularly of contaminated fish). Mercury vapor exposure can damage the lungs, leading to toxic pulmonary edema (lung injury followed by fluid accumulation and interference with gas exchange). The central nervous system is the primary target of chronic elemental mercury exposure, with injury producing the classic triad of tremor, gingivitis, and erethism (insomnia, shyness, memory loss, emotional instability, nervousness, and anorexia). Ingested elemental mercury is not absorbed well, although it can produce local gastrointestinal corrosion. Ingestion of organic mercury, particularly methyl mercury, leads to a form of central nervous system toxicity with some similarities to but also important differences from that of elemental mercury. Typical are depression, decreased intellectual ability, clumsiness (from cerebellar damage), sensory numbness, slurred speech, and spasticity of movement leading to paralysis. As with lead, increased attention is now being focused on the risks posed by low-level mercury exposure. Recent epidemiological work found an association between low levels

24. ATSDR 1988.

of mercury in urine and indicators of early kidney and neuro-psychological damage.[25]

Arsenic exposure can occur through inhalation, which is an issue in uranium mining, as well as ingestion. Arsenic is an effective acute poison at high doses, leading to severe blood, brain, heart, kidney, and gastrointestinal tract injury. Lower levels of exposure can cause skin eczema and darkening, muscle wasting, and painful peripheral nerve lesions. Of greatest concern with low levels of exposure is cancer, particularly of the skin, lungs, bladder, and liver, as well as other organs.

In addition to the radiological hazard it poses, uranium poses a significant risk of kidney damage. Acute damage can be severe, but also is largely reversible. Low-level exposure may possibly cause kidney damage in the form of tubular dysfunction. Compounds of uranium that are commonly found in the nuclear weapons industry also carry the risk of kidney damage as well as other toxicities. One such example is uranium hexafluoride, which, in the presence of water, hydrolyzes to uranyl fluoride and hydrogen fluoride, a severe irritant (see discussion below).

Exposures to beryllium, a hard metal used extensively in the nuclear weapons industry, occur principally through inhalation of beryllium particles or oxides. Beryllium's main toxic effects take the form of inflammation and immunological sensitization manifesting, for instance, in fibrosis of the lung (with resultant shortness of breath) and enlargement of internal lymph nodes. The mechanisms are poorly understood. The spectrum of health effects is similar to those seen in other chronic granulomatous diseases, particularly sarcoidosis. A mildly elevated risk of lung cancer may also exist.

Acids

The nuclear weapons industry involves the heavy use of acids, such as nitric acid, that can cause intense irritation and destroy cells upon contact. These compounds are highly reactive, and their effects are generally nonselective and limited to the sites of exposure. Acids easily burn the skin and eyes. Inhaling vapors, gases, or dusts of acids in their anhydride form can cause lung injury. Scarring can complicate recovery from acid skin burns, and severe pulmonary inhalation injury can lead to fibrosis, restrictive lung disease, or chronic obstructive airway disease.[26] As a rule, acids are not absorbed, and their action stops at the sites of exposure.

25. Rosenman et al. 1986.

26. Linden 1992.

The uniquely toxic properties of hydrofluoric acid are noteworthy, however, particularly because it is commonly used in nuclear weapons production. In contrast to other acids, even small amounts of hydrofluoric acid can penetrate the skin, underlying lipid barriers, muscles, and even bones. It penetrates tissues rapidly and deeply, but the effects are delayed; it may be several hours before intense pain develops at the site of the burn. The tissue is gradually destroyed, and the affected part may ultimately become gangrenous. Hydrofluoric acid can also cause pulmonary edema (excess body fluids in the lungs) and other respiratory damage. Hydrofluoric acid distributes throughout the body. The fluoride ion can dissociate from the hydrogen ion and combine with calcium, magnesium, and other positively charged ions to increase the risk of cardiac arrhythmias. Depending on the exposure level, burns from hydrofluoric acid can be much more severe than they first appear and can even lead to fatal metabolic disturbances.[27] Some of these properties are also shared to a degree by other fluorine-containing compounds found in the nuclear weapons industry, notably chlorine trifluoride.

Organic Compounds

The making of nuclear weapons uses a wide variety of organic compounds including solvents and chelating agents. Some are or have been used in large quantities, including toluene, carbon tetrachloride, benzene, acetone, methyl ethyl ketone, EDTA, HEDTA, and tributyl phosphate.

Most solvents consist of a single compound, but some are mixtures of compounds. Given their chemical nature, the body readily absorbs solvents through inhalation, ingestion, and skin contact. The body excretes or metabolizes most solvents fairly quickly, so they don't accumulate. Solvents mainly affect the nervous system, liver, kidneys, and skin. Benzene and some other solvents are known human carcinogens; others are suspected carcinogens. Heavy exposure produces, most acutely, an anesthetic effect manifested by dizziness and lightheadedness. Prolonged exposure can produce persistent, potentially irreversible impairment of cognitive (intellectual) function, including impaired memory and concentration, and disturbances of mood and affect such as depression. Some solvents can cause degeneration of peripheral nerves, leading to paralysis or numbness. Liver injury and kidney toxicity occur during excretion and as the body attempts to detoxify the original compound. Skin injury occurs as solvents dissolve the protective fat compounds in skin, leading to dermatitis and increasing the risk of infection.

27. Krenzelok 1992.

An important issue arising from lower levels of exposure to solvents, particularly with respect to people living near nuclear weapons plants, is groundwater contamination and possible links with cancer. Volatile organic compounds (VOCs) can migrate from hazardous waste sites through soil, leading to detectable levels in groundwater. As with many toxic agents, it is not known whether low levels of VOCs cause human cancer. A few epidemiological studies of drinking-water contamination by VOCs have found small but statistically significant increases in the risks for leukemia and cancers of the bladder, colon, and rectum.[28] Public concerns underlie the current conservative assumption for purposes of regulation that exposure to known and suspected carcinogens must be reduced to the lowest level possible.

ENVIRONMENTAL AND OCCUPATIONAL EPIDEMIOLOGY

It is important to understand the process by which scientists actually try to assess and quantify the relationship between toxic exposures and disease using environmental and occupational epidemiology. Epidemiology studies the occurrence and distribution of disease among populations. Environmental and occupational epidemiology focuses specifically on the relationship of environmental and occupational "exposures" to the development of disease. Although infectious agents, nutrition, and other personal lifestyle factors such as smoking and drinking can be considered external exposures, environmental and occupational epidemiology generally refers to exposures to chemicals, dusts, physical factors (such as radiation), jobs (which may entail multiple exposures), and other stressors.

Types of Studies

A key feature of any epidemiological study is the technique of examining the relationship between an exposure and a health outcome by comparing populations. Two main approaches exist. Cohort studies compare the health outcomes of people who were exposed versus those of people who were less exposed or not exposed. Case-control studies compare the exposures of people who have a particular disease versus the exposures of people who do not have that disease. Health outcomes are specified either in terms of mortality, usually in the form of diagnoses derived from death certificates, or morbidity (illness). In the case of cancer, morbidity usually takes the form of diagnoses compiled in cancer registries (centralized bureaus that record cases of cancer).

28. Shy 1985.

Most of the epidemiological studies discussed in this book are of the cohort type, which are usually undertaken when the exposure (in this case, to radiation) and the group of people exposed are well-defined; the chief aim is to assess whether the disease rates of the exposed individuals differ from those of unexposed individuals. The standard measure that is calculated in such studies is the risk ratio, also known as the rate ratio or relative risk. Simply speaking, the calculation takes the form of the number of adverse health events or conditions observed among a set number of individuals of the exposed population divided by the number of adverse health events or conditions observed among the same number of individuals of the unexposed population.

Comparisons of disease risks in cohort studies are generally adjusted or "standardized" to take account of important differences between the exposed and nonexposed groups. For example, it is important in making such comparisons to ensure that the age distributions of the study groups are similar, since cancer rates differ strongly with age (e.g., the elderly are more prone to most cancers, young adults to some forms of leukemia). Without this adjustment, markedly different disease rates may appear simply because two groups had differing age distributions.

Many epidemiological studies compare the number of cancer deaths that occur in a study population with the number that would be expected if the cancer mortality rates in the general population applied. This is expressed as the standardized mortality ratio (SMR), defined as

$$\text{SMR} = \frac{\text{observed number of cancer deaths among exposed}}{\text{expected number of cancer deaths among exposed}}$$

To calculate the expected number of cancer deaths among the exposed population, the age-specific cancer mortality rates of the general population are applied to the number of individuals in each age stratum of the exposed population, and the number of expected cancer deaths are then summed across all age strata. It can readily be seen that if the observed number of cancer deaths is no greater than that which would be expected in the general population with the same age distribution, the SMR will be 1.00.

In addition to the SMR, cohort studies can provide an estimate of the difference in (as opposed to the ratio of) disease rates between exposed and nonexposed persons. This is known as the *attributable risk* or *excess risk*. For example, if population A is exposed to carcinogen X and has a lung cancer rate of 15 cases per 100,000 people per year, and population B is *not* exposed to carcinogen X and has a lung cancer rate of 10 per 100,000 people per year, the SMR is $15/10 = 1.5$; the attributable (or excess) risk, however, is $15 - 10 = 5$ cases per 100,000 people per year. (In other words, the risk attributable to carcinogen X in population A is 5 cases per 100,000 people per year.)

Ideally, an epidemiological study is able to not only compare the experience of an exposed with an unexposed population but also relate this experience in term of levels of exposure. Then, the SMR, other rate ratio, or attributable risk can be expressed per unit of dose. For example, in a study of lung cancer deaths among a cohort of 2,103 uranium miners (see chapter 5), Howe et al. discovered that higher doses of radiation were associated with higher rates of lung cancer. They calculated a relative risk of 0.27 percent per working level month; that is, a rise in radiation exposure of a single WLM per person was associated with a rise in the lung cancer standardized mortality ratio of 0.27 percent. They also calculated an attributable risk of 3.10 total cancers per WLM–million person-years; that is, a rise in radiation exposure of a single WLM per person and a million person-years of observation (e.g., 33,333 workers followed for 30 years or 40,000 workers followed for 25 years) was associated with 3.10 lung cancer deaths over expectation.

A number of the environmental epidemiology studies relevant to this book are ecological studies. Ecological studies compare the disease rates of populations using average or aggregate measures of exposure. Ecological studies commonly define an individual's exposure in terms of living in a specific geographic area. Since these studies are not based on actual exposure measurements of individuals, the investigations are subject to a certain amount of error, and the results must be interpreted with caution.

Case-control studies appear less frequently in this book. These studies compare the exposures of persons suffering from a given disease with the exposures of persons from the same population who are free of disease. They are generally used to study very rare diseases. Although they are able to provide a measure of the relative risk of a disease, i.e., a risk ratio of the incidence of a disease among exposed people compared to that among nonexposed people, case-control studies are unable to provide estimates of the absolute rates of disease in either the exposed or nonexposed populations.

Measuring and Interpreting Data from an Epidemiological Study

Epidemiological studies aim at determining whether associations exist between a defined exposure and disease. Much of the methodology of epidemiology is concerned with designing these studies so that they draw valid comparisons and analyze data in a fashion that provides meaningful results.

Once an epidemiological study has established that an association exists, (i.e., in the case of a mortality study, that the SMR is greater than 1.00), it first must be recognized that an association could occur by chance. Epidemiological studies try to quantify the probability that chance played a role in the association observed by providing a statistical measure—a *p-value*—that is calculated based on probability theory. In general, there is recognition that an association is "significant" in a statistical sense if the calculated p-value for an association is below 0.05 (i.e. the probability that the association could have happened by chance is less than 5 percent). A related measure is the 95 percent confidence interval. Loosely speaking, this specifies a range around the point risk estimate in which the researchers are confident that the true risk estimate lies.

The finding of an association that is "statistically significant" (i.e., the p-value is less than 0.05) does not prove that a causal relationship exists. An association could exist because of interference by a confounder, an unmeasured factor that actually causes the disease. For example, white women are more likely than Hispanic women to get lung cancer. However, smoking confounds the apparent association between lung cancer and race—since white women are much more likely to smoke. An association might also be seen because of *bias* inherent in the study. For example, if a physician knows that the chest X ray being read comes from a patient who had heavy dust exposure, the physician may be more likely to find pneumoconiosis—an example of "observer" bias that must be avoided by making sure at the outset of the study that the observer is "blind" to exposure status.

Finally, it is important to note that even at their most sophisticated level, epidemiology studies cannot by themselves "prove" causality; rather, causality can only be assessed by taking into account a wide body of evidence, including epidemiology and laboratory experimentation, as well as evaluation of mechanisms through which causation is likely to occur.

Challenges Specific to Environmental and Occupational Epidemiology

Epidemiological studies of environmental and occupational exposures must contend with several methodological challenges. Chief among them is the difficulty of identifying who is exposed and who is not (or the extent of exposure). If exposed and nonexposed individuals are incorrectly labeled, the likelihood of detecting any true association between the exposure and the effect of interest generally decreases.[29] For studies of the nuclear weapons industry, this problem takes the form of identifying workers or community residents who were exposed and quantifying the extent of their exposure. Radiation badge readings can help quantify worker exposures, but these are not used outside a facility. Moreover, radiation badges cannot efficiently detect exposure to some forms of radiation, notably alpha particles. For ecological studies of community residents, proximity to a facility is often a

29. Barron 1977.

proxy for exposure, but this is grossly insensitive to actual patterns of exposure that depend on many unmeasured factors, such as wind direction, food and water sources, time spent in the area, and surface topography.

New methods have been developed (and continue to be developed) to better identify those who have been exposed, and to quantify the amount of that exposure. Loosely referred to as *dose reconstruction,* such methods may rely on available information on a combination of factors—such as the pattern of known environmental releases, working patterns, physical characteristics and dietary habits of those potentially exposed—to estimate the cumulative doses of radiation or other toxins to individuals. Attention is also being focused on developing *biological markers*—such as counts of chromosomally damaged lymphocytes in blood, or measures of arsenic accumulations in toenails—that can better define internal dose and/or early toxic effects of radiation and other toxins.

A second challenge is the long period between an exposure and the onset of disease. For many cancers, the latency period may be 20 years or more. A highly toxic exposure in 1993 may have put people at great risk for cancer, but a 1998 study would be unlikely to detect any resulting wave of disease.

A third challenge is the difficulty of documenting health effects among a group under study, especially those with a long latency period. Over time, people move, shift employment, marry, and otherwise become lost to follow-up. For mortality studies, death certificates often identify the cause of death incorrectly. These effects tend to decrease the chances of finding a true association.

A fourth challenge is the small sample size of the populations that are amenable to study coupled with the low background rate of many of the diseases in question. In these circumstances, epidemiological studies are unlikely to detect a small increase in the relative risk of disease. If accurate diagnosis and reporting of cases are assured, then a given number of cases could be more easily found. However, given the problems of reporting, the problems arising from small population size and low background rate are formidable.

A fifth challenge is the possible existence of interactions between environmental exposures and other factors that may influence the susceptibility of individuals to disease. For example, a toxic exposure may only cause disease among women. In addition to gender, other factors may include age, diet, smoking and other personal habits, and genetic variability. Unless these interactions are controlled for, such as by stratifying an analysis for the presence of a given factor, important associations may be missed, or spurious ones inferred.

Another issue is the proper grouping of populations according to exposure. When exposure records are poor or incomplete, improper grouping of populations may result in erroneous results both for absolute and relative risks of exposure. For instance, epidemiological studies of workers often group workers according to external gamma and beta exposure, partly because internal exposure records are often lacking. Such studies may yield misleading or erroneous results in cases where internal exposures were in fact important and not closely correlated with external exposure.

Finally, a challenge peculiar to occupational epidemiology studies is the *healthy worker effect*, which can yield apparently lower risks of disease, including cancer, for workers than for the general population. The effect occurs because the general population includes many more people at high risk of poor health, who are too sick to work, lack good medical care, have lower average socioeconomic status, have higher rates of alcohol ingestion, and so on. In view of the healthy worker effect, any study that demonstrates elevated death rates for workers relative to the general population should be viewed with particular concern, and examined carefully. Approaches for factoring in the healthy worker effect include using other occupational cohorts as comparison populations, or comparing standardized mortality ratios for cancer with the ratios for other diseases that are not theoretically linked to the exposure of interest.

In short, epidemiology is a powerful tool for assessing the relationship between environmental exposures and disease in populations. Limitations exist, as for any of the branches of science. Nevertheless, epidemiological research has been invaluable in demonstrating the toxic effects of a number of environmental hazards, including radiation. Moreover, new resources and methods are emerging that are enabling investigators to overcome these limitations. For instance, as mentioned above, biological markers (i.e., laboratory tests using blood, urine or some other type of test on individuals) are being developed that can indicate the degree of exposure to hazards, internal dose, or early biological effects of toxins. As these methods develop, epidemiology will no doubt become an even more valuable tool.

HEALTH RISKS OF MAKING NUCLEAR WEAPONS

The production of nuclear weapons potentially exposes workers and communities to a wide variety of radiation and nonradiation hazards. Unfortunately, an accurate understanding of the extent of exposures and resulting diseases would depend on a cumulative database from the countries that have produced nuclear weapons. That database is either woefully incomplete or intentionally kept from the public. Nevertheless, many generalizations are possible using available information and tenets from the preceding discussions of radiation and nonradiation hazards.

Uranium Mining and Milling

Most research on uranium mining and milling has involved occupational groups. Few investigations have examined health effects beyond the workplace, due largely to the small radiation doses received by individuals in the general population and the difficulties faced in detecting small increases in disease rates. Further, many uranium mines are in tribal areas, in the Third World, or both, where medical services tend to be poor, leading to added difficulties in disease detection. Nevertheless, the long-term total dose to the whole population due to environmental releases may be substantial, so possible health effects in the populations near mines and mills must be considered despite the speculative nature of such assessments.

Radiation Hazards of Uranium Mining and Milling The radioactive decay of uranium-238 is not a one-step process, as the decay products of uranium are themselves radioactive (see chapter 3). In all, there are thirteen main radioactive intermediate decay products in a chain that ends with nonradioactive lead (lead-206), and these decay products are almost always all present in a body of uranium ore. While uranium-235 is the fissile material of interest in uranium ore, it and its decay products contribute less than 5 percent of the radioactivity of ore. We will not discuss their health effects here. The radiological hazards of mining uranium arise mainly from radon-222 and its daughter products; hazards of milling are mainly from uranium itself. Long-term radiation hazards from mill tailings arise from a combination of radon-222, radium-226, and thorium-230.

The radiological properties of the decay products fall into three groups. Metals with long half-lives (thousands of years) dominate the first group, from uranium-238 to radium-226. The second group consists of the radioactive gas radon (radon-222) and its short-lived (minutes or less) decay products. The final group also consists of radon-222 decay products (lead-210 to polonium-210) but its constituents have somewhat longer half-lives.

• Uranium-238 to radium-226: The main radiological impact to workers comes from inhaling dust. Dust can lodge in the lung, and the radionuclides may be leached out by lung fluids. The major radiological effect arises from doses to the lung and from radionuclides that subsequently end up in other parts of the body, such as bones or thoracic lymph nodes. In the environment, these radionuclides can enter the food chain, leading to radiation doses following ingestion.

• Radon-222 and its short-lived decay products: Radon-222 gas forms from the decay of radium-226. It can diffuse into the atmosphere, although its half-life of 3.8 days limits the distance it diffuses through ore before decay. Radon is inert, so the lungs retain very little of the gas and resulting radiation doses are small. However, short-lived radon decay products, also known as radon "daughters" or "progeny," form quickly in the air. These are metals and as such can readily lodge in the lung. The first four decay products of radon, Po-218, Pb-214, Bi-214, and Po-214, are short-lived, with half-lives of less than 30 minutes; these short half-lives increase the likelihood that they will decay before the lung can clear them out, exposing the lung to alpha radiation. Thus, under most circumstances, these first four radon decay products represent the most serious respiratory health hazard in terms of radiation dosimetry.

• Lead-210 to polonium-210: These other radon decay products are metals with longer half lives. They can also lodge in the lung if inhaled, or they can be absorbed following ingestion. The ultimate fate of these metals is similar to those of the other radon decay products, with some differences in biological action.

In addition, uranium ore may contain nonradioactive materials that are hazardous to workers and, potentially, to people living near mines. These materials include molybdenum, arsenic, and vanadium. For example, molybdenum contamination can occur in groundwater near mill tailing sites.

Exposure Pathways in Mining Underground miners can be surrounded by ore, exposing them to significantly higher doses from radon decay products than in open-pit mining, where natural ventilation of the open pit prevents high radon concentrations from occurring. Uranium ore also emits gamma radiation, exposing anyone in the vicinity. The dose depends on the ore grade and a person's proximity to the ore. Careful planning, such as keeping access ways in low-grade areas of the mine, can reduce gamma doses to an extent. In some ore bodies, "non-entry" mining is feasible, with operations conducted from the periphery of the ore body as much as possible to reduce exposure. In some cases, robot equipment can reduce exposures. In practice, shielding is impractical, although the material of heavy machinery may provide some slight shielding.

Inhaling radon decay products is historically the largest source of radiation exposure to underground miners. The most important control mechanism is ventilation. Good ventilation dilutes the radon entering the mine atmosphere and also reduces the time available for the decay products to form. Conversely, radon decay product concentrations rise markedly in poorly ventilated areas.

General mine housekeeping is important: potent radon sources, such as stockpiled ore and worked-out areas, should be removed or sealed off. Respiratory protection can reduce doses from radon decay products by a factor of 10 or more. In mines where improved ventilation has reduced doses from radon decay products, the relative importance of other radiation exposures rises.

Inhaled ore dust from blasting, crushing, and mechanical handling can result in significant doses from alpha particles. Good design and work practices—for example, wetting down broken ore before moving it—can reduce the generation of dust. Dust can be diluted and removed by ventilation. Again, respiratory protection can reduce doses.

Although natural ventilation in open-pit mines helps reduce dust concentrations, some precautions are generally necessary to control doses from alpha emitters. In both hot and cold climates, if control cabins for machinery are enclosed and air conditioned, the associated filtration reduces doses to the operators.

Exposure Pathways in Milling In a uranium mill, large ore concentrations occur only in stockpiles, where there are generally few workers. Most mill operations are carried out on ore slurries, and this dilution of the ore by water results in gamma doses considerably below those from the original ore. Similarly, the relatively small amount of ore in the mill at any one time and the "open air" nature of most mills means that radon decay-product concentrations are generally very low. Where climate makes it necessary to enclose the mill, tanks containing the ore slurries can be vented externally.

However, the crushing and grinding areas of a mill can be significant dust sources if not well designed and operated. The concentrates produced by the process, ammonium diuranate (ADU) and yellowcake, have such a high concentration of uranium that inhaling small quantities can lead to significant radiation doses. Heating and drying of ammonium diurate produces dusts that can be significant sources of worker exposure. Spills must be cleaned up promptly. The packing of the concentrate into drums needs special facilities with excellent ventilation.

Waste Management in Mining and Milling The tailings contain about 85 percent of the radioactivity in the original ore, and therefore radiation exposure can result through the pathways already discussed. However, tailings are usually on the surface and occupancy of tailings areas is usually low, so exposures to workers are generally far less than those from mining. It is possible for radionuclides to leach into ground or surface water or to enter the food chain in other ways.

Tailings activity will decay at a rate determined by the half-life of thorium-230 (about 75,000 years). After 250,000 years the total activity in the tailings will be about one-tenth that of the original tailings. The enormous timeframe over which exposures from tailings might occur means that the collective population dose to people nearby could eventually be larger than the collective dose to miners. The radionuclides that pose the greatest problems in tailings are thorium-230 and radium-226. Essentially these are all in the tailings and constitute the main sources of water contamination and, for dry tailings, of doses from wind-blown dust. Further, decay of radium-226 also gives rise to radon-222 and thence to its decay products.

Liquid mill wastes can contain high concentrations of radionuclides. Depending on climate, there are two approaches to management: containment followed by evaporation or treatment to remove radionuclides followed by discharge of decontaminated water. Heavy metal exposure (molybdenum, vanadium) via groundwater is also a health threat at many sites.

In the early days of large-scale uranium milling, tailings were often left in unlined dams, or the tailings material itself was used for dam walls. Rehabilitation of these waste disposal areas was rare. Waste management practices have generally improved in the OECD countries, parallel to better regulation of occupational exposures. Procedures to limit seepage of contaminated groundwater, contain the tailings, and protect against erosion are generally required, as is covering the tailings when operations are completed. The requirements in any particular case are highly site specific, and the ways in which requirements are implemented vary from country to country. Early sites, and those in the Third World, continue to present substantial hazards requiring expensive remediation.

Health Effects of Uranium Mining on Workers The best known health effect of mining uranium is the increased risk of lung cancer due to radiation exposures during the excavation or handling of the ore. This is due principally to inhalation of short-lived radon decay products that are deposited on the bronchial walls and emit alpha radiation, damaging the genetic material of the mucosal cells. Increased rates of lung cancer among underground miners were noted over a century ago, although only in the 1920s was the true nature of the disease demonstrated and radon suggested as a possible cause.

The risk of lung cancer is dose-related, with no evidence of a threshold below which there is no risk. The nature of the dose-response relation is still debated, but the model that fits best the available data is that exposure to radon decay products increases the risk of lung cancer in proportion to the background risk of the disease. It is generally accepted that the relationship between exposure and risk is linear, with some flattening of the curve above 2,000 working level months. However, age at exposure, time since exposure, and the rate at which the dose is received may modify the effect of radon decay products on lung-cancer risk.[30] For example, several studies indicate that the in-

30. National Research Council 1988, pp. 49–52, 417.

crease in cancer risk per unit exposure increases with duration of exposure.[31] If this is so, linear extrapolation of risks from high-dose, high-dose-rate workforces may underestimate the risk of lung cancer experienced by low-dose, low-dose-rate groups, such as some modern miners.

Cigarette smoking, itself a powerful cause of lung cancer, also appears to increase the risk of lung cancer due to radon decay products. In some studies, the risk due to combined exposures can be more than the sum of the risks of smoking and radon decay products separately, a phenomenon known as synergism.[32]

It is not clear that the type of lung cancer due to exposure to radon decay products differs from lung cancer due to other causes. But several studies have shown a preponderance of small cell cancers early in follow-up of miners.[33]

Although only lung cancer has been definitely linked to uranium mining, there is some evidence of links to other diseases. A study of Czech miners reported an increase in basal cell cancers of the skin, possibly due to prolonged exposure to ore dust or to the effects of arsenic present in the ore.[34] However, this observation has not been reported for other workforces, although it must be noted that these studies have generally been based on death records, and basal cell cancers of the skin are rarely fatal.

Studies that have compared geographic areas with varying levels of background radon have suggested a link with leukemia. Henshaw et al. have proposed a mechanism specific to radon exposure.[35] While evidence of cause and effect is weak, the issue deserves attention in light of laboratory studies showing that alpha particles can be powerful inducers of chromosomal aberrations.[36] This may imply a more powerful leukemia-causing effect of alpha radiation than was thought to apply previously. The most heavily exposed workers who painted watch and instrument dials using alpha-emitting radium-226 may have experienced increased leukemia rates, in addition to radiation-induced bone malignancies.[37] Otherwise, limited experience with some people receiving high doses of ingested or injected alpha emitters have not so far demonstrated increased rates of leukemia. No studies have found a statistically significant increase in leukemia among underground

31. Howe et al. 1987; Sevc et al. 1993.

32. Samet et al. 1989; Saccamano et al. 1988.

33. Samet 1986.

34. Sevcova, Sevc, and Thomas 1978.

35. Henshaw et al. 1990.

36. Kadhim et al. 1992.

37. National Research Council 1988, pp. 225–230.

miners, including recent analysis of one of the largest cohorts of uranium miners, workers from the Western Bohemia region in former Czechoslovakia.

In addition to cancer, experiments with animals have shown that exposure to high doses of radon decay products can inflame and scar lungs, impairing respiratory function.[38] This occurs only at very high doses, beyond those experienced by any reported miner groups. Epidemiological studies of miners have reported increased deaths from nonmalignant respiratory disease and diminished respiratory function in proportion to time spent working underground.[39] Such a trend has also been observed among miners of other metals and is unlikely to be due to radiation. Among long-term miners, silica and other work-related dusts and fumes are likely to be more important contributors to chronic, nonmalignant respiratory problems than is radiation. Non-radiation health dangers to uranium miners are similar to those of mining in general. These include injury, chest diseases due to dust (and especially silica), exposure to trace contaminants (such as vanadium and arsenic), and other work-related conditions such as hearing loss. Silicosis may be a major problem in uranium mines where the silica content of the ore is high. Noise-induced hearing loss is also likely to be a common problem in all mechanized mines, unless miners have suitable protection.

Recently, questions have arisen about the effect of uranium mining on human reproduction. The effects of ionizing radiation on genetic combination and fetal development have been closely studied in other settings, but relatively little is known about the effects of exposure to radiation on miners and their offspring. Studies in Czechoslovakia and New Mexico have reported that the ratio of male to female births—one of the most sensitive measures of genetic damage—was reduced in populations involved in uranium mining. These findings are preliminary and inconsistent.[40] Further research involving New Mexico Navajo mining communities has found associations between some reproductive damage and exposure to radon decay products, but these links are weak.[41]

Elevated rates of stomach cancer have been reported in some groups of miners.[42] This may be associated with ingesting dust or diesel fumes while working underground, but the evidence is weak.

Injury is notable because it is so common. It is the most frequently reported cause of time lost from work, and most work-related deaths,

38. Cross et al. 1981.

39. Archer 1980; National Research Council 1988.

40. Wiese and Skipper 1986.

41. Shields et al. 1992.

42. Fox, Goldenblatt, and Kinlen 1981.

at least in the short term, are due to injury. Concern over the occupational health of uranium miners has often overlooked the importance of injury and, ironically, injury rates have even been called upon as a standard—an indication of acceptable risk—on which to base radiation exposure limits.[43]

Injuries in mining cover a wide range, from severe crushing trauma due to rock falls or vehicle accidents to minor soft-tissue injuries. In the mining industry generally, the frequency of severe injury had declined by a factor of two between the mid-1960s and the mid-1980s.[44] However, fatalities still occur in underground mines more often than in almost any other work setting. In Australia in the early 1980s, the fatal injury rate in mining and quarrying was 200 times greater than that in clerical occupations.[45] Reported fatal injury rates for underground mines vary from 600 to 1,000 per million people employed per year.[46]

It is often difficult to distinguish between causes of ill health that are work related and those that are part of a general lifestyle. This is a particular issue with injury. Accidents and violence are the most common cause of premature death for miners, but most of these deaths occur away from the workplace. The prevailing view holds that these events reflect the general lifestyle of people who choose to work in mines and cannot be attributed to their occupation. However, the environment of mining communities (geographic isolation, shift work, the "boom and bust" economic cycle, lack of conventional family structures) undoubtedly contributes to the hazardous behaviors that miners often exhibit away from their workplace.[47] In that sense, statistics based only on injuries that occur during working hours and at the workplace underestimate the effect of mining on the health of workers and their families.

Health Effects of Uranium Mining and Milling on the General Public Mining operations and tailings release radon into the atmosphere, while uncovered tailings are the chief sources of airborne particulates. Particulates settle relatively quickly and only affect people near uranium facilities. However, radon and the long-lived radon decay products may be distributed over great distances, depending on atmospheric conditions and the site of release.

Estimates of global uranium production and radon release quoted in UNSCEAR lead to an estimated radon release from uranium mines

43. ICRP 1985; Fry and Carter 1992.

44. Green et al. 1986.

45. Harrison et al. 1989.

46. ICRP 1985, p. 22.

47. Brown 1980.

and associated mill wastes of about 1.6 million curies (about 60 million gigabecquerels) per year. Radon has a short half-life and therefore its emissions from tailings and mine wastes would not add significantly to global radon doses. However, local effects may be significant. Further, tailings have been used in home and other construction, which produces disproportionately high doses due to the high indoor radon concentrations that result. Finally, we must consider doses of toxic materials and radiation from groundwater and other routes of illness far into the future.

Two important points relate to the risk estimates presented here: they refer to worldwide or continental averages, and they are based generally on assumptions of standard or usual practice. More relevant questions are: What are the possible effects of uranium mining and milling on local populations, especially when standard or usual practice is not followed? And what are the plausible consequences of accidental release or negligent operation?

While the aim of radioactive-waste management is to ensure that the wastes will not result in any radiation dose or more general environmental effect, this goal will not always be achieved. Waste management generally can fail in one of two ways: a catastrophic failure, such as a breach of a tailings-dam wall, or a chronic low-level release, such as slow seepage over an extended period.

There may be significant health effects among local populations from nonradioactive contaminants. Evidence of high molybdenum levels in water (24 to 60 parts per million) and soil (50 to 90 parts per million) has been reported from a site adjacent to a uranium mill in southern Colorado.[48] Animals that grazed near the mill have showed signs of molybdenosis; local residents report symptoms consistent with molybdenum toxicity, such as persistent diarrhea, fatigue, and joint pain. Formal health-effects research is needed.

Hazards of Uranium Conversion to Hexafluoride, Uranium Enrichment, and Fuel Fabrication

Uranium conversion and fuel fabrication involves the mechanical and chemical processing of uranium compounds and uranium metal. Some of these are in the form of dry powders, which can result in the discharge of uranium dust to the environment. Present-day plants are equipped with filtration equipment that should remove the uranium dust, but during the 1940s, 1950s, and 1960s, plants typically emitted far larger amounts of uranium to the atmosphere.[49]

48. Smith, M. T. 1993.

49. Eisenbud 1987, p. 181; Voillequé et al. 1993 provide an example.

Uranium metal chips and turnings can catch fire when exposed to air. If fires occur, more than normal amounts of activity may be released and worker exposures are also increased. Many plutonium production reactors use uranium metal as fuel and target rods; others use uranium dioxide.

A major hazard in both uranium conversion and uranium enrichment comes from handling uranium hexafluoride, which is highly corrosive and chemically toxic as well as radioactive. When uranium hexafluoride touches water (such as the moisture in air), it breaks down into hydrofluoric acid (a highly corrosive toxin) and fine particles of uranyl fluoride (a heavy-metal compound). Each of these compounds poses health hazards, as already noted.

Although uranium is radioactive and therefore poses some risk of carcinogenic effects, the main hazard when soluble is its chemical toxicity as a heavy metal. It had been thought that significant damage was not caused at concentrations less than about 3 micrograms of uranium per gram of kidney, but it now seems that subtle damage can occur at lower concentrations.[50] Clearly, not enough is known about the effects of chronic exposure.

Worker radiation doses from uranium conversion due to inhaling uranium may have been much higher than suspected. Infrequent urine sampling and lung-burden monitoring create large uncertainties in estimates of dose.[51]

Reactor Health Effects

Reactors are used to produce plutonium and tritium. There are some routine emissions of gaseous radionuclides from reactors. However, most routine emissions of fission products occur in the reprocessing step (see below). Reactors can also suffer various kinds of accidents during which large quantities of radioactive materials may be released. The most serious health effects from such releases would be from fission products such as iodine-131, strontium-90, and cesium-137. In chapter 8 we will discuss a reactor accident that occurred in the United Kingdom in 1957.

Hazards from Separating and Producing Plutonium and Tritium

After highly radioactive fuel rods and target rods come out of a nuclear reactor, they must be reprocessed if plutonium is to be separated. In the past, plutonium for nuclear weapons has been produced

50. Leggett 1989.

51. Franke et al. 1992; Franke and Gurney 1994.

at reprocessing plants such as Hanford in the United States and Sellafield in the United Kingdom.

In the 1940s, 1950s, and 1960s, reprocessing resulted in relatively high radiation doses to both workers and members of the public. According to UNSCEAR, the annual average effective dose equivalent per worker at reprocessing plants was between 4 and 15 millisieverts in the early 1970s, dropping early in the 1980s to 2 to 4 millisieverts.[52] Further, according to Charles et al.:

> Doses to members of the public living near reprocessing plants are higher than other nuclear installations simply because discharges from reprocessing plants are much higher. For example, Sellafield discharges 20 or so times the activity of all nuclear power stations in the United Kingdom combined, although this total is now only about 1 percent of the discharge levels of the mid-1970s. Reprocessing plants contribute about 98 percent of the collective dose from the nuclear fuel cycle to the countries of the European Community.[53]

Whether the relatively high discharges from reprocessing plants result in health problems is a controversial question. The most studied plant has been Sellafield, where an excess of childhood leukemia was discovered. The "Seascale cluster" (as this became known) stimulated other studies of incidences of leukemia or other cancers around nuclear installations. A number of clusters have been reported. Explanations remain elusive.[54] (See chapter 8 for a more detailed discussion of this issue.)

A recent study of childhood leukemias in the vicinity of France's La Hague reprocessing plant found an incidence rate close to that expected; however, the ecological design of the study and the small size of the population designated as exposed limited the power of the study to detect anything other than a very large excess risk.[55]

More than 13,000 people living near the Hanford, Washington, reprocessing plant received significant doses of radiation as a result of emissions from the facility in the 1940s. From airborne emissions alone, about 5 percent of the 270,000 residents living near the plant during and after World War II may have accumulated thyroid doses of radiation averaging 330 millisieverts over three years, and a small number of infants and children could have accumulated radiation doses to their thyroid glands as high as 29 grays.[56]

The radionuclides of principal concern in reprocessing effluents are the long-lived nuclides, principally tritium, carbon-14, krypton-85,

52. UNSCEAR 1988, 158.

53. Charles, Jones, and Cooper 1990.

54. Draper et al. 1993.

55. Viel et al. 1993.

56. Shulman 1990.

strontium-90, technetium-99, iodine-129, cesium-134, cesium-137, and isotopes of transuranic elements (plutonium-239, plutonium-240, plutonium-241, and americium-241).

Hydrogen-3 (Tritium) Tritium has a half-life of about 12 years and emits low-energy beta particles. It is an isotope of hydrogen and, hence, chemically akin to it. Although tritium occurs naturally, nuclear installations are by far the greatest source of tritium in the environment.

Tritium is generally thought to have relatively low radiotoxicity, but an important distinction separates tritium gas from organically bound tritium and tritium as a constituent of tritiated water. Tritium as a constituent of water can irradiate a large number of cells, since tritiated water is chemically the same as regular water in the body's cells. Tritiated water also crosses the placenta and could affect developing fetuses. Finally, tritium in water can become organically bound by replacing hydrogen atoms in the cells of the body. It has been suggested that organically bound tritium may be responsible for an increased incidence of Down's syndrome found near a nuclear plant in Canada and for an increased incidence of prostate cancer in U.K. Atomic Energy Authority workers.[57]

Carbon-14 Carbon distributes itself quickly in the biogeosphere. Because the half-life of carbon-14 is 5,730 years, the dose from carbon-14 introduced into the environment will be delivered for many generations. As Merrill Eisenbud has pointed out, "A special reason for concern is that carbon-14 can be incorporated in the molecules of which genes are formed. Thus, in addition to the mutations due to the energy deposited by the decay of carbon-14, mutations may result from the transmutation of carbon atoms incorporated in the genetic material."[58]

Krypton-85 Most of the fission product krypton-85 is retained in reactor fuel until released during reprocessing. Because of its inertness and half-life of 10.7 years, krypton-85 distributes uniformly throughout the earth's atmosphere within a few years after release, so collective doses are important.

Strontium-90 Strontium-90 is a beta-emitting fission product with a half-life of 28.8 years. In the body, it behaves rather like calcium and concentrates in bone.

57. Fairlie 1992.

58. Eisenbud 1987, p. 334.

Technetium-99 Large amounts of technetium-99 are produced during nuclear fission; it has a half-life of 215,000 years. As with iodine-129, dose estimation involves predicting radionuclide behavior over very long time scales.

Substantial quantities of technetium-99 are carried along with uranium during reprocessing, and hence into conversion and enrichment plants. In these situations technetium-99 is usually present in soluble form and exposures are lower than uranium compounds. However, technetium is concentrated in the food chain and technetium-99 is also discharged from reprocessing plants into the sea. There have been relatively few studies of technetium in the marine environment. It is believed that it is present in seawater mainly in the chemical form of pertechnetate, which is very mobile and can be transported very long distances. In the human body, the pertechnetate ion behaves in a similar way to the iodide ion: it concentrates in the thyroid (unlike iodide, however, it is not incorporated into hormones).

Other chemical forms of technetium (for example, organic complexes), their stability, and their behavior in the body are relatively less well understood. Nor is it known whether technetium in marine sediments can become available to marine organisms.[59]

Iodine-129 Iodine-129 is a fission product with a half-life of 16 million years. Its long half-life means that it will accumulate in the environment, become part of the iodine pool, and deliver a thyroid dose to the general population. Iodine is very mobile in the environment, so when released it will be rapidly incorporated into foods. "The highest concentrations of iodine occur in seawater and, as with carbon-14, the greatest uncertainties surround the transfer of iodine-129 to deep oceans and any sedimentation that may remove activity from biological chains."[60]

Because of its half-life, the dose from iodine-129 is delivered over a very long period of time: 0.003 percent of the dose is delivered within 100 years of release, 0.03 percent in 10,000 years, and 5 percent in a million years, leaving 95 percent of the collective dose to be delivered from 1 million years after its release.[61]

Cesium-134 and Cesium-137 These emit both beta and gamma rays and have physical half-lives of 2.06 years and 30.2 years respectively. Since the metabolism of cesium resembles that of potassium, these isotopes are predominantly found in muscle cells. They have a biological half-life in the range of 50 to 150 days.

59. Cognetti 1990.

60. UNSCEAR 1988, p. 161.

61. UNSCEAR 1988, p. 161.

Hazards of Lithium Separation Lithium-6 is used in making tritium. A principal toxic hazard associated with lithium separation activities comes from the use of large quantities of mercury, a heavy metal noted for its severe effects on the nervous system when present as methyl mercury. Small doses of mercury may cause irritability, nervousness, and headaches, while larger doses may result in convulsions, coma, and death.

Hazards from the Fabrication of Plutonium Components Plutonium-239 is an alpha emitter and, as such, poses dangers qualitatively similar to other alpha emitters. Differences in biological effects among alpha emitters arise due to a number of factors. The first is specific activity. Consider three other important alpha emitters: radium-226, thorium-230, and uranium-238. Plutonium-239 is almost two hundred thousand times more radioactive (per unit of weight) than uranium-238, about three times more radioactive than thorium-230, but only about one-sixteenth as radioactive as radium-226. Other differences in biological effects arise from differences in energies of emitted alpha particles, routes and amounts of uptake, and retention times in the body. As we have discussed, all of these radionuclides pose some dangers in the context of nuclear weapons production. The special importance of plutonium as regards its health and environmental effects arises from its relatively high specific activity and the fact that large amounts of it are processed in pure or nearly pure forms in nuclear weapons production. Thus, the dangers of processing plutonium are in some ways more comparable to those posed by handling pure radium when it was an industrial material than to the dangers of processing uranium ore or mill tailings. Indeed, early worker protection standards during the Manhattan Project were based on the experience of the uranium industry in the early part of the twentieth century.[62]

A number of potential health hazards arise from making plutonium pits for nuclear warheads, mostly due to the materials involved. The main hazards are associated with plutonium-239, plutonium-240, beryllium, americium-241, and organic compounds. For some of these, such as beryllium, risk may be mainly to workers, and not as much to offsite populations.

Plutonium-239, the principal material ingredient of nuclear warhead pits, is highly carcinogenic, chemically reactive, and flammable. It thus presents a number of dangers.

Since plutonium is retained by the body for many years, it is potentially one of the most radiologically hazardous alpha emitters. Insoluble plutonium particles (e.g., $^{239}PuO_2$), will remain in the lungs

62. Durbin 1994.

after inhalation for long periods of time. The more soluble plutonium chemical forms (e.g., $^{239}Pu(NO_3)_4$) will migrate to lymph nodes and the general circulation with eventual deposition into bone.[63] These patterns of deposition and kinetics are largely responsible for the excess lung and bone tumors that are seen in animals after inhalation of insoluble and soluble plutonium particles, respectively.

Plutonium can also enter the bloodstream through a wound, with ultimate deposition in liver or bone marrow, again raising the risk of bone cancer as well as harm to the blood-forming process.[64]

The degree of carcinogenic risks posed by plutonium and the risk of cancers other than lung and bone remain controversial. Experiments in beagle dogs have provided some quantitative estimates of the amounts of plutonium that can cause lung cancer. For instance, Bair and Thompson reported that 0.003 microcuries of plutonium-239 dioxide in particles of less than 10 microns deposited per gram of lung tissue caused bronchio-alveolar cancer in all the dogs in the experiment.[65] Plutonium has also been shown to cause damage to the blood-forming elements in mice.[66] Since 3 to 7 percent of plutonium that is deposited in human bone is located in the bone marrow,[67] some radiobiologists also consider plutonium to carry a significant risk of leukemia.[68]

On the other hand, carcinogenic risks have not been consistent over all animal studies. For instance, studies in dogs have not been particularly suggestive of a leukemia risk.[69] In addition, epidemiological studies in humans have not yet provided consistent information with which to quantify the carcinogenic risks posed by plutonium.

To protect against plutonium's health risks, extensive safety measures must be implemented. One of these is the use of glove boxes for processing and fabrication operations. Materials are manipulated through gloves installed in sockets on the wall of the enclosure. However, the gloves can tear or develop holes, increasing the risk to workers.

63. IPPNW and IEER 1992, p. 11.

64. National Research Council 1989, pp. 121–122.

65. Bair and Thompson 1974. If we extrapolate these results to the human beings, they indicate that a total lung burden of about 2 microcuries of plutonium deposited in small particles (about 30 micrograms, rounded to one significant figure) would cause lung cancer with high probability. However, this extrapolation to humans has not been validated one way or another, since such high lung burdens have not been recorded in humans in numbers large enough to enable meaningful epidemiological studies.

66. Vaughn, Bleaney, and Taylor 1973.

67. McInroy and Kathren 1990.

68. Vaughn 1976.

69. Thompson 1989.

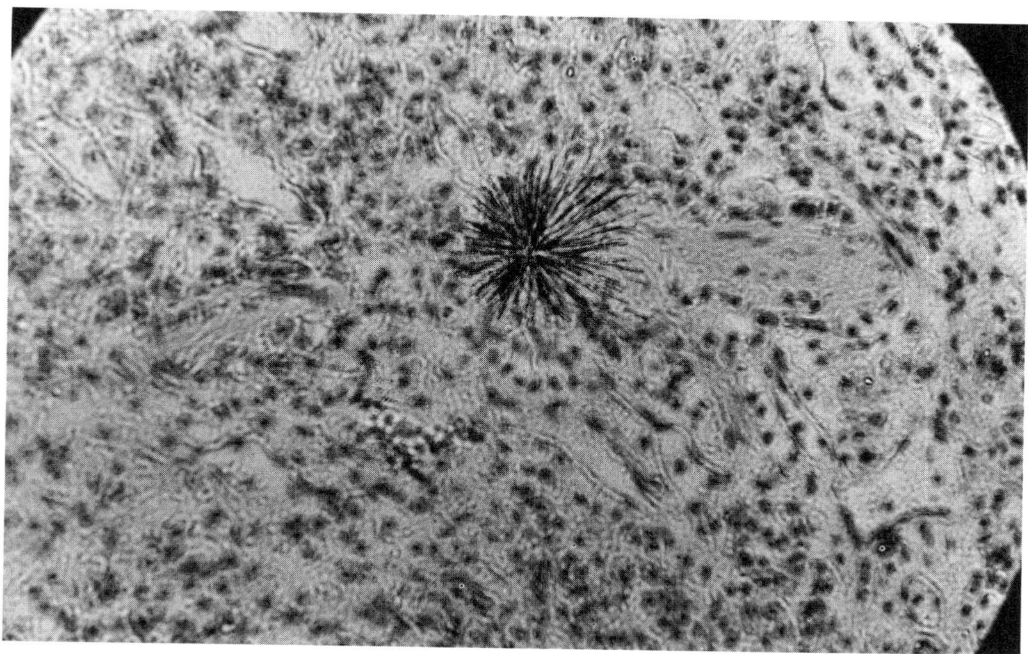

Figure 4.3 Tracks made by alpha radiation emitted by a particle of plutonium in the lung tissue of an ape, magnified 500 times. Photo by Robert del Tredici.

As we discuss in later chapters, mismanagement can reduce or eliminate the effectiveness of safety measures.

Health Effects of Plutonium on Workers A study of Rocky Flats by G. S. Wilkinson and his colleagues claims to present the "first epidemiological findings that suggest an association between exposure to plutonium and untoward health effects in humans."[70] This study classified 5,413 workers into those with a body burden of plutonium greater than 74 becquerels (2 nanocuries) and those with less than that. The death rates from all causes and some cancers (notably leukemias and lymphatic cancers) were greater in the exposed group. However, this study did not indicate an increase in lung, bone, or liver cancers.

Suggestions of plutonium-related effects appear in a study of workers at Great Britain's Atomic Weapons Research Establishment at Aldermaston.[71] In workers monitored for possible internal exposure, mortality from all cancers did not show an increase, but death rates from prostatic and renal cancers were generally more than twice the national average after a 10-year lag. However, the excess cancers were

70. Wilkinson et al. 1987.

71. Beral et al. 1988.

in a small group of workers monitored for exposure to multiple radionuclides and the excess cancers do not appear to be correlated with reported internal burdens of plutonium as such. Though mortality from lung cancer in workers monitored for exposure to plutonium was below the national average, it was some two-thirds higher than for other radiation workers. The results of this study are inconclusive so far as the specific effects of plutonium are concerned.

As discussed in an earlier IPPNW-IEER work,[72] researchers have now gathered 42 years worth of health records for 26 white males who worked with plutonium-239 during World War II at Los Alamos. Studies of their health status have been published periodically, most recently in 1991.[73] The amounts of plutonium deposited in the bodies of the subjects were estimated to range from 110 becquerels up to 6,960 becquerels. Neither the dose initially received nor the route of exposure are known with certainty. Seven subjects had died by 1990. The listed causes of death are three lung cancers, one bone cancer, one myocardial infarction, one pneumonia/heart failure, and one accident. While four of the seven deaths are due to cancer, little can be inferred from these small numbers. Interpretation is also complicated by the fact that the three people with lung cancer had smoked cigarettes. However, bone cancer is rare in humans; there is a probability of only 1 in 100 of bone cancer arising in an unexposed group of 26 men observed over 40 years. Its occurrence among a population of this size is suggestive, especially in view of plutonium's affinity for bone. Any other inference is very difficult. A clearer picture of the risks from plutonium exposure may emerge from larger and updated worker studies. A long follow-up period is important, given that the time between exposure to radiation and the appearance of cancer may be several decades.

Overall Environmental Contamination from Nonradioactive Wastes and Health Risks

"Contamination of soil, sediments, surface water, and groundwater throughout the [U.S.] Weapons Complex is widespread," summarizes *Complex Cleanup*, a 1991 report of the congressional Office of Technology Assessment (OTA) in the United States. It goes on to note:

> Almost every facility has confirmed groundwater contamination with radionuclides or hazardous chemicals. All sites in nonarid locations probably have surface water contamination.... Substantial quantities of radioactive and mixed waste have been buried throughout the

72. IPPNW and IEER 1992, p. 17.

73. Voelz and Lawrence 1991.

Table 4.1 Examples of Nuclear Weapons Site Contaminants and Mixtures

Inorganic contaminants	Organic contaminants
Radionuclides	Benzene
Americium-241	Chlorinated hydrocarbons
Cesium-134, 137	Methylethyl ketone, cyclohexanone, acetone
Cobalt-60	Polychlorinated biphenyls, select polycyclic aromatic hydrocarbons
Plutonium-238, 239	
Radium-224, 226	Tetraphenylboron
Strontium-90	Toluene
Technetium-99	Tributylphosphate
Thorium-228, 232	
Uranium-234, 238	*Organic facilitators*[1]
	Aliphatic acids
Metals	Aromatic acids
Chromium	Chelating agents
Copper	Solvents, diluents, and chelate radiolysis fragments
Lead	
Mercury	*Mixtures of contaminants*[2]
Nickel	Radionuclides and metal ions
	Radionuclides, metals, and organic acids
Other	Radionuclides, metals, and natural organic substances
Cyanide	
	Radionuclides, and synthetic chelating agents
	Radionuclides and solvents
	Radionuclides, metal ions, and organophosphates
	Radionuclides, metal ions, and petroleum hydrocarbons
	Radionuclides, chlorinated solvents, and petroleum hydrocarbons
	Petroleum hydrocarbons and polychlorinated biphenyls
	Complex solvent mixtures
	Complex solvent and petroleum hydrocarbon mixtures

Note: The contaminant list is being upgraded as new information becomes available.

1. Facilitators are organic compounds that interact with and modify metal or radionuclide geochemical behavior.
2. Information on mixture types is sparse, and concentration data are limited.

Source: U.S. Department of Energy/Office of Health and Environmental Research, Subsurface Science Program, Co-Contaminant Chemistry subprogram, "Draft Strategy Document," March 1990, as cited in U.S. Congress, OTA, 1991.

complex, many without adequate record of their location or composition.[74]

The list of specific nonradioactive contaminants found by the OTA is long and includes all the substances discussed earlier in this chapter as well as mixtures of contaminants (see table 4.1).

Unfortunately, information about the extent and magnitude of human exposure to these contaminants is extremely limited. Even in the United States, with the most open disclosure of environmental hazard information, little published data accurately quantifies off-site environmental transport of contaminants, let alone human exposure. Most investigations have concentrated on radioactive releases and contamination.

Nevertheless, off-site health effects may occur due to exposure to nonradioactive toxins. Independent observations of toxic contamination, leading to public pressure, are often the driving force behind investigations, as exemplified by the published hearings regarding the massive amounts of mercury released at the Oak Ridge Complex.[75] Notably, a pilot study carried out by the U.S. Centers for Disease Control on a small group of people living near Oak Ridge found no significant difference in urinary and hair mercury concentrations between individuals who were classified as being potentially mercury exposed and nonexposed based on responses to a questionnaire.[76] However, as is often the case in environmental epidemiology, the study was ill suited to reassure the surrounding populace. Because the period of greatest mercury releases had occurred years or even decades before the data were collected and urinary mercury only reflects relatively recent exposure, the potential for high previous exposures could not be ruled out.

74. U.S. Congress, OTA 1991, p. 23.

75. U.S. DOE 1987; U.S. Congress, House 1983.

76. Rowley, Turri, and Paschal 1986.

5 Uranium Mining and Milling for Military Purposes

Katherine Yih, Albert Donnay, Annalee Yassi, A. James Ruttenber, and Scott Saleska

Until World War II, uranium was regarded as little more than a substance used to color ceramics and glass, a by-product of radium production. However, since the discovery of nuclear fission in 1938, the international nuclear industry has produced more than 1.7 million metric tons of uranium in about 30 countries.[1] Of this quantity, approximately one-fourth has been used to make highly enriched uranium, almost all of which is in military stockpiles or weapons or has been used for military purposes, including naval reactors.[2] Additional quantities of low-enriched uranium have been used as fuel in naval reactors and in some plutonium production reactors. The IAEA estimates that about 360,000 metric tons of natural uranium, or about 20 percent of the world's production, has been used for military purposes.[3]

In most countries involved in nuclear materials production, uranium mining has been the most hazardous step, in terms of radiation doses and numbers of people affected. At least until the mid-1960s, both the United States and the Soviet Union ignored unequivocal early evidence of radiation hazards, failing to implement ventilation and other straightforward radiation protection measures in their zeal to beat each other in the nuclear arms race. Examples of negligence toward uranium workers and the public are plentiful in the case of uranium mining by the other nuclear weapons states as well.

Indigenous, colonized, and other dominated peoples have been disproportionately affected by uranium mining worldwide. In the United

1. Quantities of uranium are in metric tons of uranium oxide (U_3O_8), following industry convention.

2. This calculation is based on an assumption of a worldwide inventory of highly enriched uranium of 2,300 metric tons. As noted in chapter 3, it takes about 180 metric tons of natural uranium to make one metric ton of weapons-grade highly enriched uranium. There is considerable uncertainty in the total, which arises from diverse estimates of former Soviet stockpiles (Makhijani and Makhijani 1995, chapter 7).

3. Underhill and Muler-Kahle 1993.

States, Canada, and Australia, traditional aboriginal lands turned out to hold deposits of uranium to which the governments or government contractors laid claim. Gilbert Oskaboose, a Canadian Indian, has observed:

> White people came here a long time ago, took all the furs, trapped all the beaver out, and the otter and the mink, things like that.... They went away and they left us with the bush and the rocks. It wasn't too much later they came back again. They call that logging. Cut down all the trees—white pine, red pine, cut it all down. And they left us on the bare rocks. Then they discovered uranium here. And the old man said, "Now the sons-a-bitches are back for the rocks."[4]

Indigenous peoples have suffered notable health effects as a result of working in or possibly even living near the mines. In defiance of a UN decree designed to preserve the natural resources of one of the world's last remaining colonies for the benefit of its people, Great Britain bought uranium from a South African-controlled company in pre-independence Namibia for several years, some of it for military purposes. And accusations that the Soviets forced people to labor in uranium mines are frequently heard in eastern Germany, the Czech Republic, and Russia.

Chris Olgiati, a BBC producer who visited the French-run Arlit mine in Niger in the early 1980s, summed up one of the central ironies of uranium mining: "Some of the poorest people on earth labor in one of the deadliest environments to power the electric train sets and fuel the bombs of the world's richest nations."[5]

It is important to add that, seeing uranium mining as a fact of life, many aboriginal people have shifted their focus from opposing it to trying to gain greater control over it and get more direct economic benefits from it.

HISTORICAL OVERVIEW

Once uranium from a particular mine or mill finds its way to the world market, it is difficult to trace it through the various transformations and transactions to its end use. Thus, it is not always possible to accurately determine which uranium mining and processing activities have been for military purposes and which for nuclear power. Neither can the health and environmental consequences of mining at particular sites be accurately partitioned between military and civilian.

Nevertheless, some general distinctions can be made based on known periods of nuclear weapons manufacture by various countries.

4. Gilbert Oskaboose, Serpent River Band, as quoted in Canada, National Film Board 1991, cover.

5. WISE International Networking Bulletin for the Safe Energy Movement, Amsterdam, June 1982, cited in Moody 1992, p. 721.

In the case of the United States and United Kingdom, uranium procurement for nuclear weapons extended from World War II to the mid-1960s. After the mid-1960s, the vast bulk of the uranium trade was for power plants, while uranium for new weapons tended to be taken from military stockpiles accumulated in the earlier period or from obsolete weapons upon dismantlement.[6] However, some of the 8,600 metric tons of Namibian uranium, production of which began only in 1976, apparently went to British military uses. France, which developed the bomb later than the United States and the United Kingdom, may have continued procuring uranium for military purposes beyond the mid-1960s; in any event, military and civilian nuclear activities in France are so intertwined that it is appropriate to consider any uranium procured by that country as potentially military. However, the large size of France's civilian nuclear power program means that the vast majority of the uranium use in France (on a cumulative basis) has been in this civilian program.

In sum, most of the uranium produced until the early 1960s went into weapons or military stockpiles. Some uranium produced after that time (in countries other than the United States), but a rapidly declining fraction, also was used in weapons. Although production in Gabon, Niger, and Namibia began after the West's massive weapons buildup, it is safe to say that some of their uranium has gone to the weapons programs of France and the United Kingdom, as explained above.

The Soviet Union used East German and Czech uranium in its nuclear weapons program at least in the early years; mines in these countries were already open, having been used by the Nazis. The Soviet arms buildup, initially slower than that of the United States, persisted well into the 1980s (see figure 5.1), but it is not clear how much uranium each country contributed to it. There is also considerable uncertainty about how much uranium has accumulated in the military stocks of the former Soviet Union.

We must also recall that in some countries, notably France, Russia, and the United Kingdom, uranium used in civilian reactors is reprocessed to produce plutonium. These stocks of separated plutonium can be used for military purposes.

Health and safety precautions and environmental protection measures in all producing countries were lax in the early years, when production for military purposes was predominant. For example, the United States failed to establish protections against known occupational hazards of uranium production, notably radon gas, until after the weapons-related uranium production boom was over in the early 1960s. Mining overburden containing elevated levels of uranium and its radioactive decay products blights many mining areas. This has

6. Neff 1993.

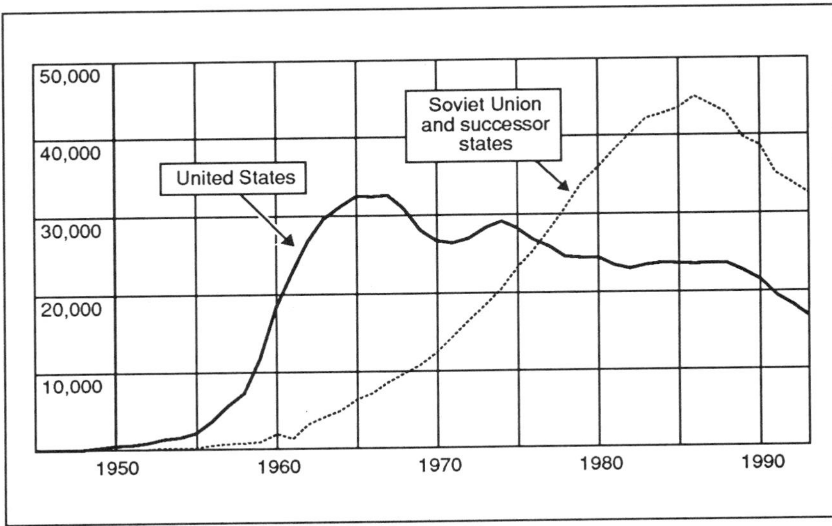

Figure 5.1 Line graph of U.S. and Soviet nuclear warhead buildup. Source: Norris and Arkin 1993.

created an enormous and persistent legacy of environmental problems at many sites that are now abandoned, being remediated, or producing uranium only for commercial power. Thus, the fraction of the total environmental and occupational hazards arising from mining and milling for weapons production is undoubtedly considerably more than would be indicated simply by the proportion of uranium produced for that purpose.

Figure 5.2 presents the total uranium production of all countries with a cumulative production through 1991 of 20,000 metric tons or more. The ups and downs in production by the nonsocialist world over time are shown graphically in figure 5.3.

Uranium ore production in the nonsocialist world peaked in 1959, to be surpassed only 20 years later as demand increased for fuel for commercial power reactors. Weapons-related uranium trade also peaked in 1959, with international shipments totaling almost 20,000 metric tons of uranium.[7]

In the early 1960s, U.S. and U.K. weapons procurements began to decline. Early nuclear power programs had not yet created an offsetting demand for uranium; moreover, there were already large stockpiles of it. As a result, uranium mine development and production in the nonsocialist world went into a slump, with worldwide prices and production dropping nearly 60 percent by 1966.[8]

7. Neff 1984, p. xxi.

8. Neff 1984, pp. xxi, 171.

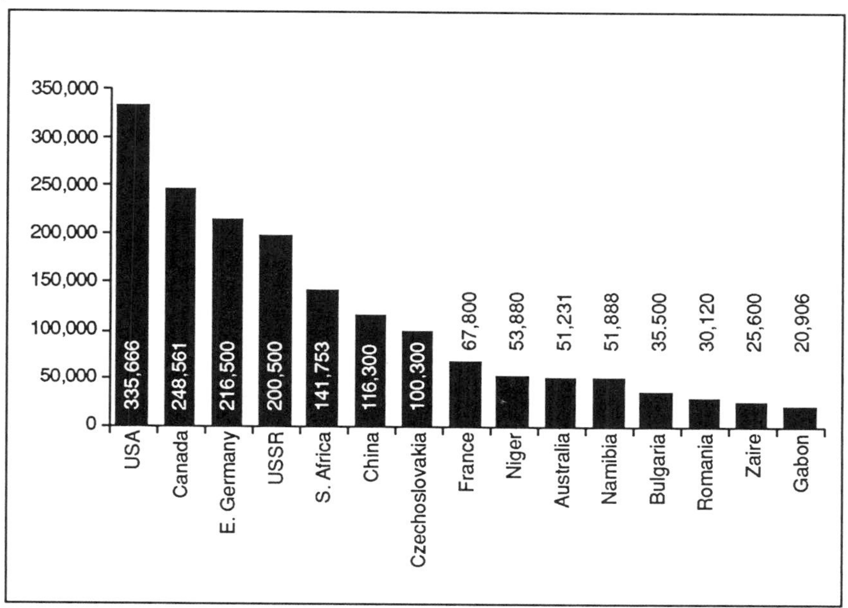

Figure 5.2 Cumulative uranium production in metric tons, 1945–1991. Source: Uranium Institute 1992, p. 43.

In the late 1960s, the prospects for nuclear power began to look up due to a surge in orders. The new demand caused a resurgence in the uranium industry in the United States and elsewhere starting around 1968 and lasting until the early 1980s (see figure 5.3). Mirroring the earlier boom-and-bust cycle, rapid expansion of the uranium industry in the late 1970s led to excess production and excess capacity. Prices dropped sharply after 1979, and new procurements from primary producers fell off. Uranium production in the nonsocialist world, having reached new heights in 1980–1981, again declined. It continues to fall in the 1990s, especially with the entry on the world market of uranium from the former Soviet Union and Eastern Europe.

In total, the West and South produced 1,017,000 metric tons of uranium through the end of 1991, while the Soviet Union, Eastern Europe, and China produced 720,000 metric tons.[9] No uranium was exported from the West to the East.[10] Only small amounts of uranium were traded in the other direction, although the annual amount has increased since the late 1980s. As of October 1992, cumulative exports to the West by the formerly socialist countries totaled about 13,000 metric tons.[11]

9. Uranium Institute 1992, p. 19.

10. Uranium Institute 1991, p. 55.

11. Uranium Institute 1992, p. 30.

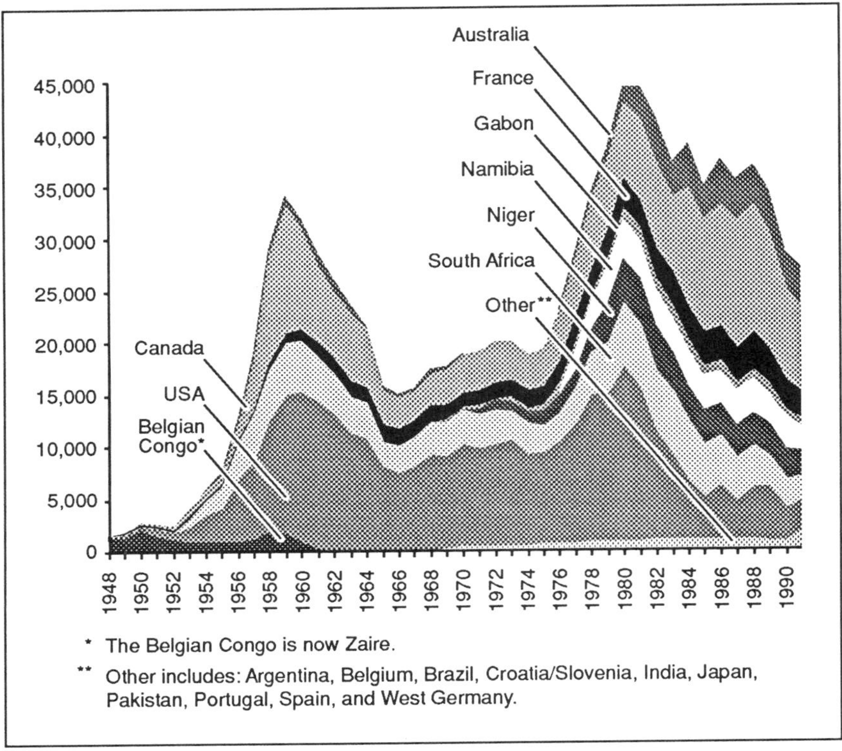

Figure 5.3 Historical uranium production in metric tons. Sources: Uranium Institute 1992, p. 43; Neff 1984; OECD and IAEA, Paris, as cited in Neff 1984.

The rest of this chapter discusses uranium production and resulting conditions in each of the most significant producing countries historically (in descending order of cumulative production): the United States, Canada, East Germany, the Soviet Union, South Africa, China, Czechoslovakia, France, Niger, Australia, Namibia, and Gabon. They are grouped below by the nuclear state to which they have supplied the most uranium. We consider the United States and United Kingdom together since they have obtained uranium largely from the same sources, dating back to the Manhattan Project. In fact, in 1944 the two countries formed a joint corporation, called the Combined Development Trust, to acquire uranium and thorium supplies for what was then the joint military program.[12]

URANIUM SOURCES FOR U.S. AND U.K. WEAPONS

The uranium used by the U.S. weapons program in the early years, including uranium for the bomb dropped on Hiroshima, came princi-

12. Hewlett and Anderson 1990, pp. 285–288.

pally from the Shinkolobwe deposit in the Belgian Congo (now Zaire). The Belgian company Union Minière du Haut Katanga had mined this deposit for radium since the early 1920s, but Canadian sources eliminated the demand for radium from Africa and led to the temporary closing of the Shinkolobwe mine. However, the United States imported uranium via Belgium (where the ore had been refined for radium production) as early as in 1937;[13] in 1939 an official of Union Minière, recognizing the possible military uses of uranium, arranged for stockpiled ore from the mine to be shipped to the United States. Uranium from the Belgian Congo ultimately constituted 80 percent of the total uranium used in the Manhattan Project.[14]

As noted above, until the mid-1960s, uranium was used almost exclusively for nuclear weapons.[15] Although nuclear energy for civilian uses seemed promising, such applications on a large scale remained several years distant, the main interest at the time being to build bombs. As the 1946 U.S. Bureau of Mines' *Minerals Yearbook* put it, "In the present state of world affairs, the most important use of uranium would seem to be in making atomic bombs";[16] it noted in 1949, "As the second decade following the discovery of uranium fission unfolded, it was evident that development of the socially beneficial aspects of atomic energy would continue to be secondary to military considerations."[17]

An intensive global hunt for uranium ores began in 1946. David Lilienthal, chair of the U.S. Atomic Energy Commission (AEC), conveyed something of the tenor of the U.S. prospecting effort in 1948: "The Atomic Energy Commission of the United States is responsible for one of the most extensive and intensive searches for a mineral that has ever been conducted, and on a world-wide basis."[18]

After the first Soviet nuclear test explosion in 1949 and the beginning of the war in Korea, the United States dramatically stepped up nuclear weapons production in 1950. "Urgency was the keynote of the atomic energy program in 1950," declared the *Minerals Yearbook*.[19] The United States expanded its weapons program to an unprecedented level in the early 1950s.

The uranium for the new U.S. weapons initially came from the Belgian Congo, the United States, and Canada. By 1953, South Africa had

13. Neff 1984, p. 171.

14. Owen 1985, p. 3.

15. Owen 1985, p. 36.

16. U.S. Bureau of Mines 1946, p. 1206.

17. U.S. Bureau of Mines 1949, p. 1248.

18. U.S. Bureau of Mines 1948, p. 1272, statement made on 17 December.

19. U.S. Bureau of Mines 1950, p. 1257.

begun shipping uranium from its Rand gold deposits to the United States, and mines at Radium Hill and Rum Jungle, Australia, had been developed. By the late 1950s, the United States' main sources of uranium other than domestic mines were Canada (principally), South Africa, the Belgian Congo, and Australia.[20] Starting around 1958, when uranium imports were 60 percent of supplies, the United States rapidly reduced its imports of uranium; imports were only 20 percent of supplies by 1965.[21]

The United Kingdom got its uranium principally from the Belgian Congo, Australia, Canada, South Africa, and Namibia.[22]

United States

After World War II, the U.S. Government guaranteed prices, gave generous investment allowances, and offered other financial incentives to encourage uranium mining in the United States. The government was the sole purchaser of the ore. The states of the Colorado Plateau were the center of exploration activities: Arizona, Colorado, New Mexico, Utah, and Wyoming.[23]

A "uranium rush" gripped the U.S. West in the 1950s. Stories spread of amateur prospectors making millions of dollars overnight, exciting wild hopes in others. "School teachers, insurance brokers, used car salesmen and shoe clerks around the nation converged on the Colorado Plateau in droves," according to Utah writer Raye Ringholz.[24] Ambitious entrepreneurs opened mines and milling facilities.

The domestic procurement program was a success. U.S. uranium production climbed sharply through the 1950s. "The AEC had turned the tap and engendered a flood," writes Ringholz. By the mid-1950s, there were 600 uranium producers, mostly small, on the Colorado Plateau and 8,000 workers in the mines and mills. Mills were not keeping up with ore production.[25] In 1956, 12 private uranium mills were operating; by 1962, there were 26.[26] Production peaked temporarily at 14,457 metric tons in 1960 (to be surpassed only in the late 1970s with the boom generated by contracts for civilian nuclear power plants).[27] By the early 1960s, the uranium market was saturated, and

20. U.S. Bureau of Mines 1953, 1954, 1955, 1958.

21. U.S. Bureau of Mines 1958, 1965.

22. U.S. Bureau of Mines 1952; Neff 1984; see section on Namibia.

23. Owen 1985, p. 112.

24. Ringholz 1989, p. 77.

25. Ringholz 1989, p. 153.

26. Owen 1985, p. 112.

27. Neff 1984, p. 315.

Table 5.1 Location of U.S. Surface and Underground Uranium Mines

State	Surface	Underground	Total
Alaska	0	1	1
Arizona	135	190	325
California	13	10	23
Colorado	268	1,008	1,276
Idaho	2	4	6
Montana	9	9	18
Nevada	9	12	21
New Jersey	0	1	1
New Mexico	38	177	215
North Dakota	13	0	13
Oklahoma	3	0	3
Oregon	2	1	3
South Dakota	111	30	141
Texas	54	0	54
Utah	391	806	1,197
Washington	15	0	15
Wyoming	242	38	280
Total	1,305	2,287	3,592

Source: U.S. EPA 1983.

the AEC had stockpiled more than enough uranium for its military needs.

With the resurgence of the uranium market in the 1970s, a new U.S. production high of 16,800 metric tons was reached in 1980. Thereafter, production declined, falling to 3,036 metric tons in 1991,[28] as demand fell worldwide and competing producers emerged (see figure 5.3).

Colorado and Utah led the country in uranium production in the 1950s and early 1960s, with New Mexico and Wyoming gaining first place in the boom of the 1970s.[29] Table 5.1 shows the number of open-pit and underground mine sites by state as of 1983 (cumulative). Mines have been progressively closed since that time.

U.S. uranium mining has disproportionately affected indigenous people. Deposits are located on the lands of the Navajo, Hopi, Laguna Pueblo, Zia Pueblo, Spokane, Northern Arapaho, Lakota, Acoma, and Jemez.[30] Impoverished and with few other employment opportunities, many indigenous Americans took jobs in the mines, unknowingly

28. Uranium Institute 1992, p. 23.

29. Owen 1985, p. 113.

30. Rogers 1992.

risking their health and lives. Furthermore, overburden from mines continues to pose radiation hazards in some of these areas.

History of U.S. Radiation Protection Efforts and Health Effects on Miners In the early days of military procurement, uranium ore was mined by drilling and bringing the rock to the surface to be broken and milled (see chapter 3). Underground conventional mining is the most dangerous mining method for workers due to exposure to radon decay products. In 1983, 70 percent of uranium mined in the United States was extracted by conventional methods.[31] With the decline in mining and the introduction of new methods, this proportion has fallen. By 1988, nonconventional methods—such as solution mining (where a solvent is injected into underground ores and the uranium-bearing liquid is pumped to the surface) and heap leaching—accounted for 47 percent of production.[32]

During the whole military procurement boom of the 1950s and early 1960s, credible technical reports of high concentrations of radon in the mines and appeals for mandatory radiation limits and effective ventilation were assiduously ignored by the AEC and other federal agencies. The experiences of miners in the uranium-rich mines of Germany and Czechoslovakia were well known by World War II. Official inaction was abetted by a loophole in the radiation protection law. The AEC was responsible for radiation protection only after the ore had been removed from the ground. Mining safety and health were state responsibilities.[33] However, the states had scant experience in radiation protection in the late 1940s and the 1950s.

Starting in the late 1940s and into the 1960s, high levels of radon and its decay products (radon progeny) in poorly ventilated mines exposed U.S. uranium miners to high lung doses of alpha radiation—which were predominantly due to radon decay products rather than to the radon itself. Exposures in the mines of the Colorado Plateau were first measured between 1948 and 1950 by the U.S. Public Health Service (USPHS).[34] These surveys confirmed that concentrations of radon and its decay products were similar to those in European mines, for which a link between radon and lung cancer had been made in the early 1940s. Based on these data, the USPHS began efforts to reduce radon decay products concentrations in mines—only to meet with resistance from mine operators and the Atomic Energy Commission.[35]

31. Cochran et al. 1987a, p. 122.

32. U.S. DOE 1989a, p. 129.

33. Eisenbud 1987, p. 175.

34. Wagoner et al. 1964; Holaday 1969.

35. Ringholz 1989, pp. 46–51.

In 1949, the Colorado State Health Department and several uranium mining and milling companies made a formal request for a health study of U.S. miners and millers. In 1950, the USPHS, in cooperation with other governmental agencies, began a study of miners and millers in the Colorado Plateau area. A cohort was established comprising miners and millers who had volunteered for at least one physical examination and provided personal identifying data to facilitate follow-up of their health status.[36]

In 1951, officials of the USPHS, the National Institutes of Health, and the AEC held a meeting in Washington, D.C. to discuss early measurements of high levels of radon in mines. Duncan Holaday of the USPHS emphasized the feasibility of lowering the high radon concentrations in the mines through ventilation. The USPHS then decided to follow the European recommended standard of 100 picocuries (3.7 becquerels) per liter of air, but the setting of any mandatory standard was deemed to require further study.[37]

In May 1952, two years into the study, Holaday and his colleagues produced an interim report describing the experience of the German and Czechoslovakian mines, where prolonged exposure to 1,500 picocuries (55.5 becquerels) of radon per liter of air had led to high lung cancer mortality in miners, and again warning of the dangers in the Colorado Plateau mines. Of these Colorado Plateau mines, 48 had a median of 3,100 picocuries (115 becquerels) of radon per liter of air, with a maximum reading of 82,800 picocuries (3,060 becquerels) per liter; 18 had a median of 4,000 picocuries (150 becquerels) of radon decay products per liter, with a maximum of 120,000 picocuries (4,440 becquerels) per liter. The report presented recommended measures to correct hazardous conditions, including ventilation. About 2,000 copies of the report were distributed to the AEC, the U.S. Bureau of Mines, state bureaus of mines, state health agencies, and the mining companies.[38]

Conditions in the mines did not improve, and it became clearer that inaction on the part of the AEC was due to more than ignorance or bureaucratism. The AEC's foremost concern was uranium output, and publicity about radiation hazards to miners was seen as a threat to production and even characterized as subversive. Jesse Johnson, director of the Raw Materials Procurement Division of the AEC, cautioned against wide distribution of the report:

> The report might become the basis for press and magazine stories which could adversely affect uranium production in this country and abroad.... There is no doubt but that we are faced with a problem

36. Wagoner et al. 1964.

37. Ringholz 1989, pp. 93–94, 97.

38. Ringholz 1989, p. 169.

which, if not handled properly, could adversely affect our uranium supply.... The possibility exists that communist propagandists may utilize any sensational statements or news reports to hamper or restrict uranium production in foreign fields, particularly at Shinkolobwe [Belgian Congo]."[39]

At Holaday's insistence, the Surgeon General wrote to the chief of the AEC Division of Biology and Medicine in Washington in October 1958, urging the AEC to take some concerted action and pointing out that, since the U.S. government was the sole purchaser of uranium ore, the AEC could require the mines to adhere to recommended radiation standards. This had been suggested in 1948, and Holaday had continued to advocate it throughout the ensuing decade. Once again, the AEC argued it had no jurisdiction over the mines, by virtue of its licensing regulations and the Atomic Energy Act.[40]

In the meantime, the USPHS study continued. Prior to 1954, about 1,200 workers were examined, and in 1954, a more aggressive plan for worker recruitment was implemented; by 1962, 5,370 workers had been enrolled.[41] The study involved triennial physical exams, an annual census of workers in the uranium mining industry, and the collection of information from cohort members by mail. An important outcome of this study was that death certificates recorded the underlying cause of death in addition to the proximate cause.

By 1960, nine uranium miners had died of lung cancer—a finding that was interpreted to confirm the similarity between U.S. and European hard rock miners with regard to risk for lung cancer from exposure to radon progeny. These preliminary results from the epidemiologic study helped to prod state and federal agencies into finally addressing excessive radon exposures in mines. After 1961, levels of radon and its decay products decreased considerably.[42]

The first published research paper on the USPHS study reported on cancer mortality for 5,370 miners and millers who were followed through the end of 1962.[43] This study noted an excess mortality from accidents (mainly mine-associated) and for cancer of the respiratory system (SMR = 3.6) for all white uranium miners, and a much greater mortality from respiratory cancer for white underground miners with 5 or more years of underground experience (SMR = 10.0). No excess mortality was detected at that time for white uranium millers, or for 1,103 nonwhite (mainly American Indian) millers and miners.

39. U.S. Atomic Energy Commission 1951, as quoted in Ringholz 1989, p. 170.

40. Ringholz 1989, p. 202.

41. Wagoner et al. 1964.

42. Holaday 1969.

43. Wagoner et al. 1964.

Considerable illness and suffering could have been prevented by imposing radiation limits and improving ventilation in 1948 or 1950, when the dangers of uranium mining were already evident. In 1964, Holaday summed up his experience with the health and safety study undertaken in 1950:

> It is now fourteen years since the uranium study was started.... The study was undertaken with the belief that all that was required was the evaluation of environmental conditions in the industry and comparison of the results of these studies with the data on human experience which was available in the literature [the German and Czechoslovakian experience]. Measures to control the exposures of the workers to toxic materials could then be recommended. This belief was a delusion. It required ten years and the accumulation of a number of deaths to convince the authorities that real hazards existed in the uranium mines.[44]

The next report on the USPHS study was for mortality through September 1967, and was restricted to 3,414 white and 761 nonwhite underground uranium miners.[45] The excess of deaths from respiratory cancers in the white miners was higher than previously reported (SMR = 6.1) and showed a strong trend with exposure to radon and its decay products. The study also showed that miners who smoked had a risk for respiratory cancer that was ten times the risk for those who did not. A relatively small increase in respiratory cancer (SMR = 1.5) was noted for nonwhite miners.

In 1971, the United States adopted a radiation exposure limit for miners of 4 working level months (WLM) per year—equivalent to a lifetime radiation exposure limit of 120 working level months.[46] (A working level month is equal to one working level for a working month of 170 hours.)

The USPHS cohort continued to be followed, with periodic reports in the scientific literature.[47] The first published evidence of a substantially increased risk for lung cancer in American Indian miners (SMR = 4.2) was reported by Archer et al. (1976). This finding was described in more detail by Gottlieb and Husen (1982), who noted that of 17 Navajo men admitted to the Shiprock Indian Health Service Hospital for lung cancer between February 1965 and May 1979, 16 had been uranium miners and 14 did not smoke.[48] Their results were independently confirmed by Samet et al.[49] By 1978, white underground miners in the

44. Holaday 1964, quoted in Ringholz 1989, p. 206.

45. Lundin et al. 1969.

46. Fleming and New 1981.

47. Archer et al. 1973; Archer et al. 1976; Whittemore and McMillan 1983.

48. Gottlieb and Husen 1982.

49. Samet et al. 1984.

United States had suffered 205 respiratory cancer deaths, compared to an expected value of 40.[50]

A 1988 study of the entire USPHS cohort by Hornung and Meinhardt,[51] which is now being followed by the National Institute for Occupational Safety and Health, reported on mortality through the end of 1982 for 3,346 members of the cohort who had worked at least one month in an underground uranium mine. They used statistical models to estimate risk per unit dose of cumulative exposure to radon and its progeny, and found significant dose-response relations, with predicted relative risks ranging between 0.9 and 1.4 per 100 working level months. Hornung and Meinhardt also found that cigarette smoking and exposure to radon decay products exerted a synergistic effect on lung cancer, but concluded that the relation was becoming additive over the later follow-up years. This study also found that relative risk increased with age at initial exposure to underground uranium mining, and that risk decreased dramatically in the years following the cessation of exposure.

Roscoe et al.[52] studied a subset of the USPHS cohort composed of 516 white, nonsmoking underground miners. They found an SMR of 12.7 for lung cancer, demonstrating that radon decay products are carcinogens in the absence of smoking exposure.

In 1977, the University of New Mexico established a cohort of New Mexico miners with at least one year of mining experience before the end of 1976. This group was exposed to radon decay products during the 1960s and 1970s, when exposures were limited by state and federal regulations. In spite of lower exposures for this cohort compared with the USPHS cohort, Samet et al. (1989) found, in a case-control study of 65 miners, that lung cancer risk increased with cumulative exposure to radon decay products. They also estimated risks per unit exposure to both radon decay products and cigarette smoking that are similar to those found in other studies.

Nonmalignant respiratory disease has also been studied in uranium miners. Studies of the USPHS cohort identified spirometric evidence of obstructive lung disease in workers in the higher radon decay product exposure groups.[53] Mortality from nonmalignant respiratory disease was also elevated in this cohort,[54] with a fivefold excess reported in the most recent follow up by Waxweiler et al.[55]

50. Robinson 1980, pp. 19–21.

51. Hornung and Meinhardt 1988.

52. Roscoe et al. 1989.

53. Archer 1973.

54. Archer 1976.

55. Waxweiler et al. 1981.

Samet et al. (1984) examined 192 long-term miners for evidence of nonmalignant pulmonary disease. They found no evidence for marked decline in lung function with short durations of mining and concluded that only lengthy mining experience would lead to clinically important disease. They noted, however, that cigarette smoking would speed the development of airflow obstruction. Eight percent of this group had radiographic evidence of simple pneumoconiosis. Samet et al. concluded this was most consistent with silicosis, and noted that free silica has been detected in New Mexico mines.

In 1980, the National Institute for Occupational Safety and Health (NIOSH) embarked on a reevaluation of the exposure standards for miners, based on studies indicating a twofold excess risk of lung cancer mortality at and below 120 working levels of lifetime exposure to radon decay products.[56] As of 1993, federal standards remain at 4 WLM per year.

Since the first report of Wagoner et al. (1964), which found no increase in cancer mortality, there have been few studies of uranium millers. NIOSH studied a group of Colorado millers and detected evidence of renal toxicity.[57] No other studies of U.S. millers have been published.

Effects on the U.S. Environment and the Public Field measurements by the U.S. Environmental Protection Agency and contractors have found exposure rates from overburden materials to vary from 0.2 to 1.1 micrograys per hour, with a mean of about 0.4 micrograys per hour. Exposure rates from low-grade ores ranged from 0.8 to almost 10 micrograys per hour, averaging 2 micrograys per hour. Ambient radiation in mining areas ranged from 0.1 to 0.85 micrograys per hour, with an average of about 0.2 micrograys per hour.[58]

Shields et al. studied birth data for 13,329 Navajo children born between 1964 and 1981 at the Public Health Service/Indian Health Service Hospital in Shiprock, New Mexico, which is in a region where there are many tailings areas and mine waste dumps. They found a statistically significant tendency for the offspring of mothers living near mine dumps and mill tailings areas to experience adverse health outcomes such as birth defects, still births, and deaths from illnesses during infancy, but this did not seem to be related to reported duration of exposure.[59]

The milling of conventionally mined uranium generates tremendous volumes of wastes (see chapter 3). In the United States, as of the late

56. Fleming and New 1981.

57. NIOSH 1981.

58. U.S. Environmental Protection Agency 1989; S. Cohen and Associates, Inc. 1989.

59. Shields et al. 1992.

1980s, some 220 million metric tons of mill tailings had accumulated from uranium production for nuclear weapons and nuclear power,[60] accounting for over 95 percent of the volume of all radioactive waste from all stages of the nuclear fuel cycle.[61] Accumulations at individual mill sites range between 0.5 million and about 30 million metric tons.[62] (While the specific activity of tailings is low, especially compared to high-level reprocessing wastes and spent fuel, which contain large amounts of fission product radioactivity, the wastes are very long-lived, since thorium-230—the parent of radium-226 and all its decay products—has a half-life of about 75,000 years.)

Mill tailings in the United States have been used for construction purposes, posing a risk to building occupants. For example, between 1952 and 1966, several hundred thousand metric tons of mill tailings were taken from the Climax Uranium Company's mill tailings pile to Grand Junction, Colorado, for use as construction material.[63] Thousands of properties including homes were contaminated, and inhabitants were exposed to excessive levels of radon, putting them at increased cancer risk, according to the Colorado Department of Health.[64] The Colorado Department of Health estimated that 10 percent of the occupants of several thousand of the homes whose foundations incorporated tailings received the equivalent of 553 chest X rays per year due to gamma emissions from the tailings.[65] Assuming 5 to 10 millirems (0.05 to 0.1 millisieverts) whole-body equivalent exposure per chest X ray, this would amount to a whole-body exposure of several rem (tens of millisieverts) per year.

Radioactive pollutants from tailings piles can reach rivers by runoff or when tailings dams break. Of 22 tailings piles investigated by the U.S. Environmental Protection Agency in 1982, half were near rivers or streams. Elevated levels of radium-226 and chemical contaminants from milling were found downstream of Shiprock, New Mexico; Durango, Colorado; and other milling sites of the 1960s.[66]

60. U.S. DOE 1989a, p. 133. (licensed sites). To the amount of 116.9 cubic meters for 1988 are added the mill tailings at inactive sites listed on p. 152; an average tailings density of 1.6 metric tons per cubic meter is assumed.

61. Based on the total volume of all radioactive waste (including spent fuel, high-level waste, transuranic waste, low-level waste, and uranium mill tailings) from all sources (both commercial and military) produced in the United States since the 1940s. Based on U.S. Department of Energy records as compiled in Saleska et al. 1989, appendix C.

62. U.S. DOE 1990a, pp. 130–131.

63. U.S. DOE 1989a, p. 150; U.S. DOE 1988a, p. 156.

64. Carter 1987, p. 13.

65. Lipschutz 1980, p. 137.

66. Wilson 1985; U.S. Congress 1971.

Between 1955 and 1977, 15 tailings dams broke, releasing their contents into the wider watershed areas.[67] The most notorious instance of this occurred in July 1979, when a dam holding waste water in a mill tailings settling pond at the United Nuclear Corporation mill near Church Rock, New Mexico, gave way and released about 94 million gallons (about 360 million liters) of mill tailings liquid into the Rio Puerco. Some 60 miles (about 100 km) of the river, which cuts through Navajo grazing lands in New Mexico and Arizona, were contaminated.[68] The spill contained about 35 curies (about 1,300 gigabecquerels) of radioactivity, most of which was from thorium-230.[69]

At virtually all U.S. mill sites, tailings have contaminated the groundwater, according to the U.S. Environmental Protection Agency.[70] This contamination may be permanent. According to the U.S. Nuclear Regulatory Commission, "the chance of returning an aquifer to premining water quality is minimal."[71]

A plume of contamination in groundwater caused by the tailings pile at Tuba City, Arizona, has been reported to be moving 8.5 meters per year toward the drinking water sources of Tuba City and Moenkopi. The Moenkopi Wash is the Hopi people's only source of water for irrigation.[72] The contaminant plume from the tailings pile at Church Rock, with elevated levels of arsenic and thorium, is believed to have reached the Navajo Reservation.[73]

In addition, milling wastes pose the risk of contaminating the food chain. In 1986, New Mexico's Environmental Improvement Division found high radionuclide levels in cattle raised near uranium mines and mills in the state. The mean concentration of uranium in the kidneys of cattle reached 11 picocuries (about 0.4 becquerels) per kilogram of tissue, over eight times the level found in the controls. Mean concentrations of radium-226 in the femur bones of cattle were as high as 7,100 picocuries (about 263 becquerels) per kilogram, almost 50 times the comparable concentration in the controls. The report concluded that "ranchers who raise and regularly consume these cattle ... may be receiving doses that are excessive."[74] Radionuclides have also been measured in a variety of wild plants and animals raised in the vicinity

67. Polsgrove 1980.

68. Carter 1987, p. 12.

69. Based on radionuclide concentrations provided in U.S. Centers for Disease Control 1980, table 1, p. 3.

70. U.S. Environmental Protection Agency 1983, p. 45941, col. 2.

71. U.S. NRC 1985a.

72. Gilles, Reed, and Seronde 1990; Lambrecht 1991.

73. U.S. NRC 1985.

74. Lapham and Samet 1986, table 2, pp. vi, 31.

of several mill tailings piles near Grants, New Mexico. Radium concentrations in the muscles of rabbits captured near tailings piles, for example, have ranged from 4 to 30 times the normal background level. People living on food produced near a tailings pile could receive a dose to their bones of up to 1 millisievert per year from radium, increasing their risk of bone cancer.[75]

Cleanup Prospects in U.S. Mines and Mills Mine wastes and mill tailings may be the most neglected of all radioactive waste materials. There are as yet no regulations governing disposal of mine wastes. As for mill tailings, Congress first enacted legislation in 1978 that mandated their cleanup and required that companies producing the tailings pay the cost of cleanup. Many mill tailings piles remain exposed and untreated.

The 1978 legislation directs the Department of Energy to conduct cleanup operations at inactive, abandoned mill tailings sites. Twenty-five inactive or abandoned uranium processing sites located in 10 states are currently subject to cleanup operations under DOE's Uranium Mill Tailings Remedial Action Project (see figure 5.4).[76] DOE presently considers surface cleanup complete at 15 of the 25 sites.[77] Companies responsible for the cleanup have often resisted implementing remediation. In New Mexico, for example, pressure from mill companies to maintain state standards at a level substantially weaker than the federal government requirements prompted the state to rescind its regulatory authority to the federal Nuclear Regulatory Commission to avoid a battle in court.[78]

Some limited funds for remediation of mill sites that were active after 1978 and supplying the government with uranium were authorized by Congress as part of the Energy Policy Act of 1992. Table 5.2 lists the sites designated for cleanup, with production, quantity of tailings attributable to government contracts, and status of operations and of tailings, as of the end of 1991. Of the 190 million metric tons of tailings at licensed mill sites, 50.9 million are attributable to the U.S. government. About half are sites that the DOE is responsible for remediating, while DOE makes contributions to the other half.

75. National Research Council 1986, pp. 107–108. This report notes that for the people studied, the 1-millisievert dose represents a substantial overestimate, because locally raised food sources probably do not constitute a large fraction of their diet. The report also notes, however, that the population studied was white, whereas other groups not studied, like Native Americans, might rely more on locally grown food. One millisievert per year is currently the maximum allowable exposure for radiation other than background and medical exposures.

76. U.S. DOE 1988, p. 154.

77. U.S. DOE 1992, pp. 171, 172.

78. Robinson 1994.

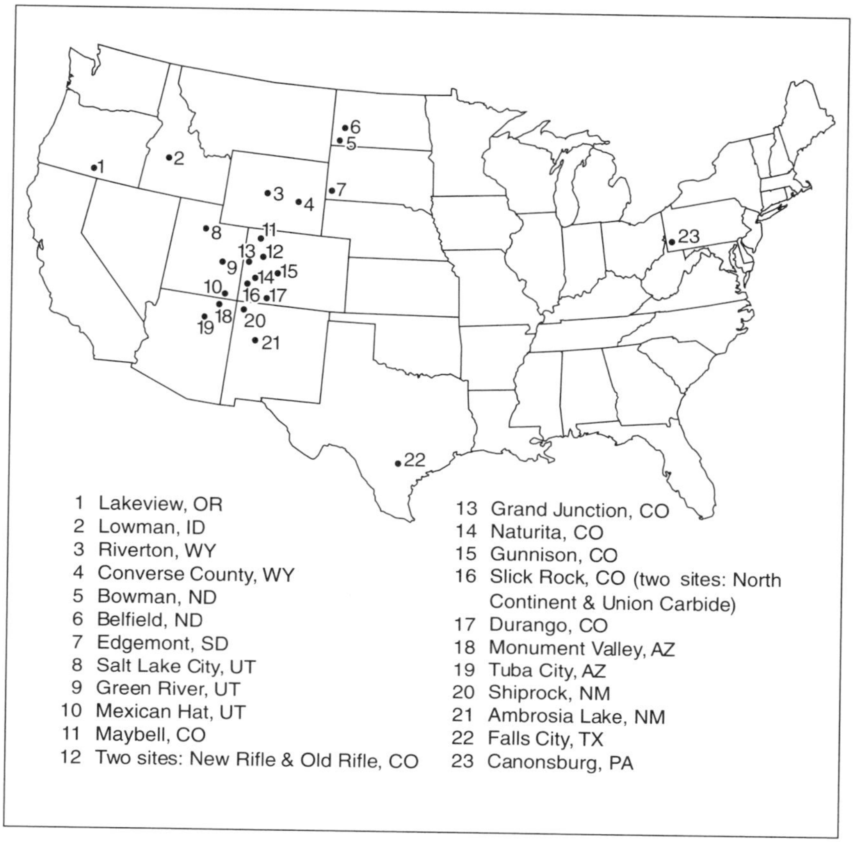

Figure 5.4 Map of sites in the Uranium Mill Tailings Remedial Action Project. Source: U.S. DOE 1992, p. 165.

DOE has begun the long process of cleaning up or securing contaminated mill tailings sites.[79] DOE's Albuquerque, New Mexico, field office is responsible for remediation at 24 government sites and roughly 5,000 properties in their vicinity.[80] The estimated cost of cleaning these sites for fiscal years 1992 through 1996 is $510 million.[81]

In the Grand Junction Remedial Action Project, a pilot project that was operated between 1973 and 1987 by DOE and the Colorado Department of Health, some 140,000 metric tons of mill tailings and contaminated materials were removed from some 600 properties in the

79. For a description of UMTRAP sites, *see* U.S. DOE 1992, pp. 171–173.

80. Under the 1978 Uranium Mill Tailings Radiation Control Act, states must pay 10 percent of cleanup costs at sites within their borders; DOE pays the remaining 90 percent. Four sites are on Indian lands; the federal government pays for all cleanup on these. U.S. DOE 1993a, p. II-68.

81. U.S. DOE 1993a, p. II-74.

Table 5.2 Tailings at Conventional U.S. Uranium Mill Sites Due to Government Contracts, as of December 1992

Location	Operator	Capacity (metric tons of ore per day)	Capacity (metric tons of U per year)[1]	Status of Operations[2]	Status of Tailing[3]	Mass of Tailings (Million metric tons)[4]
Colorado						
Canon City	Cotter	1,090	330	SD 1987	Wood-chip covering	0.3
Uravan	Umetco	1,180	1,000	DEC	PART ST	5.2
New Mexico						
Grants	Anaconda	5,440	3,000	DEC 1987	PART ST	8.0
Grants	Quivira	6,350	62	SD 1985	Fenced	9.1
Grants	Homestake	3,080	1,000	DEC	UNSTABLE	10.4
South Dakota						
Edgemont	TVA	680	N/A	DEC 1983	PART ST	1.5
Utah						
Moab	Atlas	1,270	600	DEC	UNSTABLE	5.4
Washington						
Ford	Dawn Mining	410	500	SD 1982	Wood-chip covering	1.1

Wyoming						
Gas Hills	American Nuclear	860	N/A	DEC 1988	UNSTABLE	1.9
Gas Hills	Pathfinder	2,540	N/A	SD 1988	UNSTABLE	2.4
Jeffrey City	Western Nuclear	1,540	N/A	DEC 1988	INTERIM	3.0
Natrona	Umetco	1,270	500	DEC 1987	UNSTABLE	1.9
Shirley Basin	Petrotomics	1,360	N/A	DEC 1985	UNSTABLE	0.7
Total						50.9

1. Source: Nuclear Engineering International Publications 1991, pp. 153–156.
2. SD = shut down, DEC = decommissioning now or previously decommissioned (date given).
3. UNSTABLE = unstabilized, INTERIM = interim stabilization, PART ST = partially stabilized.
4. Mass of total tailings under government contract. Most sites have additional tailings that do not fall under government responsibility.

Source: U.S. DOE 1994, pp. 152–153. Only sites with government portion of cleanup costs are included.

area.[82] Overall, 4,150 local sites or buildings are included in the current remediation plan, which is expected to cost $155 million between 1994 and 1998.[83]

According to rules issued by the EPA,[84] mill owners had six years beginning in 1986 (subject to extensions) to phase out the use of existing large tailings piles. The deadline was missed in most cases and has been set back to the end of 1997.[85] As of the end of 1993, less than half of the total tailings in the United States had been stabilized and isolated. The U.S. DOE projects that it will cost $2 billion to take care of just the approximately 22 million metric tons that it is directly responsible for remediating. The original cost estimate was $350 million. Surface cleanup is projected to be complete in 1998, and groundwater cleanup completion is projected for 2014.[86]

Canada

Canada was the first country to mine and refine uranium on a large scale, and it has been the world's largest exporter of uranium throughout most of the history of uranium production. Today, northern Saskatchewan, which has among the richest ores in the world,[87] produces more uranium than any other place on earth.

Uranium was first produced in Canada by the Eldorado Gold Mining Company as a by-product of radium mining at Port Radium on Great Bear Lake in the Northwest Territories. Radium mining operations at Port Radium, as well as a privately owned radium refinery built at Port Hope, Ontario, in 1932, were shut down in 1940 due to a lack of demand for radium.[88]

In 1942, at the secret request of the U.S. government, the Canadian government reopened the Port Radium mine and the Port Hope refinery to produce uranium for the U.S. weapons program. The refinery at Port Hope processed all the uranium for the Manhattan Project. The uranium was extracted from previously accumulated ore from Port Radium and from ore concentrates from the Belgian Congo that had been stored on Long Island, New York.[89]

82. U.S. DOE 1988a, p. 156.

83. U.S. DOE 1991b, p. 12; U.S. DOE 1993a, p. II-9.

84. See U.S. EPA 1983 and *Federal Register*, vol. 52, p. 36000, 24 September 1987.

85. *Federal Register*, vol. 59, p. 5676, 7 February 1994.

86. Chernoff 1993.

87. Estimated ore reserves of 130,000 metric tons have an average grade of 14 percent uranium oxide. U.S. Bureau of Mines 1988, p. 183.

88. Neff 1984, p. 142; Edwards 1993.

89. Neff 1984, p. 142; Edwards 1993.

In 1943, the Canadian government decreed its complete control over exploration and mining of radioactive substances and, in 1944, acquired Eldorado, creating a crown corporation called Eldorado Mining and Refining, Ltd. (later Eldorado Nuclear, Ltd.). However, in 1947, the Atomic Energy Control Board, created by legislation the previous year, loosened restrictions on private involvement in uranium prospecting and development in the Yukon and the Northwest Territories.[90]

The massive uranium exploration and procurement effort launched by the United States after World War II created a boom in the Canadian uranium mining and milling industry. On the north shore of Lake Athabasca in northern Saskatchewan, Eldorado established its Beaverlodge mine, along with a mill, a transportation system, and a town called Uranium City. In 1952, the U.S. Atomic Energy Commission contracted to buy all of Eldorado's production over the following ten years.[91] By the early 1950s, financial incentives offered by the U.S. and U.K. weapons programs had spurred a large-scale exploration effort, similar to the one taking place in the United States, and production by Eldorado and several private companies rose steeply. Uranium in the 1950s came from the Elliot Lake (eleven mills) and Bancroft (three mills) areas of Ontario, from the Northwest Territories (two mills), and from northern Saskatchewan (three mills). Exports to the United Kingdom began in 1958.[92]

In November 1959, the U.S. Atomic Energy Commission announced that it would not enter into new contracts for Canadian uranium. This was a rude shock to Canada's 17 uranium mining companies, threatening the existence of 28 mines and the communities of Uranium City, Elliot Lake, and Bancroft.[93] The United States and United Kingdom agreed to stretch out deliveries under existing contracts scheduled to expire in 1962 and 1963, until 1966 in the case of the United States and until 1971 in the case of the United Kingdom.[94] Nevertheless, by 1962, only 4 uranium companies were still in business in Canada, and the uranium towns were deserted.[95] To cushion the industry, the Canadian government intervened, creating an expensive government-funded stockpile, which allowed the main producers to continue operations.

Canadian annual uranium production, which had reached a high of 12,200 metric tons in 1959, dropped to a low of 2,800 metric tons by

90. Neff 1984, pp. 142–143.

91. Edwards 1993.

92. Neff 1984, pp. 143, 145.

93. Edwards 1993.

94. Neff 1984, p. 143.

95. Edwards 1993.

1967, due to termination of U.S. and U.K. military contracts and a protectionist bar on imports by the United States, whose own production had fallen sharply in the same period.[96]

Virtually all the uranium produced and exported by Canada until the mid-1960s was sold to the United States and, to a lesser extent, to the United Kingdom for nuclear weapons production.[97] The last shipment of Canadian uranium to the United States explicitly for military purposes was in 1966. The last shipment to the United Kingdom explicitly for military purposes was in 1972.[98] Since that date, all uranium exported from Canada was supposed to be used for peaceful purposes.

As in other countries, Canada's uranium industry experienced a comeback in the 1970s through the early 1980s and another decline after that, due to the rise and fall in demand for fuel for civilian nuclear power reactors and the eventual dissolution of an international price-fixing uranium cartel to which Canada belonged.

The Beaverlodge area in Saskatchewan, which once had ten mines producing uranium ore and feeding three mills, by 1964 had only the Eldorado Mines/Mill facility operating. It, too, closed in 1982. In 1988, Eldorado Nuclear Limited, a federal crown corporation, merged with the Saskatchewan Mining Development Corporation to form Cameco (Canadian Mining and Energy Corporation), a crown-owned uranium company. Shares have been sold to the private sector, although the Saskatchewan government is still a part owner.

In the early 1990s, activity related to uranium mining in Canada has focused on decommissioning the Elliot Lake mines and developing the extraordinarily rich deposits in the Athabasca Basin of northern Saskatchewan. Six projects, planned to develop 800 million pounds (about 360,000 metric tons) of new U_3O_8, have been proposed. Two of these are expansions within established production complexes. One is at Cluff Lake, about 100 kilometers south of Lake Athabasca, operated by Cogéma Resources, Ltd., of France (a subsidiary of the French atomic energy commission). It has processed uranium ore since 1980. The other, at Rabbit Lake on the western shore of Wollaston Lake, has operated since 1975 and is owned by Cameco Corporation. A third project, proposed at McArthur River, is a joint venture of Cameco, Cogéma, and Uranerz. Two of the projects have already been developed to the test-mine stage: one is at Cigar Lake, with the highest-grade ore in the world, and the other is Mid-West Joint Venture (MJV). The sixth project is the McLean Lake proposal by Total Minatco. Cogéma recently bought out Total Minatco's uranium interests and is

96. Neff 1984, p. 143.

97. Edwards 1993.

98. Edwards 1985.

thus the sole proponent for three of these and codeveloper in two of the others.

These projects are all in the process of evaluating their potential environmental impact. The MJV proposal was recently rejected by the Federal/Provincial Panel examining the environmental impact statements. In its October 1993 report, the panel also recommended a delay of at least five years in the development of the McLean Lake Project and granted only conditional approval to the expansion at Cluff Lake.[99] Neither the provincial nor the federal levels of government is bound by the panel's recommendation. The Saskatchewan government accepted the panel's recommendations regarding Cluff Lake and MJV, but decided to grant approval to the McLean Lake project without requiring a five-year delay.

Presenters at environmental impact assessment hearings claimed that Cogéma's interest in Canada's uranium resources is to stockpile Canadian uranium at bargain prices. Concern was also expressed that some of the uranium ore proposed for mining would be destined for use in weapons production. Of ongoing concern is that France has a large nuclear power industry and yet refuses to separate its civilian nuclear power from its weapons program.

An opinion poll in Saskatchewan indicated a 75 percent approval of uranium mining but widespread (67.4 percent) opposition to the use of Saskatchewan uranium in nuclear weaponry. The fact that no proven method exists for preventing the use of Canadian uranium for military applications troubles many Canadians. It is noteworthy that while the issue of nuclear weapons proliferation is outside the mandate of the panel reviewing the uranium mining proposals, the extent of concern expressed regarding this topic motivated the panel to bring this to the attention of government.[100]

As in the United States, many Canadian uranium deposits are located on traditional lands of indigenous peoples—the Inuit, Cree, Metis, and especially the Dene of the Athabasca region, where the current developments are being proposed.

Health Effects on Canadian Workers Studies on the health effects of uranium mining in Canada have been conducted in Ontario, Saskatchewan, and the Northwest Territories.

In Ontario, there were 12 mines in Elliot Lake at the peak of mining (1954–1958) and 4 in the Bancroft region. Total employment at Elliot Lake rose from about 500 at the end of 1954 to 10,000 at the end of 1958. By the end of 1974, only two uranium mines were operating,

99. Joint Federal-Provincial Panel 1993.

100. Joint Federal-Provincial Panel 1993, p. 26.

both at Elliot Lake (Denison and Rio Algom), employing about 1,600 people.

Of the health and safety problems that emerged in Elliot Lake, most important were dust, with its consequent risk of silicosis and chronic bronchitis, and radiation, causing lung cancer. Details of the dust conditions during that time were presented by a Royal Canadian Commission, known as the Ham Commission.[101] The ores at Elliot Lake had a far higher silica content (60 to 70 percent) than those at Bancroft (5 to 15 percent).

More than half of the 1,634 workers employed at the end of 1974 were in characteristically dust-exposed occupations. According to the Ham Commission, 36 (3.6 percent) had radiological classification of silicosis and an additional 53 (5.3 percent) had radiological classification of "dust effect" at that time.[102] Workers Compensation Board statistics to the end of 1974 recorded 143 cases of silicosis among men who had been exposed to uranium dust in Ontario, and 325 new cases of "dust effect" (suggestive of pneumoconiosis) were also identified. The early cases of confirmed pneumoconiosis in Elliot Lake had unusual characteristics. The nodules were more widely dispersed throughout the lungs than usual, and silicosis developed at a younger age. The Ham Commission also noted that there was evidence to link the conditions in Elliot Lake uranium mines with an increased incidence of chronic bronchitis.

Before 1968 Canada had no required limit for radiation exposure in uranium mines. From 1955 to the end of 1967, the guideline accepted by the Mines Accident Prevention Association in Ontario was 12 working level months; this level was adopted in the code on radiation exposure in 1967. The level was reduced to 8 WLM per year for 1973, to 6 WLM in 1974, and to 4 WLM in 1975.

While the Bancroft area produced fewer victims of silicosis than Elliot Lake, the same could not be said about lung cancer, which occurred throughout Ontario uranium mines. The high occupational health risk associated with radiation exposures in these mines has been extensively discussed and studied. The lung cancer experience of Ontario uranium miners has, in fact, been the subject of several major epidemiological investigations since 1974.[103]

Mortality data from 50,201 Ontario miners, including about 15,000 who had worked exclusively in Ontario uranium mines, were examined for 1955–1986 by linking the cohort to the Canadian National Mortality Database. The first measurements of the levels of radon decay products were made in 1958–1960, and earlier levels were

101. Royal Commission 1976.

102. Royal Commission 1976, p. 32.

103. Muller et al. 1983, 1986; Kusiak et al. 1991, 1993.

extrapolated from these. After 1968, each miner's daily time card and measurements of the level of radon decay products were used to estimate their exposure. Including all miners who worked at least two weeks in a uranium mine, but excluding miners who also worked in a uranium mill or in an asbestos mine or in a uranium mine outside Ontario, studies showed an excess of deaths from lung cancer (291 cases observed versus 171.8 expected, with a standardized mortality ratio of 1.69 and 95 percent confidence limits of 1.50–1.89). Based on the inclusion and exclusion criteria, this is generally thought to be a conservative estimate.

The excess of lung cancer in Ontario uranium miners who had also mined gold was considerably larger than that among other uranium miners. The results indicated that lung cancer mortality was related not only to the levels of exposure to radon decay products but also to arsenic and other carcinogens to which these miners may have been exposed. A survey of smoking habits of uranium miners and additional subanalyses revealed that smoking cannot account for the excess lung cancer mortality. According to Ontario Workers' Compensation Board statistics, 116 claims for lung cancer in Elliot Lake miners had been allowed by the early 1990s.

Geoffrey Howe et al. studied a cohort of 2,103 workers employed at Eldorado's Port Radium uranium mine in the Northwest Territories between 1942 and 1960, when the mine closed. The observation period was from 1950 (the earliest year that cause of death was available from the Canadian National Mortality Data Base) to 1980, during which 57 of the miners died of lung cancer. The relative risk of dying of lung cancer was 2.30 for the whole cohort ($p < .0001$; 95 percent confidence interval: 1.74–2.94). In other words, these miners were 2.3 times more likely to die of lung cancer than the general population, and the difference was highly statistically significant. There was a highly statistically significant linear dose-response relationship between exposure and increased risk of lung cancer, from which the relative and attributable risk coefficients were estimated at 0.27 percent per working level month and 3.10 per working level month per million person-years, respectively.[104]

Compared to the case in Ontario, much less is known about the health of Saskatchewan uranium miners despite two inquiries (Cluff Lake and Key Lake inquiries) that examined the continuation of uranium mining in the province. In their study of Beaverlodge miners, Howe et al. followed a cohort of 8,487 workers employed at this northern Saskatchewan uranium mine operated by Eldorado between 1948 and 1982. The observation period was from 1950 to 1980. A total of 65 miners in the cohort died of lung cancer in the period, compared

104. Howe et al. 1987.

to 34 expected. The relative risk of dying of lung cancer was 1.9 for the whole cohort ($p < .0001$). There was a highly significant linear relationship between exposure and increased risk of lung cancer, with estimates for the relative and attributable risk coefficients of 3.28 percent increase in the SMR for lung cancer per working level month and 20.8 excess lung cancer deaths per working level month per million person-years. The study adds to the concerns about whether the current standard for occupational exposure to radon decay products of 4 working level months per year in Canada and the United States is in fact adequate. The authors noted that silica exposures were always very low and the observed effects were not likely to be due to dust.[105]

It is notable that the risk coefficients from Beaverlodge (Saskatchewan) were greater by a factor of 10 or more than those estimated for miners at Port Radium (Northwest Territories) and in Ontario. This may be due to the exposure rate—Port Radium miners were exposed to much higher concentrations of radon decay products, in part because they were at work in the mines earlier, before ventilation and other safety measures had been introduced. This and other studies suggest that a low dose rate can lead to a greater cancer risk per unit of exposure than a high dose rate (see chapter 4). However, many experts feel that it was more likely due to inaccuracies in exposure measurement. The role of concomitant occupational exposures to other substances also cannot be ruled out as a cause for higher risk among the Beaverlodge miners.

A consulting firm contracted by the Atomic Energy Control Board (Canada's nuclear regulatory agency) reportedly has completed a follow-up study of the Beaverlodge cohort. However, the results have not been made public. The panel examining the proposed new uranium-mining developments have called for the establishment of an ongoing epidemiological study of Saskatchewan uranium miners, as has been conducted for Ontario miners. It is likely that more information will be forthcoming.

In the meantime, there is considerable concern regarding the proposed developments in view of the very high concentration in the ore of both arsenic and nickel as well as uranium. For example, typical core samples at MJV contained 1.08 to 9.62 percent arsenic, 0.94 to 4.8 percent nickel, and 0.25 to 11.8 percent uranium.[106] Arsenic and nickel are both toxic, and uranium is toxic and radioactive, posing significant risks (see chapter 4). Of particular concern is the possible combined effect resulting from occupational exposure to high concentrations of both uranium and arsenic. As noted by the panel, the potential health risk when all three exposures (arsenic, nickel, and uranium) are pres-

105. Howe et al. 1986.

106. Joint Federal-Provincial Panel 1993, p. 37.

ent needs more study. Although remote-controlled mining methods have been proposed and ventilation levels are considerably better than in the past, a large potential exists for an unacceptable level of occupational health risk.

Canada is contemplating adopting the new ICRP-60 recommendations, which would change the worker dose from 50 millisieverts per year to a cumulative 100 millisieverts over five years. The uranium mining industry has noted that this may force companies to adopt job rotation schemes; hence, the collective dose, and the consequent overall cancer risk, may rise. Therefore, the panel recommended implementing ICRP-60 before approval of any additional mines and also adopting measures to ensure that collective doses are not permitted to rise.[107] The government has not yet responded to these recommendations.

Effects on the Canadian Environment and the Public By the late 1970s, the uranium mines that operated in the 1950s and 1960s in the Elliot Lake area of Ontario had contaminated 80 kilometers of the Serpent River system, including 10 local lakes. At that time, the International Joint Commission on the Great Lakes declared the Serpent River, which flows into Lake Huron, the largest single contributor of radium contamination to the Great Lakes. Local water supplies were also contaminated with radium—at three to four times the drinking water standard. There have been more than 30 breaches of tailings dams in the Elliot Lake area.[108]

Government and uranium mining company officials acknowledge that serious degradation of the environment occurred but note that the situation is improving. The government is stocking Quirke Lake with trout, although this lake was the major receptacle for tailings effluent from the northern mine in Elliot Lake. Radium-226 levels are reportedly approaching the drinking water standard, and officials claim that survival and reproduction of fish are good. Both operators in Elliot Lake have submitted decommissioning plans, and an environmental impact assessment panel is being formed to make further recommendations.[109]

As noted by Dirschl and colleagues, 191 spills were reported to the Saskatchewan Spill Control Program, 125 of which involved radioactive materials.[110] Within 6 months of start-up in 1983, the mill at Key Lake, Saskatchewan had half a dozen radioactive spills, including the release of 100 million liters into a nearby bog when an overfilled

107. Joint Federal-Provincial Panel 1993, p. 17.

108. Canada, National Film Board 1991, p. 15; Dropkin and Clark 1992, pp. 106–107.

109. Joint Federal-Provincial Panel 1993a, p. 5.

110. Dirschl, Novakowski, and Burgess 1992.

Figure 5.5 Uranium tailings pile (Stanrock Tailings Wall), Elliot Lake, Ontario, Canada. Photo by Robert del Tredici.

containment dam failed in January 1984.[111] This spill contained significant concentrations of nickel, uranium, zinc, arsenic, and radium-226.[112]

Wollaston Lake in Saskatchewan has suffered several instances of radioactive contamination. From 1975 to 1977, almost two million liters of untreated waste were released into it.[113] In November 1989, at Cameco's mine at Rabbit Lake, two million liters of radioactive liquid, containing significant concentrations of arsenic and radium-226, spilled into a creek feeding Wollaston Lake.[114] This massive spill went undetected for more than 16 hours.[115] This was despite the fact that instrumentation and visits every two hours were in place to detect spills.[116] Soon thereafter, in January 1990, a radioactive spill of 90,000 liters occurred at the same mine site.[117]

111. Canada, National Film Board 1991, pp. 15, 18.

112. Dirschl, Novakowski, and Burgess 1992.

113. Miller 1993, p. 76.

114. Canada, National Film Board 1991, p. 17; Dirschl, Novakowski, and Burgess 1992.

115. *Nuclear Fuel*, 5. February 1990, p. 15; Dirschl, Novakowski, and Burgess 1992.

116. Dirschl, Novakowski, and Burgess 1992.

117. Canada, National Film Board 1991, p. 17.

Pipeline-related spills represented a high percentage of all reported spills. The provincial government noted that the development of new mines at Rabbit and Cluff Lakes had resulted in recent additions to the pipeline networks, increasing the risk of pipeline spills. In a letter to Cameco Corporation, dated 11 December 1989, provincial authorities said, "The Rabbit, Key and Cluff Lake operations are aging and spills due to pipeline wearing are becoming more common." Despite new containment and detection procedures, pipeline spills continued to occur, accounting for three-quarters of the volume released into the environment in 1990–91.[118] The experts examining the impacts concluded that the spills must be regarded as potentially serious contributors to a gradual buildup of chemical and radioactive loadings in the waters, sediments, and organisms of the region. They also noted that water in the vicinity of the mines has indeed suffered significant environmental degradation and that increased salinity and heavy-metal contamination have altered species composition in downstream aquatic ecosystems. Many Indian residents of Wollaston Lake, justifiably afraid to eat the fish and game they once lived on, reportedly had to buy expensive food from grocery stores.[119] The extent of the impact on air quality in this area is not known.

Some environmental impact studies have been conducted around the two major abandoned sites near Uranium City. For example, it has been noted that there has been no recovery of vegetation on the exposed tailings of the Lorado uranium mill, and the sediments in Beaverlodge Lake have high levels of radionuclides.[120] D. W. Waite et al. found high levels of radionuclides in sediments in Langley Bay, which received the tailings from the Gunnar uranium mine. The effects on the fish in the area have not been fully studied.[121]

P. Thomas and colleagues studied radiation in the Arctic food chain.[122] They found worrisome levels of polonium-210 and lead-210 in the bones and organs of caribou at Snowdrift, north of decommissioned and abandoned uranium mines at Uranium City, Saskatchewan. Government experts looked at the health risk to someone eating caribou daily. They calculated a dose to a consumer of 1.41 millisieverts per year. Using ICRP-60 risk coefficients, an increased lifetime cancer risk of 0.5 percent was estimated.[123] Further study, using newer data, is underway.

118. Dirschl, Novakowski, and Burgess 1992.

119. Dzeylton 1984.

120. Ruggles et al. 1978.

121. Waite et al. 1988.

122. Thomas et al. 1993.

123. Lawson 1993.

Cleanup Prospects for Canadian Mines and Mills Tailings management is regulated by the Atomic Energy Control Board and appropriate provincial authorities. The regulations require the design, construction, maintenance, and monitoring of an engineered facility for storing tailings, as well as treatment of effluents and limitation of access to the site. While various new methods are being developed to prepare tailings for abandonment, several principles generally hold true: tailings must be immobilized and covered; off-site levels of radioactive pollution from the tailings should not be significantly greater than natural background radiation; and no human intervention should be necessary to maintain the integrity of the containment.

While the management of tailings during a mill's operational phase has improved since the mid-1970s, the effectiveness of the regulations has been questioned, particularly in view of the spills mentioned above and the extremely long-lived radioactivity of tailings. Effective containment of the approximately 100 million metric tons of radioactive wastes from abandoned uranium mines and mills around the country, whose radioactivity will not diminish appreciably for tens of thousands of years, seems unlikely. And once the tailings have been abandoned, especially when the owner or operator goes out of business, it is not clear what entity is responsible for managing the tailings over such long time periods.[124]

The cleanup of Elliot Lake, as noted above, is to be the subject of an impending major review. Eldorado's Beaverlodge operation in Saskatchewan has been decommissioned and is in the post-decommissioning monitoring phase. However, the Gunnar and Lorado sites were abandoned in the 1960s, when no environmental regulations were in place regarding decommissioning. The original companies that owned and operated the properties have been dissolved. Prospects for cleanup of these sites in the near future appear slim.

South Africa

Uranium was discovered in the gold reefs of the Witwatersrand in 1922,[125] but it was not until the late 1940s that uranium exploration in South Africa began in earnest. In 1952, the country started producing uranium, destined mainly for the U.S. and U.K. nuclear weapons programs.[126] In 1959, production reached 4,960 metric tons.[127] With declining procurement of uranium for weapons by the United States

124. Canada, National Film Board 1991, pp. 11, 17.

125. Lloyd 1980, p. 33.

126. Owen 1985, p. 109.

127. Neff 1984, p. 315.

and United Kingdom in the 1960s, South African uranium output dropped temporarily. Production rose again in the late 1970s in response to demand from the nuclear power industry and high prices, and a new high of 6,146 metric tons was reached in 1980.[128]

There are several modes by which uranium is produced in South Africa. By far the greatest quantity has come from gold mines, especially those of the Witwatersrand basin, mainly as a by-product but also by recovery from tailings. There is one open-pit copper mine, Palabora, where uranium is produced as a by-product. The three mines intended primarily for uranium production have not been successful due to the depressed market: Afrikaner Lease stopped construction on a new processing plant in 1982 (after an investment of $180 million), and West Rand Consolidated stopped production in 1982.[129] The average ore grade is very low, typically between 0.01 and 0.02 percent uranium, which is why it is only economically feasible to produce uranium in conjunction with gold or copper or by recovering it from gold mill tailings.[130]

Table 11.1 shows the many facilities where uranium has been mined or recovered in South Africa, with opening and closing dates. Those mines opening in the 1950s can be assumed to be the main ones supplying the U.S. and U.K. weapons programs.

South African law forbids the release of details about uranium contracts, but apart from U.S. and U.K. procurement for weapons, Japan, Germany, the United States, France, Taiwan, Belgium, and Spain have all been major customers.[131]

Production is on the decline, with only 1,687 metric tons produced in 1991.[132] Sanctions against the apartheid state were undoubtedly one reason, but low world demand for uranium and reduced South African gold sales have also been instrumental. Decreased demand led to the layoff of 80,000 goldmine workers in 1989 and 1990.[133]

Health Effects on South African Workers Since uranium is produced in South Africa mainly as a by-product, with gold of much greater economic significance, health effects and risks among workers involved in uranium production cannot be ascribed to uranium production alone.

128. OECD Nuclear Energy Agency and IAEA 1990, p. 43.

129. Owen 1985, pp. 109–110; Neff 1984, pp. 180–181.

130. Huisman 1990, p. 29; Owen 1985, p. 110.

131. Owen 1985, p. 110.

132. Uranium Institute 1992, p. 23.

133. U.S. Bureau of Mines 1990, vol. 3, p. 202.

Although limited direct data are available on health effects, as recently as late 1992 the mining industry did not appear to have widely accepted the importance of limiting radiation exposure. For example, the abstract of a December 1992 paper in the *Journal of the Mine Ventilation Society of South Africa* begins, "The introduction of the concept of radiation dose limitation into the South African mining industry seems inevitable from the increasing international evidence concerning the effects of nuclear radiation and new trends in radiation legislation."[134] A 1980 paper states, "At uranium plants associated with the mining industry it is found that radiological surveying is infrequent if done at all." At that time at least, assessment of mine air quality was based on limited air sampling.[135] The problem of mine air quality in South Africa may be more variable than in other countries since ore quality is low in many instances, as noted above.

Effects on the South African Environment One assessment claims that although lack of dam stability was the major cause of radionuclides reaching the larger environment in the past, "modern methods of construction have overcome this" in present-day operations.[136] However, another report maintains that under South African climatic conditions of dry winter months with gusty winds and wet summers with intense rain showers, all release mechanisms could be important: overflow of dissolved and suspended material into surface water, seepage of dissolved and leached material into groundwater, suspension of fine particles by wind action, and radioactive emanations (mainly radon-222) from the tailings with subsequent dispersion of radon-222 and its decay products by the wind.[137]

Over the course of a century of mining in the Witwatersrand, more than a billion metric tons of uranium-containing wastes from gold mining have been dumped in a highly populated stretch of terrain near Johannesburg. The population of this area was estimated at five million in 1980.[138] Some of these gold mining tailings were so concentrated with uranium that it was deemed worthwhile to extract it from them.[139]

There are a number of laws to control releases from tailings residues: the Air Pollution Prevention Act, the Health Act, the Mines and Works Act, and the Water Act. Although some chemicals are specified, no

134. Van der Linde 1992.

135. Kruger 1980, p. 454.

136. Lloyd 1980, p. 41.

137. van As 1980, pp. 5, 6.

138. Lloyd 1980, p. 33.

139. van As 1980, p. 2.

radionuclides are mentioned.[140] The Chamber of Mines' 1979 *Handbook of Guidelines for Environmental Protection* does not mention the radiological impact of tailings at all.[141] No formal regulations regarding exposure of the public from mine and mill tailings exist; international (ICRP) dose limits are occasionally referenced.[142]

There are indications that any industry response to public concern about the hazards of tailings is a relatively recent phenomenon. For example, a 1982 paper states, "Intense public concern regarding the environmental and health effects of uranium tailings has forced a re-evaluation of past disposal practices." The paper goes on to consider the options for ensuring that tailings are impounded in an environmentally acceptable manner.[143] Apparently, such measures were only proposals at that stage.

Australia

Australia mined uranium ores for radium as early as the 1920s. Commercial production of uranium began in 1954, after a period of development and exploration. The mines of this period were Radium Hill, South Australia; Mary Kathleen, Queensland; and Rum Jungle, Northern Territory; with minor production at sites on the Alligator River, Northern Territory.[144]

Radium Hill was an underground mine in the semidesert of eastern South Australia. Ore was concentrated at the mine site and the concentrate transported by rail to Port Pirie, also in South Australia, for extraction of the uranium. The mine was opened in 1954 (although some development occurred earlier) and closed in 1961. Approximately 1,000 metric tons of uranium were produced.

Rum Jungle was an open-pit mine approximately 80 kilometers south of Darwin, Northern Territory. This is a high rainfall tropical area, with resulting implications for runoff and leaching, particularly since the actual mine was on the banks of a river. A mill and extraction plant were adjacent to the mine. Operations commenced in 1954; the mine closed in 1963, although milling of stockpiled ore continued until 1971. Production totaled approximately 3,000 metric tons. Copper was also extracted from the ore, and lead and zinc were also present.

The Mary Kathleen mine is in central Queensland, a dry hot area. Operations started in 1956 and continued, with an interruption from

140. van As 1980, p. 11.

141. South Africa, Chamber of Mines 1979, as cited in van As 1980, p. 2.

142. van As 1980, p. 12.

143. Robertson 1982.

144. Movement Against Uranium Mining 1991, p. 2; Fry and Morison 1982.

1963 to 1976, until 1982. Mining was by open pit, and a mill and extraction plant operated on the site. About 8,000 metric tons of uranium were produced.

Australia's annual uranium production reached 1,200 metric tons of uranium in 1961. The closure of these mines in the 1960s was caused by depletion of reserves and the collapse of prices. Until these mines closed in the 1960s, production was purchased by the Combined Development Agency, an Anglo-American defense purchasing agency, and the United Kingdom Atomic Energy Agency. Total production during this initial period was approximately 8,000 metric tons, with approximately 6,000 metric tons exported to the CDA and UKAEA and 1,730 metric tons retained in a government stockpile.[145]

In 1970, a new phase commenced with the introduction of a commercial export policy and the discovery of major new deposits. Deposits with reserves amounting to over 300,000 metric tons were discovered between 1970 and 1973. However, a new government limited development of these discoveries by requiring a major inquiry. As a result of the inquiry, the government approved only two deposits for operation, Nabarlek and Ranger, both in Northern Territory. These began production in 1980 and 1981, with annual production of approximately 1,500 and 2,000 metric tons, respectively. Production at Nabarlek ceased in 1988, when reserves were exhausted. Ranger production reached approximately 2,500 metric tons per year but has recently fallen in response to depressed markets.[146]

In 1977, a large copper-uranium ore body was discovered at Olympic Dam in South Australia. Production began in 1988, with initial production of approximately 850 metric tons per year, subsequently rising to 1,200 metric tons per year (together with over 50,000 metric tons of copper per year).

In 1973, Australia signed the Nuclear Non-Proliferation Treaty and, in 1974, concluded a Safeguards Agreement with the International Atomic Energy Agency. This agreement is implemented in part by government-to-government agreements with customer countries prohibiting any military use of Australian-produced uranium.[147] All exports since that time have been for commercial power production. Of the approximately 56,000 metric tons of uranium produced in Australia as of the end of 1992, not more than about 6,000 (produced before 1965) have been available for nuclear weapons production; the actual quantity so used is not known. Annual production is shown in figure 5.3.

145. Neff 1984, pp. 112–113.

146. Supervising Scientist for the Alligator Rivers Region 1991, p. 50; Supervising Scientist for the Alligator Rivers Region 1992, p. 65.

147. Australia, Department of Primary Industries and Energy 1991, pp. 2, 3, 36.

Australia possesses approximately 30 percent of the Western world's uranium reserves but currently accounts for only about 15 percent of annual production. Increased production is constrained by the policy of the present government not to allow the development of further mines and by the present depressed uranium market.[148]

Health Effects on Australian Workers The only known study of health effects on Australian miners was conducted on Radium Hill miners. A significant excess of lung cancer mortality was found: 54 deaths were observed, against 28 to 47 expected. Further excess lung cancer deaths are expected as this population ages.[149]

All other mines were open cut, and radiation exposures generally, and radon exposure in particular, would be very much less than in the (underground) Radium Hill mine.

Effects on the Australian Environment and Prospects for Clean-up The only wastes remaining at Radium Hill are the tailings from concentration of the ore. These have a low radionuclide content and have been covered by several meters of soil. As a result, dose rates and radon emanation rates are comparable to the surrounding areas. The low rainfall of the area and the extreme insolubility of the mineralization have prevented any significant seepage.

The residues from extraction of uranium from the Radium Hill concentrate are stored in dams on the outskirts of Port Pirie. Despite the unpromising location—tidal mud flats—the dams have retained their integrity, and there is no indication of seepage. Measurements of radon and radioactive dusts in the town are within the bounds of natural variation. The tailings, covered with several meters of slag (from an adjacent smelter) and topsoil, have been revegetated.

The processing residues at Rum Jungle included waste rock and overburden piles, mill tailings, and heap-leach residues, together with the open-pit excavations themselves. These were not rehabilitated on closure, and leaching resulting from heavy monsoon rains damaged the environment. The main concerns were the heavy-metal and sulfide contents, the latter enhancing leaching by the generation of sulfuric acid through oxidation. In 1982, rehabilitation of the site was undertaken, with collection and burial of contamination in the overburden heap, stabilization of the heap itself, and diversion of runoff away from contaminated areas.

After Mary Kathleen closed in 1982, extensive rehabilitation was carried out. Residual solids from the evaporation ponds were buried in the tailings dam, together with other contaminated materials. The

148. Australian Department of Trade and Resources 1981.

149. Woodward et al. 1991.

tailings dam was contoured and covered with two meters of soil, clay, and waste rock. The dam wall was protected from erosion by waste rock and boulders, and then partially revegetated. Seepage control measures were implemented, including collection and evaporation of liquid wastes and diversion of runoff water away from rehabilitated areas.

Namibia

Although uranium deposits were discovered in 1928 in the Rössing area of what is now Namibia, it was not considered economically viable to exploit them until the 1960s, when South Africa found itself in need of revenue for a dam project. Generous concessions were granted to the British company Rio Tinto Zinc (still the major shareholder), construction began in 1973 as the world uranium market began to revive, and this huge open-pit uranium mine began producing in 1976.[150] In addition to the open-pit mine, the Rössing enterprise has an acid-leach plant and a plant to produce the sulfuric acid used in the leach process.[151]

As of the end of 1991, Namibia's Rössing mine had produced 51,231 metric tons of uranium, making it the eleventh largest producer in the world in terms of cumulative production. In addition to the United Kingdom, major customers have included France, Japan, Iran, Germany, and Taiwan.[152]

A total of 8,600 metric tons of nonsafeguarded Namibian uranium was imported by the United Kingdom under two contracts, one for 7,500 metric tons, signed in 1970, the other for 1,100 metric tons, signed secretly in the mid-1970s.[153] All of it was presumably for military purposes—as pointed out in 1985 by Tony Benn, U.K. Minister of Technology at the time of the signing of the first contract and a member of the 1974–1979 Labor Government: "It is important to remember that the interest of the Rössing mine to the Government was that it was not under safeguards and uranium from it, unlike uranium from Canada, could be used freely for nuclear weapons."[154]

From the beginning, Rössing had a particularly blighted reputation. In part, this was due to its political significance as a strategic foreign holding (partly South African-owned) whose production and profits

150. Moody 1992, p. 655.

151. Smit and Brent 1991.

152. Owen 1985, p. 106; Huisman 1990, p. 23; Moody 1992, p. 662.

153. Campaign Against the Namibian Uranium Contracts 1986, p. 50.

154. Letter from Tony Benn to Greg Dropkin, 26 March 1985, reproduced in Campaign Against the Namibian Uranium Contracts 1986, p. 52.

were exported and whose taxes helped uphold the occupying South African forces in a colony fighting for its independence. Before independence in 1990, the mining directly violated UN Decree No. 1, adopted by the General Assembly in 1974, which prohibited the extraction of natural resources from Namibia without the consent of the UN Council for Namibia.[155]

Furthermore, labor relations and living and working conditions were notoriously bad, especially in the 1970s. According to Roger Moody:

> Every aspect of management at Rössing has given cause for alarm and protest since the mid-1970s; ranging from housing standards at the black township of Arandis [built to house non-white workers], to gross discrimination in wage awards, suppression of legitimate trade unions, and a patronizing attitude to employees and poor standards of health and safety.[156]

By the 1980s, Rössing had responded to international criticism, introducing better services and amenities, which now are comparable to other mines and mills in southern Africa.[157]

Production at the Rössing mine has declined sharply and in 1993 was at half capacity. A third of the workforce was abruptly dismissed in 1992.

Health Effects on Namibian Workers Occupational health hazards persisted at Rössing at least well into the 1980s. These include dangerously high levels of silica dust and radiation.

In open-pit mining, although radon is dissipated much more readily than in underground chambers, exposure to silica dust is a serious health hazard.[158] A worker described conditions in the open pit in a letter smuggled out of Namibia in 1979 as follows:

> Working in open air, under hot sun, in the uranium dust produced by grinding machines, we are also exposed to the ever-present cyclonic wind which is blowing in this desert. Consequently our bodies are covered with dust and one can hardly recognize us. We are inhaling this uranium dust into our lungs and many of us have already suffered the effect. We are not provided with remedies and there is no hospital to treat us. Our bodies are cracking and sore.[159]

By 1982, Rössing had adopted an exposure standard (threshold limit value) for respirable dust of 0.5 milligrams per cubic meter, based on the recommendations of the American Conference of Governmental Industrial Hygienists. A company report on environmental control

155. Singham 1980.

156. Moody 1992, p. 664.

157. Moody 1992, p. 664.

158. Erickson 1981.

159. As quoted in Moody 1992, p. 666.

Figure 5.6 Namibian uranium miners, 1991. Photo by Orde Eliason, Impact Visuals.

from July 1985 found a maximum dust level of 1.48 milligrams per cubic meter and a mean of 0.76 milligrams per cubic meter.[160]

These exposures are probably not characteristic of all operations. Dust is typically worst where ore is crushed and handled. Workers at ore-handling sites are supposed to wear face masks, but these resemble surgical masks; they are not true respirators and cannot effectively screen out the fine particles that pose the greatest hazards. (See figure 5.6.) In the open pit, many workers have breathing problems, and some show signs of fibrosis or of impaired bronchial tubes.[161]

Regarding radiation, Greg Dropkin and David Clark reviewed company documents from the 1980s and calculated cancer risk based on internationally accepted dose-risk factors:

> Throughout the period the Rössing industrial hygiene standard for airborne uranium was nearly 6 times the ICRP Derived Air Concentration for natural uranium, and 36 times the limit implied by current scientific evidence. In 1982, measured levels of airborne uranium frequently exceeded the inadequate Rössing standard, even reaching 88 times the limit implied by the [US] National Academy of Sciences (1990)....

160. Dropkin and Clark 1992, p. 46.

161. Dropkin and Clark 1992, pp. 34, 45, 47.

Applying the US National Academy of Sciences cancer risk estimates, about 17 excess cancer deaths are to be expected on the basis of current company data [in a workforce of 2,400]. Excess *lung* cancer deaths due to radon are estimated at 5 among the Rössing workforce, a low figure by comparison with underground uranium mining.[162]

Dropkin and Clark noted that workers in the "Final Product Recovery" area, which includes the roasting ovens where yellowcake is heated to uranium oxide, are subjected to especially high exposures. Although numbers vary depending on which cancer risk estimates are used and what lifetime radiation exposure is assumed, they concluded, based on the 1990 U.S. Academy of Sciences risk estimates, that out of just 21 employees in that area, 1 or 2 could be expected to die of radiation-induced cancer.[163]

Reinhart Zaire, a Namibian researcher, has recently noted a statistically significant excess of brain cancer among white mineworkers at Rössing, compared to a control group of South African whites. (Whites have been monitored more consistently than blacks; also, blacks have tended to be short-term contract workers, making follow-up more difficult.)[164] Brain cancer had not previously been observed to be associated with uranium mining. Peripheral lymphocytes and sperm of black workers are being analyzed for possible cytogenetic indicators of radiation damage, with the approval and cooperation of the Mineworkers Union of Namibia (MUN).[165]

The existing agreement between the MUN and Rössing (signed in 1988) barely addresses health, safety, or environmental concerns and only commits the company to "discuss and clarify" these matters with the union.[166] According to a 1989 statement by Ben Ulenga, general secretary of the MUN, the company had complete control over radiation monitoring at the mine, and the union played no part in determining the nature or extent of radiation hazards.[167]

Effects on the Namibian Environment In a dry environment such as that of Rössing (with a mean annual rainfall of only 3 centimeters and a mean evaporation rate of 0.72 centimeters per day),[168] radioactive dust from mining operations and mill tailings can present a special problem, despite the practice of sprinkling the tailings to

162. Dropkin and Clark 1992, pp. 35–36.

163. Dropkin and Clark 1992, p. 103.

164. Clapp 1993.

165. Clapp 1993.

166. Dropkin and Clark 1992, p. 119.

167. News releases cited in Moody 1992, p. 666.

168. Smit and Brent 1991.

keep dust levels down. At times, a large part of the tailings area has not been covered by water.[169]

A second environmental problem is waterborne radioactivity from the tailings reservoir. The tailings are disposed of by pumping the slurry to a disposal dam, which covers 650 hectares. The tailings area is about 250 meters higher than the Khan River, only about 6 kilometers away.[170] In 1980, Rössing scientists noted signs of seepage downstream of the collection system in two gorges that empty into the Khan River, which runs westward to the coast and empties at Swakopmund, near Walvis Bay. By computer modeling, it was calculated that some 45 percent of the seepage from the tailings dam was bypassing the collection system, mainly into fractured bedrock in the western half of the dam, resulting in significant seepage into the gorges.[171]

Although Rössing has built walls and trenches and implemented other measures to monitor and contain the seepage, in 1991 two Rössing scientists, M. T. R. Smit and C. P. Brent, reported evidence of contamination of the Khan River by seepage from uranium tailings,[172] confirming earlier reports.[173]

A third environmental problem is the high demand for water at many stages of the operations and the possible lowering of the water table and rivers in a large area around the mine as a result. Smit and Brent declare Rössing the largest consumer of water in the coast region, pointing out that it uses twice the amount as Swakopmund, whose population numbers about 18,000. At its peak in 1980, Rössing consumed 28,000 cubic meters of fresh water per day. Acting out of "moral as well as commercial interest," Rössing took steps to reduce potable water consumption by half over the course of the 1980s, mostly by recycling contaminated water from its operations and tailings areas. Smit and Brent write that "significant water savings have been effected by reducing the wetted area on the tailings dam to about 30% of its former size," but they do not discuss the implications in terms of radioactive dust or radon emissions.[174]

After a progressive decrease in fresh (not total) water use from 1980 to 1985, its use jumped back up to about 22,000 cubic meters per day in 1987, since "the tailings pond had been depleted to a very low

169. Moody 1992, p. 669.

170. Smit and Brent 1991.

171. "Design of Waste Disposal," a Rössing document cited in Dropkin and Clark 1992, p. 110.

172. Smit and Brent 1991.

173. Moody 1992, p. 669.

174. Smit and Brent 1991, p. 192.

level,"[175] and extra water was need to keep down airborne contamination from tailings. The use of fresh water has declined again since 1987, to 12,000 cubic meters per day in 1990. This is still not an insignificant quantity, especially in view of the rising demand placed by growth in population and industry downriver at the coast.

Cleanup Prospects for Namibian Mines and Mills It is unclear what, if anything, will happen regarding environmental cleanup. Rio Tinto Zinc, which runs the Rössing mine, has been under no legal obligation to conduct an environmental impact statement on its operations and had not done so until recently, when it contracted with CSIR (then a South African parastatal agency, now a private consulting firm) for that purpose. In August 1992, CSIR delivered its environmental impact statement to the company, which passed on a copy to the government. Somewhat beholden to Rio Tinto Zinc, a large investor in Namibia, the Namibian government, as of spring 1993, had not yet consented to give a copy to the Mineworkers Union of Namibia.[176]

URANIUM SOURCES FOR SOVIET WEAPONS

The Soviet Union, which embarked on its nuclear weapons program in 1942, got uranium from its own territory, as well as from eastern Europe and China. Until the end of 1990, the whole uranium production of East Germany, Czechoslovakia, Hungary, and Bulgaria went to the Soviet Union, where it was processed into reactor fuel, used for military purposes, or stockpiled. Only Romania kept a small inventory.[177] In the case of East Germany at least, 6,300 metric tons of uranium in the form of reactor fuel was reimported for use in Soviet-made VVER reactors.[178] In an agreement that lasted from 1956 until the Sino-Soviet split in 1960, the Soviets agreed to provide China with "full-scale assistance" for its nuclear weapons program in exchange for uranium.[179]

East Germany and Czechoslovakia were by far the largest foreign suppliers of uranium to the Soviet Union. Cumulative production through 1990 was 216,500 metric tons for East Germany and 98,500 metric tons for Czechoslovakia. No information is available on what

175. Smit and Brent 1991, p. 192.

176. Moody 1993.

177. Uranium Institute 1991, p. 70.

178. Madel 1990, p. 2.

179. Lewis and Xue 1988, pp. 61–62.

fraction of this uranium production went to nuclear weapons as opposed to nuclear power.

East Germany

In the years immediately following World War II, East German mines (which the Nazi government had operated during the war to supply its nuclear research program) supplied uranium for the Soviet nuclear weapons program.[180] Even after mining started in the Soviet Union, East Germany continued to serve as the major source of uranium. The most important uranium deposits were in the Erzgebirge ("Ore Mountains") in Saxony and the Thüringian Forest in Thüringia.[181] The first of the deposits to be exploited were in the Erzgebirge, which defines the border with Czechoslovakia. These mountains have been an important mining area since the Middle Ages, originally for silver.[182]

In an urgent quest for uranium for nuclear weapons, the Soviet Union started mining the Erzgebirge in 1946. The Soviet enterprise Wismut (meaning "bismuth," disguising its uranium-mining activities), founded in 1946, ran the operations.[183] In 1950, Wismut opened new mines in the Katzhütte area of Thüringia state and one near Wernigerode in the Harz Mountains.[184]

Allegations of forced labor have been made, especially in anti-communist Western accounts of the period.[185] In conflicting accounts, high salaries and access to housing and consumer goods were effective in attracting workers.[186]

Within a few years, Wismut turned the southern part of East Germany into the largest uranium mining region in Europe. The pace of work in the 1950s has been called "feverish."[187] According to one source, at times more than 100,000 workers were employed under unsafe conditions.[188] In 1950, the British Control Commission reported that 300,000 Germans had been drafted as miners and worked on an intensive, 24-hour basis aimed at maximum output of uranium "re-

180. Zaloga 1991, p. 178.

181. U.S. Bureau of Mines 1957.

182. Schüttmann 1993; Toro 1991a.

183. Schüttmann 1993, pp. 363–364.

184. U.S. Bureau of Mines 1950.

185. Gordon 1953, quoted in U.S. Bureau of Mines 1953, p. 1233; *Newsweek*, 8 November 1954, pp. 45, 46, as quoted in U.S. Bureau of Mines 1954; Kahn 1993, p. 449.

186. Kahn 1993, p. 449.

187. Kahn 1993, p. 448.

188. Schüttmann 1993, p. 364.

gardless of wastage in manpower and material."[189] Another source says that by the mid-1950s, Wismut had 150,000 employees, most of whom worked in underground mines and some in the yellowcake factories.[190] The peak year was 1968, during which 8,000 metric tons of uranium were produced.[191] In total, about 400,000 people have worked in Wismut mines, producing a total of about 220,000 metric tons of uranium.[192]

One author identifies 24 mines and 10 processing facilities.[193] A representative of Wismut states that, all told, there were about 400 "industrial sites"—including exploration shafts, mine shafts, underground mines, open pits, milling plants, waste dumps, and tailings ponds.[194]

By 1989, production was down to 3,900 metric tons; there were 39,000 employees that year, of whom 22,000 were involved in uranium mining and milling. The mines and mills are now closed.

Health Effects on East German Workers Despite the radiation protection principles that had been recognized and adopted in the region after decades of epidemiological research into lung cancer among East German miners, Soviet authorities disregarded the existing recommendations in their rush to build the bomb, operating both old and new mines without radiation protection measures until 1954. Miners drilled without wetting down the surfaces or the dust, and ventilation was inadequate.

When Wismut became a joint Soviet-East German enterprise in 1954, conditions in the mines improved somewhat. Wismut introduced drills that automatically watered down dust and installed new ventilation systems. (According to a 1993 article in *Science*, "The workers often turned off the ventilation to cut down on noise, and more regular use of safety measures came about only gradually.")[195] The company began taking monthly radiation measurements. The range of radiation exposures of workers in the mines varied considerably over the years. A summary of exposures is given in table 5.3. For comparison, cancer of the lung or bronchial cells was considered a work-related illness

189. A description issued on 23 August 1950, as quoted in U.S. Bureau of Mines 1950, pp. 1269–1270.

190. Kahn 1993, pp. 448–449.

191. Madel 1990, p. 5.

192. Krause 1991.

193. Kahn 1993, p. 448 (map).

194. Madel 1990, p. 2.

195. Kahn 1993, p. 449.

Table 5.3 Range of Annual Radiation Exposures per Worker in Wismut Mines

Time	Radon and Progeny Products (WLM/yr)
Before 1956	30 to 300
1956–1960	10 to 100
1961–1965	5 to 50
1966–1970	3 to 25
1971–1975	1 to 4

Source: Krause 1991.

Table 5.4 Cumulative Worker Exposures Officially Accepted during Various Periods as Indicators of Work-Related Illness

Time	Radon and Progeny Products (cumulative WLM)
until 1974	450 WLM
1974	250 WLM
1976	200 WLM
1990	discussion if > 100 WLM, accepted if > 150 WLM

Source: Krause 1991.

if the following cumulative exposures had been received, listed in table 5.4.

After the collapse of the East German government in 1989, medical records on 450,000 uranium workers were tracked down by an East German scientist with West German support. These records on workers in the uranium industry in the Erzgebirge are collected by Wismut-associated hospitals and clinics; they have not yet been fully analyzed, but various articles on the subject offer roughly comparable mortality values. One reports that, according to the records of the organization providing insurance to the miners, about 15,000 of the 450,000 miners died of silicosis and about 6,000 of lung cancer.[196] Another value for deaths from lung cancer through 1989 is 5,132.[197] Yet another article estimates the occurrence of thousands of cases of pneumoconiosis and at least 9,000 fatal cases of lung cancer with the claim that these diseases and deaths could have been avoided had radiation protection measures been in place.[198] None of these three sources specifies whether these cases of lung cancer also included those possibly induced by smoking.

196. Kahn 1993, p. 449.

197. Toro 1991a.

198. Schüttmann 1993, p. 363.

The Wismut medical service has claimed that miners who started work after 1960 have not shown an excess of lung cancer. New cases of lung cancer are expected to arise, particularly among older miners who worked in the 1940s and 1950s,[199] and the Federal Office of Radiation Protection estimates that the final number of lung cancer cases will reach 10,000 to 15,000.[200] Again, these presumably include cases due at least in part to smoking.[201]

The German government plans large-scale epidemiological studies, using the Wismut archive on the 450,000 workers as a starting point and following up on living miners. Various problems bedevil the effort: questions about the reliability of the existing records, the accuracy with which miners' radiation doses can be estimated, and the influence of confounding factors such as high background radiation, smoking, and exposure to nonradioactive toxic materials.[202]

Effects on the East German Environment and the Public East German industry created notoriously severe pollution and other environmental degradation, and Wismut's uranium mining and milling operations were no exception. These activities have contaminated an area estimated at between 1,000 and 1,400 square kilometers.[203]

Lakelike basins of mill tailings and associated waste up to 2.5 square kilometers in area dot the region.[204] Waterways and groundwater have been polluted with radioactive and other toxic substances. Caustic solutions involved in separating uranium from ore ended up in large reservoirs, from which the contents were gradually fed into local rivers and streams. As with other tailings areas, the waste basins contain arsenic, uranium, and radium.[205] Contaminants from waste basins have leached into groundwater.[206]

199. Schüttmann 1993, p. 364.

200. Kahn 1993, p. 449.

201. These numbers raise questions. If smoking and therefore lung cancer rates are similar to the West's, about 5 percent of the workforce (450,000), or 20,000 to 25,000, would be expected to develop lung cancer without even taking into account the effects of mining.

202. Dickman 1991.

203. Toro 1991a; Kahn 1993; Bundesamt für Strahlenschutz, Salgitter (1991) calls 1,400 square kilometers of Sachsen, Thüringen, and Sachsen-Anhalt "areas of suspected radioactive contamination," based on remote sensing data and existing local measurements.

204. Toro 1991a; Toro 1991 mentions 18 waste ponds associated with the Wismut operation.

205. *New York Times*, 19 March 1991.

206. Deutsches Atomforum *e.v.* 1991; Toro 1991a.

The Crossen-Oberrothenbach facility is one of the worst sites in terms of groundwater contamination. The *Minerals Yearbook 1990* reported that waste from uranium mines had been discharged directly into the Elbe River drainage basin.[207] A spill, apparently similar to the accident at Church Rock, New Mexico, occurred near the village of Oberrothenbach in the Erzgebirge in 1964 when a floodgate to a mill tailings impoundment failed. A large amount of radioactive tailings sludge spilled into a small creek flowing through the town, raising the water level by 1 meter for more than a day. The radioactive contamination is believed to have worked its way eventually to the Elbe river.[208] As of 1991, the Crossen-Oberrothenbach facility had extensive seepage problems. The major concern at the facility is a large 60-meter-deep pool, composed of uranium mining and milling wastes that are about one-third under water. "We have to take an inventory just to find out what is in there," one federal official said.[209]

Like the Crossen-Oberrothenbach area, several other facilities have serious groundwater contamination problems, or the potential for contamination. In Königstein, about eight kilometers from Dresden, two of the four aquifers of the area have been contaminated. The hydrogeologic factors are poorly understood, though, and it is uncertain how severe the contamination may become. At the Seelingstädt facility, water is seeping through vast tailings piles at the rate of about 15.8 cubic meters per hour. The major contaminant is sulfate, and the tailings piles overlie fault zones. Some of the seepage is collected around the waste piles, but much of it escapes collection; some of the seepage has entered the cellars of nearby homes. Seepage and runoff from both the Seelingstädt and Schmirchau complexes (described below) feed into tributaries of the White Elster River, resulting in calcium buildup in the river water.[210]

Local air quality has been affected as well. Near Oberrothenbach, a tailings pile with a dam, used as a dump by a nearby yellowcake factory, reportedly contains large amounts of arsenic, lead, cadmium, sulfuric acid, low-level radioactive wastes, and other toxic substances.[211] On windy days, mill tailings dust from the exposed area would be so thick that the town could barely be seen from a nearby hilltop.[212] Villagers had to stay inside to avoid irritation of the nose

207. U.S. Bureau of Mines 1990, p. 119.

208. Beleites 1988, p. 28.

209. *Nuclear Fuel* 1991, p. 5.

210. Robinson 1991, pp. 6–9 and p. 12.

211. Kahn 1993, p. 448.

212. Beleites 1988, p. 33.

and throat.[213] In 1986, residents urged the mine operators to take steps to reduce the dust levels, but the director of mining operations replied that no special measures were needed because "Oberrothenbach is, after all, not a vacation spot."[214] As of 1992, the tailings ponds were covered with water to reduce dust and radon emissions. This approach is only a stopgap measure though, since the water is severely contaminating the groundwater.[215]

Another site with serious air quality problems is the Schmirchau facility, an open-pit complex. Black particles from "pyramid" waste piles, which are about 100 meters high, are distributed by wind throughout the countryside. The particles are visible on roads and fields, and have resulted in the sheep of the area appearing gray, rather than white. Mine venting has also affected air quality in the region, since by 1989 the ventilation rates were 80 cubic meters per minute of fresh air to each worker.[216]

As in other countries, mill tailings were used as construction material, a practice apparently not prohibited until 1980. More recently, tailings have been used in road building.[217]

Radon concentrations in indoor air at certain sites in the Erzgebirge are the highest in the world. In the country as a whole, thousands of homes have radon levels above the maximum allowable limit of 250 becquerels per cubic meter of air (about 6.75 picocuries per liter).[218] In Schneeberg, the average value is 8.1 picocuries per liter (300 becquerels per cubic meter),[219] and 5 percent of the houses have "extraordinarily high" radon levels.[220] In 1.2 percent of the buildings, the concentration exceeds 405 picocuries per liter (15,000 becquerels per cubic meter) based on short-term measurements. The highest level measured was 3.1 nanocuries per liter (115,000 becquerels per cubic meter), found in the cellar of a Schneeberg house.[221] These levels are similar to levels in some unventilated or poorly ventilated underground mines. However, as Schüttmann points out, this contamination cannot be attributed solely to Wismut uranium mining but is also due to the general

213. Kahn 1993, p. 450.

214. Beleites 1988, p. 33.

215. Robinson 1994a.

216. Robinson 1991, p. 11.

217. Beleites 1988, p. 25.

218. Kahn 1993, p. 450. The U.K. National Radiological Protection Board recommends no more than 200 becquerels per cubic meter, and the U.S. EPA recommends action to be taken at levels equivalent to 4 picocuries per liter (about 150 becquerels per cubic meter).

219. Schüttmann 1993, p. 364.

220. *New York Times*, 19 March 1991.

221. Schüttmann 1993, p. 364.

radioactivity of geological deposits and the centuries of mining in the region, which left old shafts and mining debris upon which houses were sometimes built.

Based on epidemiological data from uranium miners and the assumption of a linear dose-effect relationship, one would expect an above-average mortality from lung cancer among the inhabitants of this region of high indoor radon concentrations. However, according to Schüttmann, none of the investigations carried out as of early 1993 had shown increased lung cancer mortality for nonminers in the region, a finding he considers surprising and in need of explanation. More extensive studies are being conducted on radon concentrations in homes and lung cancer mortality in one area.

Cleanup Prospects in East Germany It is estimated that remedial work will take 10 years and cost Germany about 10 to 15 billion marks (or about 6 to 9 billion dollars at the early 1994 exchange rate). The Wismut Corporation put off programs for improving the safety of the affected sites and for restoring the landscape, leaving no reserve funds for these activities.[222]

The estimated time and cost for remediation seem optimistic. Some 1,600 kilometers of tunnels, about 140 mine and ventilation shafts, and 6 underground caverns must be filled and sealed. In the meantime, in opencast and underground mines, naturally occurring acids (formed when sulfide-bearing ores meet water and oxygen) as well as leftover acid from *in situ* leaching extraction threaten to contaminate groundwater and nearby creeks. Mounds of radioactive waste rock, some huge, must be isolated from groundwater, contained to prevent erosion by water or wind, and either hauled away to some waste dump or covered and replanted.

Moreover, there are 18 tailings ponds to decontaminate or isolate and cover. Drying out the ponds may take at least a decade, in part because of the fineness of the tailings. As drying proceeds, the waste must be protected from the rain to avoid leaching. The recommended covering would be about 3 meters thick, made of layers of clay, gravel, and dirt.[223]

The Former Soviet Union

Within the vast area of the former Soviet Union, almost every type of uranium deposit known in the world, including many unique formations, has been reported and exploited.[224] Uranium for both nuclear

222. Schüttmann 1993, p. 364.

223. Toro 1991a, pp. 42–43; Dayton 1991, p. 45.

224. Cochran et al. 1989, p. 90.

power and weapons programs has come from over 50 sites in many regions, including several parts of Russia, Ukraine, Kazakhstan, Kirghizstan, Uzbekistan, and Tajikistan (see chapter 7 and table 7.1).[225]

Many uranium mining areas were developed with forced labor. A list of 32 such sites appears in *The First Guidebook to Prisons and Concentration Camps of the Soviet Union* under the heading of "extermination camps."[226] According to this 1982 book, these are places "where prisoners, forced to work under dangerously unhealthy conditions for the Soviet war machine, face a virtually certain death."[227] The labor camps supporting these mining operations reportedly ranged in size from 1,000 to 5,000 prisoners.

According to a 1990 report in the Soviet journal *Atomnaya Energiya*, mining methods include open pits, underground drilling, and *in situ* underground leaching.[228] A study of the Soviet uranium industry by NUEXCO International Corporation claims that the Soviet Union also recovered uranium from waste water generated during petroleum refining, a source that is not used in the United States.[229]

Uranium ores are processed at major production centers to yield yellowcake from ore by milling, leaching, and ion-exchange methods.[230] Other metals found with uranium in some deposits, such as molybdenum, gold, and copper, are also recovered and refined at these production centers.

A 1991 organizational chart of the Ministry of Atomic Power and Industry of the former Soviet Union (which was reorganized into the Ministry of Atomic Energy in 1992) lists nine facilities under the Department of Uranium Mining and Ore Refining Industry. Three are in Russia: the Almaz Industrial Association in Lermontov near the Caucasus Mountains, the Zabaikal'sky Mining and Refining Complex in Pervomaisky, and Malyshevsk's Mining Utility in Malysheva. Two are in Kazakhstan: the Tsellinny Mining and Chemical Complex in Stepnogorsk and the Pricaspiysky [Caspian] Mining and Smelting Complex in Aktau (formerly Shevchenko). The four remaining sites, in Ukraine, Tadjikistan, Uzbekistan, and Kirghizstan, respectively, are the Vostochny [East] Mining and Chemical Complex in Zheltyye Vody ["Yellow Waters"], the Vostochny Rare Metal Complex in Tchkalovsk, the Navoi Mining Chemical Complex in Navoi, and the South Poly-

225. Lindberg 1992; Shifrin 1982.

226. Shifrin 1982, pp. 30–35.

227. Shifrin 1982, pp. 30–35.

228. As cited in Levine 1991, p. 10.

229. As cited in Levine 1991, p. 10.

230. Szymanski 1992, pp. 7–10.

metals Complex in Bishkek (formerly Frunze).[231] The latter was reportedly shut down in 1991.[232]

Papers presented by Russian scientists at international meetings of the OECD Nuclear Energy Agency and the International Atomic Energy Agency in 1991 and 1992 identify three other production centers: the Priargunsky (Argun) Mining and Chemical Complex in Krasnokamensk, Siberia; the Agrokimservis Company in Dneprodzerzhinsk, Ukraine (also reported to produce heavy water, zirconium, and hafnium);[233] and the Vostochny Rare Metal Complex, shut down in 1991, in Khudzhand (formerly Leninabad), Tadjikistan.[234]

Another major uranium production center, the Chemical-Metallurgical Industrial Association, operated in Sillamaë, Estonia.[235] Various other sources report unnamed production centers in the Stavropol region of the Caucasus[236] and in Uzbekistan at Toypeta and Uchkuduk.[237]

Soviet production is generally reported in the form of uranium concentrates, which typically contain 85 to 95 percent U_3O_8. Output of uranium concentrates at each of the 10 major processing sites in the former Soviet Union ranged from 400 to 4,000 metric tons per year, although no site-specific numbers are available. Total production of uranium concentrate for 1991 was reported at 13,500 metric tons—50 percent from Kazakhstan and 23 percent from Russia—with cumulative production estimated at between 240,000 and 440,000 metric tons.[238] Two of the primary production centers had closed as of 1992, and production in that year is estimated to have fallen to 10,500 metric tons.[239] Even the largest production center, the Prikaspiisky plant on the Mangyshlak peninsula in Kazakhstan, has cut back on uranium mining and processing.[240]

The uranium mining industry has been shifting to solution mining methods.[241] This method is to account for 40 to 50 percent of ura-

231. MINATOM 1991.

232. Syzmanski 1992, p. 11.

233. Potter and Cohen 1992, p. 27.

234. As cited in Szymanski 1992, p. 11.

235. Potter 1993, p. 11.

236. Ruzicka 1992, p. 19.

237. Potter and Cohen 1992, p. 30.

238. Szymanski 1992, pp. 9–10.

239. Szymanski 1992, pp. 10–11.

240. ECOTASS 1991.

241. Szymanski 1992, p. 7.

nium production by 1995.[242] Solution mining can be carried out at depths of up to 800 meters. Its cost is reportedly just one-tenth that of underground mining.[243]

Effects on the Soviet Union's Environment and the Public Based on the waste disposal practices of other nuclear facilities in the Soviet Union and other countries, operators of uranium mining and milling operations can be assumed to have released or buried radioactive wastes on site. The type and toxicity of wastes produced at each site varies with the mining methods used. Underground and open-pit mining have presumably left the largest quantities of mill tailings, while *in situ* underground leaching operations using sulfuric acid produce toxic liquid wastes.

A study of radiation levels throughout Estonia provides some estimates of the waste volumes collected at the Chemistry and Metallurgy Industrial Association plant in Sillamyae, which recovered and processed uranium from oil shale for the Soviet Union from the late 1940s until 1980. Conducted by Estonian officials and the Swedish laboratory Studsvik AB, the study describes an open radioactive waste storage reservoir only 15 meters from the Gulf of Finland, 20 meters above sea level.[244] Approximately 190 meters square, this reservoir leaks constantly, and the dam on its inland side has broken in the past. Only a 3-meter dam holds the reservoir on the Baltic side, where increased levels of radioactivity have been detected.[245] According to the company, this one impoundment holds 5 to 6 million metric tons of waste, including 1,000 metric tons of uranium, 500 of thorium, up to 30,000 of calcium fluoride, 600,000 of calcium sulfate, and 900,000 of oil shale ash particulates.[246]

Vladimir Nosov, the plant's laboratory manager, admits, "Of course, there is increased radioactivity in the sea outside the dam, but some kilometers off the coast, the levels are normal."[247]

The Estonian government announced plans in 1990 to dismantle and decontaminate the uranium production facilities and convert the site to the production of consumer products.[248]

As in the United States and East Germany, uranium mill tailings were sometimes used for a variety of housing and construction projects

242. As quoted in NUEXCO 1991a, p. 27.

243. Szymanski 1992, pp. 7, 10.

244. As reported in Sandback 1992 and Isherwood 1992.

245. Monterey Institute of International Studies 1992.

246. As reported in Sandback 1992 and Isherwood 1992.

247. As quoted in Isherwood 1992.

248. Kaazik 1990.

in the former Soviet Union. In the Ukrainian city of Zheltyye Vody, where uranium was mined and milled for 40 years, tailings were used to build foundations for homes and roads. The radioactive roadways were torn up and resurfaced in 1987, and a park and pioneer camp were decontaminated, but the city council is pressing for more to be done. It wants the republic's Council of Ministers to declare Zheltyye Vody an ecological disaster zone.[249]

Cleanup Prospects in the Former Soviet Union No other uranium mining areas of the former Soviet Union appear to have undertaken any comprehensive environmental decontamination or restoration projects.[250] Mining officials in Russia and the other former Soviet republics have little financial incentive or ability to begin cleaning up on the scale required, especially at a time when prices for raw uranium are low. The United States has agreed in principle to pay Russia for some of its highly enriched uranium (500 metric tons over 20 years),[251] but it appears unlikely that a significant fraction of the funds would be used for remediation, given many competing demands for the money.

Czechoslovakia

Czechoslovakia was the third largest supplier of uranium to the Soviet Union, with a cumulative production through 1990 of 98,500 metric tons. The first operating uranium mines were at Jachymov and Joachimsthal, in the Erzgebirge, at the northwest edge of the country. During German occupation from 1939 to 1944, 110 metric tons of pitchblende concentrates with 40 percent recoverable U_3O_8 were produced. The mines were reportedly "working normally" after the withdrawal of Russian troops in 1945.[252]

During the 1950s, several new deposits were found and exploited, most notably a high-grade deposit at Pribram, southwest of Prague, as well as deposits at Jihlava and other sites in Moravia.

By the end of 1990, the workforce had been reduced to about 12,000. Eight operational mines and three mills produced 2,400 metric tons that year. By 1991 production was down to 1,800 metric tons.[253]

Health Effects on Czechoslovakian Workers A follow-up study of several cohorts of miners in Czechoslovakia has been under way

249. Klimov and Shtengelov 1992.

250. Szymanski 1992, p. 11.

251. National Academy of Sciences 1994, p. 31.

252. U.S. Bureau of Mines 1945, 1946.

253. U.S. Bureau of Mines 1990; Uranium Institute 1992, p. 23.

since 1970. A recent publication details findings regarding the oldest cohort, those miners who began underground work between 1948 and 1957 (inclusive) and who had worked underground for at least four years.[254]

There were 4,042 miners in this group, excluding those no longer in Czechoslovakia and those for whom insufficient data existed. Non-occupational exposure to radon (for example, in residences) did not appear to be important among these miners on the whole. Although no information on the smoking habits of the whole cohort is available, in a sample of 300 miners, the proportion of smokers appeared about equal to the proportion of males smoking nationally. There was no reason to believe that the miners smoked more than males nationwide.

As of 1985, the observed number of lung cancer deaths was 574, almost five times the expected number of 122. The analysis suggested that the relationship between lung cancer risk and exposure was non-linear, with a greater risk per Working Level Month at low exposures, as noted in the Canadian studies above. Rates of mortality from other causes were not significantly different from national rates.

Reportedly, exposures were reduced considerably after the first years of mining by means of ventilation. Also, according to the above study, there was never extensive dry drilling in the Czechoslovak uranium mines, so silicosis has not been frequent among uranium miners. The few cases of silicosis per year are for the most part in miners who had previously worked in non-uranium mines.[255] By the 1980s, personal dosimeters were coming into use, and the intake of radon and its decay products was reportedly declining.[256]

Effects on the Czechoslovakian Environment and Public Radioactive dust related to uranium activities is the most severe problem at the sites of greatest handling and processing of the ore—mines, processing plants, and transportation routes. Data collected by Czech scientists indicates that rates of deposition of uranium-containing radioactive dust in some residential areas are greater than deemed safe by Czech authorities.[257]

A survey of radon in inhabited houses in Jachymov found that 62 percent of the homes had radon in concentrations greater than the "intervention limit" of 200 becquerels per cubic meter (5.4 picocuries per liter), 22 percent had more than 600 becquerels per cubic meter, and 3 percent more than 2,000 becquerels per cubic meter (the latter two categories are included in the 62 percent). Of 567 measurements

254. Sevc et al. 1993.

255. Sevc et al. 1993.

256. Vancl 1985.

257. Smetana and Jech 1992.

of drinking water in eastern Bohemia, 18 percent showed more than 1350 picocuries (50 becquerels) of radon per liter of water.[258] As one reerence point, the U.S. standard for drinking water limits radon to 300 picocuries (about 11 becquerels) per liter. As in other historic mining areas, however, part of this may be attributable to the natural radioactivity in the region or contamination from the mining of other minerals.

URANIUM SOURCES FOR FRENCH WEAPONS

France is second only to the United States in cumulative spending on domestic and foreign uranium exploration.[259] To develop its nuclear arsenal and fuel its massive nuclear power program, France has relied heavily on uranium deposits in its own national territory and in its former colonies in Africa. The first uranium used by the French nuclear weapons program came from the Belgian Congo, but already in the late 1940s, France, seeking independence in uranium, found and began developing its own deposits.[260] French prospecting in Africa turned up additional deposits in Madagascar, Gabon, and Niger. Most of their production has gone to France,[261] contributing to the expansion of its nuclear arsenal.[262]

France has come to rely increasingly on overseas sources, as is evident from figure 5.7. By 1989, imports constituted more than half of France's consumption.[263] Besides the increasing proportion of French uranium coming from Gabon and Niger over four decades, France also gets uranium from Canada, Australia, and the United States.[264]

The Commissariat à l'Energie Atomique (CEA), through Cogéma and its foreign subsidiaries, owns a major share of most of its overseas suppliers. These include two uranium mining companies in Niger, at least one in Gabon, five in Canada (one wholly owned by Cogéma), three in Australia (one wholly owned by Cogéma, although Cogéma does not actually produce uranium in Australia), two wholly owned subsidiaries in the United States, and five in France.

258. Benes 1991, pp. 169–173.

259. Owen 1985, p. 102.

260. Barrillot 1991, part 2, section 2.11.

261. Owen 1985, pp. 105, 109; Neff 1984, pp. 172, 205; U.S. Bureau of Mines 1961, 1963, 1965, 1966, 1967, 1968, 1970; Huisman 1990.

262. Neff 1984, p. xxi.

263. Owen 1985, p. 102.

264. Barrillot 1991, part 2, section 2.13.

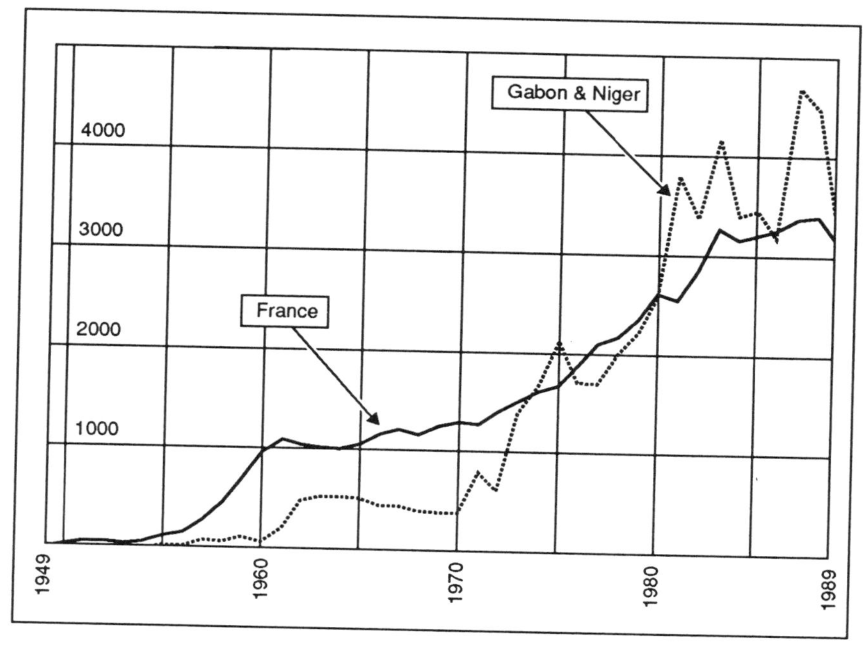

Figure 5.7 French production of uranium in metric tons. Source: Barrillot 1991, p. 31.

France

Most of France's uranium mines are located in the regions of Brittany, the Massif Central, the Vendée, and the Haute-Vienne (see table 9.1).

France has exploited both surface and underground mines. Poorer ores are piled up on site and treated by heap leaching. Other ores are processed in conventional uranium mills.

Although many uranium mines have been opened in France, their deposits have turned out, in general, to be small and relatively low-grade, with few prospects for expansion. Indeed, domestic production is now falling.

With the world market for uranium declining, Cogéma announced in 1991 that it was seeking to reduce its production costs by closing its Le Chardon and La Commanderie mines. The number of miners employed by Cogéma has fallen by more than a third in recent years, from 3,300 at the end of 1988 to 2,100 at the end of 1990.[265]

Uranium Milling in France Société Industrielle de Minerais de l'Ouest (the Industrial Company for Western Ores), or SIMO, a wholly owned subsidiary of Cogéma, built the first industrial-scale facility in France for concentrating uranium ore into yellowcake in 1957 in Gétigné in the Loire Atlantique region. SIMO built France's

265. Barrillot 1991, part 2, section 2.13.

second and largest facility in 1958 in Bessines, Haute-Vienne, with a capacity of 1,500 metric tons per year. In 1960, SIMO built the Bois Noirs plant in St.-Priest-la-Prugne, Loire, which operated for 20 years, producing 6,400 metric tons of uranium.[266]

Cogéma built specialized plants in 1961 and 1965 to complete processing of ores mined in Africa. The first, in Gueugnon, Sâone-et-Loire, treats ore from Gabon; between 1955 and 1961, it processed French ores mined in the Morvan.[267] The second, at the Le Bouchet plant, acquired by CEA in March 1946, processed only uranothorianite ore that had already been sorted in the washing plants of the Madagascar mines. It closed in 1971. Heap-leaching facilities to recover uranium have been built adjacent to mines in La Chapelle-Largeau, Vendée; Fanay, La Crouzille; Margnac, La Crouzille; and Langogne, Lozère.

SIMO had a monopoly on the yellowcake market for domestically mined uranium until 1979, when Total Compagnie Minière (TCM) built a 500-metric-ton-per-year plant called Mailhac-sur-Benaize in Jouac, Haute-Vienne. After some further expansion, TCM agreed in 1993 to exchange all its French and foreign uranium operations for a 10.8 percent share of Cogéma, making it Cogéma's first private shareholder. Under the deal, Cogéma also paid TCM 2.52 billion francs (about $500 million) for 4.3 percent of its stock. Cogéma is now the world's largest uranium supplier and the only company involved in every stage of the nuclear fuel cycle, from uranium mining to reprocessing.[268]

Health Effects on French Workers In the early 1980s, the CEA and Cogéma undertook a follow-up study of mortality among more than 2,000 mine workers. Periodic assessments have been made, some of which have been reviewed by Roger Belboech;[269] the following discussion relies on his work.

The assessment as of the end of 1985, as reported by Dr. Chameaud of Cogéma, found elevated standardized mortality ratios (SMRs), especially where the cause of death was lung cancer or cancer of the larynx. Table 5.5 summarizes Cogéma's data on standardized mortality ratios with respect to various causes of death for the whole group, as well as for two subgroups: those who began working in the mines before 1956 (mean observation time: 28.6 years) and those who started work between 1956 and 1972 (mean observation time: 23.7 years). No confidence intervals or tests of statistical significance were provided by Belboech.

266. Cogéma 1991, p. 40.

267. Cogéma 1991, p. 27

268. WISE 1993b.

269. Belboech 1991, pp. 6–9.

Table 5.5 Standardized Mortality Ratios in French Uranium Miners

Causes of Death	Full Group	Miners Starting Work Before 1956	Miners Starting Work 1956 to 1972
All causes	1.14	1.23	1.02
All cancers	1.30	1.44	1.12
Lung cancer	2.26	2.71	1.83
Cancer of the larynx	2.00	2.45	1.83

Source: Belboech 1991, p. 6.

It is necessary to account for smoking in order to carry out an accurate dose-response analysis or to estimate the amount of risk attributable to radiation exposure among miners, as pointed out by Tirmarche and fellow researchers of the CEA and Cogéma.[270] This was not done. However, Belboech argues that miners do not differ substantially from the general population in their smoking habits. Thus, the greater incidence of death from lung and larynx cancers observed among miners may well be due to mining. However, synergistic effects between smoking and radon inhalation are known to exist, and this complicates and renders more uncertain any evaluation not explicitly controlled for smoking.

French legislation on radiation exposure limits has tended to be somewhat lax. In the 1980s, when the U.S. occupational limit for internal radiation exposure was 4 working level months, France's limit was effectively 9 times this. Only in July 1989 did France pass legislation limiting occupational exposure to external and internal radiation to 50 millisieverts per year, in agreement with ICRP recommendations of the time. Statements by officials of the CEA's Institute of Nuclear Protection and Safety published in 1984 may explain the tardiness: "One can see that the increase in stringency of the future regulation will lead to some miners being put in a situation of exceeding the individual dose [limits] in most of the underground galleries."[271]

As noted earlier, in 1990 the ICRP revised its recommended radiation exposure limits downward to an occupational limit of 20 millisieverts per year averaged over five years. Like the United States and many other countries, France has not yet revised its exposure limits in accordance with this.[272]

Effects on the French Environment The 63,000 metric tons of uranium produced in France through January 1991 have created an

270. Tirmarche et al. 1988, as cited in Belboech 1991, p. 6.

271. *Gazette Nucléaire* 1991h; translated from the French.

272. *Gazette Nucléaire* 1991h.

estimated 20 million metric tons of conventional radioactive mill tailings, 17 million metric tons of tailings treated by heap leaching, and 26 million metric tons of finely ground tailings from uranium ore concentration plants.[273] Although the radioactivity of the untreated portion is not known, the heap-leaching piles were estimated to still contain about 1,400 metric tons of natural uranium and about 1500 curies (54 terabecquerels) of radium-226 in 1991; the tailings from production plants hold about 3,000 metric tons of uranium and about 19,000 curies (700 terabecquerels) of radium-226.[274] Presumably, the amounts of radioactivity from thorium-230 and radium-226 in the tailings were approximately equal, since that is generally a characteristic of uranium ore.

Although the bulk of this radioactivity is concentrated at a few sites—Gétigné in Loire-Atlantique; Bessines in Haute Vienne; St. Priest-la-Prugne in Loire; and Bersac in Haute Vienne—there are at least 15 sites where wastes are stored, in unlined pits, open reservoirs, and underground mines, and only one of these sites is properly licensed. Nine contain more than 1,000 curies (37 terabecquerels) of radium-226, which means that by law they should be regulated as "Basic Nuclear Installations." These sites have avoided regulation, however, through a legal provision that allows mining companies to characterize their mill tailings as "natural uranium."[275]

Tailings not used to backfill mines are normally supersaturated with water to form a mud-like slurry that is poured into settling basins for decantation or evaporation. Overflow from these basins is treated with barium chloride (to form insoluble barium-radium sulfate), effectively removing radium from the overflow and lessening its impact on streams and rivers.[276]

One such dump is located near Bessines in Haute-Vienne. Known as Lavaugrasse, the 25-hectare site contains about 5.7 million metric tons of tailings from uranium mills in Bessines and Le Bouchet. The amount of long-lived radioactivity from radium-226 is estimated at about 3,780 curies (about 140 terabecquerels).[277] The tailings basin is mostly covered with water, which is held back by a dike also built from tailings. Water from the site drains into the Gartempe River. Gamma radiation readings taken 30 centimeters above exposed tailings at Lavaugrasse in 1991 detected 4.5 micrograys per hour, about 30 to 45 times natural background radiation.

273. *Gazette Nucléaire* 1992a; Davis 1992, p. 1.

274. *Gazette Nucléaire* 1992a.

275. Davis 1992, p. 1.

276. Davis 1988, p. 96.

277. *Gazette Nucléaire* 1992a.

Approximately 5.8 million additional metric tons of uranium mill tailings from Bessines and Le Bouchet are stored in an open pit at the abandoned Brugeaud uranium mine in Haute-Vienne. The wastes contain an estimated 3,375 curies (about 125 terabecquerels) of radium-226, as well as high levels of lead and arsenic.[278]

Regulations Governing French Mines and Public Health Environmental regulations regarding protection of the public and the environment from the effects of uranium mining in France went into effect on 9 March 1990. These regulations impose certain constraints on uranium mining, but are not retroactive and cannot be applied to mines already in operation. The maximum allowable dose to a member of the public is 5 millisieverts (500 mrem) per year, compared to the maximum limit of 1 millisievert (100 mrem) per year recommended by the ICRP. Levels of radon and radioactive dust need only be measured once a year. Furthermore, there are loopholes. Article 9 of the regulations, concerning containment of liquid effluents, and Article 10, concerning the minimum distance required between ventilation discharge sites and any home, can be circumvented with an authorization from the local authorities. And Chapter 3, Article 8 of the decree entrusts the mine administration with assessing the impact on the public of its own solid-waste management, with no independent evaluation of the possible risks.[279]

Although environmental impact statements are required for mines opening after 1977, the restoration of abandoned mines is not explicitly required by national law but rather left to the prefectures to impose or not.

Niger and Gabon

In 1946, France's newly established Commissariat à l'Energie Atomique (CEA) launched "the first modern uranium exploration program,"[280] sending people with Geiger counters to French colonies in Africa in search of uranium.[281] CEA's reconnaissance effort spanned all of Francophone Africa, from Mauritania to the Congo.[282] Discoveries were made in Gabon in 1956 and Niger in 1965, and production began at the Mounana deposit in Gabon in 1961 and the Arlit deposit in Niger in 1971. France has been and remains the dominant buyer of

278. *Gazette Nucléaire* 1991, 1991a, and 1992a.

279. *Gazette Nucléaire* 1991i.

280. Tona 1986, p. 172.

281. Moody 1992, p. 178.

282. Tona 1986, p. 172.

uranium from these two countries and is said to have developed its nuclear arsenal in part from this supply.[283] Japan, Spain, Germany, and Italy also buy Niger's uranium,[284] and Japan and Italy have bought small amounts of Gabon's.[285]

Published information on health and environmental conditions at the mines and mills of Gabon and Niger is scarce. However, one can infer that conditions have not been good. According to one researcher, "None of the governments of the [African] uranium producing countries have an environmental policy related to uranium mining and milling."[286] The French state company Cogéma extracts uranium in Gabon and Niger, yet is not required to submit to the health and environmental regulations it is supposed to abide by in France.[287] Occasional journalistic accounts have appeared that chronicle the unsafe conditions. For example, in the early 1980s a BBC television crew visited the mine at Arlit. There it found 15- and 16-year-old Tuareg boys emerging dust-covered from the mines.[288]

URANIUM SOURCES FOR CHINESE WEAPONS

Uranium for China's nuclear weapons comes from domestic sources. Although the Chinese searched for uranium as early as 1934, their uranium prospecting campaign was officially initiated with the signing on 20 January 1955 of a secret Sino-Soviet protocol creating a joint uranium-prospecting organization. By the end of 1956, more than 20,000 people were involved in prospecting. The first mines were established in the contiguous southern provinces of Hunan, Jiangxi, Guangxi, and Guangdong. The Chengxian uranium mine, in the steamy malaria belt of southern Hunan Province, was partially operational in 1960, making it the first to open.[289] Two other early mines were the Dapu mine in Hunan Province and the Shangrao mine in Jiangxi Province. The Soviet Union designed all three mines.[290]

As uranium ore became available, mills were constructed to process it. Plans for the mass production of oxides were developed in the late 1950s. Hengyang Uranium Hydrometallurgy Plant, near the city of Hengyang on the Xiang River in Hunan Province, was the first such

283. Neff 1984, p. xxi.

284. Huisman 1990, p. 11.

285. Owen 1985, p. 105.

286. Huisman 1990, p. 2.

287. Huisman 1990.

288. In Moody 1992, p. 721; traditionally, the Tuaregs are nomads.

289. Lewis and Xue 1988, pp. 76, 85, 86; Fieldhouse 1991, p. 53.

290. Lewis and Xue 1988, p. 78.

plant to be constructed. Production began in 1962, although it was not until late 1963 that an acceptable level of purity was achieved and the plant went into mass production. A uranium mill was also built at Shangrao, Jiangxi Province. Eventually, processing plants were built at most of the mines in order to avoid the problems and expense of transportation and reduce the contamination of ores.[291] Hengyang was still operational as of 1991, using the acid leach process, with a production capacity of 1,100 to 1,400 metric tons per year.[292]

By 1963, the Chinese had eight uranium mines in operation,[293] including (in addition to the above) Maoshan and Zhushan in Guangxi Province, Xiazhuang in Guangdong Province,[294] and Linxian in Hunan Province.[295] By 1967, two major mining and metallurgical complexes had been built, one in Fuzhou, Jiangxi Province, and the other in Guangdong Province.[296]

In the 1980s, 26 uranium mines were operating in 24 provinces and autonomous regions.[297] NUEXCO lists eight operating mills: Hengyang and Zhuzhou in Hunan, Shao Kuan in the Jiangxi-Guangdong uranium belt, Hengjian in Zhejiang, Guizhou in Guizhou, Huxian in Shaanxi, Lianshanguan in Liaoning, and Urumqi in Xinjiang.[298]

Health Effects on Chinese Workers As in other countries, the emphasis in the late 1950s and early 1960s was on finding and extracting uranium, not on safety. Radon was a serious problem in the Chenxian mine from the beginning. Also, lack of water for drilling made for high levels of airborne dust and attendant health problems such as silicosis. Sustained and effective efforts toward reducing radiation hazards to miners were made at the national level only two decades after the start of mining.[299]

Methods of gathering uranium other than underground mining were used during the Great Leap Forward of 1958–1959, with their own implications for health. By 1958, geological teams had found several small, shallow deposits of good quality uranium ore. In mid-1958 the government announced, "The whole people should engage in uranium

291. Lewis and Xue 1988, pp. 85, 95, 97.

292. Low figure from Nuclear Engineering International 1991, p. 151; high figure from NUEXCO 1991, p. 22.

293. Lewis and Xue 1988, p. 87.

294. U.S. Bureau of Mines 1965, p. 983.

295. Fieldhouse 1991, p. 55.

296. Lewis and Xue 1988, p. 80.

297. Lewis and Xue 1988, p. 268.

298. NUEXCO 1991, pp. 18, 19, 21.

299. Lewis and Xue 1988, p. 86.

mining." Masses of people were mobilized, as was characteristic of the Great Leap Forward, in prospecting and in collecting uranium from shallow deposits near the surface. Peasants were provided with scintillation and Geiger counters and instructed in simple methods for mining surface ores. As of September 1959, in Hunan Province alone tens of thousands of peasants had joined the prospecting effort, which covered 95 percent of the province.[300] Most peasants were presumably unprotected radiologically. It was the Chinese version of the U.S. uranium boom on the Colorado plateau.

This mass prospecting movement, which lasted until 1961, produced the first 150 metric tons of uranium concentrates, and it was this uranium that went into China's first bomb, rather than uranium from underground mines, which were fully operational only in 1962 or 1963.[301]

The Sino-Soviet break and the departure of Soviet technicians in 1960 led to large number of small-scale efforts to mine uranium. Although there may have been isolated efforts by some officials to provide information about health protection measures,[302] it seems doubtful that there was much public awareness in that period of the health risks posed by uranium's decay products.

Effects on the Chinese Environment Lewis and Xue report that the decentralized surface mining during the mass mobilizations of the late 1950s "produced serious environmental pollution because of the near absence of environmental protection equipment."[303] We have no more specific information about environmental effects of uranium mining in China.

300. Lewis and Xue 1988, pp. 87, 88.

301. Lewis and Xue 1988, pp. 87–88.

302. Lewis and Xue 1988, p. 90.

303. Lewis and Xue 1988, p. 88.

6 The United States

Arjun Makhijani, A. James Ruttenber, Ellen Kennedy, and Richard Clapp

In late 1938 and early 1939, experiments in Germany and the United States confirmed that uranium could be split into fragments by bombarding it with neutrons; fission, in other words, was possible. The potential military implications of this discovery were clear and were discussed in the press. Internationally, scientists and governments feared that Germany, with its advanced scientific capabilities in this area, would outpace others in making an atom bomb for use in a war it was already pursuing relentlessly in Eastern Europe. As a result, scientists in the United States, Britain, and elsewhere began intense investigations of fission.

Leo Szilard, an emigrant scientist from Hungary, persuaded Albert Einstein to sign a letter to President Franklin D. Roosevelt asking for a research effort on nuclear fission. On 11 October 1939 Alexander Sachs, a vice president of a Wall Street firm, Lehman Corporation, presented this letter to President Roosevelt. As a result of this meeting, the president formed the Advisory Committee on Uranium, headed by a government scientist, Lyman J. Briggs. By doing so Roosevelt took the first major step to establishing a new role for the U.S. government: large-scale funding of basic scientific research for military purposes.[1]

By 1940, Germany had invaded Belgium and France and was bombing London. There had been progress on the understanding of chain reactions (the first chain reaction was accomplished on 2 December 1942) and even on nuclear reactor design concepts. The Committee on Uranium was brought under the umbrella of the National Defense Research Committee (NDRC), headed by Vannevar Bush, a former vice president of the Massachusetts Institute of Technology and president of the Carnegie Institution. In the same year, a parallel British effort, code-named the MAUD Committee, established a liaison with the United States. The subsequent bomb-building project was

1. Hewlett and Anderson 1990, pp. 14–20. This official history of the 1939–1946 period is the source for dates and events in this introductory section, unless otherwise mentioned.

conducted largely as a joint U.S.-British effort, with the British in a junior role.[2]

By mid-1941, the research project was transferred to an organization with greater authority, namely the Office of Scientific Research and Development (OSRD). The OSRD was part of the Office of Emergency Management in the Executive Office of the President, and according to the official history of the period, its mission was "to serve as a center for mobilizing the scientific resources of the nation and applying the results of research to national defense."[3] At this time, the project acquired a new name—the S-1 Section.

Central to this new effort was research on the enrichment of uranium and experiments to prove that a nuclear chain reaction could be sustained. Another line of crucial research also developed in mid-1941. In July, Emilio Segré and Glenn Seaborg established that the element plutonium had a higher probability than natural uranium of fissioning with fast neutrons (plutonium was the fissile material used in the first atom bomb). President Roosevelt made a decision on 9 October 1941 to commence full-scale research on the possibilities of building atomic weapons.[4] The United States had not yet entered the war.

Within a year after the Japanese bombing of Pearl Harbor (7 December 1941), which precipitated the U.S. entry into the war, there was no doubt that a nuclear bomb could be made. A chain reaction had been achieved. Electromagnetic separation devices for uranium enrichment had been built. And there had been enormous progress toward designing reactors (called atomic piles) to produce plutonium as well as a chemical separation process to extract plutonium from irradiated uranium fuel.

The next step was to build the factories for enriching uranium and for producing and separating plutonium and the facilities for designing and testing an atom bomb. In September 1942, Leslie R. Groves, a colonel from the Army Corps of Engineers, was appointed to head up a vastly expanded construction and engineering effort to produce the materials, create the designs, and actually fabricate and test the first atom bombs. He was promoted to Brigadier General at that time.

Three sites were crucial to the bomb-building effort. Two of them, Hanford and Oak Ridge, are discussed in more detail below. The site for the third, the Los Alamos Scientific Laboratory, near Santa Fe, New Mexico, was selected by General Groves in 1942. It began operations in 1943. Some of the laboratory was sited on lands traditionally used by

2. Sherwin 1987, chapter 1.

3. Hewlett and Anderson 1990, p. 41.

4. Hewlett and Anderson 1990, p. 52.

Pueblo Indians. The official history of the period does not mention the Pueblo use but refers to the land situation as follows:

> It would have been difficult to find a place more inaccessible than this mesa with its steep rock walls and execrable roads. Plenty of space for safe testing lay deep in adjacent canyons. Most of the land required was public domain, while the rest, valuable only for grazing, would cost little.[5]

The first nuclear weapon was exploded in the New Mexico desert on 16 July 1945. The bomb's plutonium was made at Hanford; the bomb itself was designed and fabricated at Los Alamos.

FROM WORLD WAR II TO THE COLD WAR

It is generally assumed that atomic weapons were used on Japan because they were not ready at the time the war against Germany ended. While it is true that atomic weapons could not have been used on Germany for this reason, a decision not to target Germany had been made long before. The first discussion of targeting on record appears to have taken place on 5 May 1943 at a meeting of the Military Policy Committee, which was established to coordinate the military and scientific aspects of the project. (However, its members included no senior military commanders, most of whom were not even aware of the existence of the Manhattan Project.)[6] General Groves recorded the discussion on targeting as follows:

> The point of use of the first bomb was discussed and the general view appeared to be that its best point of use would be on a Japanese fleet concentration in the Harbor of Truk [an island in the Pacific Ocean]. General Styer suggested Tokio [sic] but it was pointed out that the bomb should be used where, if it failed to go off, it would land in water of sufficient depth to prevent easy salvage. The Japanese were selected as they would not be so apt to secure knowledge from it as the Germans.[7]

Germany was not targeted because of its presumed greater knowledge of nuclear physics; Japan was targeted because it would probably not learn as much from an unexploded bomb. Tokyo or other Japanese cities were not targeted at the time because the loss of a nuclear weapon to the enemy was seen as unacceptable. This is possibly the first example of nuclear deterrence in history. A potential nuclear power, Germany, seems to have deterred another potential nuclear power, the United States, from targeting a weapon against it.

5. Hewlett and Anderson 1990, pp. 229–230.

6. Makhijani and Kelly 1985, chapter 5.

7. Groves 1943.

(See chapters 11 and 12 for a summation of the proliferation effects of deterrence strategy.)

Americans tend to associate the end of World War II and the surrender of Japan with the atomic bombings of Hiroshima and Nagasaki in August 1945. The bombings were greeted with relief by most Americans. An invasion of Japan that many people supposed would be necessary to end the war, and that they feared would be a brutal affair, was not required.

U.S. leaders later fed this idea with grandiose statements about the role the bombings played in ending the war and supplied very large estimates of the number of lives saved. According to Harry Truman's memoirs, General George Marshall, head of the Joint Chiefs of Staff, told the President that the bombings of Hiroshima and Nagasaki had saved half a million lives.[8] Winston Churchill put the number at one million American and half a million British lives.[9] Such pronouncements generally omitted the role of the Soviet Union's entry into the war against Japan, which came between the two atomic bombings. This lack of perspective in the comments of western leaders had an important role in convincing the U.S. public that nuclear weapons had been central to ending World War II, that they were proper instruments of war more generally, and, therefore, that they ought to have a large role in the postwar U.S. arsenal.

Accompanying official statements was silence about the fact that most of Japan's leaders were ready to surrender by mid-July 1945—as U.S. leaders knew at the time.[10] Nor do the published estimates correspond to military estimates made during the war. Rather, they apparently originated in a comment by an unidentified economist friend of Secretary of War Henry Stimson. But when Stimson asked the Joint Chiefs of Staff for an evaluation, they concluded in June 1945 that "the estimated loss of 500,000 lives due to carrying the war to a conclusion under our present plan of campaign is considered entirely too high." General Marshall expressed "general agreement" with this staff evaluation and forwarded it to Stimson.[11]

The military's own projection, made in June 1945, of the number of American deaths in the final campaign against Japan ranged from 25,000 to 46,000.[12] This figure is supported by a look at the war's actual figures. The total number of Americans killed in action in both Europe and the Pacific during the entire war was about 300,000, with

8. Truman 1955, p. 417.

9. Churchill 1953, pp. 638–39.

10. Alperovitz 1985; Makhijani and Kelly 1985, pp. 24–27.

11. Handy 1945; Marshall 1945.

12. U.S. Joint Chiefs of Staff 1945, p. 7.

almost two-thirds of this number in Europe and the Mediterranean area. The fierce Pacific operations under MacArthur between 1 March 1944 and 1 May 1945 resulted in the deaths of 13,742 U.S. soldiers.[13]

In fact, there is evidence that the military disputed a proposed statement of the Secretary of War in September 1945 that the atomic bomb saved 200,000 American lives, but decided not to take issue with it.[14] And in time the numbers escalated, as a range of half a million to a million became a part of popular mythology. Relief over the end of the war merged with the atomic bombings to start what Robert Alvarez, a Deputy Assistant Secretary in the Department of Energy, has called "America's romance with the atom."[15]

Congress and the Manhattan Project

Pulling exaggerated estimates out of thin air was not an idle exercise, born of elation at the war's end. It related to a great fear held by many of those responsible for the Manhattan Project: their immense effort could come to naught if the war ended before the United States could demonstrate that the bomb was a decisive weapon. This, in turn, was linked to the post–World War II role of nuclear weapons. As Tufts University historian Martin Sherwin has remarked, "The race for the bomb ... changed from a race against German scientists to a race against the war itself."[16]

Congress was a prime audience for the bombings. During the war, Congress had known only that the Manhattan Project was a sinkhole for money and other resources. Many members of Congress were concerned, and some were upset, that a project that seemed to contribute nothing to the war effort had so much funding—and no accountability to them.

One of those who was unhappy was then-Senator Harry Truman, a member of the Special Committee Investigating the National Defense Program. In 1944, he wanted to send a general as his own representative to investigate the huge project near Pasco, Washington, that was to become the Hanford plant and the source for plutonium for the Nagasaki bomb. Stimson resisted. "It may be necessary for the Committee to consider the appointment of a subcommittee to investigate the project," an irritated Truman responded to the Secretary of War. "On your urgent request, that usual procedure will not be adopted at this time. The responsibility therefor and for any waste or improper

13. Hull 1945.

14. For more on the decision to drop the atomic bomb on Japan, see Makhijani and Kelly 1985.

15. Robert Alvarez, personal verbal communication to Arjun Makhijani, 1988.

16. Sherwin 1987, p. 145.

action which might otherwise be avoided rests squarely upon the War Department."[17]

In February 1945, Undersecretary of War Robert P. Patterson informed Stimson about a "growing restlessness and impatience among Members of Congress about the size and cost of the project and also on account of the fact that they can find out nothing officially about the nature of it."[18] Secrecy won the day, and almost all members of Congress as well as military commanders were not aware of the purpose of the Manhattan Project.[19] As a result, leaders of the Manhattan Project felt tremendous pressure to demonstrate to Congress that the bomb would be decisive. General Leslie Groves, head of the effort, quoted a special assistant to Undersecretary of War Patterson to this effect: "[If] the project succeeds, there won't be any [congressional] investigation. If it doesn't, they won't investigate anything else."[20]

A further motivation for promoting the role of the atomic bomb in ending World War II was postwar funding, which would relate directly to perceptions of the success of the Manhattan Project. Leo Szilard, who had a large role in initiating the U.S. atom bomb project and who sought to stop the arms race after the war, recalled a comment of Secretary of State James Byrnes: "How could you get Congress to appropriate money for atomic energy if you do not show results for the money which has been spent already?"[21]

The Buildup of U.S. Nuclear Weapons

Thus the atomic blasts that killed at least 200,000 people in Hiroshima and Nagasaki helped lay the political foundation for the U.S. nuclear weapons complex. In the years following World War II, the United States designed and built vast numbers of nuclear weapons, ranging from compact battlefield weapons to immense thermonuclear bombs. By 1967, the arsenal reached a peak of 32,000 nuclear warheads, ranging from a few kilotons to megatons. The size of that arsenal declined slowly to about 21,000 warheads by 1990. Since then, arsenal reduction has accelerated due to the end of the Cold War, bilateral treaties, and mutually reciprocated unilateral U.S.-Russian measures to reduce stockpiles. The U.S. nuclear arsenal in 1993 was estimated at 16,750 warheads.[22]

17. Truman 1944.

18. Patterson 1945, p. 2.

19. Makhijani and Kelly 1985, chapter 2.

20. Jack Madigan, as quoted in Groves 1962, p. 360.

21. As quoted in Sherwin 1987, p. 202.

22. Norris and Arkin 1993e.

In the 1950s, U.S. military planning drew from the experience of World War II strategic bombings and from the nuclear attacks on Hiroshima and Nagasaki. Thus, in the mid-1950s, the Strategic Air Command, headed by General Curtis LeMay, had a targeting plan that would use 750 nuclear bombs in a simultaneous strike on the Soviet Union: "Virtually all of Russia would be a smoking, radiating ruin at the end of two hours."[23] After the launch of Sputnik in 1957 started the missile race, there was a proliferation of targets and with it the demand for a larger quantity of nuclear weapons. As the targets multiplied, the United States raced to produce thousands of nuclear weapons for massive bombing. Between 1959 and 1961, warhead production reached a rate of about 5,000 to 6,000 a year.[24]

In the 1970s, the United States shifted its emphasis from producing ever more nuclear weapons to making them more accurate. While the size of the arsenal edged down between the mid-1960s and the late 1980s, the demand for new nuclear weapons continued. For example, the Pentagon developed missiles that could carry more than one nuclear warhead, calling for lighter and more compact nuclear weapons. These weapons could be independently targeted, so one missile could destroy many different targets. Known as "multiple independently targetable reentry vehicles" (MIRVs), these weapons initiated a new era in nuclear weapons strategy because they increased the possibility that a preemptive first strike could destroy an enemy's land-based missiles. At the same time, new battlefield nuclear weapons (called neutron bombs) were developed to destroy enemy troops with limited damage to property.

With the demand for new weapons, the United States increasingly dismantled old ones to recycle their weapons-grade plutonium, highly enriched uranium, and tritium. Dismantling primarily took place at the Pantex plant near Amarillo, Texas and also at the Y-12 plant at Oak Ridge, Tennessee. The Rocky Flats plant in Colorado reworked plutonium, while the Y-12 plant reworked highly enriched uranium.

Since 1989 dismantling weapons has outpaced building new ones. No new weapons are being produced as of 1993, though design work continues. This shift is due to the disintegration of the Soviet Union, the end of the Cold War, treaties to reduce strategic nuclear arsenals, and reciprocal unilateral decisions by both the United States and the former Soviet Union to reduce their arsenals of tactical nuclear weapons. As a result, Pantex's main function is to dismantle nuclear weapons and store their plutonium triggers. Oak Ridge dismantles and stores the highly enriched uranium parts.

23. War planning document as quoted in Rosenberg 1982.

24. Cochran et al. 1987a, p. 19.

The United States also stopped producing new plutonium and tritium for nuclear weapons in the late 1980s. At that time, the Department of Energy shut the last weapons-related reactors because of health, environment, and safety concerns. And despite very large expenditures dedicated to trying to restart some of these reactors, notably the K-Reactor that produced weapons materials at the Savannah River Site in South Carolina, the United States has scrapped plans to start up old reactors or to build new ones for producing plutonium or tritium. (Russia continues to produce plutonium—see chapter 7).

Institutional Background

Until 1946, the U.S. Army was responsible for producing nuclear weapons under the Manhattan Project. That year, Congress passed the Atomic Energy Act, also known as the McMahon Act, establishing the Atomic Energy Commission (AEC). This was to be a civilian body, in part because it seemed preferable to have civilian control of nuclear weapons production. Such control was also deemed desirable in view of the contemplated widespread civilian uses of atomic energy. The Joint Committee on Atomic Energy was set up to exercise congressional oversight over the agency. Both opened their doors in 1947.

The Department of Defense was, of course, the "customer" for nuclear weapons. Therefore, a joint council of the AEC and the Department of Defense, called the Military Liaison Committee, was established to determine the numbers and types of weapons that the AEC would produce for the military. This body is now called the Nuclear Weapons Council.

The AEC carried out testing of nuclear weapons, but this was not exclusively an agency affair. The armed forces were involved on a large scale, with their personnel being the responsibility of the Armed Forces Special Weapons Project, established in 1947.[25] This body is now called the Defense Nuclear Agency.

The AEC owned and regulated the nuclear weapons plants. It also promoted and regulated civilian nuclear power. Both situations created conflicts of interest that were a principal factor behind the AEC's production-first attitude and its downplaying of its health and environmental responsibilities.

The rise of environmental awareness in the 1960s and early 1970s and the coming to light of serious safety issues in nuclear reactor design in the same period created a demand for separating AEC's regulatory function for civilian power from its weapons-related functions. The 1973 oil crisis also heightened the need for an agency with a

25. Hewlett and Duncan 1990, p. 131.

broad energy mandate. With these issues in mind, Congress passed the Energy Reorganization Act in 1974, establishing the Energy Research and Development Administration (ERDA) and the Nuclear Regulatory Commission (NRC). ERDA would manage nuclear weapons production and energy programs (including research and development on nuclear power), while the NRC would regulate civilian nuclear power programs and license facilities. However, this did not resolve the conflict of interest inherent in the self-regulation of the weapons program. (The legacy of this split on the civilian regulatory side is beyond the scope of this work.)

In 1977, President Carter created the cabinet-level Department of Energy out of ERDA and gave it a broad mandate for all energy programs, including energy efficiency and renewable energy sources, in addition to the more traditional role of civilian nuclear energy research and development. However, a central function of DOE and the largest single portion of its budget continued to be nuclear weapons production and associated design and testing activities.

Until the mid-1980s, DOE claimed it was exempt from all laws regulating civilian industry on the grounds that its responsibilities derived from the Atomic Energy Act (and amendments). However, a lawsuit forced DOE to bring its facilities in compliance with the Resource Recovery and Conservation Act. In 1992, Congress further applied the operating norms of civilian industry to weapons facilities by passing the Federal Facilities Compliance Act, requiring federal facilities to comply with civilian laws pertaining to environmental protection. This act allows for grace periods to bring them into compliance.

The Future of the U.S. Weapons Complex

The U.S. nuclear weapons complex has three principal functions today. The first is to attend to the complicated and expensive environmental legacy of past nuclear weapons production and testing. DOE is planning or implementing a variety of options for cleaning up and storing wastes from the weapons complex, including two waste repositories. Most notable are the Waste Isolation Pilot Project in New Mexico for disposing of transuranic wastes and the Yucca Mountain project in Nevada to dispose of high-level waste. The latter, which is still in the characterization phase, is also meant to hold high-level waste from civilian nuclear reactors.

The second major task is maintaining whatever arsenal political and military leaders decide that U.S. policy requires. This function has come to be called "stockpile stewardship." Existing weapons will require regular checks for safety and integrity. Tritium, which decays with a half-life of 12.3 years, will be restocked to keep existing weapons operable.

Finally, the DOE must dismantle nuclear weapons no longer required for the arsenal, store the fissile materials, and implement decisions regarding their ultimate fate.

Some expenditures of DOE's Office of Environmental Restoration and Waste Management appear to be thinly disguised production activities, including producing plutonium from irradiated fuel and plutonium scrap. Despite the essential shutdown of nuclear warhead production, DOE continues to spend about $5 billion a year on nuclear weapons design, testing, and production out of its budget for the nuclear weapons complex of about $11 billion a year. (About $6 billion a year is being spent on Waste Management and "environmental restoration.") The department maintains a number of plants on standby status.

Experimentation on Human Subjects[26]

Radiation experiments conducted on human subjects epitomize the history of the nuclear weapons complex in the United States, especially during the period from the 1940s through the 1960s. We include a brief discussion of some of the experiments here since they have often been overlooked in histories of the U.S. atomic age. In addition, this account may inspire researchers in other countries to investigate allegations about experiments in their own countries.

U.S. Secretary of Energy Hazel O'Leary announced on 7 December 1993 that the nuclear establishment had conducted radiation experiments on humans since the 1940s. It was a stunning admission—the first time that the head of a nuclear weapons establishment stood before the people it was pledged to protect to admit that it had been experimenting upon them.

It is apparent that other agencies beyond the Department of Energy were involved in human radiation experiments.[27] For example, the Department of Defense funded whole-body irradiation of terminally ill charity patients in the University of Cincinnati Hospital during 1960–1972.[28] Some of the experiments were to determine how well troops could function in the nuclear battlefield; others were for studying the metabolism of radionuclides; yet others were for designing radiation weapons. The latter were discussed as far back as the Manhattan Project; they are designed to create temporarily high radiation fields to kill or debilitate enemy soldiers.

26. Makhijani 1994.

27. Also involved were the Department of Defense (DOD), the National Aeronautics and Space Administration (NASA), the Department of Veterans Affairs, formerly the Veterans Administration, and the Central Intelligence Agency.

28. *Congressional Record—Senate*, S 16371, 15 October 1971.

There had been sporadic publicity about some of the experiments since the early 1970s. In 1986, the staff of a committee then chaired by Congressman Edward Markey had even written a report about them.[29] But these did not attract widespread attention. It was Secretary O'Leary's high-profile annoucement that, in effect, opened the Pandora's box of U.S. radiation testing on humans to public view.

Purpose of Experiments Table 6.1 shows a list of many of the human radiation experiments categorized according to the goals of the funding agencies. Some experiments may have had more than one purpose; for example, some involving external exposure of sick people to radiation were purportedly to treat cancers. The objectives of the experiments will not be entirely known until more documentation is made available.

In general, there were five broad purposes of the experiments. First, scientists hoped to develop instrumentation for spying on the Soviet nuclear weapons complex. By intentionally releasing radioactive material to the environment, scientists hoped to track similar materials released by the Soviets to learn more about plutonium production in the Soviet Union and other details about the Soviet nuclear weapons complex. Second, some experiments were conducted to develop radiation weapons. These experiments also involved the intentional release of radioactive materials to the environment. Third, several experiments were designed to determine the effects of radiation on the ability of military personnel to function on the nuclear battle field and, in some cases, the effects of radiation on astronauts. These experiments involved, for example, whole-body irradiation of subjects, or the exposure of pilots to mushroom clouds from nuclear tests. Fourth, some experiments attempted to determine the effects of occupational exposure to radiation on workers in the U.S. nuclear weapons complex. And finally, a large number of experiments were designed to study the metabolism of radioactive materials.

The scope of experiments included in the table is by no means exhaustive. In general, the table encompasses experiments related to the development of nuclear weapons, including the effects to workers and the general population. Most of the experiments discussed here include those funded by the Atomic Energy Commission or the Manhattan Project. We have excluded experiments whose relation to the development of nuclear weapons is unclear. Finally, we have not listed the broad implications of the participation of armed forces personnel in nuclear weapons testing, apart from a specific case explicitly and officially listed as an experiment. A presidential advisory committee is investigating the scientific and ethical aspects of the human radiation experiments.

29. U.S. Congress, House 1986.

Table 6.1 Some Examples of Radiation Experimentation on Humans

Date	Description	Institution[a]
Experiments to Develop Instrumentation for Spying on the Soviet Nuclear Weapons Complex		
1949	Intentional release of iodine-131 to environment	AEC, Air Force[1]
1950	Intentional release of radioactive material to the environment	Los Alamos Lab., Air Force[1]
Experiments to Develop Radiation Weapons		
1948	Intentional release of lanthanum-140 to environment	AEC[1]
1949–52	Intentional release of tantalum-182 and possibly other radioactive material to the environment	Army, AEC, Air Force[1]
Experiments to Determine the Effects of Radiation on the Ability of Military Personnel to Function on the Nuclear Battlefield and/or Effects of Radiation on Astronauts		
1943–44	Whole-body irradiation by X rays	Univ. of Chicago[2]
1953	Exposure of hands to radioactive material	Foster D. Snell (consulting firm), Monsanto[2]
1960–71	Whole-body irradiation by X rays	Univ. of Cincinnati[3]
1960–74	Whole-body gamma irradiation	Oak Ridge Institute of Nuclear Studies (TN)[3]
Early 1970s	Neutron and ion beam irradiation	Lawrence Berkeley Laboratory[2]
Occupational Exposure to External Radiation		
1945	Exposure of skin to beta rays	Clinton Lab. (Oak Ridge, TN)[2]
1947	Exposure of fingers to radioactive material	Univ. of Chicago[2]
1955	Exposure of skin to radium-224	New York Univ.[2]
1963–71	Irradiation of the testicles of prisoners by X rays	Pacific Northwest Research Foundation, Univ. of Washington[2,b]
Experiments to Determine Metabolism of Radioactive Materials[c]		
1943–47	Polonium injections	Univ. of Rochester[2]
1945–47	Plutonium injections	Manhattan District Hospital (Oak Ridge), UCSF, Univ. of Rochester Univ. of Chicago[2]
1946–47	Injections of U-234 and U-235 uranium nitrate to induce renal injury	Univ. of Rochester[2]
Late 1940s	Adminstration of radioactive iron to pregnant women	Vanderbilt Univ.[5]
1946–56	Ingestion of radioactive iron and calcium	MIT, Harvard[4]
1950, 1952	Exposure of skin to tritium; also some by ingestion and inhalation	Los Alamos Scientific Lab.[2]
1953–57	Uranium injections	Mass. General Hospital (Boston), ORNL[2]

Table 6.1 (cont.)

Date	Description	Institution[a]
? (results published 1959)	Calcium-45 and strontium-85 injections	Columbia Univ., Montefiore Hosp. (Bronx, New York)[2]
1960s	Uranium-235 and manganese-54 ingestion	Los Alamos Sci. Lab.[2]
1961–63	Ingestion of real and simulated fallout from nuclear tests	Univ. of Chicago, Argonne National Laboratory[2]
1961–65	Radium and thorium injections/ ingestion	MIT[2]
? (results published 1962)	Ingestion of lanthanum-140	Oak Ridge Institute of Nuclear Studies[2]
1963	Phosphorus-32 injections	Battelle Memorial Institute (Richland, Washington)[2]
1962–65	Intentional release of iodine-131 to environment/ingestion	ORNL, National Reactor Testing Station (ID)[2]
1965	Technetium-95 (metastable) and technetium-96 injections/ingestion	Pacific Northwest Lab.[2]
1965–73	Inhalation of argon-41/ingestion of various radioactive isotopes	AEC[2]
1967	Promethium-143 injections/ingestions	Hanford Environmental Health Foundation, Battelle Memorial Institute[2]
? (results published 1968)	Lead-212 ingestions/injections	Univ. of Rochester[2]

Note: Categories are those considered most appropriate from publicly available evidence. The purpose is not always explicitly stated and in this case represents judgments made by IEER staff.

a. AEC = U.S. Atomic Energy Commission, UCSF = University of California, San Francisco, MIT = Massachusetts Institute of Technology, ORNL = Oak Ridge National Laboratory
b. These experiments were possibly related to radiation weapons development.
c. Some of these experiments may fit into other categories, and some may have had military applications.

Sources are indicated by superscript numerals:
1. U.S. GAO, *Examples of Post World War II Radiation Releases at U.S. Nuclear Sites, GAO/RCED-94-51FS*, November 1993.
2. U.S. House of Representatives, "American Nuclear Guinea Pigs: Three Decades of Radiation Experiments on U.S. Citizens," November 1986.
3. *Congressional Record—Senate*, S 16371, 15 October 1971.
4. Rosenberg, Howard, "Informed Consent," *Mother Jones*, Sept./Oct. 1981, pp. 31–37, 44.
5. Hahn, Paul F., et al, *Journal of Obstetrics and Gynecology*, vol. 61, no. 3, March 1951.

Lack of Informed Consent In addition to their connection to warfare and weapons development, the U.S. radiation experiments on humans were disturbing for their failure to obtain informed consent from many of the subjects involved. The concept of informed consent, though not as fully developed and institutionalized as it has been since the 1970s, was an understood and valued criterion for responsible research in the United States. As Bette-Jane Crigger of the Hastings Center has pointed out, the concept of informed consent first surfaced in the United States as early as 1914 in *Schloendorf v. Society of New York Hospital*, and reached full development in a widely read book by Joseph Fletcher, *Morals and Medicine*, published in 1954.[30] Moreover, in 1947, the concept of informed consent and the ethical guidelines for human experimentation were stated explicitly at the Nuremburg Trials.

One of the best indicators that the spirit of informed consent was abused is the profiles of the type of human subjects chosen to receive the radiation. In case after case, the subjects were in a compromised or powerless position. For example, some subjects were prisoners, others poor, pregnant, young children, elderly, or mentally retarded (or believed to be mentally retarded). Some were soldiers or military personnel who felt they had to follow orders. Many were people of color with curtailed legal rights or access to education.

Some of the experimenters bypassed the issue of informed consent by experimenting on themselves. A journalist, Charles Mann, reported in 1994 that 5 of 31 widely publicized experiments were performed by researchers on themselves.[31]

Nameless Subjects Radiation experimentation on humans for the purposes of the nuclear weapons complex has been well documented for some time. As noted above, in 1986 Congressman Edward Markey of Massachusetts released a report called "American Nuclear Guinea Pigs," documenting many of the radiation experiments on U.S. citizens and calling for further investigation.[32] Yet at the time, the report went largely unnoticed.

There are several reasons why the experiments generated a public outcry in 1993–1994. First, the Department of Energy is slowly trying to redefine itself according to post–Cold War reality, thanks in large part to Secretary O'Leary. Further, as noted above, the 1986 Markey report released information about *nameless* human subjects. But when a reporter from the *Albuquerque Tribune*, Eileen Welsome, uncovered the

30. Crigger 1994.

31. Mann 1994.

32. U.S. Congress, House 1986.

identities of some of the subjects, the matter took on a different aspect. Putting faces and names on the subjects is integral to understanding and redressing the experiments.

A Question of Ethics Many of the scientists who conducted the experiments are still alive and most are now scrambling to defend their actions. The experimenters, however, are finding slippery footing as they search for justification for their actions.

Many of the human experimenters now claim that not enough was known about radiation to recognize that it might harm the subjects of the experiments. While it is true that risk estimates of exposure to low levels of radiation have increased over the decades, the dangers of radiation exposure were well known during the era of experimentation, which extended into the early 1970s.[33] In addition, the bombing of Hiroshima and Nagasaki clarified the hazards of radiation. And since the late 1950s, radiation standards for workers have been set at 5 rem per year, the same as they are today.

Even the dangers from lower levels of radiation—comparable to downwind fallout from atmospheric testing—were well recognized. For instance, Colonel Stafford L. Warren, chief of radiological safety, studied the widespread fallout produced by the very first nuclear weapons test in New Mexico on 16 July 1945 and recommended that no tests be conducted within 150 miles of human habitation.[34]

Despite the experimenters' protestations that the doses were low and therefore not dangerous, many of the experiments were designed to induce harm. Among these was the irradiation of the testicles of prisoners, carried on till 1971. The irradiation levels ranged up to 600 rads, known to be very dangerous even during the Manhattan Project. Another example was the injection of uranium salts into subjects at the University of Rochester in 1946 and 1947 to determine the levels that would produce injury to the kidneys.[35]

33. For instance, the amounts of plutonium injected in the human experiment conducted by Los Alamos and the University of Rochester School of Medicine and Dentistry during 1945–47 ranged from 0.095 to 5.9 microcuries, which were about 2.4 times to 14.7 times the "tolerance dose," of 0.04 microcuries set for workers in 1944 in the Manhattan Project; the average dose was 0.35 microcuries, or almost nine times the "tolerance dose," according to Patricia Durbin of the Lawrence Berkeley Laboratory. She stated in testimony that this standard for "tolerance dose" was established at a level at which "no clinically detectable biological damage would result" during an exposed worker's entire lifetime. The standard was based mainly on animal studies and on analysis of the deaths of radium dial painters in the early part of this century. In other words, doses greatly in excess of those thought not to produce damage were given to all the plutonium injection experimental subjects. Durbin 1994.

34. Warren 1945.

35. U.S. Congress, House, 1986, p. 2.

Complex 21[36]

While DOE has no plans to build new nuclear weapons, a third function—designing future weapons—could continue under the rubric of Complex 21. Planned by the Office of Environmental Restoration and Waste Management, Complex 21 will enable production to resume and continue at some level until the middle of the twenty-first century. The size and composition of this new nuclear weapons complex and its relation to existing facilities is a matter of considerable debate and controversy. Some, including the Clinton administration, argue that deterrence requires several thousand weapons. Others say that a hundred, one, or even just the ability to make weapons may suffice. Given that the existing arsenal is larger than the range of debate, current policy is leaning towards a program of "stockpile stewardship" and maintenance of design capability rather than building new weapons.

The location of Complex 21's large operations, should the United States proceed toward new production, is somewhat less vague than the size of Complex 21, partly because several major facilities either have closed or almost certainly will shut in the near future. Proposed locations for new facilities are the Savannah River Site, Oak Ridge, Pantex, and the Los Alamos National Laboratory. In addition, the nuclear weapons establishment would like to keep the Nevada Test Site open. The manufacture of nonnuclear components for nuclear weapons is slated for consolidation at Kansas City, Missouri.

A parallel debate concerns the extent and nature of design capabilities that will be part of Complex 21. A number of labs carry on research and development, but the main centers are three national laboratories: Los Alamos and Sandia in New Mexico and Lawrence Livermore near San Francisco, California. Regardless of the size of the future nuclear arsenal, the design requirements depend on political and military judgments about the advisability of introducing new kinds of nuclear weapons, as well as on the political response to the financial pressures to continue spending on design and testing. Even a comprehensive test ban treaty may not eliminate new design activity since U.S. weapons labs are developing techniques for proving the design of new weapons in laboratory settings complemented by computers.

QUANTITY AND QUALITY OF DATA

There is far more publicly available information on the health and environmental problems occasioned by U.S. nuclear weapons production than for any other country. For this reason, it is possible to discuss

36. Albright et al. 1992, pp. 3, 7–8, 13–16.

more vividly both the state of the data and the actual problems created by nuclear weapons production. The Freedom of Information Act and investigations by journalists, environmental activists, and scientists have brought crucial information to light. Research by victims, congressional committees, and environmental organizations has also increased the volume and quality of public information. Court actions against weapons contractors and the U.S. government have contributed to making information available. In all, the diverse array of data on the U.S. nuclear weapons complex includes millions of pages of documentation.

Even so, the record is far from complete. Much data regarding weapons production and the use of hazardous materials like chlorine trifluoride for uranium enrichment, along with materials-accounting information for plutonium and highly enriched uranium and the documents relating to experiments, such as those relating to the purpose of a 1948 release of iodine-131 at Hanford, are still secret, even after the Cold War. Yet this secrecy, too, is eroding under public pressure to disclose the truth of the nuclear legacy. Dozens of citizens groups around the country are actively pressuring DOE to disclose information regarding health and environmental dangers from past operations *and* any planned future operations related to production and cleanup. Since 7 December 1993 the flow of this information has increased due to the openness and declassification initiative taken by Secretary O'Leary. This includes data on production of fissile materials.

While the quantity of available data is enormous, quality is another matter. It is well established that the AEC and DOE both regarded their prime purpose to be building and testing nuclear weapons. The best talents and energies went into design, testing, and production, with few incentives for talented people to work on environment and health. In fact, the government regarded those who raised health or environmental concerns as obstructions to production. This attitude severely affected the quality of data and analyses of worker health and safety, environmental impacts, and the health of civilians. Neither the AEC nor DOE monitored many sources of radioactive emissions, including sources that generated considerable amounts of radioactive waste. For example, the scrubbers and the associated stacks at the Fernald, Ohio, uranium processing plant were major sources of uranium emissions, but the plant conducted no direct continuous monitoring of the stacks, making only crude estimates that seriously understated the case.[37]

Among the most serious quality problems with data are those relating to records of worker exposures. Internal dose records seem to be particularly deficient. For instance, at the Fernald, Ohio, uranium

37. For details on this study, see Makhijani and Franke 1989, pp. 17–22, table 8.

processing plant mentioned above, no doses from lung burdens were actually entered into the dosimetry records of workers. Thus, dose records of workers implicitly indicate that internal doses were zero. However, the lung burden and urine data recorded for these same workers tell a different story.

An independent dose reconstruction from urine data (complemented by limited lung burden data) indicates that from 1952 to 1962, over half the workers were exposed to more than the 15 centisievert limit for lung dose that was the prevailing occupational standard in every year but one, and that over 10 percent of the workers exceeded this limit through the end of the 1960s.[38] This means that many plant workers were at considerably greater risk than indicated by official dose records, which focus on external doses.

Plant emissions of many nonradioactive hazardous materials were also poorly monitored when they were recorded at all. For example, the plants that manufactured uranium hexafluoride near Gore, Oklahoma, and at Portsmouth, Ohio, emitted hydrofluoric acid to the environment. These emissions were not monitored in the stacks and only sporadic measurements, if any, were made of hydrofluoric acid releases to the environment. Similarly, uranium enrichment used fluorine gas and chlorine trifluoride, two highly toxic chemicals. Emissions records for these chemicals are spotty. There are almost no data on worker exposure to such materials.

Almost no records exist of worker exposures to other potential hazards, such as heavy metals and beryllium, which were present at many stages of the processing and production of nuclear weapons. The same is true for organic solvents. This lack of data is highly relevant both to assessing the nature and variety of health effects that workers experienced from such agents and to understanding the effects of the exposure of workers to radioactivity.

The data on environmental monitoring are similarly deficient. Many pollutants, radioactive and nonradioactive, were not monitored, making it difficult to trace the environmental impact of production. As a result, it is impossible to accurately assess the impacts on ecosystems, at least until the advent of the environmental laws of the early 1970s. Since that time, the requirements imposed on nuclear weapons plants to secure water pollution permits have improved the record of discharges to the water. Groundwater monitoring has occurred on-site and off-site for nuclear weapons plants, providing some basis for judging the extent of aquifer contamination and the rate of its spread. The discharges of many nonradioactive hazardous materials as well as radioactive materials to the air still appear to be recorded with less certainty than emissions to the water.

38. Franke and Gurney 1994.

Another area of documentation is the dumping of radioactive and hazardous wastes, on-site and off-site; but this body of documentation is also incomplete. Nuclear weapons plants created large quantities of contaminated wastes. Dumps established on-site for discharging this waste created many health hazards. For example, dumping dry waste into pits often threw a good deal of dust into the air. The quantity of materials discharged into on-site dumps or sent off-site to radioactive- or hazardous-waste dumps and incinerators is not well known. In addition, for some categories of wastes, such as buried wastes, there are no adequate records of the quantity and nature of the material dumped. This makes it difficult to make realistic assessments of the character and cost of a cleanup or of future health effects and environmental pollution. Many of these problems are in the process of being corrected. As noted above, since the mid-1980s, the DOE has been increasingly subject to civilian environmental laws. Characterization of wastes dumped on many sites will become more detailed over the next several years.

Compared with other countries, more data is available on accidents and maintenance problems. Analyses of the probabilities of accidents and reports describing many past accidents have been made public for several U.S. nuclear weapons plants. Some computer printouts recording problems with maintenance and in other areas are also now available.[39] Still, as in the other areas of information, a considerable amount of data is missing and the quality of available data is often unreliable. For instance, one of the largest potential threats to public health and the environment is the risk of an explosion in any of the several dozen tanks containing high-level radioactive waste at Hanford. A number of organic materials, ferrocyanides, and other chemicals are in these tanks in addition to the high-level radioactive wastes they were supposed to hold. These chemicals have reacted with one another. Because instruments to monitor the tanks often don't work or have not been calibrated in decades, it is difficult to determine their true contents and condition.

Extensive programs are underway to remedy some of the problems resulting from decades of faulty practices, of inadequate record keeping and insufficient environmental measurement and monitoring. But it will take years and considerable expenditures of effort and money to assess the total cleanup and waste management costs from manufacturing nuclear weapons. Comparable efforts are in order to improve assessments of health damage to workers and off-site residents from past operations. It is unlikely that the true extent of the damage will ever be known accurately.

39. See for instance the appendix tables in Makhijani, Alvarez, and Blackwelder 1986.

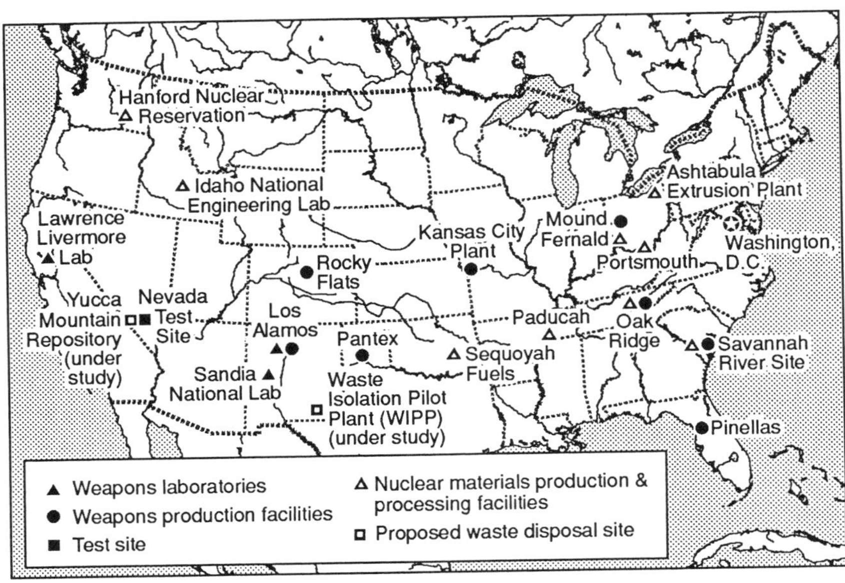

Figure 6.1 Nuclear weapons testing and production sites in the United States.

NUCLEAR WEAPONS–RELATED ACTIVITIES AND FACILITIES

Seventeen major facilities in the United States have taken part in the research, development, production, and testing of nuclear weapons and nuclear weapons materials. In addition to the principal sites, there are many smaller facilities that date back to the Manhattan Project, and some that have devoted only a small portion of their efforts to making weapons (see table 6.2 and figure 6.1). Figure 6.2 shows the flow of materials in the U.S. nuclear weapons complex during the period of weapons production.

In effect, the U.S. nuclear weapons production complex has been shut down since the late 1980s. DOE production facilities are either on standby or they are closed and their operations have been transferred to the jurisdiction of the department's Office of Environmental Restoration and Waste Management. DOE's five-year plan, published in June 1993, only lists the 86 sites with environmental restoration and waste management activities. In addition, it enumerates 23 sites where remediation in relation to uranium mill tailings is taking place.[40]

At many sites, the United States abandoned production activities in the 1940s and 1950s. Under the Formerly Utilized Sites Remedial Action Program (FUSRAP), mandated by Congress in 1974, DOE must identify and clean up sites from the early years of weapons production and the atomic energy program. DOE adds to this list as it discovers more formerly used sites and as more sites go from production to

40. U.S. DOE 1993, pp. D-1 to D-3.

Table 6.2 Location of Nuclear Weapons–Related Activities and Facilities in the United States

Official Facility Name	Location	Operating Status[1]	Corporate Owner or Mangement and Operating Contractor[2]	Source[3]
R&D Weapons Laboratories (only the largest are listed)				
Lawrence Livermore National Laboratory	Bordering east side of Livermore, California	OP	University of California[4]	a, h, o
Los Alamos National Laboratory (Manhattan Project Site Y)	Los Alamos, New Mexico	OP	University of California[4]	a, b, h
Sandia National Laboratory	Kirtland Air Force Base near Albuquerque, New Mexico	OP	AT&T's Sandia Corporation (1945–93)[4] Martin Marietta ESI (1993–)[4]	a, h
Oak Ridge National Laboratory at Oak Ridge Reservation (Manhattan Project Site X)	About 16 km southwest of Oak Ridge, Tennessee	OP	Metallurgical Lab. of UOC (1943–45)[4] Monsanto Chemical Corp (1945–48)[4] Union Carbide (UCCND) (1948–84)[4] Martin Marietta ESI (1984–)[4]	a, h, o, p
Uranium Mining[5]				
The DOE and its predecessor agencies purchased uranium from hundreds of domestic and foreign mines.				
Uranium Milling and Production (production of yellowcake from uranium ore)[6]				
Canon City (cap = 330t U/yr)	Colorado	NO 1987	Cotter	l
Uravan (cap = 1,000t U/yr)	Colorado	DE	Umetco	l

Table 6.2 (cont.)

Official Facility Name	Location	Operating Status[1]	Corporate Owner or Mangement and Operating Contractor[2]	Source[3]
Grants #1 (cap = 3,000t U/yr)	New Mexico	DE 1987	Anaconda	1
Grants #2 (cap = 62t U/yr)	New Mexico	NO 1985	Quivira	1
Grants #3 (cap = 1,000t U/yr)	New Mexico	DE	Homestake	1
Edgemont	South Dakota	DE 1983	TVA	1
Moab (cap = 600t U/yr)	Utah	DE	Atlas	1
Ford (cap = 500t U/yr)	Washington	NO 1982	Dawn	1
Gas Hills #1	Wyoming	DE 1988	American Nuclear	1
Gas Hills #2	Wyoming	NO 1988	Pathfinder	1
Jeffrey City	Wyoming	DE 1988	Western Nuclear	1
Natrona (cap = 500t U/yr)	Wyoming	DE 1987	Umetco	1
Shirley Basin	Wyoming	DE 1985	Petrotomics	1
Conversion of Yellowcake (U_3O_8) to Uranium Oxide (UO_3), Tetrafluoride (UF_4), and Hexafluoride (UF_6) (in preparation for uranium enrichment)				
Sequoyah Fuels Corporation Plant	Gore, Oklahoma	NO, DE	Kerr-McGee (1970–88) General Atomic (1988–)	h
Destrehan Street Plant	St. Louis, Missouri	NO 1958	Mallinckrodt Chemical (1943–58)[4]	h

Weldon Spring Plant near St. Louis	Weldon Spring, Missouri	NO 1967	Mallinckrodt Chemical (1958–67)[4] RC = Jacobs Engineering Group	h, l
Fernald Environmental Management Project (formerly Feed Materials Production Center)	Fernald, Ohio	NO, DE	National Lead of Ohio (1951–85)[4] Westinghouse MCO (1986–92)[4] RC = FERMCO[4]	h
Uranium Enrichment				
Paducah Gaseous Diffusion Plant	about 16 km west of Paducah, Kentucky, on Ohio River	NO[7]	Union Carbide (1952–84)[4] Martin Marietta ESI (1984–)[4] RC = Martin Marietta ESI[4]	h, o
K-25 Plant (gaseous diffusion) at Oak Ridge	about 21 km west of Oak Ridge, Tennessee	NO 1987, DE	Union Carbide (1947–84)[4] Martin Marietta ESI (1984–)[4]	h, o
Y-12 Plant (electromagnetic separation) in S-50 Plant, Oak Ridge Reservation	3.2 km from Oak Ridge, Tennessee	NO, DE	Tennessee Eastman (1943–47)[4]	a, h, o
Portsmouth Gaseous Diffusion Plant (Portsmouth Uranium Enrichment Complex)	Piketon, Ohio	NO[8]	Goodyear Atomic (1956–86)[4] RC = Martin Marietta ESI (1986–)[4]	d, h, o
Uranium Processing and Fuel Fabrication (of enriched uranium metal)[9]				
Feed Materials Production Center (Fernald)	Fernald, Ohio, 27.4 km northwest of Cincinnati	NO 1989, DE	National Lead of Ohio (1953–85)[4] Westinghouse MCO (1986–92)[4] RC = FERMCO[4]	a, h, o

Table 6.2 (cont.)

Official Facility Name	Location	Operating Status[1]	Corporate Owner or Mangement and Operating Contractor[2]	Source[3]
Oak Ridge Reservation	24.1 km from Knoxville, Tennessee	OP	Union Carbide (1945–84)[4] Martin Marietta ESI (1984–)[4]	a, h, p
Ashtabula Extrusion Plant	Ashtabula, Ohio	NO, DE	Reactive Metals (1952–)[4]	h, p
Production of Heavy Water				
Heavy Water Plant, Savannah River Site	About 20 km south of Aiken, South Carolina	NO 1982	DuPont DNC (1950–89)[4] Westinghouse SRC (1989–)[4]	h, d, p, o
Production of Tritium				
Tritium Production Reactors, Savannah River Site	About 20 km south of Aiken, South Carolina	NO	DuPont DNC (1950–89)[4] Westinghouse SRC (1989–)[4]	h, d, o, s, t
Recycling of Tritium				
Savannah River Site	About 20 km south of Aiken, South Carolina	OP	DuPont DNC (1950–89)[4] Westinghouse SRC (1989–)[4]	d, h, o, s
Plutonium Production From:				
Military Production Reactors				
Hanford Reservation	Bordering and north of Richland, Washington	NO, DE	DuPont DNC (1943–46)[4] General Electric (1946–64)[4] Isochem (1965–67)[4]	a, h, o

			Atlantic Richfield (1967–76)[4] Rockwell HO (1977–87)[4] RC = Westinghouse HC, UNC Nuclear, and Battelle Pacific NWL	
X-10 pilot graphite reactor at Manhattan Project Site X, Oak Ridge Reservation	24.1 km west of Knoxville, Tennessee	NO	Metallurgical Lab. of UOC (1943–45)[4] Monsanto Chemical (1945–47)[4] Union Carbide (1947–84)[4] Martin Marietta ESI (1984–)[4]	h, p, o
Savannah River Site	About 20 km south of Aiken, South Carolina	NO	DuPont DNC (1950–89)[4] Westinghouse SRC (1989–)[4]	a, h, s, t, o
Civilian Power Reactors				
No plutonium recovered from civilian fuel for military purposes.		NA		
Reprocessing Plants				
Hanford Reservation	Bordering and north of Richland, Washington	NO, DE	DuPont DNC (1943–46)[4] General Electric (1946–64)[4] Isochem (1965–67)[4] Atlantic Richfield (1967–76)[4]	h, o
Savannah River Site	About 20 km south of Aiken, South Carolina	NO[10]	DuPont DNC (1950–89)[4] Westinghouse SRC (1989–)[4]	f, h, o, s

Table 6.2 (cont.)

Official Facility Name	Location	Operating Status[1]	Corporate Owner or Mangement and Operating Contractor[2]	Source[3]
Idaho Chemical Processing Plant at the Idaho National Engineering Laboratory (formerly National Reactor Testing Station)	About 67.6 km northwest of Idaho Falls, Idaho	NO	American Cyanamid (1951–71)[4] Allied Chemical (1971–80)[4] Exxon NIC (1980–83)[4]	a, h, f, o
West Valley	West Valley, New York	NO	Nuclear Fuels Services (1966–72)	f, p
Fabrication of Plutonium Metal Components				
Pu-239 Pits and Triggers				
Rocky Flats Plant	Golden, Colorado, about 25.7 km northwest of Denver	NO, DE	Dow Chemical (1952–75)[4] Rockwell International (1975–89)[4] EG&G RFI (1990–)[4]	a, h, w, o
Hanford Works	Richland, Washington	NO 1965	General Electric (1949–65)[4]	q
Pu-238 Radioisotope Thermal Generators				
Mound Laboratory	Miamisburg, Ohio	NO 1993	Monsanto Research (1948–88)[4] EG&G MAT (1988–)[4]	d, h, o

Other Warhead Components				
Beryllium Processing				
Rocky Flats Plant	Golden, Colorado, about 25.7 km northwest of Denver	NO, DE	Dow Chemical (1952–75)[4] Rockwell International (1975–89)[4] EG&G RFI (1989–)[4]	h, p, o
Electronic Components				
Kansas City Plant	Kansas City, Missouri	OP	Bendix KCD (1949–)[4]	h, o
High-Explosives Testing, R&D				
Pantex Plant	About 27 km northeast of Amarillo, Texas	OP	Procter & Gamble (1952–56)[4]	h, o
Site 300, Lawrence Livermore National Laboratory	Bordering east side of Livermore, California	OP	University of California[4]	o, p
Lithium-6 and Lithium-Deuteride Production				
Y-12 Plant at Oak Ridge Reservation	About 3.2 km from Oak Ridge, Tennessee	NO 1963	Union Carbide (1945–84)[4] Martin Marietta ESI (1984–)[4]	p
Neutron Generators				
Pinellas Plant	St. Petersburg, Florida	OP	General Electric (1957–92)[4] Martin Marietta ESI (1992–)[4]	a, h
Assembly and Disassembly of Nuclear Weapons				
Pantex Plant	About 27 km northeast of Amarillo, Texas	OP[11]	Procter & Gamble (1952–56)[4] Mason & Hanger SM (1956–)[4]	y, h
Buffalo Works	Buffalo, New York	NO 1952	ACF Industries (1945–52)[4]	h

Table 6.2 (cont.)

Official Facility Name	Location	Operating Status[1]	Corporate Owner or Mangement and Operating Contractor[2]	Source[3]
Clarksville Modification Center	Clarksville, Tennessee	NO 1965	Mason & Hanger SM (1958–65)[4]	y, q
South Albuquerque Works	Albuquerque, New Mexico	NO 1967	ACF Industries (1952–56)[4]	q
Burlington AEC Plant (formerly Iowa Ordinance Plant)	Burlington, Iowa	NO 1975	U.S. Army/AEC (1947–51) Mason & Hanger SM (1951–75)[4]	y, h, q
Medina Modification Center	Medina, Texas	NO 1966	Mason & Hanger SM (1958–66)[4]	y, q
Testing of Nuclear Devices[12]				
"Peaceful" Nuclear Exposions (PNEs)				
Rio Blanco gas simulation test site (1 UT)	Rifle, Colorado	1973	one test of 3 simultaneous detonations	p, m
Rulison gas simulation test site (1 UT)	Grand Valley, Colorado	1969		p, m
Gasbuggy gas stimulation site (1 UT)	Farmington, New Mexico	1967		p, m
Gnome-Coach test sites (1 UT)	Carlsbad, New Mexico	1961		p, m
32 other PNEs conducted at the Nevada Test Site	104.6 km northwest of Las Vegas, Nevada	1961–73	EG&G Reynolds EE, Holmes & Narver, and Fenix & Scisson (1951–)[4]	p, m, x, c
Weapons Testing, inside USA				
Trinity Site (1 AT)	near Alamogordo, New Mexico	NO 1945	Manhattan Project (1945)	a, m, q

Nevada Test Site (100 AT + 793 UT)	104.6 km northwest of Las Vegas, Nevada	NO 1992	EG&G, Reynolds EE, Holmes & Narver (1951–)[4]	d, h, m, q, x
Amchitka (3 UT)	Amchitka Island, Alaska	NO		m, q
Central NTS (1 UT)	Central, Nevada	NO		m, q
Shoal Test (1 UT)	Fallon, Nevada	NO		m, q
Tonopah Bombing & Gunnery Range (5 AT)	at Nevada Test Site, administered by Sandia Lab.	NO		m, q
Tatum Dome test site (2 UT)	Hattiesburg, Mississippi	NO		m, q
Johnston Island Area (12 AT)	Johnston Island, U.S. Territory, Pacific	NO		m, q
Weapons Testing, outside USA				
Hiroshima, Japan (1 AT)		1945		m
Nagasaki, Japan (1 AT)		1945		m
Enewetak Proving Grounds (43 AT)	Enewetak Atoll, Marshall Islands	1958	Holmes & Narver (1948–58)[4]	h, m, q
Bikini Test Site (23 AT)	Bikini Island, Marshall Islands, Pacific	NO		m, q
Pacific Ocean (4 UT)	Open ocean, Pacific	NO		m, q
Christmas Island (24 AT)	Christmas Island, mid Pacific ocean	NO		m, q
South Atlantic Ocean (3 AT)	S. Atlantic, 1,770 km southwest of Capetown, South Africa	NO		m, q

Table 6.2 (cont.)

Official Facility Name	Location	Operating Status[1]	Corporate Owner or Mangement and Operating Contractor[2]	Source[3]
Permanent Repositories for Radioactive Waste				
Waste Isolation Pilot Plant (WIPP)	Carlsbad, New Mexico	US 1981	Westinghouse WIPP	h
Yucca Mountain Repository	104.6 km northwest of Las Vegas, Nevada; adjacent to Nevada Test Site	US	TESS	h

1. Status abbreviations:
DE = decommissioning begun or completed
NA = no information available
NC = not completed (construction suspended or abandoned)
NO = not operating and either will not resume or unlikely to resume, with closure year indicated if known
OP = operating at some capacity
UC = under construction
UR = under remediation (by DOE contractor)
US = under study; R&D work in progress and fate still uncertain

2. Abbreviations:
AEC = Atomic Energy Commission
Batelle Pacific NWL = Batelle Pacific Northwest Laboratory
Bendix KCD = Bendix Kansas City Division of Allied Signal
DOE = Department of Energy and/or its predecessor agencies
DuPont DNC = E. I. du Pont de Nemours and Company
EG&G = Edgerton, Germeshausen & Grier
EG&G MAT = EG&G Mound Applied Technologies
EG&G RFI = EG&G Rocky Flats, Inc.

Exxon NIC = Exxon Nuclear Idaho Company
FERMCO = Fluor Daniel Environmental Restration Management Corporation
INEL = Idaho National Engineering Laboratory
Martin Marietta ESI = Martin Marietta Energy Systems, Inc.
Mason & Hanger SM = Mason & Hanger Silas-Mason
Metallurgical Lab. of UOC = Metallurgical Laboratory of the University of Chicago
RC = remediation contractor
Reynolds EE = Reynolds Electrical & Engineering, a subsidiary of EG&G
Rockwell HO = Rockwell Hanford Operations
TESS = TRW Environmental and Safety Systems
Westinghouse HC = Westinghouse Hanford Company
Westinghouse INC = Westinghouse Idaho Nuclear Company
Westinghouse MCO = Westinghouse Materials Company of Ohio
Westinghouse SRC = Westinghouse Savannah River Plant
Westinghouse WIPP = Westinghouse Waste Isolation Pilot Plant

3. Sources:
a = U.S. Congress, OTA 1991
b = National Research Council 1989
c = Norris and Arkin 1993
d = Coyle et al. 1988
f = Nuclear Engineering International 1991
h = Schwartz 1993
k = U.S. DOE 1993
l = U.S. DOE 1992
m = IPPNW and IEER 1991
o = U.S. DOE 1993a
p = Cochran et al. 1987b
q = Cochran et al. 1987a

Table 6.2 (cont.)

s = Energy Research Foundation 1993
t = Energy Research Foundation 1993b
u = Hewlett and Holl 1989
w = Rocky Mountain Peace Center 1992
x = U.S. DOE 1993d
y = Lemert 1979.

4. Facilities of the U.S. Department of Energy and its predecessor, the Atomic Energy Commission, owned by the U.S. government but run by private management and operating contractors. Not all historical contractors are included, nor are subcontractors.
5. For more on U.S. domestic uranium mining sources, see Nuclear Engineering International 1991.
6. Only mills for which DOE has accepted some responsibility for cleanup are listed. All were conventional mills owned and operated by private companies.
7. The Paducah gaseous diffusion plant now provides enrichment services only for commerical nuclear power.
8. Portsmouth produced high enriched uranium in building X-326. It still produces low-enriched uranium for nonmilitary purposes in buildings X-330 and X-333.
9. The following facilities supplied fuel for naval reactors: Nuclear Fuel Services, Erwin, TN; Nuclear Materials and Equipment Corp., Apollo, PA; United Nuclear Corporation, Hematite, MO; Corporation, Hematite, MO; Babcock and Wilcox, Apollo, PA. See Cochran et al. 1987a, p. 71.
10. The only plutonium reprocessed at Savannah River as of 1993 is Pu-238 for use in radioisotope thermal generators.
11. Pantex is now involved primarily in nuclear weapons disassembly.
12. AT = atmospheric or surface nuclear test; UT = underground or underwater nuclear test. The Nevada Test Site was the only nuclear test site used after 1973.

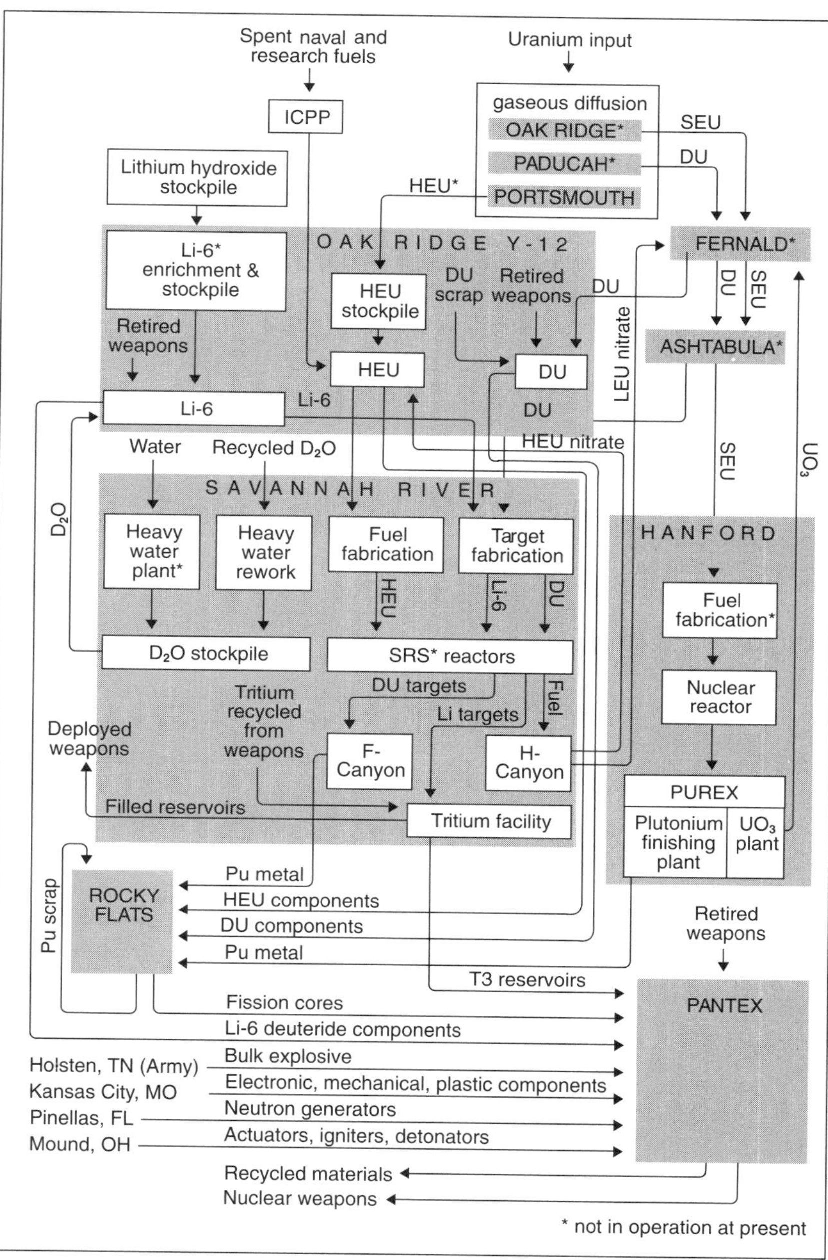

Figure 6.2 Flow of materials through the DOE nuclear weapons complex. Adapted from National Research Council 1989, pp. 104–105.

cleanup and waste management status. The 1993 five-year plan includes 33 FUSRAP sites, of which 28 related to the Manhattan Project or to the AEC (see figure 6.3).[41]

Research and Development Laboratories

The DOE operates three principal research laboratories: the Los Alamos National Laboratory, the Lawrence Livermore National Laboratory, and Sandia National Laboratory (in two locations, one in Livermore and one in Albuquerque).

Los Alamos National Laboratory (LANL), established in 1943, conducts research on warhead design. The 45-square-mile site is about 25 miles northwest of Santa Fe, New Mexico. LANL designed and built the world's first nuclear weapon. DOE projections indicate spending over $1 billion in fiscal years 1994 through 1998 to clean up contamination at this site.[42] The site was established partly on land taken from Pueblo Indians with the promise that it would be returned after World War II (evidently it was not).

Lawrence Livermore National Laboratory (LLNL) in California was added as a second nuclear weapons design center in 1952. LLNL also designed and tested fusion (thermonuclear) devices. It has an estimated cleanup cost of over $357 million for fiscal years 1994 through 1998.[43] The University of California administers both LANL and LLNL.

Sandia National Laboratory, south of Albuquerque, New Mexico, was founded in 1945. Sandia is located on 1,141 hectares of Kirtland Air Force Base. In addition, Sandia was in charge of two off-site test areas: Tonopah Test Range in Nevada and the Kauai Test Facility on Kauai Island, which is within the U.S. Navy's Pacific Missile Range. Sandia was managed by an AT&T subsidiary until September 1993, when DOE transferred management to Martin Marietta Corp. Sandia designs nonnuclear components for nuclear devices. DOE estimates that five years of cleanup activities at Sandia will cost about $310 million.[44]

Smaller DOE weapons research labs are the New Brunswick Laboratory in New Jersey and the Savannah River Laboratory in South Carolina. Cleanup at New Brunswick cost over $1 million in FY93.[45] The Savannah River Laboratory, now called Savannah River

41. U.S. DOE 1993a, p. II-143.

42. Kimball, Siegel, and Tyler 1993, p. 8.12.

43. Kimball, Siegel, and Tyler 1993, p. 6.6.

44. Kimball, Siegel, and Tyler 1993, p. 8.13.

45. Kimball, Siegel, and Tyler 1993, p. 4.18.

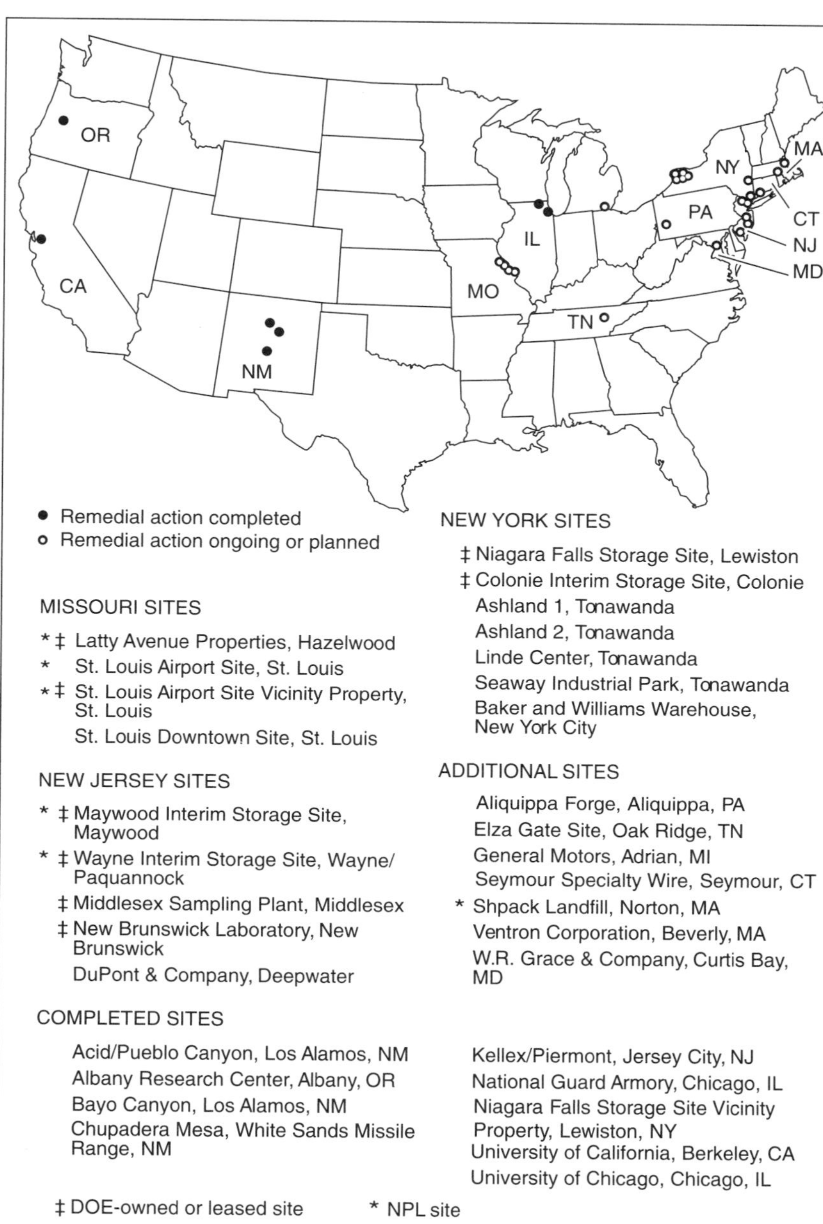
OR
CA
NM
MO
IL
TN
PA
NY
MA
CT
NJ
MD
● Remedial action completed
○ Remedial action ongoing or planned
MISSOURI SITES
* ‡ Latty Avenue Properties, Hazelwood
* St. Louis Airport Site, St. Louis
* ‡ St. Louis Airport Site Vicinity Property, St. Louis
St. Louis Downtown Site, St. Louis
NEW JERSEY SITES
* ‡ Maywood Interim Storage Site, Maywood
* ‡ Wayne Interim Storage Site, Wayne/Paquannock
‡ Middlesex Sampling Plant, Middlesex
‡ New Brunswick Laboratory, New Brunswick
DuPont & Company, Deepwater
NEW YORK SITES
‡ Niagara Falls Storage Site, Lewiston
‡ Colonie Interim Storage Site, Colonie
Ashland 1, Tonawanda
Ashland 2, Tonawanda
Linde Center, Tonawanda
Seaway Industrial Park, Tonawanda
Baker and Williams Warehouse, New York City
ADDITIONAL SITES
Aliquippa Forge, Aliquippa, PA
Elza Gate Site, Oak Ridge, TN
General Motors, Adrian, MI
Seymour Specialty Wire, Seymour, CT
* Shpack Landfill, Norton, MA
Ventron Corporation, Beverly, MA
W.R. Grace & Company, Curtis Bay, MD
COMPLETED SITES
Acid/Pueblo Canyon, Los Alamos, NM
Albany Research Center, Albany, OR
Bayo Canyon, Los Alamos, NM
Chupadera Mesa, White Sands Missile Range, NM
Kellex/Piermont, Jersey City, NJ
National Guard Armory, Chicago, IL
Niagara Falls Storage Site Vicinity Property, Lewiston, NY
University of California, Berkeley, CA
University of Chicago, Chicago, IL
‡ DOE-owned or leased site
* NPL site

Figure 6.3 Map of FUSRAP sites. Source: U.S. DOE 1992, p. 166.

Technical Center, is part of the Savannah River Site, which altogether will cost about $4 billion for five years of cleanup.[46] In addition, several labs serve other DOE needs. These sites include Ames Laboratory (Iowa State University); Argonne National Laboratory (University of Chicago); Brookhaven National Laboratory (Brookhaven, New York); Hanford Engineering Development Laboratory (Washington state); Idaho National Engineering Laboratory (Idaho Fall, Idaho); Knolls Atomic Power Laboratory (Schenectady, New York); Lawrence Berkeley Laboratory (University of California, Berkeley); Oak Ridge National Laboratory (Oak Ridge, Tennessee); Mound Laboratory (Miamisburg, Ohio); and Pacific Northwest Laboratory (Richland, Washington).

Uranium Chemical Conversion and Refining

DOE produced uranium metal principally at the Feed Materials Processing Center in Fernald, Ohio; the Y-12 plant in Oak Ridge, Tennessee; and the Weldon Spring plant near St. Louis, Missouri. In addition, in the 1970s the Sequoyah Fuels plant in Gore, Oklahoma, began operation to produce uranium hexafluoride, mainly for civilian nuclear power but also for military purposes.

A variety of contractors were responsible for the plants. National Lead of Ohio managed Fernald until 1985, when Westinghouse Materials Company took over until 1989, when the plant was shut down. Union Carbide ran Oak Ridge until 1984, when Martin Marietta took over as principal contractor. Kerr-McGee Corporation owned and operated Sequoyah Fuels until 1988, when General Atomics bought the facility.

Fernald carried out a number of different conversion processes until it closed in 1989. The main end product was uranium metal.

The Y-12 plant at Oak Ridge fabricated the uranium components for nuclear warheads, including those made from highly enriched uranium.

Both the Fernald and Oak Ridge plants are Superfund sites. (These are sites that are estimated by the U.S. Environmental Protection Agency to be among the most contaminated in the country and are covered under a special law, popularly called "Superfund.") DOE estimates of cleanup costs over the 1994-1998 period alone are $2 billion for Fernald and $511 million for the Y-12 plant.[47]

The privately owned Sequoyah plant is regulated by the NRC. It was shut down in 1992. The Sequoyah plant converted yellowcake to uranium hexafluoride using the "wet process" (see chapter 3). This product then went to one of the three enrichment plants. There was a

46. Kimball, Siegel, and Tyler 1993, p. 7.17.

47. U.S. DOE 1993a, p. II-68.

serious accident at this plant on 14 January 1986 when a uranium hexafluoride cylinder was accidentally overfilled and then heated. The cylinder ruptured and released essentially all of the 14 metric tons of uranium hexafluoride it contained. One worker, who was heating the cylinder, died of pulmonary edema caused by inhalation of hydrofluoric acid fumes. (As mentioned previously, uranium hexafluoride reacts with moisture to produce uranyl fluoride and hydrofluoric acid.) All of the 42 workers on the site as well as about 100 others who were near the plant at the time of the accident were hospitalized for treatment and observation. Most suffered temporary kidney damage from the uranium intake.[48]

Two other plants conducted the preliminary conversion from yellowcake to uranium hexafluoride. The Metropolis, Illinois, plant used the "dry process." Managed by Allied Corporation, it started operating in 1959 and now supplies the civilian power industry.[49] The Weldon Spring plant in Missouri opened in 1958 and was operated by Mallinckrodt Chemical Company until 1967, when the plant was shut down and operations were consolidated at Fernald. Cleanup costs for Weldon Spring will run over $3 billion for 1994 through 1998.[50]

Uranium Enrichment

Nuclear warheads require highly enriched uranium. The AEC designed its enrichment system with this in mind, but no highly enriched uranium has been produced for warheads since 1965. All highly enriched uranium produced from 1965 onward has been for research reactors and to fuel naval vessels. In fact, the weapons program has a surplus of highly enriched uranium.

Of the several ways to enrich uranium, the United States has used only gaseous diffusion since 1946. Between 1944 and 1946, it used three enrichment processes at Oak Ridge: electromagnetic separation at the Y-12 plant, thermal diffusion at the S-50 plant, and gaseous diffusion at the K-25 plant.[51] Y-12 produced the world's first kilogram of enriched uranium, using an electromagnetic separation device known as the Calutron isotope separator. (The name "Calutron" derives from the University of California Cyclotron, where the process was invented.) The S-50 plant, which enriched natural uranium to 0.86 percent uranium-235, supplied Y-12 during its initial period of operation. In September 1945, S-50 was shut down in favor of production at the new K-25 gaseous diffusion plant. Enrichment by electromagnetic separa-

48. U.S. Congress, House 1987, pp. 6, 9.

49. Cochran et al. 1987a, p. 125.

50. U.S. DOE 1993a, p. II-187.

51. Hewlett and Anderson 1990, pp. 123–173.

tion ended the following year, because gaseous diffusion was much cheaper.[52]

Plans to enrich uranium with the gas centrifuge method, based on work at a Bayway, New Jersey, plant operated by Standard Oil, were never realized. In addition, a large gas centrifuge plant was built in the 1970s on the same site as the Portsmouth gaseous diffusion plant, but it was abandoned before going into commercial operation.

Thus, since 1946, three gaseous diffusion plants have handled uranium enrichment: Oak Ridge, Portsmouth, and Paducah, Kentucky. These plants have a combined capacity of 27.3 million separative work units per year. Oak Ridge and Portsmouth produced all the highly enriched uranium in the United States. Martin Marietta Energy Systems is the primary contractor at all three sites. In 1993 the government turned over Paducah and Portsmouth to a new government, non-DOE corporation, the U.S. Enrichment Corporation, because all U.S. enrichment is now for civilian purposes.

The Oak Ridge enrichment operations at K-25 ended in 1987. The site is severely contaminated, and DOE estimates cleanup costs will be more than $1.6 billion for fiscal years 1994 through 1998.

The Paducah plant, constructed in 1954, produces low-enriched uranium for civilian power plants. In 1959, 450 gallons (about 1,700 liters) of trichloroethylene-contaminated uranium were buried at the site in 30-gallon drums. By 1984, most of these drums were missing; possibly the drums corroded and the contents leaked away. Furthermore, large quantities of DOE uranium wastes are buried at Paducah. In 1984 and 1985, more than 300 pounds (about 135 kilograms) of uranium were released into surface water.[53] The DOE estimates cleanup costs for 1994 through 1998 to be $170 million.

The Portsmouth Uranium Enrichment Complex began operation in 1954. Highly enriched uranium for nuclear weapons and naval reactors was produced in the X-326 building. Production of highly enriched uranium ended in 1992.[54] Most of the waste generated by the plant is buried on-site or lies in an on-site waste treatment pond. Trichloroethylene and other solvents used in plant operations have contaminated the aquifer beneath the site. As a result, Portsmouth is on the Superfund list. DOE estimates the five-year cleanup costs to be $163 million. Portsmouth now operates only to meet civilian enrichment needs. It may also be used in the future to convert ("downblend") military highly enriched uranium back to low enriched uranium for civilian uses.

52. Benedict, Pigford, and Levi 1981, p. 815.

53. Coyle et al. 1988, p. 77.

54. U.S. Congress, OTA 1993, p. 101.

Uranium Processing and Fabrication

For weapons production, enriched uranium is converted from gaseous uranium hexafluoride to either uranium oxide (a powder) or uranium metal. Depleted uranium is converted to metal to make casings for warheads or target elements for plutonium production in reactors.

The Feed Materials Production Center at Fernald had primary responsibility for converting depleted uranium and low-enriched uranium into metal. The Ashtabula, Ohio, plant extruded the depleted and low-enriched uranium from Fernald into metal billets and tubes. Ashtabula opened in 1952 and no longer operates. Plans for its cleanup are included in the plans for the Feed Materials Production Center. Machining uranium metal was also carried on at many other sites. For instance, a small plant in Oxford, Ohio operated from late 1952 to early 1957 under subcontract to the Feed Materials Production Center.

Depleted uranium and low-enriched uranium were sent to the Savannah River Site for manufacture into targets for plutonium production. The fuel and target elements were nickel-plated and enclosed in aluminum cans in the 313-M building. Low-enriched uranium metal cores also went to Hanford to make rods for the dual-purpose N-reactor. There, the uranium metal cores were extruded into zirconium-clad rods for use as both fuel and targets in the N-reactor. Hanford's Fuel Fabrication Facility (in the "300 area") was responsible for extrusion. The 300 area has undergone some environmental restoration.[55] The Rocky Flats plant fabricated depleted-uranium tampers (casings for the fissionable materials in nuclear weapons) and manufactured highly enriched uranium components until 1965, when the Y-12 plant began making all such components.

Highly enriched uranium goes into fuel rods for some reactors and into nuclear warheads. Highly enriched uranium metal, also known as Oralloy, was stored at the Oak Ridge Y-12 plant for use in Savannah River and other DOE research reactors. Highly enriched uranium metal was also recovered from the spent fuel of research reactors at the Idaho Chemical Processing Plant, which is the reprocessing plant at the Idaho National Engineering Laboratory. This recovered highly enriched uranium metal was used in Savannah River Site reactors as driver fuel. INEL also recovered highly enriched uranium from naval fuels, processed them into the oxide form, and sent the product to the Oak Ridge Y-12 plant to be processed into fuel for Savannah River Site reactors. The acidic high-level radioactive waste resulting from reprocessing was calcined; it is stored in stainless steel tanks in concrete vaults. The prime contractor for INEL is Westinghouse Idaho Nuclear Com-

55. U.S. DOE 1993a, p. II-190.

pany. INEL's projected cleanup cost is $1.8 billion in fiscal years 1994 through 1998.[56]

Military Reactors for Plutonium and Tritium Production

The United States produced plutonium and tritium for warheads in dedicated "production" reactors, except for the dual-purpose Hanford N-reactor. In all, the United States has operated 14 plutonium production reactors. No reactor, however, has operated for this purpose since 1988. These reactors are discussed briefly in the sections on Hanford and the Savannah River Site later in this chapter.

Future U.S. plutonium production is uncertain. Officially, there is to be no more military plutonium processing, but it may be recovered under the guise of waste management or environmental restoration programs at Hanford, Savannah River Site, or the Idaho National Engineering Laboratory. There are more than 30 metric tons of plutonium in the form of residues, oxides, metal, and process materials in the U.S. nuclear weapons complex.[57] The U.S. government has not yet officially declared this material a waste.

Chemical Separation of Plutonium and Uranium (Reprocessing)

The United States has operated eight reprocessing plants. Reprocessing ended in 1988. Hanford had five reprocessing plants (T, B, U, Redox, and PUREX). The Savannah River Site had two reprocessing plants (F and H canyons). Beginning in 1953, the Idaho Chemical Processing Plant, using the PUREX process, recovered uranium from spent fuel from naval propulsion reactors and from research and test reactors. It was shut in 1992, having operated for a total of 10 months in the 1980s. Its current functions are to develop technology to immobilize calcine wastes and to prepare spent fuel for repository disposal. In addition, some military reprocessing was done at the privately-owned reprocessing plant in West Valley, New York, which operated from 1966 to 1972.

Manufacture of Plutonium Components

The Rocky Flats plant manufactured pits (also known as triggers or cores) for warheads. Pits are plutonium-239 spheres that are surrounded by uranium or beryllium tampers that reflect neutrons back into the core. Both metallurgical and chemical processes are part of pit

56. Kimball, Siegel, and Tyler 1993, p. 5.8.

57. U.S. DOE 1993d, p. 23.

manufacturing. All pit manufacturing stopped in January 1992. Rocky Flats is severely contaminated and estimates of 1994-1998 cleanup costs are on the order of $1.8 billion.

Manufacture of Other Components

• *Beryllium:* The Rocky Flats plant manufactured beryllium components until 1992.

• *Deuterium:* Both the Oak Ridge Y-12 plant and the Savannah River Site processed deuterium.

• *Lithium-6 and Lithium Deuteride:* Large-scale lithium separation activities for weapons purposes took place between 1950 and 1963 at the Y-12 plant, which also produced and sold 99.97 to 99.99 percent pure lithium-7, and continues to do so. Production of lithium-6 ended in 1963. Presumably, existing stockpiles and recycling the contents of retired warheads have met all weapons requirements since that time.[58]

• *Nonnuclear Components:* Other components vital to making nuclear weapons were produced in Kansas City, Missouri; Pinellas, Florida; and Mound, Ohio. Bendix managed the Kansas City plant, which produced mechanical, electrical, and plastic components for nuclear weapons. The plant will cost about $107 million to clean up from 1994 through 1998. Pinellas, managed by General Electric, manufactured neutron generators and electronic and mechanical components. Cleanup at Pinellas is estimated by the DOE to cost about $46 million for fiscal years 1994 through 1998.[59] The Mound, Ohio, plant was managed by Monsanto Research Corp. It produced cable assemblies, detonators, actuated timers and firing sets, surveillance testing equipment, and command-disable explosive components. (Tritium recovery and other functions were also carried out at Mound.) Cleanup at Mound is estimated at $290 million for fiscal years 1994 through 1998.

Assembly and Disassembly of Nuclear Weapons

Parts are shipped to the Pantex plant in Texas for assembly. Pantex also makes the chemical high explosives for the warheads. The ingredients for assembly are these high explosives, the lithium deuteride components from Oak Ridge, the pits from Rocky Flats, the tritium reservoirs from Savannah River, and nonnuclear parts.

Pantex originally operated under the U.S. Army. It was transferred to AEC control in 1963. Since 1956, Pantex has been managed by Mason and Hanger-Silas Mason Company of Lexington, Kentucky.

58. Cochran et al. 1987a, p. 91.

59. Kimball, Siegel, and Tyler 1993, p. 7.15.

Testing of Nuclear Devices

In the first years after World War II, the United States conducted all its nuclear tests in the Marshall Islands in the Pacific. In 1951, the Nevada Test Site was opened about 80 miles northwest of Las Vegas. Tests have also occurred at other sites in the United States.[60]

DESCRIPTION OF KEY SITES

Because of the large number of nuclear weapons–related sites in the United States, this book discusses in detail only seven key facilities: Fernald, Hanford, the Nevada Test Site, Oak Ridge, Pantex, Rocky Flats, and the Savannah River Site. These sites played a central role in producing and testing nuclear weapons and exemplify a range of production issues. Some of them may have a role in a proposed new weapons production complex. Furthermore, the principal steps in manufacturing, processing, assembly, and testing took place at these seven sites (though some were not exclusively at these sites). They are also the locations of most of the severe contamination problems in the nuclear weapons complex, although many other sites—such as the Idaho National Engineering Lab and the many uranium processing plants—pose considerable challenges in terms of both environmental contamination and waste management.

Fernald Uranium Processing Plant, Ohio

Originally called the Feed Materials Production Center, Fernald opened fully in 1953; some start-up operations were begun in 1951 and 1952. Perhaps owing to its name and a pattern on the plant's water tower resembling the Purina Company checkered logo, many residents thought the center processed or manufactured pet food. The more colloquial name derives from the nearby small town of Fernald, Ohio, about 20 miles northwest of Cincinnati. Since 1989 when production stopped, the plant has been called the Fernald Environmental Management Project.

Fernald is one of the smaller sites in the nuclear weapons complex, comprising about 1,150 acres (see figure 6.4). National Lead of Ohio operated Fernald through December 1985. Westinghouse Materials Company of Ohio operated it through 1992.

Fernald's main function was converting a wide variety of uranium-containing raw materials, such as ores and residues in scrap, into

60. For a detailed discussion of testing, see IPPNW and IEER 1991. On 7 December 1993 Secretary of Energy Hazel O'Leary announced that the U.S. had conducted 204 secret underground nuclear tests.

Figure 6.4 Location of Fernald Plant.

uranium metal. It shipped uranium metal products to Rocky Flats in Colorado; the Savannah River Site in South Carolina; Oak Ridge, Tennessee; and Hanford, Washington. Fernald handled a variety of enrichments, ranging from depleted uranium of about 0.2 percent uranium-235 to enriched uranium with about 10 percent uranium-235. Fernald stopped production operations in July 1989.

Production at Fernald Fernald processed a variety of uranium-bearing materials in nine plants. Plant 1 prepared feed materials, which were sampled, weighed, and then directly fed to plants 2 and 3. In Plant 2 uranium was dissolved in nitric acid to produce uranyl nitrate. The uranyl nitrate was then calcined in Plant 3 to yield uranium trioxide. This material was sent to Plant 4, which first converted uranium trioxide to uranium dioxide, then fluorinated it with hydrofluoric acid to make uranium tetrafluoride (also known as green salt). Green salt was

Figure 6.5 Target element cores at Fernald. Photo by Robert del Tredici.

reduced to metal by reacting it with magnesium metal flakes in crucibles in Plant 5. This metal was then cast into ingots or billet shapes. Plant 6 machined the metal. Plant 7 operated for only 18 months in the 1950s, and produced uranium hexafluoride. Plant 8 processed residues and scrap for refinery feed. Plant 9 processed metal pieces that were too large to be fabricated in other parts of the Fernald facility.

Fernald also had several storage areas and a pilot plant. In addition, it processed large quantities of low-enriched uranium and natural uranium in the 1950s and 1960s. In the 1980s, production increasingly focused on depleted uranium. In all, Fernald produced about half a million metric tons of uranium metal.[61]

Waste Generation and Environmental Contamination at Fernald
Fernald generated a variety of waste products (see table 6.3). These include low-level radioactive waste, toxic waste, mixed waste (chemically hazardous and radioactive waste together), water treatment sludges, fly-ash from the steam plant, and general waste and refuse.[62] Production wastes were sent to six pits, or stored in drums, two silos, and scrap piles. Many buildings and pieces of equipment

61. Coyle et al. 1988, p. 117.

62. U.S. DOE 1991, Attachment No. 1, p. ES-1.

Table 6.3 Summary of Hazardous Substances Released to the Environment from the Feed Materials Production Center, Fernald, Ohio

Contaminant	Air	Soil	Surface Water	Groundwater	Sediment
Radionuclides	Radon	Radon		Cesium-137	Technetium-99
	Radon decay products	Uranium		Gross alpha	Uranium
	Thoron[1]			Gross beta	
	Uranium[2]			Neptunium-237	
				Potassium-40	
				Ruthenium-106	
				Strontium-90	
				Thorium-232	
				Uranium	
Metals		Lead	Chromium	Barium	
				Chromium	
Inorganic compounds	Hydrogen fluoride[3]		Cyanide	Chlorides	
				Fluorides	
				Nitrates	
				Sulfates	
Volatile organic compounds (VOCs)	Perchloro-ethylene[4]	Perchloro-ethylene[4,5]	Perchloro-ethylene[4,5]	Perchloro-ethylene[4,5]	
		Trichloro-ethane[5]	Trichloro-ethane[5]	Trichloro-ethane[5]	
Miscellaneous	Particulates	Asbestos	PCBs[6]	PCBs[6]	
		PCBs[6]			

1. Although believed present, inappropriate methods have been used to detect presence and contamination potential.
2. Approximately 96 metric tons of this radioactive contaminant had been released up to mid-1986.
3. An unspecified amount of this contaminant was released to the air from the uranium reduction plant (used for reducing UF_6 and UF_4) in January 1986.
4. This VOC is also know as tetrachloroethylene or tetrachloroethane.
5. The presence of potential contamination associated with this pollutant has not been fully determined.
6. PCBs = polychlorinated biphenyls.

Source: U.S. Department of Energy, Office of Environmental Audit, "Environmental Survey Preliminary Report—Feed Materials Production Center, Fernald, Ohio," DOE/EH/OEV-1-P, March 1987, as cited in U.S. Congress, OTA, 1991.

are highly contaminated. Uranium was the main pollutant emitted to the air, water, and soil.

Fernald's six waste pits were dumping grounds for vast amounts of mixed wastes, estimated to contain a total of about 5,000 metric tons of uranium. One pit contains as much as 5 percent uranium and another as much as 9 percent. (Commercial uranium ore is about 0.1 to 0.2 percent uranium.) The pits also contain radium-226, technetium-99, and a variety of nonradioactive pollutants, including barium compounds. Surface soil samples in the waste storage area have from 2.2 to 1,790 picocuries (about 0.081 to 66.2 becquerels) per gram of radioactivity.[63]

Two silos, known as the K-65 silos, contain uranium-processing residues from making the first atomic bombs. Since these bombs relied on ores from the Congo, the residues belonged to Belgium (specifically, the Afrimet-Indussa Company). In the 1950s, the silo contents were coveted for the radium-226 content in the waste (radium-226 was considered valuable from a commercial point of view in the 1950s). However, radium has lost all commercial value, since its luminous properties can be obtained from far less dangerous materials. It is now considered a dangerous pollutant, and K-65 wastes have taken on an onerous character. They are estimated to contain about 1,650 curies (about 61 terabecquerels) of radium-226. In 1983, the United States agreed to assume ownership of this waste if Belgium would allow Pershing 2 missiles on Belgian soil.[64]

Thousands of drums containing radioactive material, notably thorium, sit at Fernald. These materials were generated by other facilities and shipped to Fernald for temporary disposal. DOE plans to ship the 55-gallon (about 200 liter) drums to the Nevada Test Site for disposal as low-level waste. By July 1993, Fernald had shipped 1,621 drum equivalents of waste to Nevada. About 10,000 drum equivalents awaited approval for shipping.

DOE estimates that Fernald holds a total of 343,000 cubic meters of low-level radioactive wastes. This waste contains uranium and thorium as the primary radioactive contaminants.[65]

Fernald's release of uranium to the air has been highly controversial. Release estimates by DOE's contractors (National Lead of Ohio and Westinghouse) range from 97 to 135 metric tons emitted between 1951 and 1986.[66] However, independent estimates reveal that Fernald's

63. U.S. DOE 1991, p. ES-3.

64. Coyle et al. 1988, pp. 118–119.

65. U.S. DOE 1992, pp. 117, 129; U.S. DOE 1991, Attachment No. 10, p. 1. As of 1991, 1,094 metric tons of thorium were on-site.

66. Makhijani 1988, p. 7.

discharges to the environment were systematically underestimated. In 1989, IEER made a preliminary estimate that uranium releases were between 270 and 1,400 metric tons.[67] The Radiological Assessment Corporation, under contract to the federal Centers for Disease Control and Prevention (called the Centers for Disease Control prior to 1993), is conducting a detailed independent look at Fernald's discharges of radioactivity. Its preliminary estimates range from 370 to 600 metric tons of uranium released.[68]

Since this question became an issue in 1985 when the plant changed contractors, DOE and its contractors have admitted that releases were higher than their previous estimates. That was a crucial year because in late 1984 publicity about a relatively large accidental release of uranium had led to a number of disclosures about previous operational and environmental problems. Some people living near the plant filed a class action lawsuit in early 1985 against then-contractor National Lead of Ohio. The lawsuit was settled by DOE on the contractor's behalf for $78 million in 1989. Another lawsuit, filed by Fernald workers, was settled by the DOE in 1994. The U.S. government is to provide the workers with medical monitoring for life and other compensation.

Plans for the disposal or cleanup of wastes at Fernald are not clear. DOE has funded a pilot project to concentrate some contaminated wastes and solidify them into glass marbles. Known as the Minimum Additive Waste Stabilization (MAWS) Program, the effort will apply a variety of technologies—including vitrification, soil washing, and ion-exchange water treatment—to treating low-level and mixed wastes. The program began in 1992.

The MAWS process is not without its problems, however. Because some contaminants will evaporate into the air when wastes are added to the molten glass, the process relies on high-efficiency particulate air filters to remove fine particles in this "off-gas." However, the filters are delicate and, according to the Environmental Protection Agency, "can sustain structural damage relatively easily."[69] This problem can be largely addressed by installing additional filters in a series, so that the failure of one set does not result in large releases to the environment. One important problem is that DOE has not disclosed how it will dispose of the approximately half a million metric tons of radioactive marbles that would result from even partial conversion of Fernald's on-site contaminated materials.

67. Makhijani and Franke 1989, p. 2.

68. Voillequé et al. 1993, p. 39.

69. U.S. EPA 1991, p. (4–3). See also Goldfield 1988.

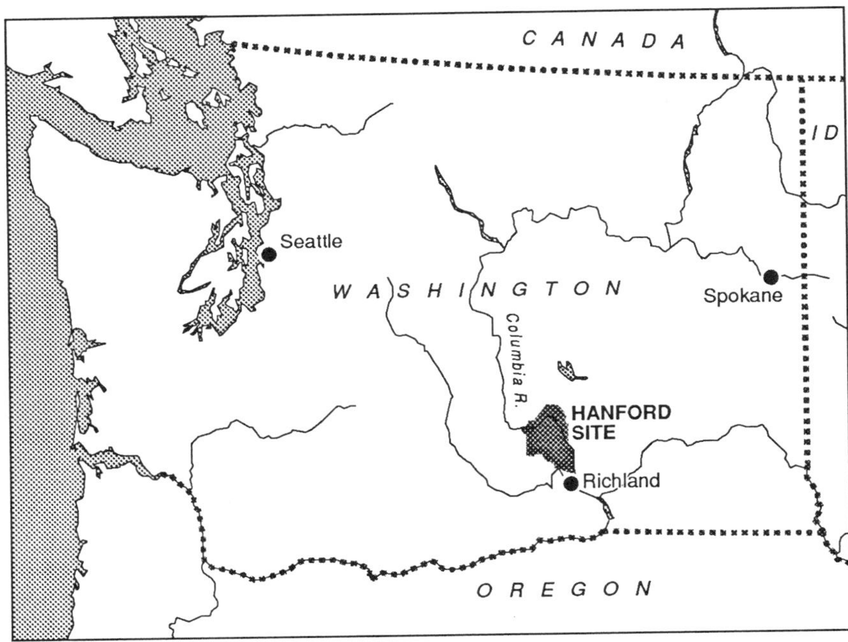

Figure 6.6 Location of Hanford Site.

Hanford, Washington

During World War II, the Manhattan Project selected a site outside Richland, Washington on which to build the world's first large-scale plutonium production facility. The location offered two advantages: the remoteness helped to maintain the secrecy the project required, and the Columbia River, which ran right through the site, could provide the plutonium production reactors with the copious quantities of water that they needed to operate. Hanford is one of the largest sites in the U.S. nuclear weapons complex, covering about 1,460 square kilometers (see figure 6.6). Hanford plutonium went into the first nuclear weapon, which was exploded at Alamogordo, New Mexico, on 16 July 1945. The Nagasaki bomb also contained Hanford plutonium.

E. I. Du Pont de Nemours and Company was the initial contractor at Hanford. As was typical of nuclear weapons facilities, the lines between contracting company and the military were frequently blurred. For example, the army's officer in charge of receiving plutonium from Du Pont was a company employee on leave to the military for the Manhattan Project.[70] Since World War II, Hanford has had many contractors and managers. Among the more recent contractors are Rockwell Hanford Operations, Westinghouse Hanford Company, Battelle

70. Sanger and Mull 1989, pp. 157–158.

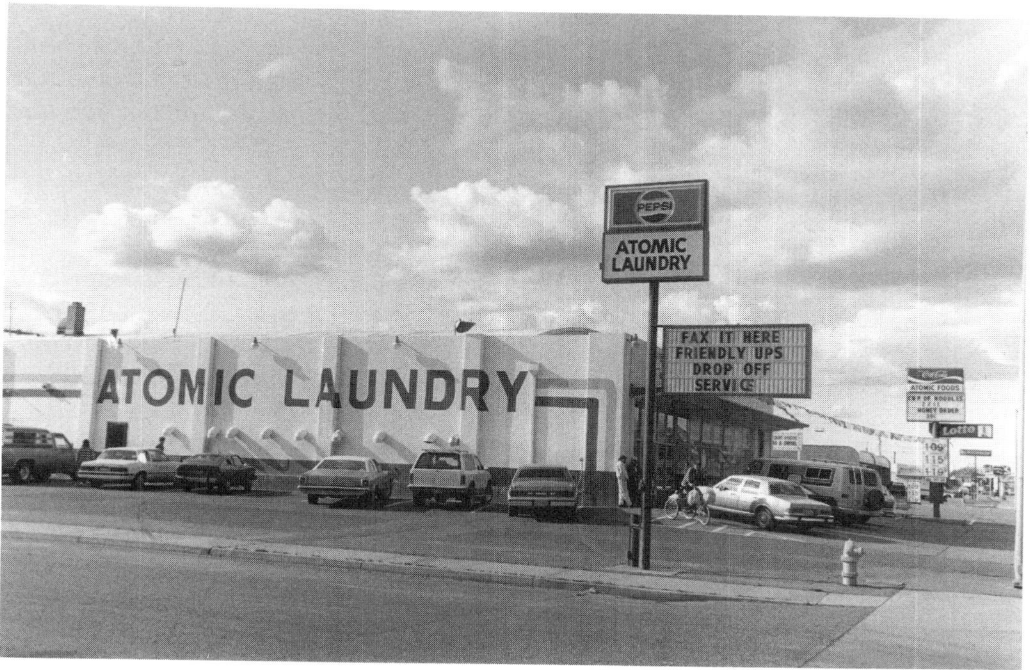

Figure 6.7 Atomic Laundry, Pasco, Washington, one of the Tri-Cities adjoining the Hanford Nuclear Reservation. Photo © James Lerager.

Memorial Institute, UNC Nuclear Industries, J. A. Jones Construction Services, Kaiser, BSC Richland, and the Hanford Environmental Health Foundation.[71]

Between 1943 and 1945, several operations facilities were built at Hanford, including the B, D, and F plutonium production reactors and the T, B, and U plants for chemically extracting plutonium from irradiated fuel. The reactors are water-cooled and graphite-moderated. The reprocessing plants used a bismuth-phosphate process based on experimental work by Glenn Seaborg and his associates. This approach to reprocessing had no provision for recovering uranium, which was disposed of with other waste products.[72] All three reprocessing plants were shut down or closed by 1958.

Between 1947 and 1963, Hanford added six more reactors. Reactors H, DR, C, KE, and KW were similar to their predecessors. The sixth, the N-reactor, completed at the end of 1963, was a thermal reactor for producing plutonium for nuclear weapons and breeder reactors. It also produced heat for generating steam, which the AEC and its successor agencies sold to the Washington Public Power Supply System to

71. Cochran et al. 1987b, p. 16.

72. Cochran et al. 1987b, p. 15.

produce electricity. Serving both military and civilian functions, the N-reactor is the only dual-purpose reactor the United States has built. It is a 4,000-megawatt thermal graphite-moderated reactor cooled with pressurized light water. Located in the 100-N area of the Hanford Reservation, it began operating on 31 December 1963. Electricity generation began in 1966 in an 860-megawatt electrical generating facility next to the reactor.[73] The N-reactor was placed on cold standby in 1988.

Hanford had two other reprocessing plants in addition to those built during World War II. The REDOX (REDuction OXidation) plant, opened in 1951, attempted to make extraction more efficient, reduce the amount of waste generated, and add the ability to extract uranium left in irradiated fuel along with the plutonium created in the reactors. The process was developed by Argonne National Laboratory and then tested in a pilot plant at Oak Ridge in 1948 and 1949.[74]

Hanford's longest-lasting reprocessing plant was PUREX (Plutonium URanium EXtraction), which represented a turning point in reprocessing technology when it was built. PUREX, which relies on solvent extraction to recover plutonium, uranium, and even neptunium, is more efficient than earlier methods. It was shut down in 1972, then restarted and upgraded in 1983. PUREX is not permanently shut down, even though the United States has produced no plutonium since 1988. However, any plutonium separation at the PUREX plant will, in the future, be designated as part of the "cleanup" program to process irradiated fuel from the N-reactor, which is now stored under water in deteriorating fuel rods.[75]

PUREX extracted plutonium and uranium in nitrate form. Both nitrates are liquids, which are far more risky to transport than solids. As a consequence, Hanford has facilities to convert plutonium into plutonium oxide and uranium into uranium trioxide. The plutonium-solution plant at Hanford also converts plutonium oxide into a metal through a hydroflourination process similar to the one used at Rocky Flats. The plutonium-finishing plant is also known as the Z plant.

Production at Hanford Hanford reactors mainly produced weapons-grade plutonium. The total from its eight graphite-moderated light-water reactors (which were shut down from the 1960s until 1971, then reactivated) was 49.1 metric tons of plutonium. The N-reactor also produced an estimated 8.1 metric tons of fuel-grade plutonium; during the 1980s, some of this fuel-grade plutonium was

73. National Research Council 1987, p. 2.

74. Benedict, Pigford, and Levi 1981, p. 459.

75. Saleska and Makhijani 1990.

mixed with super-grade plutonium for weapons.[76] The N-reactor produced weapons-grade plutonium in 1964 and 1965 and from 1982 to 1986, making an estimated 3.5 metric tons of it.[77] The N-reactor also produced tritium, another ingredient in nuclear bombs, but it was far more effective as a plutonium producer, and that was its primary mission between 1973 and 1987. The N-reactor produced fuel-grade plutonium during a portion of this period. The N-reactor stopped production in 1987 due to various safety concerns.[78] It was not formally shut until 1988, when it was placed on cold standby. It is not expected to operate again.[79]

N-reactor fuel is unlike commercial nuclear fuel, which consists of small uranium dioxide pellets stacked inside 13-to-15-foot-long thin fuel rods. Each N-reactor fuel rod consists of two concentric short, stubby cylinders of continuous uranium metal, not oxide.

There are about 2,100 metric tons of unreprocessed, irradiated N-reactor fuel stored underwater in two basins called K-East basin and K-West basin. Most of this N-fuel (1,750 metric tons) contains fuel-grade plutonium—that is, the plutonium in this irradiated fuel is about 12 percent plutonium-240, too much to be used as the primary fission component in nuclear warheads. However, 350 metric tons of N-fuel are low-burnup and contain weapons-grade plutonium. Much of the fuel is old (especially the fuel-grade material), some of it having been stored in water basins for up to 20 years or more.[80]

N-fuel has been handled very roughly. It has typically been removed simply by pushing it out of the reactor's horizontal fuel tubes and allowing it to drop into the storage pool. The resultant impact can dent the fuel element, causing a small defect in the cladding. Such small defects apparently provide a place for oxidation reactions to begin; if

76. "Super-grade" plutonium contains 2 to 3 percent plutonium-240. "Weapons-grade" plutonium is less specific but is used to describe plutonium generally containing less than 7 percent plutonium-240. "Fuel-grade" plutonium contains 7 to 18 percent plutonium-240, while "reactor-grade" contains over 18 percent plutonium-240 (Albright, Berkhout, and Walker 1993, p. 15).

77. Albright, Berkhout, and Walker 1993, pp. 33 and 35.

78. Among the concerns were acute aging phenomena, brought on by prolonged irradiation. In particular, radiation-induced nonuniform expansion of the graphite moderator was placing increasing stress on graphite cooling tubes, several of which had fractured, distorting the control rod system. The second major concern was radiation-induced embrittlement, exacerbated by hydrogen embrittlement, of the very systems subjected to stress from graphite expansion. Also of concern was the realization after Chernobyl that the understanding of what would happen if the N-reactor suffered a major loss of coolant was inadequate. Other uncertainties related to the reactor's ability to confine radionuclides during a severe accident. These issues are summarized in National Research Council 1987, pp. 25–28, 35–49, 49–55.

79. Saleska and Makhijani 1990, p. 16.

80. Saleska and Makhijani 1990, pp. 21, 23.

the fuel stays in water for a very long time, corrosion can be extensive.[81]

An estimated 3 to 7 percent of the cladding is damaged as a result of the rough handling and corrosion.[82] Thus, irradiated uranium metal is exposed directly to water in the cooling pools, allowing reactions producing uranium oxide and possibly some quantities of uranium hydride to occur. This has led to high levels of contamination in the water, which must be continually filtered to remove the radioactive fission products.[83]

Waste Generation and Environmental Contamination at Hanford A wide variety of hazardous substances are present in air, soil, surface water, groundwater, and sediments at Hanford (see table 6.4).

Hanford bears an unhealthy distinction as the U.S. site storing the largest volume of liquid high-level radioactive waste. As of 1992, Hanford stored 256,000 cubic meters of high-level radioactive wastes, containing 374 million curies (about 13.8 million terabecquerels) of radioactivity.[84] Some of the high-level wastes were processed to remove cesium-137 and strontium-90, which were then calcined and are stored as powders. Organic materials used for such processing were dumped into the waste tanks.

Between 1954 and 1957 ferrocyanides were added to some tanks to precipitate cesium-137 out of solution and add room for more-radioactive waste.[85] Large volumes of other chemicals were dumped into tanks over the decades, greatly increasing the already large volume of high-level radioactive wastes. The presence of other chemicals complicates the content of the tanks of liquid radioactive waste.

Hanford stores high-level radioactive waste in 177 tanks, ranging in volume from about 500,000 gallons (about 1.9 million liters) to over 1.1 million gallons (about 4.2 million liters) except for sixteen 55,476-gallon (about 210,000 liters) tanks.[86] Of these, 149 older tanks rely on one shell of carbon steel for the principal containment. The tanks have outer concrete cans underneath to catch leaking liquids; 68 of these tanks are declared leakers or potential leakers. The newest tanks, type V, have not leaked.[87]

81. Swanson 1988, pp. 3.8–3.9; Saleska and Makhijani 1990, pp. 23–24.

82. Saleska and Makhijani 1990, p. 23.

83. Dabrowski 1979, p. 4.

84. U.S. DOE 1992, p. 43.

85. IPPNW and IEER 1992, p. 63.

86. U.S. Congress, OTA 1991, p. 44.

87. U.S. Congress, OTA 1991, p. 44. Two new tanks have been added to the potential list of leakers since this report was published in 1991.

Table 6.4 Summary of Hazardous Substances Released to the Environment at the Hanford Site

Contaminant	Air	Soil	Surface Water	Groundwater
Radionuclides	Argon-41[1] Radon-222[1] Strontium-90[1]	Cesium-137 Ruthenium-106		Cesium-137 Gross alpha Gross beta Iodine-129 Plutonium-239 Plutonium-240 Radium Strontium-90 Tritium
Metals				Barium Cadmium Chromium Mercury
Inorganic compounds	Ammonia[1,2]			Fluorides Nitrates
Volatile organic compounds (VOCs)	Carbon tetrachloride[1]			Carbon tetrachloride[1] Chloroform Dichloromethane[1] Hexone[1] Methylcyclohexane[1] Perchloroethylene[3] Phthalates[1] 1,1,1-Trichloroethane
Miscellaneous		Pesticide rinsate[1] Untreated wastewater[1,4]	Untreated wastewater[1,4]	Coliform Kerosene[1] Oil Pesticide rinstate[1] Temperature[5] Untreated waste-water[1,4]

1. The present or potential contamination associated with current and past discharges of this pollutant has not been fully determined.
2. Ammonia is released into the air by the plutonium uranium extraction (PUREX) facility located at the Hanford Site.
3. This VOC is also known as tetrachloroethylene or tetrachloroethene.
4. The direct discharge of untreated sanitary wastewater and of process wastewaters containing radioactive and nonradioactive hazardous materials into the soil may have contaminated the soil and groundwater at the site.
5. Changes in ambient groundwater temperatures have been caused by effluent cooling waters.

Source: U.S. Department of Energy, Office of Environmental Audit, "Environmental Survey Preliminary Report—Hanford Site, Richland, Washington," DOE/EH/OEV-05-P, August 1987, and "Environmental Restoration and Waste Management Five-Year Plan for the Hanford Site—Predecisional Draft," April 1989, as cited in U.S. Congress, OTA, 1991.

Figure 6.8 High-level radioactive waste storage tank under construction at the Hanford Site, November 1984. Photo by Robert del Tredici.

Three to four dozen of the 177 tanks of high-level waste present a risk of fire or explosions due to the variety of chemicals in them. The chemicals that have been dumped into the tanks may react with one another; the presence of radioactivity also alters these chemicals and their reactions. Both these processes create new flammable compounds.

There are several kinds of risks posed by these tanks. For example, the tanks to which ferrocyanides were added also contain a variety of organic chemicals, nitrates, and nitrites. Ferrocyanides and nitrates are an explosive mixture if dry and will explode if hot enough. Fortunately, these tanks contain a great deal of water, so DOE considers the risk of explosion to be low. But DOE estimates are based on an incomplete knowledge of the tank contents, notably the identity of organic materials that are present and how that might affect the explosive properties of the mixture.

DOE has found that some tanks—the most well-known is Tank 101-SY—generate hydrogen gas, which has a relatively high explosive potential. This hydrogen had tended to accumulate in the tank due to the peculiar physical and chemical properties of its contents. In 1989, pressured by citizens' groups, the state of Washington, the U.S. Congress, and independent researchers, DOE acknowledged the seriousness of the potential for tank explosions and the need for a thorough investigation into the contents of the tanks. Public pressure also mounted after the disclosure of the disastrous 1957 explosion in a high-level waste tank at a Soviet nuclear weapons plant (see chapter 7). The contents of tank 101-SY are now stirred so as to vent the hydrogen continuously, thereby preventing a buildup to potentially explosive levels.

The production of plutonium at Hanford involved the discharge of large quantities of low-level liquid radioactive wastes directly into soil—roughly 800 billion liters over the last half century.[88] The processing of plutonium and its conversion to oxide and metal forms produce a considerable amount of waste contaminated primarily with transuranic elements. The most important are plutonium-239, plutonium-238, neptunium-237, plutonium-241, and americium-241. Hanford has three types of transuranic waste. About 10,180 cubic meters are in "retrievable storage"—that is, they are stored in drums. Another 109,000 cubic meters are buried—the wastes might have been packaged at one time but were dumped into the soil. Finally, there are 31,960 cubic meters of contaminated soil.[89]

As a result of large-scale dumping into the soil at various locations, Hanford is one of the most contaminated sites—perhaps the most contaminated—in the U.S. nuclear weapons complex. It also is where the risks of catastrophic tank explosions appear to be greatest.

Groundwater at Hanford has been contaminated with a variety of radioactive and nonradioactive hazardous materials. These include cesium-137, iodine-129, plutonium-239, plutonium-240, strontium-90, cadmium, chromium, mercury, fluorides, nitrates, carbon tetrachloride, chloroform, and various other organic compounds. Some of these groundwater contaminants have spread. According to Office of Technology Assessment of the U.S. Congress, "Tritium and nitrate contamination has been found in plumes totaling 122 square miles [316 square kilometers]. Other pollutants have been detected in more localized groundwater areas at levels that exceed drinking water standards."[90]

As noted above, two water-filled basins—K-East and K-West—store about 2,100 metric tons of irradiated N-fuel. These basins were originally used for the interim storage of fuel from the now-retired K reactors. DOE estimates that the K-East basin leaked about 15 million gallons (57 million liters) between 1974 and 1979, contaminating soil with an estimated 2,500 curies (92.5 terabecquerels) of mostly strontium-90 and cesium-137. The department has increased mitigation and

88. Alvarez and Makhijani, 1988, p. 46.

89. Makhijani and Saleska, 1992, p. 20.

90. U.S. Congress, OTA 1991, p. 150.

Figure 6.9 Ditch for low-level radioactive liquid waste disposal at the Hanford Site, 1984. Photo by Robert del Tredici.

monitoring activities, but leaks are always possible. In the absence of secondary confinement systems for the basins, leaks will discharge radioactivity directly into the soil.[91]

Indefinite continuation of basin storage poses a number of problems. First, degraded fuel continues to deteriorate, and there is no primary confinement for these leaking fuel elements. Second, the concrete in the K clearwells (adjacent to the basins and holding backup cooling water) has degraded. Third, a new roof built over the K-West storage basin may not be able to withstand an earthquake.[92]

Nevada Test Site

The Nevada Test Site (see figure 6.10) was established in 1951 so the AEC and the military could test nuclear weapons more cheaply and rapidly.[93] The location was selected even though prevailing winds blowing west to east meant that tests would carry radioactivity over almost the whole United States. Furthermore, the Western

91. Saleska and Makhijani 1990, pp. 21, 24.

92. Saleska and Makhijani 1990, p. 56.

93. IPPNW and IEER 1991, chapter 4.

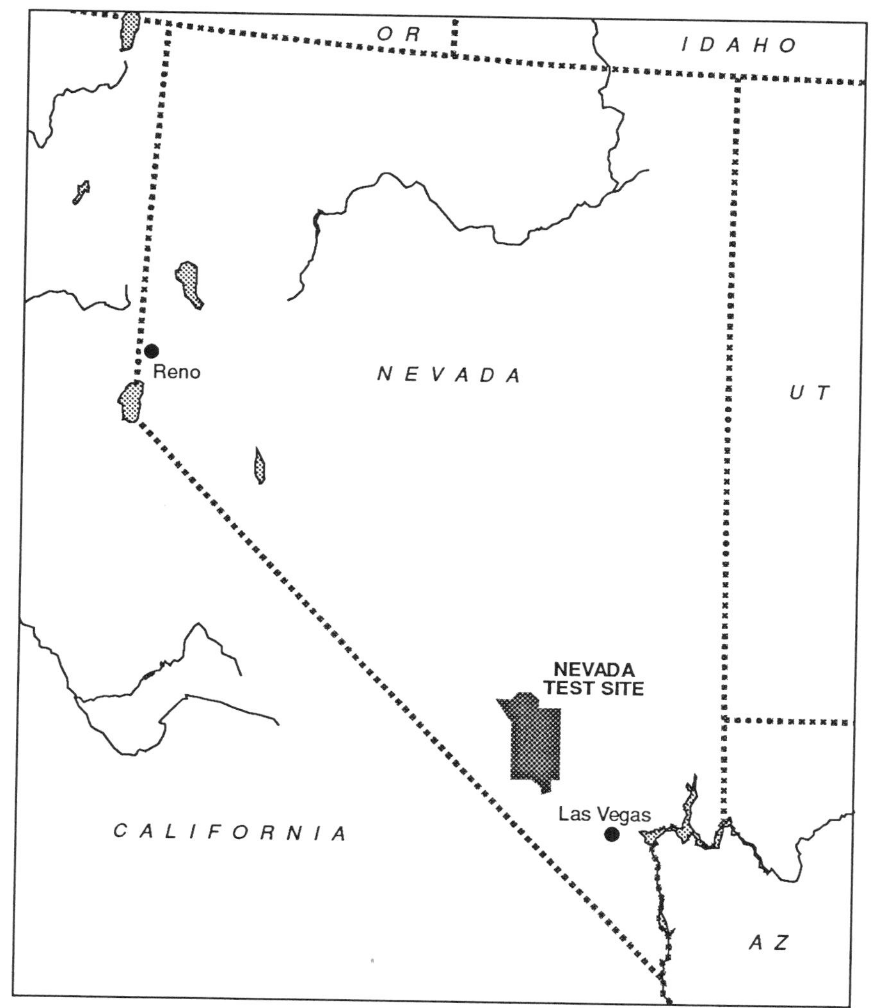

Figure 6.10 Location of Nevada Test Site.

Shoshone Nation states that it has never ceded this land to the federal government.[94]

Through 1992, the United States conducted 100 atmospheric tests and 825 underground tests at the Nevada Test Site and 126 tests elsewhere (mostly in the Pacific).[95] A testing moratorium, started in September 1992, was extended in July of 1993.

Waste Generation and Environmental Contamination at the Nevada Test Site Because of the many underground and near-

94. For a discussion of the land claims dispute, see Jaimes 1992, pp. 172–175.

95. IPPNW and IEER 1991, p. 51; U.S. DOE 1993d provides the updated total used here.

surface explosions conducted here, the Nevada Test Site is highly contaminated (see table 6.5). Underground nuclear tests have left an estimated 2.8 million curies (about 104,000 terabecquerels) of strontium-90, 4.5 million curies (about 167,000 terabecquerels) of cesium-137, and 124,000 curies (about 4,600 terabecquerels) of plutonium underground.[96] Groundwater contamination, presumably due to these test residues, has been documented at the site, but DOE expects to take several years to fully characterize the extent of contamination.[97]

About 420,000 cubic meters of radioactive waste have been buried at the site. As of 1991, this low-level radioactive waste contained about 9.8 million curies (about 363 million gigabecquerels).[98] DOE has not listed the specific amount of soil contaminated by transuranic solid waste, although it does list almost 600 cubic meters of retrievable transuranic wastes in storage. However, many parts of the Nevada Test Site have substantial amounts of transuranic contamination from plutonium dispersion tests that used a conventional explosive.

Aboveground nuclear tests spread radioactive fallout far beyond the site, reaching the East Coast and beyond. IPPNW and IEER have estimated that about 6 million curies (about 222 million gigabecquerels) of cesium-137 and on the order of 4 million curies (148 million gigabecquerels) of strontium-90 remain from atmospheric testing.[99]

Many workers at the test site have claimed that working conditions were dangerous and that exposures to radioactivity caused a variety of illnesses. Some of them filed a lawsuit to press their claims, but unlike the Fernald workers, they did not prevail.

Oak Ridge, Tennessee

Oak Ridge was one of the major materials production facilities built during World War II. Its principal function for the Manhattan Project

96. We have assumed 0.1 megacurie per megaton of explosive yield for strontium-90 and 0.16 megacurie per megaton for cesium-137. We also assumed that one-third of this radioactivity had decayed by 1993. We have escalated the estimated yield of 37 megatons for U.S. underground tests to 42 megatons to reflect the previously secret tests that were announced on 7 December 1993. We have also assumed a residue of 150 curies per nuclear test for plutonium-239. For a more detailed discussion of methodology and calculations, see IPPNW and IEER 1991, chapters 2 and 3.

97. U.S. Congress OTA 1991, p. 158.

98. U.S. DOE 1992, pp. 129–130.

99. Estimates are based on the total fission yield in megatons from U.S. atmospheric tests (72 megatons) and the total fission yield of global atmospheric tests (217.2 megatons). See IPPNW and IEER 1991, pp. 35 and 45. These figures are approximately corrected for radioactive decay.

Table 6.5 Summary of Hazardous Substances Released to the Environment at the Nevada Test Site

Contaminant	Air	Soil	Surface Water	Groundwater	Sediment
Radionuclides	Krypton-85 Plutonium-239 Tritium Xenon-133	Americium-241 Antimony-125 Beryllium-7 Cadmium-109 Cesium-137 Cobalt-60 Europium-152[1] Europium-154[1] Europium-155[1] Gross alpha Plutonium-238[1] Plutonium-239[1] Plutonium-240[1] Radium-226 Rhodium-106 Strontium-90 Uranium-235 Uranium-238 Yttrium-90[1]	Cobalt-60[1] Gross beta Plutonium Tritium	Antimony-125 Barium-140 Beryllium-7 Cadmium-109 Cerium-141 Cesium-137 Cobalt-60 Europium-155 Iodine-131 Iridium-192 Krypton Lanthanum-140 Plutonium-238 Plutonium-239 Plutonium-240 Rhodium-106 Ruthenium-103 Sodium-22 Strontium-90 Tritium	Cesium-137 Plutonium-239 Plutonium-240
Metals		Cadmium Silver	Chromium[1] Lead[1]	Lead	
Volatile organic compounds (VOCs)	Acetylene Benzene Hydrochloric acid Hydrofluoric acid Nitric acid Perchloric acid Toluene	Acetone[1] Chlorobenzene[1] Methylene chloride[1] Xylenes		Methylene chloride	
Miscellaneous	Gamma radiation[1]	Acids Caustics Chlorinated solvents Fission activation products Gamma radiation[1]		Gamma radiation[1]	

1. The present or potential contamination associated with current and past discharges of this pollutant has not been fully determined.

Source: U.S. Department of Energy, Office of Environmental Audit, "Environmental Survey Preliminary Report—Nevada Test Site, Mercury, Nevada," DOE/EH/OEV-15-P, April 1987, and "Comments on Site Summary" submitted by DOE on 18 June 1990, as cited in U.S. Congress, OTA, 1991.

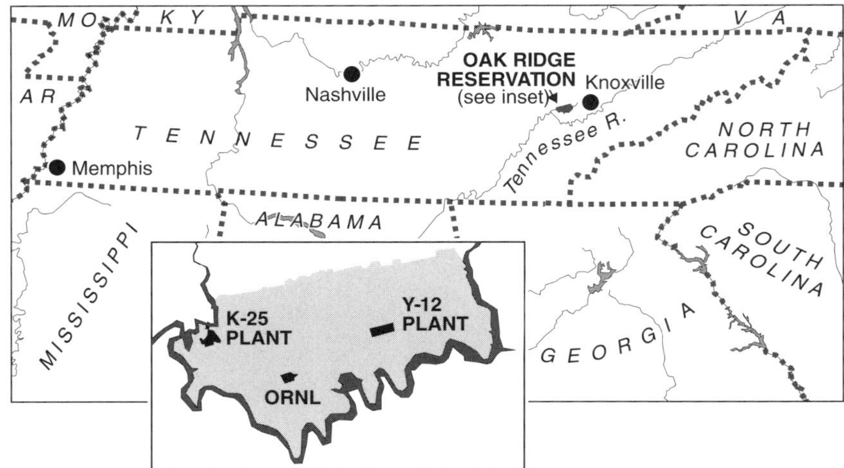

Figure 6.11 Oak Ridge Reservation. Source: U.S. DOE 1992b, vol. 1, p. 2-5.

was to produce enriched uranium. Materials from Oak Ridge went into the bomb dropped on Hiroshima on 6 August 1945.

The Oak Ridge Reservation, composed of three separate sites located within a few miles of one another (see figure 6.11), is about 20 miles from Knoxville, Tennessee. The sites correspond to Oak Ridge's three functions. A reactor in the X-10 section produced plutonium. The K-25 plant enriched uranium for warheads. The Y-12 plant's main purpose was manufacturing uranium components for nuclear weapons. For a short time, it also enriched uranium using the electromagnetic process.

During the Manhattan Project and in the years that followed, nuclear reactors at Oak Ridge produced various nuclear weapons materials. Hanford's large-scale REDOX plutonium-processing plant was tried on a pilot scale at Oak Ridge, and a prototype of Hanford's full-scale graphite-moderated reactor was built at Oak Ridge. Oak Ridge also made lithium-6, a basic ingredient for thermonuclear explosions. The process used mercury to separate lithium-6 from lithium-7.

Production at Oak Ridge With the end of U.S. nuclear weapons production, corresponding activities at Oak Ridge have stopped. However, two of the main facilities involving uranium enrichment and lithium production were shut even before the halt of weapons production. The gaseous diffusion plant went on standby in 1985.

The lithium-6 production plant closed in 1963.[100] In February 1993 DOE requested funds to decommission it.

The United States has produced highly enriched uranium at the K-25 plant at Oak Ridge and the X-326 building of the uranium enrichment

100. Cochran et al. 1987b, pp. 66, 75.

plant near Portsmouth, Ohio. DOE has not released figures on how production was partitioned between these two facilities, but a large proportion certainly occurred at Oak Ridge. It was built first, and production rapidly increased in the 1950s, while Portsmouth only went on-line in late 1954.

Until 1964, virtually the entire U.S. enrichment capacity was devoted to enriching uranium for nuclear weapons and naval reactors. Only about 3 percent of the total enrichment was for civilian light-water reactors. The total production of highly enriched uranium through 1992 was 994 metric tons, at which time it was stopped.[101] Current inventories of highly enriched uranium are lower since some of it has been used up in naval reactors, in nuclear weapons tests, and in research reactors.

Y-12 manufactured the highly enriched uranium components and subassemblies for nuclear weapons; it also dismantles and stores them. While the exact quantity is not known, Y-12 fabricated several hundred metric tons of highly enriched uranium into weapons components.

Waste Generation and Environmental Contamination at Oak Ridge

DOE and its contractors for the Oak Ridge Reservation have identified 600 contaminated sites that they say may require further investigation and remediation (see table 6.6).[102] The initial official estimate for radioactivity released from Oak Ridge between 1950 and 1968 was 240 curies (about 8.9 terabecquerels). Uranium released to the water accounts for about 15 curies (about 555 gigabecquerels), 24 to 34 curies (about 888 to 1,258 gigabecquerels) of uranium are buried, and about 15 curies were released to the air. In addition, an estimated 8 curies (about 296 gigabecquerels) of thorium originated in the Y-12 plant and are buried on site. An estimated 90 curies (about 3.3 terabecquerels) of technetium-99 have been released to the water and 10 curies (about 370 gigabecquerels) to the air.[103] However, the official estimates made by DOE and its contractors are often far too low, as is shown by recalculations IEER and others have made for releases from Fernald and other sites. Estimates of releases from Oak Ridge are now being revised in studies conducted under the supervision of the State of Tennessee.

During the 1940s and early 1950s, the X-10 site (now the site of the Oak Ridge National Laboratory) produced lanthanum-140, a beta- and gamma-emitting radioisotope with a half-life of about 40 hours. Large quantities of lanthanum-140 were produced in the graphite pilot reactor at X-10. The lanthanum-140 was then chemically separated from

101. DOE 1994b, p. 52.

102. U.S. Congress OTA 1991, p. 159.

103. Coyle et al. 1988, pp. 73, 74.

Table 6.6 Summary of Hazardous Substances Released to the Environment at the Oak Ridge Reservation

Contaminant	Air	Soil	Surface Water	Groundwater	Sediment
Radionuclides	Questionable[1]	Americium-241	Americium-241	Antimony-125	Americium-241
		Cesium-137	Cesium-137	Cesium-137[2]	Cesium-137
		Cobalt-60	Cobalt-60	Cobalt-60[2]	Cobalt-60
		Curium-244	Curium-244	Europium	Curium-244
		Plutonium-238	Gross beta	Gross alpha	Europium
		Plutonium-239	Strontium	Gross beta	Plutonium-238
		Radium-228	Tritium	Plutonium	Plutonium-239
		Strontium-90		Ruthenium-106	Strontium-90
		Uranium-232		Strontium[2]	Uranium-232
		Uranium-233		Technetium-99	Uranium-233
		Uranium-234		Thorium-232	Uranium-234
		Uranium-235		Tritium[2]	Uranium-235
		Uranium-238		Uranium-232	Uranium-238
				Uranium-233	
				Uranium-234	
				Uranium-235	
				Uranium-238	
Metals	Lead[2]	Mercury	Chlorine	Arsenic	Chromium
Inorganic compounds	Questionable[1]			Barium	Lead
				Cadmium	Mercury
				Chromium	
				Lead	
				Mercury	

Volatile organic compounds (VOCs)	Questionable[1]			Acetone Benzene Carbon tetrachloride Chloroform 1,1-Dichlorethylene Trans-1,2-Dichloroethylene Dimethyl phthalate Ethylbenzene Methylene chloride Naphthalene 1,1,2,2-Tetrachloroethane Toluene 1,1,1-Trichloroethane Trichloroethylene Xylene	Undefined VOCs[2]
Miscellaneous		Stored petroleum products[2]	Fecal coliform Total suspended solids	Endrin Stored petroleum products[2]	PCBs[3]

1. Although radionuclide and chemical releases to the air are in compliance, the facility's lack of documentation and quality control regarding reported emission estimates, as well as the inappropriate design and calibration of air samplers, are of concern.
2. The present or potential contamination associated with current and past discharges of this pollutant has not been fully determined.
3. PCBs = polychlorinated biphenyls.

Source: U.S. Department of Energy, Office of Environmental Audit, "Environmental Survey Preliminary Report—Oak Ridge National Laboratory (X-10), Oak Ridge, Tennessee," DOE/EH/OEV-06-P; "Comments on Site Summary" submitted by DOE on 18 June 1990; and Thomas Wheeler, Oak Ridge Reservation, personal communication, 9 July 1990, as cited in U.S. Congress, OTA, 1991.

the fuel slugs. The initial process was installed in 1945 and upgraded in the late 1940s. By the time upgrades were finished, Oak Ridge could produce batches of 10,000 curies of lanthanum-140.[104] The process had a number of problems, including ruptures of radioactive slugs that released large quantities of radioactivity.

According to a 1992 estimate by Martin Marietta, the operations of the graphite reactor during the 1940s and 1950s annually released 182,500 curies of argon-41 (about 6,750 terabecquerels), 550 to 1,200 curies (about 20 to 44 terabecquerels) of xenon-133, and 290 to 880 curies (about 11 to 33 terabecquerels) of iodine-131. Martin Marietta estimates the off-site radiation doses from these releases as less than 10^{-6} millisieverts (10^{-4} millirem) per year from xenon-133, between 0.02 and 0.06 millisieverts (2 and 6 millirem) per year from iodine-131, and about 0.009 millisieverts (0.9 millirem) per year from argon-41.[105] These doses are based on release estimates and other assumptions that have yet to be independently verified.

Draft estimates published in 1993 for upper-bound releases to the air from the X-10 facility appear to be far too high. The study counted all materials in the fuel rods not shipped out of the plant as having been released to the air. In other words, radionuclides stored in drums, released to streams, or disposed of in any way were all counted as potential air emissions. This was done to provide screening calculations for maximum possible release estimates. For instance, it estimated potential releases of cesium-137 as 280,000 curies (10,360 terabecquerels) in the worst single year for that radionuclide (1952). It estimated potential cumulative strontium-90 releases during 1944–1956 as about 1.3 million curies (about 48,000 terabecquerels), or about 6 times the strontium-90 in the fallout plume from an explosion in the waste tank at Chelyabinsk-65 in 1957 in the Soviet Union. It raised the iodine-131 potential release estimate from several hundred curies in the Martin Marietta estimate cited above to almost 600,000 curies (22,200 terabecquerels) from 1944 to 1956.[106] For materials other than noble gases and perhaps to some extent iodine isotopes, these screening estimates will probably be revised downward substantially when more refined calculations are done.

The immense range of release estimates for the X-10 reactor and associated chemical processing illustrates the nature of uncertainties confronting serious analyses of what actually happened in terms of doses and health effects as a result of nuclear weapons production. The main problems relate to secrecy, inadequate documentation, and the difficulties of reconstructing day-to-day events that took place decades

104. Unger 1951, p. 5.

105. O'Donnell 1992, attachment 1.

106. ChemRisk 1993, pp. 33–34 and appendix D.

ago. The worst periods of releases in the United States (as in the Soviet Union) were in the 1940s, 1950s, and 1960s. Assuming the release of the entire inventory of iodine-131, which affects the thyroid, the estimated dose would be several hundredths of a sievert (several rem) per year.[107]

Radionuclide contamination resulted from an accident at Oak Ridge—the "red oil" explosion of 20 November 1959. According to the official report, about 600 milligrams of plutonium spread "rapidly over a limited area south and east of buildings; Graphite Reactor Building most contaminated." The report states that some employees were contaminated with plutonium internally, but "no one received more than 2% of the life-time body burden of plutonium or an overexposure to sources of ionizing radiation either at the time of the incident or during subsequent cleanup operations."[108]

DOE estimates that about 440,000 cubic meters of low-level radioactive waste have been disposed of at Oak Ridge through 1991. In addition, about 41,000 cubic meters of "mixed low-level waste" were at the reservation in 1991.[109]

Mercury from Lithium Separation Discharges of large amounts of mercury, now estimated to be three-quarters of a million pounds (340,000 kilograms), occurred during lithium separation. Much of the mercury was discharged into the East Fork of Poplar Creek.[110] This has resulted in extensive leaching of contamination to ground and surface waters.[111] A 1985 study estimates that about 170 grams of mercury were discharged per day. Mercury concentrations in samples taken from a nearby pond averaged 3 micrograms per liter and ranged up to 99 micrograms per liter,[112] far above EPA's prescribed limits on mercury in surface water.[113] Mercury concentrations in uncontaminated surface waters are 0.02 to 0.05 micrograms per liter.[114]

Pantex Plant, Texas

The Pantex Plant near Amarillo, Texas, began operations in 1942. During World War II, its main function was to load conventional

107. O'Donnell 1992.

108. King and McCarley 1961, p. 2.

109. U.S. DOE 1992, pp. 117, 212.

110. U.S. DOE 1993d, p. 47.

111. Ahearne 1990.

112. Turner et al. 1985, pp. 7, 10.

113. U.S. EPA 1985.

114. Turner et al. 1985, p. 10.

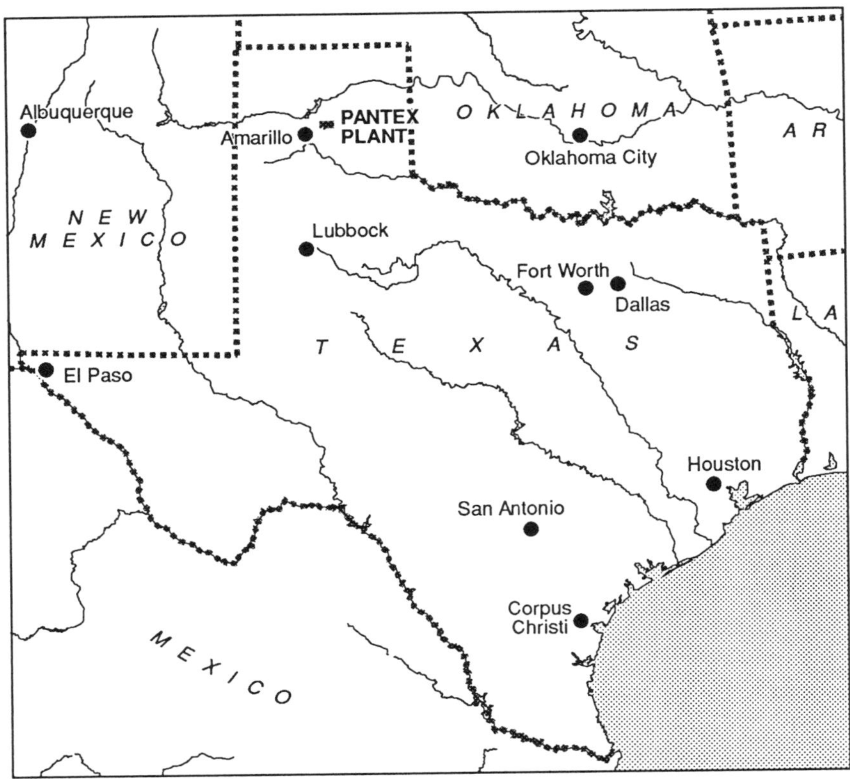

Figure 6.12 Location of Pantex Plant.

ordnance bombs and shells with explosive materials.[115] This factory was decommissioned in 1945. In 1949, the government sold the plant to Texas Technological College (now Texas Tech University) for one dollar. The army acquired the site in 1951 at the request of the AEC, so that the AEC could build a facility to assemble and disassemble nuclear weapons. Procter and Gamble was the operating contractor. The Mason and Hanger-Silas Mason Company, contracted to rehabilitate the facility, took over operating the plant in 1956 when Procter and Gamble declined to renew its five-year contract. In 1963, the AEC assumed full control of the site. In 1984, and again in 1989, several thousand additional acres were leased from Texas Tech as a security buffer.[116]

Pantex lies in the center of the Texas Panhandle for which it is named, in a rich agricultural area that produces grain crops and nearly 20 percent of all U.S. grain-fed beef (see figure 6.12). Large areas of the

115. TNT and Amatol were loaded into bombs and medium caliber shells. Pantex also fabricated tetrytol boosters during the War (West 1991, pp. 2–3).

116. U.S. DOE 1993b, p. 2.

site are still used for farming and raising cattle.[117] Beneath the region is the Ogallala Aquifer, the largest, most heavily used aquifer in the United States.

With the construction of Pantex and other major facilities in the early 1950s, a large-scale manufacturing operation for nuclear weapons was in place. Between the early 1950s and the mid-1960s, the U.S. nuclear weapons complex assembled well over 20,000 nuclear weapons. After that time, materials production declined, and the process of reworking materials from old weapons into new increased. Pantex enabled this by disassembling warheads and recovering components like plutonium and highly enriched uranium. (Further disassembly of highly enriched uranium components occurred at Oak Ridge.) The components were reworked at other sites in the weapons complex.

Pantex officially stopped assembling weapons in the early 1990s. However, it continues to maintain existing weapons systems, and also dismantles them. In some cases, disassembled weapons may be refurbished and reassembled for subsequent deployment. Since the early 1990s, activity at Pantex has centered on dismantlement of warheads and the storage of plutonium pits.

Production at Pantex Pantex has been the main facility that put together nuclear weapons components into the final product. Pantex assembled almost all of the over 60,000 nuclear weapons produced in the United States. It has also been responsible for disassembling nuclear weapons, but the exact number involved is not clear because the lines between assembly and disassembly are blurred. While some weapons may be permanently dismantled, others may be taken apart, according to Beverly Gattis, to "1) evaluate a random sample of a warhead type to see how the parts [or] systems are behaving over time, or 2) replace a part, update, [or] do maintenance. A disassembled weapon might be refurbished and reassembled to be deployed."[118] DOE has implied that some 50,000 nuclear weapons were permanently disassembled between 1945 and 1992 at Pantex,[119] but in 1993, DOE admitted that probably only 10,000 to 15,000 were actually permanently disassembled.[120]

Pantex has facilities for fabricating the nonnuclear high explosives that compress the plutonium trigger of a nuclear weapon. When a weapon is disassembled, the high explosive is removed to avoid an

117. Smith 1991.

118. Gattis 1993.

119. U.S. DOE 1992a, p. 1-1.

120. Gattis 1993; Walton 1993.

Table 6.7 Summary of Hazardous Substances Released to the Environment at the Pantex Plant

Contaminant	Air	Soil	Surface Water	Groundwater	Sediment
Radionuclides[1]		Gross alpha			
		Gross beta			
		Plutonium			
		Thorium			
		Tritium			
		Uranium			
Metals[1]		Beryllium	Chromium		
		Chromium	Copper		
		Copper	Lead		
		Lead	Silver		
		Silver			
Inorganic compounds[1]		Barium oxide	Cyanide		
		Hydrogen cyanide			
		Hydrogen fluoride			
		Sulfuric acid			
Volatile organic compounds (VOCs)[1]		Acetone		Acetone	Acetone
		Benzene		Benzene	Benzene
		Carbon tetrachloride		Carbon tetrachloride	Carbon tetrachloride
		Chloroform		Chloroform	Chloroform
		Dimethylformamide		Dimethylformamide	Dimethylformamide
		Ethyl acetate		Ethyl acetate	Ethyl acetate
		Methylene chloride		Methylene chloride	Methylene chloride
		Methyl ethyl ketone		Methyl ethyl ketone	Methyl ethyl ketone
		Methyl isobutyl ketone		Methyl isobutyl ketone	Methyl isobutyl ketone

	Tetrachloroethane	Tetrachloroethane	Ketone
	Tetrahydrofuran	Tetrahydrofuran	Tetrachloroethane
	Toluene	Toluene	Tetrahydrofuran
	Trichloroethylene	Trichloroethylene	Toluene
			Trichloroethylene
Miscellaneous	2,4-D[1,2]		
	Dioxin[1]		
	Gasoline		
	PCBs[1,3]		
	TNT[1]		

1. The present or potential contamination associated with these pollutants has not been fully determined.
2. 2,4-D = (2,4-dichlorophenoxy) acetic acid.
3. PCBs = polychlorinated biphenyls.

Source: U.S. Department of Energy, Office of Environmental Audit, "Environmental Survey Preliminary Report—Pantex Facility, Amarillo, Texas," DOE/EH/OEV-06-P, September 1987, as cited in U.S. Congress, OTA, 1991.

accidental detonation. The high explosives are burned in the open air in an area known as the Burning Ground.

Waste Generation and Environmental Contamination at Pantex Pantex has released both radioactive and nonradioactive hazardous materials into the environment (see table 6.7).

DOE reports only 134 cubic meters of low-level radioactive waste buried at Pantex.[121] This relatively small volume is explained by the fact that the plant ships its waste to other facilities.

One of the main sources of environmental concern due to Pantex is the potential contamination of the Ogallala Aquifer, which is about 150 meters deep in the area of Pantex. To date, no contamination has been detected in the aquifer. However, a number of crucial water systems at Pantex *have* been contaminated, and some evidence points to possible future contamination of the Ogallala.

Pantex has on-site a continuous system of "perched" aquifers, comprising shallow (about 77 meters), local zones of water.[122] In 1993, DOE reported that groundwater sampling in one of the perched aquifers (under the 110-hectare Zone 12 region) indicated the presence of various solvents, heavy metals, and high explosives, but the department maintained that perched aquifers are "distinctly separate" from the Ogallala.[123] However, a 1993 study found that "all recharging ground water that is perched" will eventually migrate "downward ... to the Ogallala aquifer."[124] Moreover, residents near the plant use water from the perched aquifer for drinking or agriculture, although it is not used for these purposes on site.[125] Furthermore, a 1988 DOE study found that the release of waste chemicals to unlined waste pits from 1954 to 1980 posed a risk of migration into groundwater, which would contaminate aquifers used for local water supplies. The study ranked this chemical contamination risk at Pantex among the greatest environmental hazards in the overall U.S. nuclear weapons production complex. The chemicals involved include toluene, acetone, tetrahydrofurane, methanol, dimethylformamide, methyl ethyl ketone, and ethanol.[126]

Environmental monitoring has documented 67 contaminants near ditches and *playas* (ephemeral lakes) on-site and in the buffer zone.[127]

121. U.S. DOE 1992, p. 129.

122. Gustavson 1993, p. 1; U.S. DOE 1993c.

123. U.S. DOE 1993c.

124. Mullican, Fryar, and Johns 1993, p. 1.

125. Gattis 1993.

126. U.S. DOE 1988, pp. ES-2, 2-74, A-119.

127. U.S. Army Corps of Engineers 1991.

Both toxic chemicals and metals appear in some groundwater samples. Nearby sediments indicate uranium-235 contamination above background levels. The largest *playa* provides irrigation water to cropland. Contamination of the *playas* could result if process effluents and rainwater run off through contaminated soils and into the *playas*.[128]

There is also evidence of uranium releases to the environment at Pantex. According to a 1985 DOE report, uranium in vegetation samples at Pantex exceeded background by 70 times, and uranium concentrations in the kidneys of jackrabbits on the site were four to six times greater than background.[129]

At least one radioactive release occurred when a 1989 accident opened the tritium reservoir of a nuclear warhead. About 40,000 curies (about 1.5 terabecquerels) of tritium were released, exposing one worker to an estimated radiation dose of 12 millisieverts (1.2 rem). The cleanup cost was about one million dollars.[130]

The plutonium pits from dismantled warheads are accumulating rapidly at Pantex. In times past, the pits were sent to Rocky Flats for reworking into new warheads. Now that new warhead production has stopped, the pits are being stored at Pantex for what the DOE has called an interim period. Since the long-term disposition of plutonium has not yet been decided, the "interim" storage of thousands of pits at Pantex could stretch out for decades. A 1994 report on plutonium by the National Academy of Sciences (NAS) concluded that it would take two or more decades to implement either of the two solutions it recommended for disposition of plutonium—burning plutonium in existing nuclear reactors in the form of mixed oxide fuel, or vitrification of plutonium mixed with high-level radioactive waste.[131] Wolfgang Panfosky, the chair of the plutonium study committee of the NAS, noted that "the world is condemned to having to baby-sit this material for at least another decade."[132]

Rocky Flats, Colorado

In 1951, the AEC announced plans to build the Rocky Flats nuclear weapons production facility to accommodate the enormous expansion of production planned for the 1950s; Los Alamos alone could not handle the anticipated growth. Rocky Flats did for the fabrication of plutonium parts what Hanford, Oak Ridge, and Savannah River had

128. U.S. DOE 1988, pp. 2-76, A-119.

129. U.S. DOE 1986, *Environmental Monitoring Report for 1985, Pantex Plant*, as cited in Coyle, et. al. 1988, p. 81.

130. U.S. DOE 1991a, p. 89.

131. National Academy of Sciences 1994, p. 13 and p. 189.

132. As quoted in the *Washington Post*, 25 January 1994.

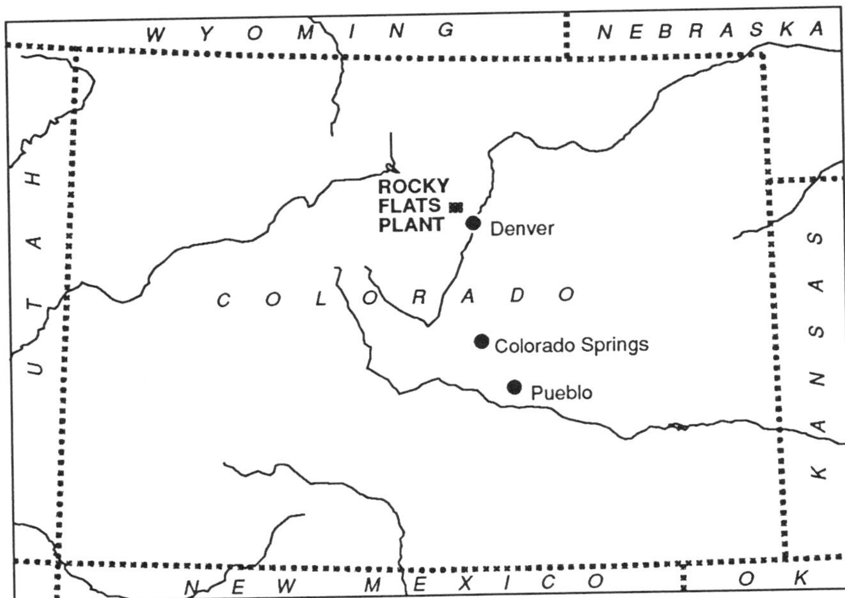

Figure 6.13 Location of Rocky Flats Plant.

done for materials production. It greatly expanded the capacity for building nuclear weapons, especially in connection with the new assembly operations at Pantex. Rocky Flats is about 16 miles (26 kilometers) northwest (and upwind) of Denver (see figure 6.13). It has been operated by Dow Chemical (1952–1975), Rockwell International (1975–1989), and EG&G (1989 to the present).

Production at the Rocky Flats Plant Rocky Flats has made virtually all plutonium pits in the U.S. arsenal. The plant has also removed plutonium from obsolete weapons and reworked it into new pit designs. The newer, more advanced weapons have about three kilograms of plutonium in the first trigger stage; older designs had four to five kilograms.

In addition to plutonium fabrication and recycling, Rocky Flats fabricated and assembled uranium, beryllium, and stainless steel into components for weapons. Americium-241, a byproduct resulting from plutonium recycling, was recovered and separated. Finally, Rocky Flats conducted some research and development on production processes.

Waste Generation and Environmental Contamination at the Rocky Flats Plant Plutonium fabrication is a significant source of transuranic waste. At Rocky Flats, this waste resulted both from processing plutonium-239 into pits and from reworking plutonium from old weapons. The latter process involved removing americium-241,

Figure 6.14 Aerial view of 700 Complex area, Rocky Flats Plant. U.S. Department of Energy photo.

which accumulates in the plutonium pits as a result of the decay of plutonium-241.

Two serious plutonium fires occurred at Rocky Flats, one in 1957 and the other in 1969. It is believed that the spontaneous ignition of plutonium scrap started the 1969 fire, which caused some $50 million in damage. Plutonium aerosols might have contaminated the Denver area if the fire had burned through the plant's roof.[133]

There is considerable controversy over the amount of plutonium emitted during these fires. Estimates for the 1957 fire are instructive not only for assessing the health and environmental impacts of Rocky Flats, but also for illustrating the uncertainties often associated with such figures. One investigation, by ChemRisk, a contractor for a Colorado Department of Health investigation of the plant, came up with a best estimate of 1 gram of plutonium, but the range was 0.03 grams to 38 grams, which indicates a twelve-hundred-fold uncertainty. Further,

133. Carter 1987, p. 66.

the investigation is continuing and the final figures may well be different.[134]

The 1969 fire marked a significant shift in the types of problems Rocky Flats presented for DOE and its contractors. Coming amid domestic political turmoil and a burgeoning U.S. environmental movement, the fire called attention to two facts about the plant. First, Denver was only 26 kilometers from this enormous plutonium-processing facility, with parts of the city's suburbs as little as 6.5 kilometers away.[135] Second, the fire raised awareness of the amount of transuranic wastes stored at Rocky Flats. The 1969 fire, as did the 1957 fire before it, significantly increased the volume of waste contaminated with plutonium. This is because a large volume of soil and equipment became so heavily contaminated with plutonium as to be rendered unusable; it had eventually to be classified as transuranic waste.

One principal political consequence of the 1969 fire was an AEC pledge to severely limit the quantity of transuranic wastes stored at Rock Flats and to establish a permanent repository elsewhere. The AEC moved the transuranic wastes to the Idaho National Engineering Lab and promised to remove the waste from Idaho by 1980.

At the time of the 1969 fire, the AEC was conducting research on radioactive waste disposal in a salt mine near Lyons, Kansas, although not to determine its feasibility as a disposal facility. After the fire, the AEC announced that Lyons would become a repository for transuranic wastes. However, a host of geological problems, including holes from previous mining and drilling, made the Kansas site unsuitable.[136] The removal of the wastes from Idaho has been repeatedly delayed. The repository location under consideration as of mid-1994 is the Waste Isolation Pilot Plant near Carlsbad, New Mexico. But that repository will not open for many years, if ever.

Production waste at Rocky Flats continues to create problems, notably in the "903" drum-storage area. About 5,000 drums of oil, most of which are contaminated with plutonium, were stored on site at Rocky Flats during the first dozen years of the plant's operation. Some drums corroded and leaked their contents into the soil. Dow Chemical Company, the contractor from 1952 to 1975, discovered the leaks in 1959 and added rust retardant. However, no drums were removed until 1967. By then, plutonium had contaminated a large volume of soil. Two and a quarter million kilograms of soil were shipped off for disposal in 1978. Although the storage area has been covered with asphalt, the soil below remains contaminated.[137]

134. ChemRisk 1994, pp. 258–259.

135. U.S. DOE 1990, pp. 2–13.

136. Lipschutz 1980, p. 119.

137. Rocky Mountain Peace Center 1992, pp. 31–32.

About 934 cubic meters of transuranic waste, containing 4,730 curies (about 175 terabecquerels) of alpha radioactivity, are stored at Rocky Flats.[138] An unknown amount of soil is contaminated with transuranic waste. DOE says that no low-level radioactive wastes are buried at Rocky Flats. Low-level radioactive wastes generated there are presumably either stored on-site or sent elsewhere for disposal. The plant still stores a large volume of mixed wastes—about 7,900 cubic meters as of 1991.[139]

The amount of plutonium at the site was until late 1993 a matter of both secrecy and speculation. In December 1993, the DOE released data on the plutonium inventory at Rocky Flats. It stated that 6.6 metric tons of plutonium metal, 3.2 metric tons of various plutonium compounds, and 3.1 metric tons "of various mixtures, etc." were stored at Rocky Flats. This total does not include "material that will be recovered during decontamination of buildings and equipment or material in waste."[140]

Rocky Flats used a variety of toxic materials, such as organic solvents and hydrochloric acid, for processing plutonium. Volatile organic compounds, which the plant used in large quantities, have been found in the aquifer.[141] One DOE study ranked this contamination as the largest off-site risk posed by Rocky Flats: the groundwater feeds into water supplies for the Denver metropolitan area.[142] Two streams and two reservoirs (both of the latter used for drinking water supply to local communities) have been contaminated with both radioactive and nonradioactive toxic materials.[143] The plant has also been a substantial source of emissions of carbon tetrachloride, a toxic ozone-depleting solvent (see table 6.8).

Kilogram quantities of plutonium have accumulated in Rocky Flats' ventilation ducts. In addition to increasing the risk of accidental releases to the environment, the accumulation of unknown quantities of plutonium and in unknown configurations poses a threat of an accidental nuclear criticality.[144] This would not mean a large nuclear explosion, but rather a release of energy and neutrons that could destroy equipment, cause considerable contamination, generate radioactive waste, and expose workers.

138. U.S. DOE 1992, p. 89.

139. U.S. DOE 1992, p. 212.

140. U.S. DOE 1993d, p. 28.

141. Coyle at al. 1988, p. 90.

142. U.S. DOE 1988, p. 2-90, A-153.

143. Rocky Mountain Peace Center 1992, pp. 36–37.

144. National Research Council 1989, p. 58.

Table 6.8 Summary of Hazardous Substances Released to the Environment at the Rocky Flats Plant

Contaminant	Air	Soil	Surface Water	Groundwater	Sediment
Radionuclides	Beryllium Plutonium[2]	Americium-241 Gross alpha Gross beta Plutonium Tritium[1] Uranium	Plutonium	Cesium-137[1] Gross alpha Gross beta Strontium[1] Tritium[1] Uranium[1]	Cesium-137 Plutonium-239 Plutonium-240
Metals		Lithium[1]		Beryllium[1] Cadmium Chromium[1] Lead[1] Manganese[1] Molybdenum[1] Nickel[1] Selenium[1] Thallium	
Inorganic compounds		Aluminum hydroxide[1] Ammonium persulfate[1] Cyanide[1] Ferric chloride[1] Hydrochloric acid[1] Lithium chloride[1] Nitrates Nitric acid[1] Sodium nitrate[1] Sulfuric acid[1]	Nitrates Sulfates	Aluminum hydroxide[1] Ammonium persulfate[1] Chloride[1] Cyanide[1] Ferric chloride[1] Hydrochloric acid[1] Lithium chloride[1] Nitrates[1] Nitric acid[1]	

Volatile organic compounds (VOCs)	Carbon tetrachloride	Acetone[1] Benzene[1] Carbon tetrachloride[1] Chloroform[1] Dichloromethane[1] Methylene chloride[1] Methyl ethyl ketone[1] Toluene[1]		Carbon tetrachloride Chloroform 1,2-Dichloroethane 1,2-Dichloroethylene Tetrachloroethylene 1,1,1-Trichloroethane Trichloroethylene
Miscellaneous	Laundry wastewater[3]	Disposed waste[4] Friable asbestos Oil sludge PCBs[1,5] Sanitary sewage sludge[4]	Disposed waste[4] Friable asbestos PCBs[1,5]	Disposed waste[4] Friable asbestos Oil sludge PCBs[1,5] Total dissolved solids

1. The present or potential contamination associated with current and past discharges of this pollutant has not been fully determined.
2. Primarily due to past accidental releases.
3. Significant releases of radionuclides into air may have occurred from 1969 to 1973 when radioactively contaminated sludges were dried at the facility's drying beds.
4. There is a potential for soil, surface water, and groundwater contamination because current practices do not prevent low-level radioactive waste improperly disposed of in landfill designed for hazardous waste.
5. PCBs = polychlorinated biphenyls.

Source: U.S. Department of Energy, Office of Environmental Audit, "Environmental Survey Preliminary Report—Rocky Flats Plant, Golden, Colorado," DOE/EH/OEV-03-P, January 1988; "Federal Facility Agreement and Consent Order—Rocky Flats Plant"; and "Report on Federal Facility Land Disposal Review," October 1987, as cited in U.S. Congress, OTA, 1991.

One of the most revealing problems in connection with the lack of government accountability on nuclear weapons production concerned an investigation into the possibility of filing criminal charges against the personnel of Rockwell, and against Rockwell itself, for violations of environmental laws. Rockwell was DOE's contractor for Rocky Flats until 1989. In 1990, a grand jury looked into whether such charges were warranted. When the evidence began to accumulate and it appeared that the grand jury would bring indictments against Rockwell personnel for serious criminal offenses, the Justice Department effectively stopped the investigation. It settled the case with Rockwell, which admitted to felony violations as a corporation and paid a fine of $18 million. While the fine was the largest in DOE history, a part of the settlement was that no criminal charges would be filed against any individual involved.[145]

Savannah River Site, South Carolina

In the 1950s, the AEC established the Savannah River Plant (later renamed the Savannah River Site [SRS]) for several tasks involved in making nuclear weapons, including expanding the agency's capacity to produce plutonium and providing facilities to make tritium on a large scale. SRS also has a heavy-water production plant, facilities for making fuel and targets, and a research laboratory. The site occupies 300 square miles (about 800 square kilometers) on the east bank of the Savannah River in South Carolina (see figure 6.15).

Five heavy-water–moderated reactors were built at SRS in the 1950s to produce plutonium and tritium. Two of these, the R and C, were shut down in 1964 and 1986, respectively. The remaining reactors (K, L, and P) were shut in 1988 for reasons of safety. Under the "restart program," the DOE planned to refurbish these reactors to restart them. Only the K-Reactor was restarted, and only for a two-month period. During the pre-startup test, almost 6,000 curies (222 terabecquerels) of tritium leaked from the reactor's cooling system. All told, the restart program cost over two billion dollars before it was canceled in March 1993.[146]

Two chemical extraction areas (200-F and 200-H, known as the "F and H Areas") contain facilities to extract plutonium and uranium from irradiated fuel. Each area has a chemical separations (reprocessing) facility (the F and H canyons) with two distinct zones, called the "hot" and "warm" canyons. In each hot canyon, spent fuel is

145. *New York Times*, 19 November 1992.

146. Energy Research Foundation 1993b.

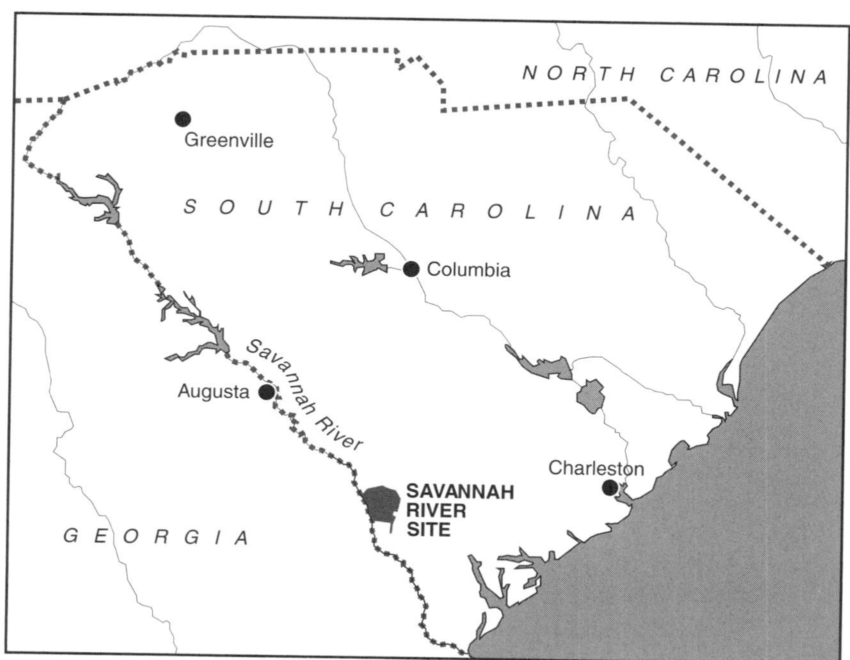

Figure 6.15 Location of Savannah River Site.

Figure 6.16 F-area reprocessing site at the Savannah River Site. Photo by Robert del Tredici.

dissolved, and fission products are separated from fuel. In the warm canyons, almost all remaining fission products are separated from the plutonium and uranium. The resulting uranium-plutonium solution is far less intensely radioactive.[147]

Plutonium and uranium leave each canyon in solutions of nitric acid (uranyl nitrate and two forms of plutonium nitrate), requiring further processing. F Area's "A-Line" converts uranyl nitrate into uranium trioxide. Through a series of processes, the "B-Lines" in both F and H canyons convert plutonium nitrate into plutonium metal.[148] These reprocessing plants started operation in 1953.

The reprocessing canyons are not officially shut down. DOE plans to restart limited operations in them and eventually phase them out. DOE restarted part of the H-Canyon, the HB-Line, in 1993 to provide plutonium-238 for the Cassini spacecraft. Since then, five workers have been contaminated with plutonium-238.[149] DOE has now agreed to prepare an environmental impact statement before restarting either canyon, although there may be certain exemptions.

The F and H areas also separated and processed tritium, beginning in 1955 and 1957, respectively. SRS recycles tritium, and the present plan is that it will continue to do so as long as nuclear weapons remain in the U.S. arsenal.

In general, the waste management system is complex at SRS. Fifty-one waste tanks are located in the F and H areas, with capacities ranging from 750,000 gallons to 1,300,000 gallons (2,800 to 4,900 cubic meters) each. Twenty-four of these tanks are of an older design and do not comply with new regulations. Other waste management facilities include tank farm evaporators (which condense radioactivity from tank farm liquids), and an effluent treatment facility (to treat waste water from the evaporators). Additional facilities and operations to process or contain waste include: extended sludge processing, waste removal projects, in-tank precipitation, a saltstone facility, a late-wash facility, a defense-waste-processing facility, a new waste transfer facility, a consolidated incinerator facility, a hazardous waste/mixed waste (HW/MW) disposal facility, M Area sludge stabilization, a transuranic-waste facility, solvent-storage tanks, nonradioactive hazardous-waste storage, and a new sanitary landfill.

E. I. Du Pont de Nemours and Company, which ran Hanford during the Manhattan Project, also had the contract for Savannah River from 1950 to 1989. The contractors as of 1993 are Westinghouse Savannah River Company, Bechtel, and Wackenhut Services, Inc. The last mentioned corporation provides security services.

147. Bebbington 1990, pp. 34, 35.

148. Bebbington 1990, p. 37.

149. Energy Research Foundation 1993.

Production at the Savannah River Site SRS produced essentially all the tritium in the U.S. arsenal. In a complex of buildings, irradiated lithium-aluminum targets were processed to separate tritium from other materials; this tritium was then purified. SRS also processed tritium recovered from weapons components at the Mound, Ohio, plant, and from stockpiled weapons components. Assuming about 4 grams per weapon and 20,000 weapons, the inventory of tritium as of 1990 was about 80 kilograms in the weapons themselves. Additional tritium is in storage. Savannah River produced an estimated 225 kilograms of tritium in all.[150]

SRS produced about 36 metric tons of weapons-grade plutonium, or about 40 percent of the total estimated to have been produced in the United States.[151] It also produced tritium, plutonium-238, and neptunium-237. The United States has also imported about one metric ton of weapons-grade plutonium from Great Britain.[152]

Waste Generation and Environmental Contamination at the Savannah River Site In terms of both volume and radioactivity, SRS has large quantities of low-level, high-level, and transuranic wastes (see table 6.9).

SRS contains about 54 percent of the total radioactivity from high-level waste at all DOE sites. Of the 51 tanks for high-level waste at SRS, 1 has been emptied and the remaining 50 contain about 128,000 cubic meters of waste. The total radioactivity as of 1991 was about 538 million curies (about 19.9 million terabecquerels).[153] Nine of the 16 single-shelled tanks from the 1950s have leaked. Some leaks from tanks and their pipe joints contaminated the soil or water around the site. The aquifers beneath SRS are shallow, and some of the underground tanks are actually partially under the water table. In the early years SRS only recorded leaks it considered significant, and no accurate record exists of high-level-waste management problems before the mid-1960s.[154]

Some SRS tanks generate substantial amounts of hydrogen. Although the tanks have ventilators to prevent this gas from building up, failure of these systems or inadvertently leaving them off can result in ignitable or even explosive levels of hydrogen. There have been two documented buildups over the lower explosive hydrogen concentration limit.[155]

150. Albright, Berkhout, and Walker 1993, p. 34.

151. US DOE 1993d, p. 21. Reactor-grade plutonium not included.

152. Albright, Berkhout, and Walker 1993, pp. 34, 43.

153. U.S. DOE 1992, p. 43.

154. Makhijani, Alvarez, and Blackwelder 1986, p. 20.

155. Makhijani, Alvarez, and Blackwelder 1986, table 5 in part 2. See also U.S. DOE 1991a, p. 79 for information on flammable gas buildup in DWPF treatment process.

Table 6.9 Summary of Hazardous Substances Released to the Environment at the Savannah River Site

Contaminant	Air	Soil	Surface water	Groundwater	Sediment
Radionuclides	Carbon-14 Iodine-129 Technetium-99 Tritium Unknown nuclides[2]	Cesium-137 Gross alpha Gross beta Iodine-129 Iodine-131 Strontium-90 Tritium	Cesium-137[1] Cobalt-60[1] Gross alpha[1] Gross beta[1] Iodine-129[1] Iodine-131[1] Strontium-90[1] Tritium Uranium	Cesium-137[1] Cobalt-60[1] Gross alpha Gross beta Plutonium-238 Plutonium-239 Radium Ruthenium-106 Strontium-90 Tritium Uranium	Cerium-243 Cerium-244 Cesium-137 Gross alpha[1] Gross beta[1] Iodine-129[1] Iodine-131[1] Strontium-90[1] Thorium-228 Tritium Uranium-235 Uranium-238
Metals	Mercury		Chromium Copper Mercury Silver	Barium[1] Cadmium Iron Lead Magnesium[1] Manganese Mercury Sodium[1] Zinc	Chromium Copper Mercury Nickel Silver
Inorganic compounds	NOx[1]		Cyanide	Chloride[1] Sulfate	Cyanide

Volatile organic compounds (VOCs)	1,1,1-Trichloroethane Unknown VOCs		Tetrachloroethylene 1,1,1-Trichloroethane Trichloroethelyne Trichloroethane	
Miscellaneous		Stored petroleum products[1]	Coal reject effluents Temperature[3]	Endrin Stored petroleum products[1] Phenol[1] Solvents[1]

1. The present or potential contamination associated with current and past discharges of this pollutant has not been fully determined.
2. Releases of other radionuclides may also have occurred but sampling equipment and monitoring procedures were inadequate.
3. Thermal impacts associated with the discharge of cooling waters to this medium include deforestation, changes in water levels, reduction of oxygen levels, and increased erosion and sedimentation.

Source: U.S. Department of Energy, Office of Environmental Audit, "Environmental Survey Preliminary Report—Savannah River Plant, Aiken, South Carolina," DOE/EH/OEV-10-P; "Comments on Site Summary" submitted by DOE on 18 June 1990; and Thomas Wheeler, Oak Ridge Reservation, personal communication, 9 July 1990, as cited in U.S. Congress, OTA, 1991.

Figure 6.17 Burial marker for low-level radioactive waste at Barnwell, South Carolina. Photo by Robert del Tredici.

In the 1980s, a new facility for vitrifying SRS's high-level waste was built. Known as the Defense Waste Processing Facility (DWPF), this facility is intended to reduce the volume of waste stored at SRS by solidifying difficult-to-handle sludges. Due to poor management and planning, growing costs and unanticipated problems have delayed the original startup date of 1989 until at least 1995.

The DWPF will concentrate the radioactivity by processing the wastes before vitrification. A large volume of "low-level" radioactive liquid containing some long-lived radionuclides will be mixed with cement and permanently disposed of at SRS. (The cement-waste mixture is called "saltstone.") In extracting cesium-137 from waste, the plant uses sodium tetraphenylborate. The residues from this extraction process degrade to form benzene when stored in tanks, increasing the risk of fire or explosion. This problem was not anticipated during project design and was one cause of the delay.

Plutonium production has also resulted in large quantities of transuranic wastes. An estimated 3,140 cubic meters are in storage; 4,530 cubic meters have been buried; 38,000 cubic meters consist of contaminated soil.[156] Until 1965, SRS buried transuranic waste in cardboard boxes and plastic bags. In 1974, restrictions on burying

156. Makhijani and Saleska 1992, p. 20.

transuranic waste were made more stringent; the more dangerous portion of transuranic waste was put in temporary drums for later shipment to a permanent repository.[157]

About 637,000 cubic meters of low-level waste are buried at SRS. One of the more notable features in the history of low-level-waste disposal at the site has been the rapid migration of plutonium into shallow aquifers. Using a computer model, SRS scientists predicted that plutonium would not migrate into groundwater for hundreds of thousands of years because ion exchange would trap it in the soil. But ion exchange only keeps plutonium in soil for long periods in the absence of complicating factors. For instance, plutonium can migrate much faster in the presence of solvents. At SRS, the solvent tributyl phosphate leaked into the soil, probably contributing to the rapid migration of plutonium in the groundwater over 12 acres (about 5 hectares) of the burial ground. By 1981, less than 20 years after the solvent entered plutonium-contaminated ground, Du Pont acknowledged that plutonium had migrated into groundwater.[158]

ENVIRONMENTAL POLICY ISSUES IN U.S. NUCLEAR WEAPONS COMPLEX

One of the principal tasks facing the U.S. nuclear weapons complex is the cleanup of the vast amounts of contamination that have resulted from over half a century of nuclear weapons production. Accomplishing this will require immense effort, technological innovation, and financial resources. In addition, decommissioning of weapons facilities and cleanup of the sites will create new waste management and disposal issues that will add to the already daunting ones that exist from weapons production. Citizen activism and the relative openness of information in the United States have resulted in a great deal of public information about policy issues related to cleanup and waste management. We therefore present this additional discussion in this chapter, since it may help inform similar efforts in other countries.

The principal challenges facing cleanup include:

- large quantities of radioactive, nonradioactive, and mixed wastes, whose management poses continuing difficulties and costs;
- widespread and partly irremediable contamination of soil and water;
- decommissioning and cleanup problems that will result in more waste generation.

Table 6.10 shows the aggregate wastes stored or buried throughout the U.S. nuclear weapons complex. It is unclear how much waste is

157. Energy Research Foundation 1993a.

158. Makhijani, Alvarez, and Blackwelder 1986, p. 78.

Table 6.10 Radioactive Waste in the U.S. Nuclear Weapons Complex, 1991

Type of Waste	Volume (thousands of cubic meters)[1]	Radioactivity (thousands of curies)	Radioactivity (thousands of terabecquerels)	Main Sites
High-level waste	395	970,000	35,890	Savannah River, Hanford, Idaho National Lab
Transuranic waste	Over 255	2,722	101	Idaho National Lab, Hanford, Oak Ridge, Savannah River, Los Alamos
Low-level waste	2,816	13,430	497	Hanford, Fernald, Nevada Test Site, Oak Ridge, Idaho National Lab, Savannah River Site
Mixed waste[2]	101	?	?	Fernald, Portsmouth, Idaho National Lab, Oak Ridge, Rocky Flats, Savannah River Site
Mill tailings[3]	118,000	?	?	Many sites

1. Figures rounded to the nearest thousand cubic meters.
2. Mixed Waste refers to low-level radioactive waste mixed with nonradioactive waste.
3. The radioactivity in tailings ranges from 26 picocuries per gram to 400 picocuries per gram (about 1 to 15 becquerels) of radium-226 for solids and 20 picocuries per liter to 7,500 picocuries per liter (about 0.74 to 280 becquerels) of radium-226 for liquids. Radioactivity levels from thorium-230 range from 70 to 600 picocuries per gram (about 2.6 to 22 becquerels). See U.S. DOE 1992, p. 155.

Source: U.S. DOE 1992, pp. 9, 87–91, 129, 130, 151, 212, 220.

buried at most sites. The table includes the volume of uranium mill tailings, although most of this has been due to producing uranium for commercial power. However, on the order of 100,000 curies (3,700 terabecquerels) each of radium-226 and thorium-230 would have been associated with producing highly enriched uranium for weapons. The burial or disposal locations of each of these types of wastes, except mill tailings, are given in Figures 6.18 to 6.21. In the case of low-level waste, only buried waste is given, thus excluding low-level waste disposed of by other means.

Cleaning up the highly contaminated nuclear weapons complex will produce large quantities of radioactive and mixed wastes, adding to already formidable quantities of existing waste. These will have to be managed and isolated from the environment so far as that is possible. The following categories of decommissioning and cleanup problems will contribute to waste volumes and radioactivity:

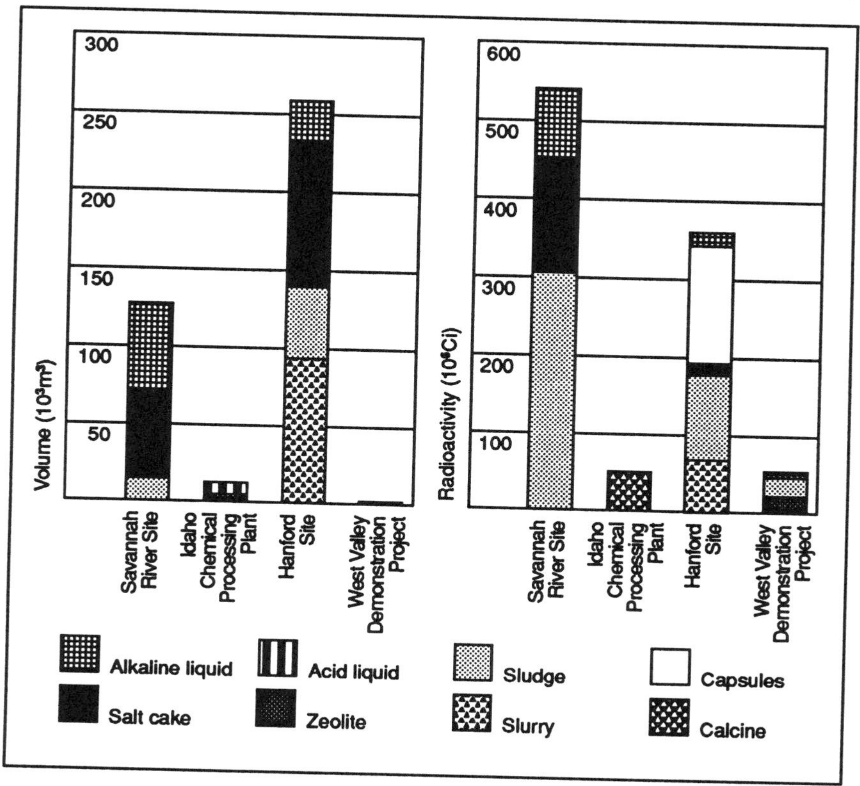

Figure 6.18 Distribution of total cumulative volume and radioactivity of high-level wastes by site and type through 1991. Source: U.S. DOE 1992, pp. 55, 56, 58.

- decommissioning highly contaminated process buildings and equipment;
- remediating disposal pits and other former dumping grounds;
- cleaning up or scraping up contaminated soil;
- decommissioning nonprocess buildings that are contaminated to varying degrees;
- cleaning up or collecting contaminated sediments;
- disposing of abandoned contaminated equipment; and
- remediating contaminated groundwater to the extent possible.

For several years, DOE has had a budget for waste management and environmental restoration of about $5 billion or more, and overall expenditures at this level or higher are expected for about 30 years. But despite these large expenditures, DOE is not on a course that would resolve the environmental legacy of nuclear weapons production. Planning for the cleanup—and implementation to date—lacks essential ingredients:

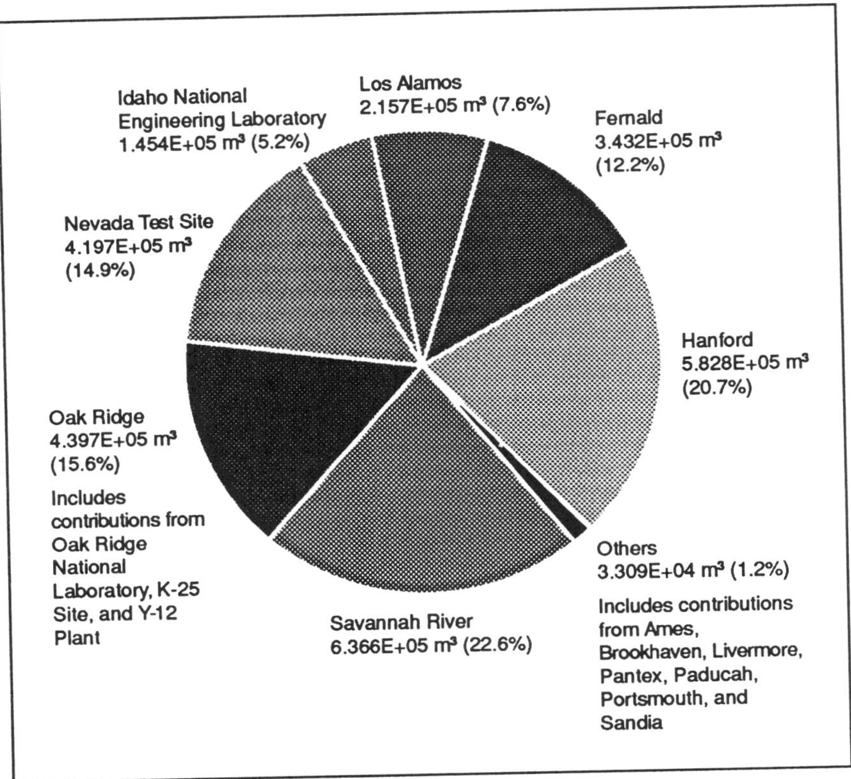

Figure 6.19 Total volume inventory of DOE buried low-level wastes through 1991. Source: U.S. DOE 1992, p. 119.

- classification of waste must be based on hazard rather than waste origin;
- cleanup standards must be based on clear health and environmental protection criteria;
- a long-term waste management plan is needed that will minimize risks to health and environment;
- clear priorities must allow for rapid interim actions to prevent serious situations from worsening and for the systematic minimization of risks to workers, people living off the site, and future generations.

Reclassification of Waste

Radioactive wastes in the U.S. are primarily classified by source (reactor spent fuel, uranium mill tailings, etc.) rather than by hazard and longevity characteristics. Thus, decisions about the disposal of radioactive waste are based to a large extent on where it originated, and do not systematically include the radioactivity level, type, half-life, and other aspects relevant to environmental and health problems. Due

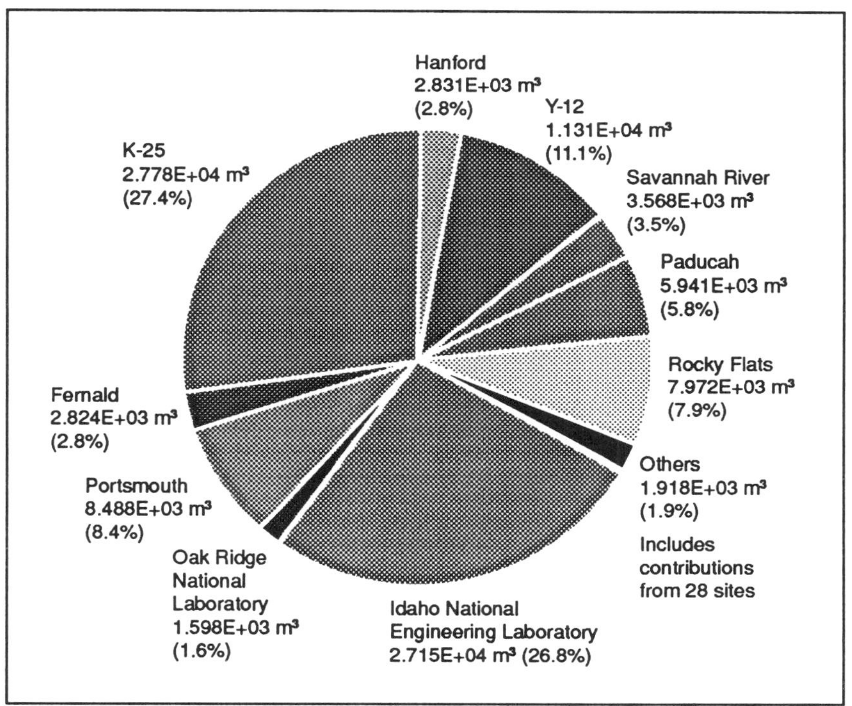

Figure 6.20 Total volume inventory of DOE mixed low-level wastes through 1991. Source: U.S. DOE 1992, p. 212.

to the uncertainty of source terms, many, if not all, DOE and contractor estimates of radionuclide releases from plants and amounts disposed of as waste need to be revisited. Official estimates cannot be accepted at face value.

The United States classifies radioactive wastes arising from the nuclear weapons program in four broad categories: high-level, transuranic, low-level, and uranium mill tailings. In addition, "mixed waste" is a hybrid containing both radioactive and nonradioactive hazardous materials.

High-level waste is either spent fuel from reactors or the highly radioactive waste containing fission products that remains after plutonium and uranium are extracted from spent fuel.

Transuranic elements have an atomic number over 92, the atomic number of uranium. For all practical purposes, they do not occur naturally. Government regulations define transuranic waste as having 100 nanocuries (3,700 becquerels) of radioactivity per gram of radionuclides with half-lives longer than 20 years. (Before 1984, the DOE definition was 10 nanocuries, or about 370 becquerels, per gram.) Thus, transuranic elements up to 100 nanocuries per gram may be disposed of in waste that is not as transuranic, but rather as "low-level"

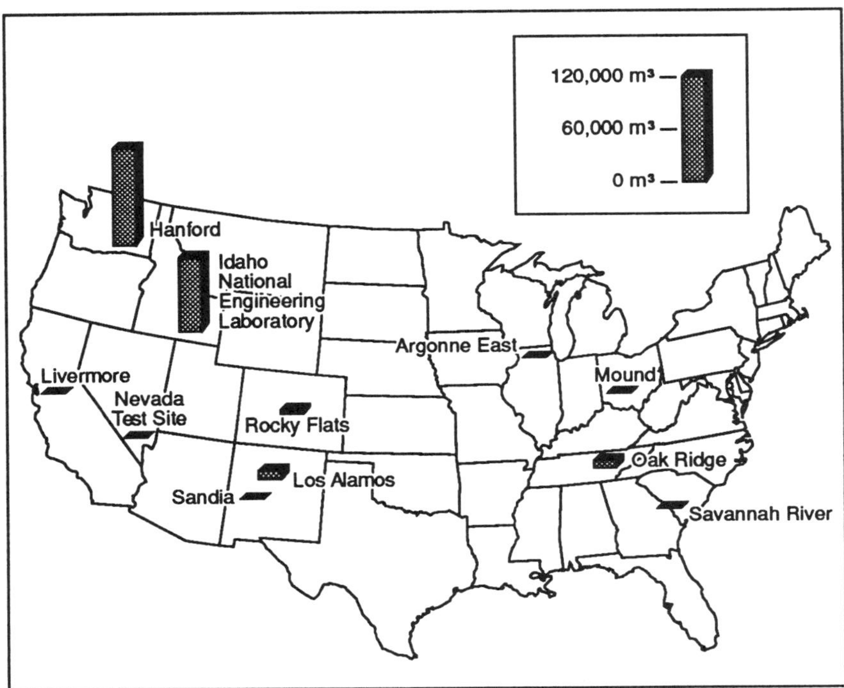

Figure 6.21 Locations of buried and stored DOE transuranic wastes through 1991. U.S. DOE 1992, p. 81.

waste.[159] Transuranic wastes emit principally alpha particles and typically have long half-lives, necessitating repository disposal.

Low-level waste is almost everything that does not fit into the high-level or transuranic waste categories. As a result, it encompasses a wide range of radioactivity levels and half-lives, from slightly radioactive trash (such as mops and gloves) to highly radioactive activated metals from inside nuclear reactors. Low-level waste is further categorized as Class A, Class B, Class C, or Greater-than-Class-C waste, where Class A is the least radioactive, on average. To further complicate the matter, low-level waste disposal regulations require that Greater-than-Class-C waste be disposed of in a manner similar to high-level waste, unless an exemption is granted.

Cleanup Standards

Cleanup standards have fallen far short of criteria that would minimize health risks and ecological harm. Often cleanup standards are either outdated or—in some cases—nonexistent. For example, a 1981 Nuclear Regulatory Commission technical paper and associated mate-

159. Makhijani and Saleska 1992, p. 17.

rials have served as a guide to some, mostly nonmilitary, decommissioning activities.[160] Generally cleanup guidelines have not been incorporated into regulations. The DOE has its own internal guidelines for its clean-up program.

Only one soil cleanup standard applies to a radionuclide. It covers radium-226 near uranium-mill tailings. There is a cleanup guideline for plutonium, but it is not mandatory. Its application has been haphazard. The cleanup guideline of 0.2 microcuries of plutonium per square meter was used at Johnston Atoll, but this figure was five times more stringent than the standard used at Enewetak.[161] Moreover, the guideline can be interpreted in various ways in terms of plutonium contamination per unit weight of soil, depending on the depth and density of soil considered.

The U.S. Environmental Protection Agency is developing cleanup and waste management standards that will apply to the DOE (and other institutions). These standards are expected to be finalized in late 1995.[162]

Long-Term Waste Management

DOE's plans for long-term waste management have involved much expenditure but little result. The department has focused its financial, technical, and institutional resources on siting deep repositories as soon as possible for transuranic and high-level civilian and military wastes. Ironically, this emphasis, guided by political expediency, has resulted in poor decisions, delays, inappropriate priorities, and wasted funds. DOE estimates for an opening date for a high-level waste repository are now much farther away than those of two decades ago.[163]

The largest expenditure of the program went to settling on a repository at a controversial site in a politically weak state, Nevada. This repository is known as Yucca Mountain for the small mountain that will house the high-level waste repository. However, the site has 32 known active faults. Volcanic and related tectonic activity could endanger the repository, and the water table under the repository could rise.[164]

Most of the waste that would go to Yucca Mountain, should the site be found suitable and licensed, would be spent fuel from civilian

160. The technical paper applies to NRC licensees that fall under the Site Decommissioning and Management Plant list (Dircks 1981).

161. IPPNW and IEER 1991, p. 86.

162. For a summary of cleanup regulations, including proposed draft standards, see U.S. EPA 1993.

163. Makhijani and Saleska 1992, p. 66.

164. Makhijani and Saleska 1992, pp. 60–61.

nuclear power plants. According to plans as of 1993, Yucca Mountain is also the disposal site for vitrified high-level waste from SRS and Hanford, but the vitrification program at Hanford is in considerable flux. The eventual outcome of putting high-level waste in a suitable form for disposal in a repository is still highly uncertain, as is the volume of waste that will arise from eventual processing of Hanford high-level waste.

DOE has also spent large sums on a plan to mix some of the high-level wastes in tanks at Hanford, containing about 32 million curies (about 1.2 million terabecquerels) of radionuclides, with cement. DOE planned to discharge the cement-waste mix (called grout) into shallow land "grout vaults" at Hanford.[165] By increasing the volume of this high-level waste, DOE planned to dispose of it as low-level waste and avoid the more expensive vitrification. This would have more than doubled the total radioactivity of low-level waste dumped at DOE sites during the entire period of nuclear weapons production.[166] However, the cement failed to set properly due to the large amounts of chemicals and radionuclides in it, and the project has, for the present, been shelved.

DOE has similar questionable plans for transuranic wastes. These wastes are slated to go to the Waste Isolation Pilot Plant (WIPP) salt repository. However, WIPP applies principally to retrievably stored wastes. The most threatening transuranic wastes in the short term are buried wastes and contaminated soil, but WIPP will have no room for most of these.[167] Dealing with buried wastes and contaminated soil has been a far lower priority for DOE than opening WIPP, despite the many indications—including pressurized brine pockets and a premature roof collapse—that WIPP may be an unsatisfactory repository.[168] In 1993, however, the DOE moved its repository program for transuranic waste (the Waste Isolation Pilot Plant) onto sounder footing by allowing EPA to proceed with laboratory tests before placing any wastes in the repository.

At the same time that DOE moves ahead with scientifically questionable programs, it is failing to attend to serious problems concerning high-level waste with vigor and imagination. These problems include how to mitigate the risks of fires and explosions, how to empty the most dangerous tanks, and how to stabilize tank contents to eliminate the threat of explosions.

165. Martin 1993, p. 9.

166. U.S. DOE 1992, p. 121. The total amount of low-level waste disposed of on DOE sites by 1992 contained 13,430,000 curies.

167. Makhijani and Saleska 1992, pp. 20, 58.

168. Makhijani and Saleska 1992, pp. 51–57.

DOE has no credible estimates of the cost of cleanup or of long-term waste management for the U.S. nuclear weapons complex. The cost will partly depend on cleanup standards, the technologies for implementing the cleanup, and the manner in which the resulting radioactive and other hazardous wastes are managed. The price could run into the hundreds of billions of dollars.[169] Even so, there is no good, proven solution for the safe, long-term management of long-lived radioactive wastes.

HEALTH EFFECTS OF U.S. NUCLEAR WEAPONS PRODUCTION

From the inception of the nuclear industry, there was interest in the health risks to scientists and production workers—particularly those risks whose effects could be noticed over short post-exposure periods. Likewise, recruits to the emerging field of health physics attempted to develop monitoring equipment and procedures and set exposure standards that would protect the health of workers and members of the public who lived around weapons facilities. Unfortunately, from the mid-1940s to the mid-1980s, weapons production goals of the U.S. nuclear weapons industry conflicted with considerations for protection of the health of workers and the public.

Before the mid-1960s, there was little effort to systematically study the health risks associated with the nuclear weapons industry. During this period only three papers were published on cancer mortality in counties near nuclear facilities,[170] and no studies had been designed to evaluate the health of workers. In 1964, the AEC began a study of the lifetime health and mortality experiences of the employees of its contractors, with Thomas Mancuso as its director. The early history of this program and the AEC's attempts to influence its research have been documented by many.[171] From the mid-1970s until 1993, the successor agencies of the AEC directed occupational and environmental research efforts that were contracted to a small group of researchers with close ties to these agencies. In the late 1980s, the Department of Energy began funding the Center for Environmental Health at the Centers for Disease Control and Prevention and state health departments to conduct research on the health risks to people who lived near weapons facilities. In 1992, the DOE began to fund the National Institute for Occupational Safety and Health to conduct epidemiologic research on worker populations.

This review describes both epidemiologic studies of workers at various nuclear weapons facilities and studies of groups of the general

169. U.S. Congress, OTA 1991, appendix C.

170. Moshman and Holland, 1949; Fadeley, 1965; Bailar and Young, 1966.

171. See Stewart and Kneale, 1991.

public who lived around these facilities. In preparing this summary, we reviewed papers published in peer-reviewed literature as well as reports of studies that have not been published in scientific journals or monographs accessible to most readers. Some were preliminary analyses which will likely be published later in scientific journals. Some were not, and a few serve as the only data on the health risks collected for certain facilities.

A number of deficiencies in the dose data and estimates must be kept in mind when reading the results of the epidemiological studies described here. There are somewhat different issues involved for off-site and worker populations. As we have already noted, for off-site populations, it is important to know the magnitude of releases and the patterns of the releases (short-term bursts versus continuous releases) in order to estimate which people may have been the most affected. This is important for identifying the populations that are selected for epidemiological study and in stratifying those populations. Further, some of the off-site populations had family members who were workers in the plants. In general, the data and the methods to factor in these issues so as to yield reliable results do not yet exist, particularly for the initial period of operation through the 1960s.

There are also serious issues of data accuracy and completeness regarding worker doses. Some of these issues emerged at a 17 March 1994 hearing held by the House Subcommittee on Oversight and Investigations of the House Energy and Commerce Committee. A number of deficiencies in worker dosimetery were discussed at the hearing by Jim Wells, Associate Director of the General Accounting Office, and by Dr. Tara O'Toole, Assistant Secretary for Environment, Safety and Health in the DOE.[172] The DOE also submitted for the hearing record a summary of its findings regarding the shortcomings of its dosimetry program. The most important of them were:

1. There is very little information regarding nonradioactive hazardous materials and there are essentially no worker dose records.

2. Records for exposure to external radiation are incomplete, unreliable, and misleading, partly due to poor calibration of measuring devices, issuance of multiple badges, and poor placement of dosimeters.

3. Internal exposure records are unreliable.

4. The electronic version of radiation records "did not accurately reflect the data on the original record."[173]

172. Wells 1994; O'Toole 1994, and notes taken by Arjun Makhijani of oral testimony by Jim Wells and Tara O'Toole before the Subcommittee on Oversight and Investigations, Energy and Commerce Committee, U.S. House of Representatives, March 17, 1994.

173. U.S. DOE 1994.

5. Methods for measuring and calculating doses varied considerably from one plant to another, so that it is not possible at this time to reliably pool information from various plants for epidemiological studies. This item must be kept in mind when reading the results of studies that have pooled data from more than one facility.

Wells also stated that in some cases when dosimeters were not returned, zeros were entered into the dose record of the worker. In effect, this means that a part of the external dosimetry record is fabricated in that the zeros entered into the records have no connection with actual dosimeter readings. Other deficiencies also exist in at least part of the dosimetry record. A review of raw data of worker records at the Fernald plant, made available to the Institute for Energy and Environmental Research as a result of a class action lawsuit filed by workers against the former contractor of the plant, National Lead of Ohio, indicates that measurements of uranium in urine as well as *in vivo* measurements of lung burdens of uranium were not translated into doses and entered into the worker's dose record as such.[174] Rather, they are maintained as part of raw data files available to individual workers, at their request, and to plant management. Thus, any study relying on dose records alone, particularly on secondary electronic or magnetic dose records, without verification with all relevant raw data in workers' files, would be liable to errors whose magnitude cannot be estimated without reference to the original records.

Some of these problems with worker dosimetry were identified in a 1992 study by the U.S. Physicians for Social Responsibility, entitled *Dead Reckoning: A Critical Review of the Department of Energy's Epidemiologic Research.* The study discusses numerous problems with dosimetry. For instance, it cites one example where doses to workers at Hanford during the late 1940s may have been seriously underestimated due to a variety of factors such as the insensitivity of early film badges, policies for issuing film badges, and changes in the frequency of processing and reading film badges.[175]

General Accounting Office Associate Director Jim Wells also testified that when workers retired because they were diagnosed with cancer, DOE did not record that cancer. However, we should note that the epidemiological studies discussed in this chapter do not rely on internal DOE records of cancer incidence or mortality. Rather they rely on death certificate records. But such records have other problems; they tend to be inaccurate regarding cases in which cancer was an underlying cause of death but not the proximate cause of death. The effects

174. Examples of the raw dosimetery data have not yet been published but are available from the Institute for Energy and Environmental Research, Takoma Park, Maryland 20912.

175. Geiger et al. 1992, p. 36.

Table 6.11 Epidemiologic Studies of Health Risks from Oak Ridge Facilities[1]

Study (Facility)[2]	Period of Exposure	Period of Follow-up	Number of Subjects[3]	Study Type	Outcome	Method of Analysis
Occupational Exposure						
Godbold and Tompkins (1979) [K-25]	1948–1972	1948–1972	852 WM 85 deaths	Historical cohort: nickel-exposed workers	Mortality from cancer and other diseases	Comparison of observed and expected deaths for exposed and unexposed groups
Cragle et al. (1984) [Y-12]	1953–1958	1953–1979	5,663 WM 1,140 deaths	Historical cohort: mercury-exposed workers	Mortality from cancer and other diseases	Cohort SMRs; SMRs stratified by mercury exposure
Cragle et al. (1984a) [K-25]	1948–1953	1948–1977	814 WM 137 deaths	Historical cohort: nickel-exposed workers	Mortality from cancer and other diseases	SMRs and directly standardized death rates for nickel-exposed and controls
Checkoway et al. (1985) [ORNL]	1943–1972	1943–1977	8,375 WM 966 deaths	Historical cohort	Mortality from cancer and other diseases	Cohort SMRs; SMRs by cumulative external dose and job categories
Carpenter et al. (1987)	1943–1977	1943–1979	89 cases	Nested case-control	Central nervous system cancer deaths	ORs for external and lung doses
Carpenter et al. (1987a) [Y-12, ORNL]	1943–1977	1943–1979	82 cases	Nested case-control	Brain cancer deaths	ORs for nonoccupational risk factors
Carpenter et al. (1988) [Y-12, ORNL]	1943–1977	1943–1979	89 cases	Nested case-control	Central nervous system cancer deaths	ORs for chemical exposures

Checkoway et al. (1988) [Y-12]	1947–1974	1947–1979	6,781 WM 862 deaths	Historical cohort	Mortality from cancer and other diseases	Cohort SMRs; SMRs stratified by cumulative external and internal doses
Wing et al. (1991) [ORNL]	1943–1984	1943–1984	8,318 WM 1,524 deaths	Historical cohort	Mortality from cancer and other diseases	Cohort SMRs; SMRs for workers with internal contamination; dose-response analysis
Environmental Exposure						
Moshman and Holland (1949)	1943–1949	1949		Ecologic/cancer mortality		
Patrick (1977)	1929–1971	1929–1971		Ecologic/cancer mortality		
Goldsmith (1989a, b)	1950–1979	1950–1979		Ecologic/cancer mortality		

Note: WM = white males; SMR = standardized mortality ratio; OR = odds ratio; unless specified otherwise, doses are from ionizing radiation.

1. Current workforce size: 16,500 (Geiger et al., 1992).
2. Discussion of facilities is given above in this chapter.
3. If two or more numbers are reported, the first is the size of the entire cohort.

of the errors introduced by the use of such data have not been taken into account in the studies.

It is not apparent what effects these deficiencies will have on the reinterpretation of the results of the epidemiological studies that are described here. However, they do mean that these analyses should be treated as very preliminary. Taking potential hazardous exposure of workers into account by looking at job classifications, for instance, could allow screening of those workers who are still alive for health effects other than cancers and for a wider range of cancers. Health monitoring of workers over the coming decades could also yield data necessitating revision of the studies described here.

Uranium Processing and Enrichment

Since 1943, uranium has been processed at a number of facilities around the United States. Most published studies are of workers exposed to ionizing radiation at various Oak Ridge operations, but exposures to toxic chemicals have also been studied (see table 6.11). Although most analyses failed to detect cohort mortality in excess of national or state rates, two studies found an excess of leukemia mortality in workers.[176] The study by Wing and his colleagues also found a statistically significant dose-response relation for all cancers that was 10 times higher than estimates for the survivors of the bombings of Hiroshima and Nagasaki.

In studies of groups exposed to elemental mercury and metallic nickel, no relations were found between these exposures and the causes of death that were examined. Case-control studies of central nervous system (CNS) cancers did not find a relation between these diseases and either crude classifications of chemical exposures or measures of exposure to ionizing radiation. Carpenter and colleagues did find a statistically significant relation between CNS cancer and employment for more than 20 years.[177] A study of Oak Ridge workers exposed to phosgene gas during the preparation of uranium for electromagnetic separation showed elevated standardized mortality ratios for all causes and for respiratory diseases in a group of workers with high exposures.[178]

Reports of cancer mortality in communities around the Oak Ridge facility have been difficult to interpret because they lack data on the extent of exposure in exposed counties and because the results from

176. Checkoway et al. 1985; Wing et al. 1991. Recent work by Wing and his colleagues taking potential confounding factors into account indicates that the earlier analysis remains valid (Wing et al. 1993).

177. Carpenter et al., 1987.

178. Polednak 1980; Polednak and Hollis 1985.

Table 6.12 An Epidemiologic Study of Health Risks to Workers at Linde Air Products[1]

Study	Period of Exposure	Period of Follow-up	Number of Subjects[2]	Study Type	Outcome	Methods of Analysis
Dupree et al. (1987)	1943–1949	1943–1979	995 WM 429 deaths	Historical cohort	Mortality from cancer and other diseases	SMRs for selected causes of death; no analysis by radiation exposure

Note: WM = white males; SMR = standardized mortality ratio.

1. Processed uranium from 1943 to 1949; total ever employed: 1,551.
2. If two or more numbers are reported, the first is the size of the entire cohort.

different studies are not consistent. One paper assessed cancer incidence without accounting for a reasonable latent period between exposure and disease.[179] Another reviewed crude rates for fetal deaths, infant deaths, and all cancer combined between 1949 and 1970 and found no evidence that they were higher in areas around Oak Ridge compared with the State of Tennessee.[180] Goldsmith compared observed and expected deaths from childhood leukemia in two counties near Oak Ridge for three decades. He found excesses for the 1950s and 1960s but not for the 1970s.[181]

A historical cohort study of one of the earliest groups to process uranium—workers at the Linde Air Products Company Ceramics Plant—showed increased standardized mortality ratios for a number of causes of death (see table 6.12). Most notable is a statistically significant increase in cancer of the larynx. The cohort mortality was not stratified by dose in this study, making it difficult to relate the detected mortality increases to radiation.

No other studies of uranium-processing facilities have been published in the open literature. The Centers for Disease Control and Prevention (including the National Institute for Occupational Safety and Health) is completing a historical reconstruction of off-site doses at the Fernald Feed Materials Production Center and conducting feasibility analyses for studies of workers and off-site radiation doses at the Idaho National Engineering Laboratory.[182] As noted above, the Centers for Disease Control and Prevention study shows far greater releases of uranium to the environment than estimates by the DOE and

179. Moshman and Holland 1949.

180. Patrick 1977.

181. Goldsmith 1989a; 1989b.

182. Ruttenber 1994.

its contractors. The dose evaluations are not yet complete. The Tennessee Department of Health is conducting a similar feasibility analysis for dose reconstruction and epidemiological studies of communities exposed to off-site releases.

Plutonium Production and Processing

The first studies of health risks incurred by U.S. nuclear weapons workers were conducted on the Hanford workforce (see table 6.13). Mancuso, Stewart, and Kneale computed mean cumulative external radiation doses for workers who died from cancer and other diseases.[183] They found radiation doses were higher for workers who died from cancer than for those who died from other causes, that cancer risk was correlated with exposure age, and that intervals between induction and death usually exceeded 10 years.[184] They concluded that radiation-associated risk was higher for multiple myeloma and cancer of the pancreas and lung than for other cancers. They also used statistical techniques to estimate the number of cancers within the workforce that were induced by radiation.

The paper by Mancuso, Stewart, and Kneale generated substantial criticism and began a debate about the health risks from low doses of ionizing radiation—a debate that has yet to be settled. Reports by DOE-sponsored researchers quickly followed.[185] These studies used more traditional methods of occupational cohort analysis, namely, standardized mortality ratios. Though these researchers differed with Mancuso, Stewart, and Kneale in their general conclusions about risks from occupational radiation exposure, they concurred on the findings of increased mortality from multiple myeloma and carcinoma of the pancreas.

Mancuso, Stewart, and Kneale expanded their analyses with dose-effect models that controlled for a number of potentially confounding factors.[186] These analyses showed a curvilinear relation between dose and response for radiation-sensitive cancers. Moreover, this relationship becomes steeper with increasing age at first exposure.

In an update of their original cohort mortality study of workers who began employment between 1944 and 1978, Gilbert et al. found evidence for a strong healthy-worker effect and small elevations in standardized mortality ratios for a number of different cancers, most of

183. Mancuso, Stewart, and Kneale 1977.

184. Stewart and Kneale 1991.

185. Gilbert and Marks 1979, 1980; Tolley et al. 1983.

186. Kneale, Mancuso, and Stewart, 1978; Stewart, Mancuso, and Kneale, 1980; Kneale, Mancuso, and Stewart, 1981.

which were not statistically significant.[187] These researchers also examined ratios of observed and expected cancer deaths for different cumulative dose categories and used a statistical test for trend to examine dose-response relations. Of the 18 categories of cancer analyzed, they found significant dose-response relations for multiple myeloma.

Kneale and Stewart published an updated analysis of the same cohort that was studied by Gilbert et al., but they restricted the cohort to workers who were monitored with film badges or thermoluminescent detectors (TLDs).[188] They analyzed the cohort data with advanced statistical risk models and concluded that people exposed to radiation after age 50 are more likely to develop cancer than those exposed before age 50. Kneale and Stewart also found no evidence for radiation being more likely to cause leukemia than solid tumors, and estimated doubling doses for radiation-induced cancer that are much greater than those estimated in the BEIR V report.

In addition to examining the relation between cancer and occupational exposures to ionizing radiation, investigators have looked into the possibility that preconception occupational exposure to ionizing radiation is associated with congenital malformations in children born near Hanford. Sever et al. determined the prevalence of congenital malformations among births in two counties close to Hanford and found a significant elevation for neural tube defects.[189] In a case-control study of neural tube defects from a different time period, they found a significant association with cumulative preconception doses to the father.

During the first two decades of plutonium reprocessing, Hanford released approximately 700,000 curies (approximately 26,000 terabecquerels) of iodine-131 to the atmosphere. Preliminary estimates of radiation doses to the exposed public are high—particularly for those who consumed locally produced dairy products.[190] Releases of other radionuclides to the atmosphere and the nearby Columbia River may have exposed regional residents by other pathways. A multi-million-dollar historical dose-reconstruction project, initially sponsored by the Department of Energy and now by the Centers for Disease Control and Prevention, is nearing completion. It is attempting to provide estimates of the doses that could have been received from these exposures, along with estimates of their uncertainties.[191] The Centers for Disease Con-

187. Gilbert, Peterson, and Buchanan 1989.

188. Kneale and Stewart 1993.

189. Sever et al. 1988; 1988a.

190. Cate, Ruttenber, and Conklin 1990.

191. Gilbert et al. 1993.

Table 6.13 Epidemiologic Studies of Health Risks Associated with Hanford[1]

Study	Period of Exposure	Period of Follow-up	Number of Subjects	Study Type	Outcome	Methods of Analysis
Occupational Exposure						
Mancuso et al. (1977)	1944–1972	1944–1972	24,939 M,F 3,710 deaths 670 cancers	Comparisons between mean radiation doses for selected cancer deaths	Mean cumulative radiation dose	Comparison of mean cumulative external radiation doses between cancers and non-cancers
Gilbert and Marks (1979)						
Gilbert, Peterson, and Buchanan (1989)[2]	1944–1981	1944–1981	44,100 M,F 7,249 deaths 1603 cancers	Historical cohort	Cancer mortality	Cohort SMRs; SMRs stratified by external dose and Pu deposition; dose-response
Kneale and Stewart (1993)[3]	1944–1986	1944–1986	35,868 badge-monitored M,F; 7,342 deaths; 1,907 fatal and nonfatal cancers	Cohort dose-response models	Cancer mortality	Cohort risk models with conditional maximum likelihood procedures

Environmental Exposure							
Sever et al. (1988)	1944–1979	1957–1980	672		Case-control	Congenital malformations	Comparison of parental preconception external radiation doses between cases and controls; dose-response analyses
Sever et al. (1988a)	1944–1980	1968–1980	454		Community-based prevalence	Congenital malformations	Number of congenital malformations per 1,000 births compared to data from monitoring programs in Washington, Idaho, and Oregon

Note: M,F = males and females; SMR = standardized mortality ratio; OR = odds ratio; unless specified otherwise, doses are from ionizing radiation.

1. Current workforce size: 13,500.
2. A number of reports by this group preceded this paper; see Gilbert (1991) for a summary of these.
3. A number of reports by this group preceded this paper; see Stewart and Kneale (1991) for a summary of these.

trol and Prevention is also funding a study of thyroid disease in people who lived near Hanford during the period of the highest iodine-131 releases.[192]

Early in the history of weapons development, plutonium was processed at the Los Alamos National Laboratory (LANL) (see table 6.14). Epidemiologists at this site have studied small cohorts of the early workers who had high plutonium exposures, and have conducted a case-control study of malignant melanoma for this group.[193] They have also evaluated the incidence of malignant melanoma in a larger cohort of workers,[194] and have conducted an incidence study for workers employed between 1969–1978, but this has not been published in the general scientific literature. No evidence of radiation-related disease was found in these studies. Likewise, an ecologic study of cancer incidence and mortality in Los Alamos County published in 1981 failed to detect differences in mortality or incidence rates between this county and counties with similar demographic features.[195] In contrast, a 1993 report released by scientists from the New Mexico Department of Health and University of New Mexico Cancer Center suggested the occurrence of a statistically significant fourfold excess of thyroid cancer in Los Alamos County in the mid–1980s, as well as a modest elevation in the incidence of brain and nervous system cancer.[196] The Centers for Disease Control and Prevention is evaluating the feasibility of a dose-reconstruction study for communities around the Los Alamos National Laboratory.

Wilkinson et al. studied white men who were employed at Rocky Flats for at least two years between 1956 and 1980, and analyzed records of cancer deaths for these workers (see table 6.15).[197] They divided the workforce into two groups based on plutonium body burden. In the group with the higher body burdens, they found higher rates of death from all causes (cancer and noncancer combined), leukemias and lymphomas (combined), cancers of the esophagus, stomach, colon, and prostate (analyzed separately), and lymphosarcoma and reticulum cell sarcoma (combined).

Rocky Flats workers were also divided into two groups based on cumulative external radiation dose as measured by the film and TLD badges they wore. The group with the higher cumulative doses had higher rates of myeloid leukemia, lymphosarcoma and reticulum cell

192. Ruttenber 1994.

193. Voelz et al. 1979; 1983; Voelz, Grier, and Hempelmann 1985; Acquavella et al. 1983.

194. Acquavella et al. 1982.

195. Stebbings and Voelz 1981.

196. Athas and Key 1993.

197. Wilkinson et al. 1987.

Table 6.14 Epidemiologic Studies of Health Risks Associated with Los Alamos[1]

Study	Period of Exposure	Period of Follow-up	Number of Subjects	Study Type	Outcome	Methods of Analysis
Occupational Exposure						
Voelz et al. (1979)	1944–1945	1944–1947	26 M	Historical cohort	Findings on physical exam and cause of death	Descriptive
Acquavella et al. (1982)	1969–1978	1969–1978	11,308	Historical cohort	Incidence of malignant melanoma	SIRs by ethnicity and sex
Voelz et al. (1983)	1944–1973	1944–1976	241	Historical cohort	Mortality from cancer and other diseases	SMRs for selected causes of death; no analysis by radiation exposure
Acquavella et al. (1983)	NA	NA	20	Case-control	Cumulative external radiation dose; Pu body burden; duration of employment; job classification	ORs for various potential risk factors
Voelz, Grier, and Hempelmann (1985)	1944–1945	1944–1982	26	Historical cohort	Findings on physical exam and cause of death	Descriptive
Environmental Exposure						
Stebbings and Voelz (1981)	1944–1974	Mortality: 1950–1969; Incidence: 1969–1974	NA	Ecologic	Incidence and mortality for selected cancers	Comparison of county rates with rates for other counties and for the U.S.

Note: WM = white males; SIR = standardized incidence ratio; SMR = standardized mortality ratio; OR = odds ratio; unless specified otherwise, doses are from ionizing radiation.

1. Current workforce size: 13,500 (Geiger et al., 1992).

Table 6.15 Epidemiologic Studies of Health Risks from Rocky Flats[1]

Study	Period of Exposure	Period of Follow-up	Number of Subjects	Study Type	Outcome	Methods of Analysis
Occupational Exposure						
Wilkinson et al. (1987)	1952–1979	1952–1979	5,413 WM 409 deaths 95 cancer deaths	Historical cohort	Cancer mortality	SMR; SMR by external and internal dose; dose-response
Brandom et al. (1990)	NA	NA	22	Cytogenetic analysis of selected workers	Sister chromatid exchange and chromosome aberrations	Chromosome abnormalities stratified by Pu burdens and chemical exposures
Environmental Exposure						
Johnson (1981)	1953–1971	1969–1971	5,747 cancers	Ecologic	Cancer incidence	Comparison of age-adjusted cancer incidence over areas defined by Pu soil concentrations
Crump et al. (1987)	1969–1981	1969–1971 1979–1981	>17,000	Ecologic	Cancer incidence	Comparison of age-adjusted cancer incidence over areas defined by Pu soil concentrations

Note: WM = white males; SMR = standardized mortality ratio.

1. Current workforce size: 6,000 (Geiger et al., 1992).

sarcoma (combined), liver cancers, and unspecified brain tumors. Only the relation between unspecified brain tumors and dose from external radiation was statistically significant. The small number of workers in the study could be responsible for elevated but statistically insignificant disease rates.

In 1990, William Brandom and co-researchers published results from a small study of chromosome abnormalities in 18 plutonium workers at Rocky Flats.[198] They divided this group according to body burden from plutonium and found more chromosome aberrations among those with high burdens. They found no differences between groups classified by exposures to chemicals. These data are hard to interpret because the study was not large enough to adjust for other factors and because it is not clear to what extent chromosomal abnormalities reflect risk for future disease.

Workers at Rocky Flats processed large quantities of beryllium, used asbestos for insulation, and were exposed to a number of chemical carcinogens. Cases of chronic beryllium disease and asbestosis (a chronic lung disease caused by exposure to asbestos fibers) have been diagnosed in Rocky Flats workers. It is likely these diseases are also present in workers at other weapons facilities.

In 1981, Carl Johnson used cancer diagnosis data for 1969–1971 from the National Cancer Institute's Third National Cancer Survey to examine the relation between cancer rates and exposures to plutonium as measured by a group of soil samples collected from the region around Rocky Flats in 1970.[199] He found increases in many cancer types for persons in exposed areas as compared with those for unexposed areas.

In 1987, Kenneth Crump and others replicated the study design used by Carl Johnson to reevaluate cancer diagnosis data for 1969–1971 and for 1979–1981.[200] Although they reconfirmed Johnson's findings, they found no association between plutonium concentrations in the soil and cancer rates after adjusting for distance from the state capital. They also noted no increase in cancer rates for all cancers combined, for radiosensitive cancers, or for cancers of the respiratory system in the region within 16 kilometers of Rocky Flats for both study periods.

Determining whether Rocky Flats was responsible for disease increases in surrounding communities is impossible with data from ecologic studies such as the ones described. An alternate approach is to carefully reconstruct the off-site releases from the plant and then estimate the resulting radiation doses to the public. The dose estimates can be used to estimate health risks directly or in planning and conducting

198. Brandom et al. 1990.

199. Johnson 1981.

200. Crump, Ng, and Cuddihy 1987.

epidemiological studies. This approach, which is being taken at Hanford and Fernald, is also being used in a study conducted by the Colorado Department of Health.[201]

Cragle and colleagues studied the workforce at the Savannah River Site.[202] They conducted a traditional historical cohort analysis and found mortality from a number of different causes to be lower than the mortality for the general public. When the workforce was divided into groups based on hourly or salaried employment, they detected increases in leukemia mortality for salaried workers.

Other Weapons Facilities

Researchers have conducted several studies of note at other sites involved in the manufacture of nuclear weapons (see table 6.16).

Two studies on the mortality of Mound workers were published in 1991. A historic cohort analysis of white men who were monitored for external radiation showed a strong healthy-worker effect for all causes of death and for all cancers.[203] The study detected elevated standardized mortality ratios for a number of specific cancers, although none were statistically significant. A dose-response analysis for leukemias detected a significant relation for all leukemias, lymphatic leukemias, and myeloid leukemias. A positive trend for leukemia rate ratio with dose was of borderline significance when one case of chronic lymphocytic leukemia was removed.

A similar analysis was conducted of Mound workers who were stratified by internal doses from polonium-210.[204] This cohort included a larger group of workers exposed at an earlier time. There was a slight healthy-worker effect for all causes, and elevated standardized mortality ratios for a number of cancers (none were significant). Dose-response analyses did not detect a relation between cancer risk and dose.

In one of the few studies of cancer incidence for workers in a weapons facility, Austin et al. detected a statistically significant increase in malignant melanoma for workers at the Lawrence Livermore National Laboratory during the period 1972–1977, using a comparison population of residents in the counties near the facility.[205] They found no difference in either cumulative external radiation dose or length of employment between the melanoma cases and a control group of

201. Ripple 1992.

202. Cragle et al. 1988.

203. Wiggs et al. 1991.

204. Wiggs, Cox-Devore, and Voelz 1991.

205. Austin et al. 1981.

Table 6.16 Some Epidemiologic Studies of Workers at Other U.S. Weapons Facilities

Study	Period of Exposure	Period of Follow-up	Number of Subjects	Study Type	Outcome	Methods of Analysis
Mound						
Wiggs, Cox-DeVore, and Voelz (1991)	1944–1972	1944–1984	4,402 WM 987 deaths 213 cancer deaths	Historical cohort	Mortality from cancer and other causes	Cohort SMRs; SMRs by polonium dose
Wiggs et al. (1991)	1947–1979	1947–1980	4,182 M 593 deaths	Historical cohort	Mortality from cancer and other causes	Cohort SMRs; SMRs by cumulative external dose
Lawrence Livermore Laboratory						
Austin et al. (1981)	1952–1977	1972–1977	5,100 W 19 cancers	Historical cohort	Incidence of malignant melanoma	Comparison of observed with expected rates; case-control analysis for cumulative external radiation and job classification
Reynolds and Austin (1985)	1952–1980	1969–1980	134 cancers	Historical cohort	Cancer incidence	Comparison of observed with expected numbers of cancers
Pantex						
Acquavella et al. (1985)	1951–1978	1951–1978	269 deaths	Historical cohort	Mortality from cancer and other causes	SMRs for cancer and other causes

Note: W = whites; M = males; SMR = standardized mortality ratio.

workers. However, workers with malignant melanoma were more frequently employed as chemists than were controls. When the study was expanded to include the years 1969–1980 and all types of cancer, they found an elevated incidence of malignant melanoma and salivary gland tumors.[206]

A mortality study of workers at the Pantex assembly plant showed a strong healthy-worker effect, with most standardized mortality ratios (SMRs) less than 1.0.[207] Slightly elevated SMRs for brain cancer and leukemia were noted, however. Analysis of SMRs by length of employment and cumulative external radiation dose did not reveal relations between these variables and mortality, but the number of deaths was not large enough to analyze with statistical techniques.

Studies of Multiple Facilities

Mortality data for workers at Hanford, Oak Ridge National Laboratory, and Rocky Flats were combined by Gilbert et al. (see table 6.17).[208] The study was restricted to workers who were monitored for external radiation exposure, and SMRs were computed for each facility and for the three facilities combined. The facility-specific and combined SMRs were also stratified by cumulative external dose and then analyzed for dose-response relations. This analysis failed to detect an association between dose and cancer for specific facilities alone or combined. A significant relation between dose and all noncancer deaths was detected for Rocky Flats, however.

Wilkinson and Dreyer analyzed leukemia data from seven nuclear facilities (including three in Great Britain and the Portsmouth Naval Shipyard in the United States).[209] They computed mortality rate ratios for two categories of cumulative external radiation dose for each facility separately and for all combined. They found a significant excess of leukemia in the groups with higher doses and a modest dose-response relation when three dose categories were compared.

In a continuation of this methodology, Wilkinson analyzed data from eight facilities—the seven mentioned above plus Mound Laboratory.[210] After stratifying for external dose, he found elevated rate ratios for all cancers, solid tumors, and leukemia for a number of the facilities separately and for all combined.

206. Reynolds and Austin 1985.

207. Acquavella et al. 1985.

208. Gilbert et al. 1989a.

209. Wilkinson and Dreyer 1991.

210. Wilkinson 1991.

Table 6.17 Epidemiologic Studies of Combined Workforce Populations

Study	Period of Exposure	Period of Follow-up	Number of Subjects	Study Type	Outcome	Methods of Analysis
Occupational Exposure						
Gilbert et al. (1989)	Hanford: 1944–1981 ORNL: 1943–1977 Rocky Flats: 1952–1979	Hanford: 1944–1981 ORNL: 1943–1977 Rocky Flats: 1952–1979	23,704 WM (Hanford) 6,332 WM (ORNL) 5,897 WM (Rocky Flats)	Historical cohort	Mortality from cancer and other diseases	Cohort SMRs; SMRs by cumulative external dose
Wilkinson and Dryer (1991)	Variable	Variable	83 deaths	Meta-analysis	Leukemia mortality	Rate ratios for published results from 7 studies combined; rate ratios for combined studies by cumulative external dose
Wilkinson (1991)	Variable	Variable	2,965 deaths	Meta-analysis	Mortality from cancer	Rate ratios for published results from 8 studies combined; rate ratios for combined studies by cumulative external dose
Alexander (1991)	Variable	Variable	78,893 M 18,460 deaths 142 brain cancers	Meta-analysis	Mortality from brain cancer; mortality from other cancers	SMRs for individual facilities; SMRs for all facilities combined

Note: M = males; WM = white males; SMR = standardized mortality ratio.

Another meta-analysis by Alexander provides a thorough review of the major workforce epidemiological studies.[211] (A meta-analysis uses special methods to combine and interpret the result of several different epidemiologic studies.) He focused on brain cancer but also reviewed data on SMRs for other cancers and all causes. Alexander's meta-analysis of data from 10 studies indicates a significant elevation in brain cancer. He also noted elevated standardized mortality ratios for a number of cancers at different weapons facilities.

In a study of communities in the United States located near nuclear facilities, Jablon and co-researchers at the National Cancer Institute compared cancer mortality for counties near nuclear facilities with cancer mortality for those further away.[212] They included nine nuclear weapons facilities in the study and computed standardized mortality ratios for exposed and unexposed county groups for periods before plant startup and after startup. They found no evidence for a relation between residence near weapons facilities and risk for cancer.

This study, however, failed to examine regional differences in cancer below the county level, and did not take into account the directionality of exposures—e.g., wind patterns and airborne plumes—in their analysis. The potential consequences—that is, the dilution of disease rates of highly exposed communities with the disease rates of surrounding nonexposed communities, as well as the possible complete misclassification of some exposed communities as non-exposed—greatly diminish the ability of this kind of ecological study to detect an effect of nuclear facilities.

Nevada Test Site

A number of epidemiologic studies have been conducted on populations exposed to fallout from atmospheric weapons tests conducted at the Nevada Test Site between 1951 and 1962. Lyon et al. first reported increases in childhood leukemia in a study of counties that received fallout from the site.[213] These results were challenged by Land et al.,[214] using the definitions of exposed counties proposed by Lyon and colleagues but mortality data from a different source.

A study of self-reported cancer incidence in a cohort of Mormons who lived in towns that received fallout from the Nevada Test Site showed increases in a number of cancers as compared with Mormons in unexposed areas.[215] Most of the findings of this study are contra-

211. Alexander 1991.

212. Jablon, Hrubec, and Boice 1991.

213. Lyon et al. 1979; 1980.

214. Land, McKay, and Machado 1984.

215. Johnson 1984.

dicted by a cancer mortality study by Machado et al. in counties that contained some of the towns studied by Johnson.[216] Machado et al. did confirm an increase in leukemia that was noted by Johnson, but they found a much lower risk.

A 1990 case-control study of Utah Mormons used environmental-pathway analysis to compute individual radiation doses for all subjects.[217] Cases were Mormons who died of leukemia; controls were Utah Mormons who did not die of leukemia. This study found a weak but nonsignificant association between bone marrow dose and all types of leukemia, and statistically significant trends in excess risk for people dying of acute leukemias discovered between 1952 and 1963 among individuals 20 years or younger at exposure.

Kerber and colleagues updated a historical cohort study of people who were children in Washington County, Utah, and Lincoln County, Nevada, from 1965 through 1970.[218] The study examined cohort members in 1985 and 1986 for evidence of thyroid disease, and it included interviews to determine food consumption during the period of atmospheric weapons tests. Data from the dietary survey was combined with data from models of fallout deposition to estimate doses to the thyroid. The researchers then stratified the cohort by thyroid dose and found a statistically significant positive dose-response trend for cancerous and noncancerous neoplasms (combined).

By using radiation doses computed for individual study subjects, the studies by Stevens et al. and Kerber et al. demonstrate important advances in the field of environmental epidemiology.[219] Both studies also incorporated the uncertainties of dose estimates into their analyses, another significant advance.

There have also been studies of cancer in members of the military who witnessed nuclear tests in the 1950s. Glyn Caldwell and collaborators at the Centers for Disease Control and Prevention published studies of leukemia and polycythemia vera (a disease characterized by excessive production of red blood cells) in a cohort of veterans who were at the Nevada Test Site in August 1957. The studies found increases in the incidence of both diseases above expected rates. Research on other cohorts of atomic veterans has not found disease elevations but cannot rule those out due to the low statistical power of the studies.[220]

216. Machado, Land, and McKay 1987.

217. Stevens et al. 1990.

218. Kerber et al. 1993.

219. Stevens et al. 1990; Kerber et al. 1993.

220. Robinette, Jablon, and Preston 1985.

SUMMARY

Compared with occupational studies in other countries, the health risks to workers in U. S. nuclear weapons facilities have been studied extensively. It is difficult to arrive at a simple summary evaluation of these health risks—as evidenced by the ongoing debate within the scientific community and among the public. The reasons include the differences between study designs, the differences from site to site in radiation and chemical exposures, and the extent to which exposures were monitored. There is also the problem of how to adjust for the healthy-worker effect in epidemiological analyses.

Despite these limitations, a number of studies of workers suggest increased risk for leukemia and brain cancer. The evidence for leukemia is perhaps more convincing because it has been linked to ionizing radiation in a number of studies of other populations, because elevated leukemia risks have been noted for a number of individual weapons facilities and in meta-analyses, and because dose-response relations have been demonstrated.

Studies of workers have used analytical alternatives to the classic cohort-wide standardized mortality ratios, which can be strongly affected by the healthy-worker effect. The alternate approaches, which stratify worker cohorts by cumulative radiation dose or a surrogate measure of dose or use statistical models to analyze the survival of workers, provide a clearer picture of work-related risk than can be obtained from analyzing SMRs for entire cohorts.

In the future, the National Institute for Occupational Safety and Health will be responsible for conducting epidemiologic studies of nuclear weapons workers. Transferring research from DOE to this agency will enable a larger number of independent scientists to participate in studies, and, one hopes, they will generate more creative approaches to data collection and analysis.

A number of challenges lie ahead for such research. The methods for estimating cumulative radiation doses need to be improved to account for discrepancies in the reporting of minimally detectable doses and missing doses for personal external dosimeters. Likewise, better methods need to be developed to assess internal doses from such radionuclides as plutonium and uranium. Estimates of external doses from neutrons need to be improved for a number of weapons facilities, particularly for the early years of operation when personal dosimeters may have underestimated doses.

Since workplace chemical carcinogens and other exposures such as smoking have also contributed to cancer risk, they need to be accounted for in order to get accurate estimates of radiation risk. Since exposures to other toxic agents such as beryllium and asbestos have also occurred, future cohort analyses could incorporate other health outcomes in addition to cancer. Future research will also rely on biological markers of exposure or early disease, such as heightened sensitivity to beryllium, for both cancer and other health effects. Combining cohorts of workers from a number of facilities and conducting case-control studies with cases from multiple facilities will also improve our knowledge of cancer risks. For many facilities, large percentages of the workforce are still alive, and their risks for cancer are still not known.

Compared with epidemiological studies in Europe, studies of communities around U.S. weapons facilities have focused less on analyses of cancer clusters, and there has been much less debate over the results from ecologic studies. Of the studies that have been conducted around weapons facilities, none has provided clear evidence for increased disease risk, although it is inappropriate to use these analyses as evidence for no risk.[221] Two studies of the relation between cancer and fallout from atmospheric tests at the Nevada Test site have provided epidemiological evidence of elevated cancer risk. Because these studies use doses computed for individual subjects, they provide good evidence for causal relations. The designs used in these studies could serve as examples for researchers in the future.

Since epidemiologic studies of community exposures from weapons facilities are severely limited in their utility for assessing disease risk, dose reconstruction studies have been proposed as alternatives for risk analysis. If conducted by independent scientists with full participation and oversight by the public, and if researchers have full access to production and environmental monitoring records, the results from dose reconstructions may prove to be quite useful. Dose reconstruction may allow a preliminary estimate of the risk of disease posed to off-site populations and enhance the ability to plan future epidemiological studies.

On the other hand, however, in light of the uncertainties about the adequacy and accuracy of dose records, as well as the revelations reviewed at the beginning of this section about the incompleteness, inaccuracy, and what appears to have been outright fabrications of data in some instances, it is essential to be cautious about the extent to which attempts at dose reconstruction may propagate errors instead of correct them. Attempts at dose reconstruction must be preceded by verification of the validity of data to be used; if such validation cannot be achieved with a reasonable amount of certainty, dose reconstruction may be a futile exercise.

When confronted with a flawed record of this magnitude, a difficult policy question must be faced: Should large amounts of scarce public resources be put into lengthy dose reconstruction, or would the workers be better served if simpler screening criteria were developed

221. Shleien, Ruttenber, and Sage 1991.

and available funds devoted to medical monitoring? The answer to such a question is primarily a political one. We must note that in the face of an official stand that grievous harm was not done, the necessary resources for medical monitoring may only be forthcoming if some dose reconstruction shows that the official record has been systematically underestimating doses and/or health effects. Further, some dose reconstruction, sufficient to determine reasonable screening criteria for each type of facility and possibly for each facility, would seem to be warranted, not only for radioactive materials but also for nonradioactive hazardous materials. But it would be unwise to invest resources in reconstruction efforts that attempt to be more precise than warranted by the quality of the data.

Two pieces of legislation may indicate the policy direction for the future. In 1988, the U.S. Congress passed a law by which veterans who had attended nuclear weapons tests and who were suffering from certain radiogenic cancers would automatically be allowed benefits on the presumption that their cancers were connected to their service. They do not have to prove this connection on an individual basis. In 1990, Congress passed legislation compensating victims of certain cancers who had been uranium miners, people living downwind from the Nevada Test Site, civilian workers at the Nevada and Pacific test sites, and atomic veterans. The process for receiving compensation appears to be cumbersome, and some controversies continue.

7 Russia and the Territories of the Former Soviet Union

Albert Donnay, Martin Cherniack, Arjun Makhijani, and Amy Hopkins

Soviet atomic physicists first considered the possibility of making a uranium bomb in 1939. Three years later, as articles on nuclear issues disappeared from U.S., British, and German scientific journals because of officially imposed secrecy, they informed Stalin that these countries already had nuclear weapons programs under way.[1] At their urging, Stalin initiated a small-scale development project in 1942, appointing Igor V. Kurchatov, a prominent researcher on nuclear fission, to lead the effort. Kurchatov's role was central and involved all major aspects of materials production and bomb design. Based in "Laboratory No. 2" in Moscow, he quickly began work on experimental atomic pile type reactors, isotope separation methods, and the design of bombs using uranium-235 and plutonium.[2]

After World War II, Stalin greatly increased the size and scope of the Soviet program. He reportedly instructed his Commissar of Munitions to "provide us with atomic weapons in the shortest possible time! You know that Hiroshima has shaken the whole world. The balance [of power] has been destroyed!"[3] Under the direction of Lavrenti Beria, head of the Soviet secret police, the program used thousands of prison laborers to quickly expand uranium mining and build production reactors, processing plants, and secret cities.[4]

By the end of World War II, the Soviet Union was importing uranium from German and Czechoslovakian mines under its control and searching for its own uranium deposits. Captured German scientists, recruited to work on isotope separation in prison labs, developed the

1. Zaloga 1991, p. 174; Holloway 1984, p. 18.

2. Zaloga 1991, pp. 174–175. There were already several centers that did research on atomic physics by the early 1930s. The main research centers were in Leningrad (now St. Petersburg) and Kharkov, Ukraine.

3. Cochran and Norris 1993, p. 14.

4. Zaloga 1991, p. 175. See Shifrin 1982 for a site-by-site discussion of Soviet labor camps.

gaseous diffusion and gas centrifuge technologies that Russia still uses. They also worked, with less success, on electromagnetic separation.[5]

P. Boltyanskaya, a Soviet participant in the project, has stated that the Soviet government signed contracts with eminent German physicists in the Soviet occupation zone, inviting them to come to the Soviet Union to work on the bomb. These Germans enjoyed "excellent living conditions and a much higher salary than their Soviet counterparts." But they could not leave the research area unaccompanied and could talk only with their co-workers.[6]

The Soviet Union's first atomic pile type experimental reactor went into operation at Laboratory No. 2 on 25 December 1946. The first reactor for producing weapons-grade plutonium was of a water-cooled, graphite-moderated design (reactor A). It came on line in June 1948 at the newly built secret production complex of Chelyabinsk-65 (formerly called Chelyabinsk-40).[7] It was also known in the West as Kyshtym, after the name of a nearby town.

The Soviet Union's first fission bombs were designed and assembled at the Arzamas-16 laboratory, in the Nizhny Novgorod region of the middle Volga.[8] The first atomic weapon was exploded on 29 August 1949, at the Semipalatinsk test site in Kazakhstan near the border with the Russian Republic.[9]

Throughout the Cold War there was considerable discussion of the role of Soviet intelligence in the early success of the Soviet nuclear weapons program. Much new information has been revealed since the end of the Cold War. Perhaps the most detailed inside perspective so far has been provided by two senior Soviet scientists intimately involved in the Soviet nuclear weapons program, Yuli Khariton and Yuri Smirnov. They confirmed that the first Soviet weapon "was based on a rather detailed diagram and description of the first American bomb, which the Soviet Union obtained through the efforts of Klaus Fuchs and Soviet intelligence"—information that the Soviet scientists received "in the second half of 1945." However, before the design could be used, Soviet scientists and engineers had to verify "that the in-

5. Zaloga 1991, pp. 175, 179.

6. Boltyanskaya 1991.

7. Zaloga 1991, p. 178. The Soviets attempted to keep secret the exact location of their nuclear weapons production sites and the "closed" cities that supported them by giving these newly built places the names of nonmilitary cities that were actually up to 100 kilometers away, combined with a postal zone number. We will use Chelyabinsk-65 throughout to refer to the site, which was called Chelyabinsk-40 until about 1990 (Cochran and Norris 1993, p. 45). Note that Chelyabinsk-65 and Chelyabinsk-70 are two separate sites far from the city of Chelyabinsk.

8. Zaloga 1991, p. 178.

9. Hewlett and Duncan 1990, pp. 362–366. This passage also provides a description of how the Soviet test was detected by the United States.

formation was reliable (which required a great number of meticulous experiments and calculations)."[10]

According to Khariton and Smirnov, the Soviet nuclear program was primarily the work of Soviet scientists, even though Soviet intelligence agencies evidently contributed to the program. Pointing to work that was going on even before Germany invaded the Soviet Union, they claim that "some members of the [Soviet] secret service have exaggerated the role the intelligence community played in the development of Soviet weapons."[11]

Khariton and Smirnov also point out a parallel in the western atomic weapons program. A British undercover agent, Paul Rosbaud, was instrumental in getting the December 1938 German uranium fission experiment by Otto Hahn and Fritz Strassman rapidly published, so that scientists in Britain and elsewhere would be aware of it.[12] Publication of these results provided a crucial impulse for U.S. investigations of fission and related events in 1939 (which are briefly described in chapter 6).

Interpretations such as those of Khariton and Smirnov are likely to be only the first round of post–Cold War efforts at understanding the history of the nuclear era. Far more definitive historical interpretations of early events both in the Soviet Union and elsewhere will probably emerge over the next decade as further information is declassified in Russia, Germany, the United States, and other countries.

Soviet work on thermonuclear weapons was initiated during 1947–1948 at the Institute for Chemical Physics. Five years later, on 12 August 1953, the Soviets exploded a thermonuclear device at Semipalatinsk.[13] Throughout the 1950s and 1960s, the Soviet Union expanded its nuclear weapons production program. A second large production complex was started in 1948 near Tomsk, Siberia, about 600 kilometers west of Krasnoyarsk. The first Tomsk reactor began operation in 1955. A third complex—built entirely underground—was begun in 1950 near Krasnoyarsk in central Siberia, about 4,000 kilometers east of Moscow, and its first reactor began operation in 1958.[14] The Soviet nuclear weapons industry grew to include dozens of uranium mining operations, eleven major complexes for producing and processing radioactive material, and two major test sites.

The Soviet nuclear arsenal is estimated to have reached its peak of 45,000 warheads in 1986, almost two decades after the U.S. arsenal.

10. Khariton and Smirnov 1993.

11. Khariton and Smirnov 1993.

12. Khariton and Smirnov 1993.

13. Zaloga 1991, p. 181, and Khariton and Smirnov 1993.

14. Cochran and Norris 1992, p. 108.

Since then, retirement and dismantlement of weapons is estimated to have reduced the arsenal to about 32,000 weapons in 1993.[15] These estimates include weapons in Ukraine, Belarus, and Kazakhstan.

Russia is scaling back its remaining nuclear stockpile significantly. The START I treaty of 1991 committed the Soviet Union (and now Russia) to reducing its total inventory of nuclear weapons to some 10,500 to 13,000 warheads by the year 2000. This inventory would be reduced further by the year 2003 to 3,000 to 3,500 strategic warheads under the terms of the START II agreement.[16] These treaties require the signatories to remove the nuclear warheads from delivery systems but not to dismantle the nuclear warheads or remove the radioactive material they contain. While no formal agreements cover the dismantling of the 16,000 to 20,000 tactical or short-range nuclear weapons that the Soviet Union had based in Russia, Ukraine, Kazakhstan, and Belarus, these republics agreed in 1991 to return all such weapons to Russian territory and had done so by 1992.[17]

In April 1992, Nuclear Energy Minister Mikhailov told an IPPNW delegation, "We dismantle more nuclear warheads than we produce, and we have decreased our arsenal by 15 to 20 percent since 1986." This suggests that some 7,000 to 9,000 warheads had already been dismantled. "The key problem," Mikhailov noted, "is where to store [the recovered] plutonium."[18]

Russia continues to produce plutonium, despite its storage problems. In January 1992, President Boris Yeltsin pledged to shut down all military plutonium producing reactors by the year 2000.[19] Russia continues to operate military production reactors at Tomsk-7 and Krasnoyarsk-26, in part because they supply heat and electricity for civilian needs. U.S.-Russian negotiations are underway to explore replacing these reactors by other energy sources or upgrading their safety.

Based on announced plans for tactical weapons and on commitments in the START II treaty, Russia is to reduce its total inventory to between 4,000 and 10,000 warheads by the year 2003, depending on numbers of tactical weapons remaining and warheads in reserve.[20] To do so, it must remove between 23,500 and 28,000 warheads from its arsenal of 32,000 (as of the end of 1993). If it dismantled all these warheads, it would greatly increase its stockpile of recovered plutonium.

15. Norris and Arkin 1993c.

16. Cochran and Norris 1993, p. 23.

17. *Washington Post*, 18 March 1993.

18. As quoted in IPPNW 1992b, p. 14.

19. Cochran and Norris 1992, p. 19.

20. Cochran and Norris 1993.

While the Soviet Union stopped producing highly enriched uranium in October 1989,[21] Russia continues to enrich uranium to low levels for use as reactor fuel. It also exports natural uranium and enrichment services.[22] In 1993, Russia concluded a 20-year agreement to sell the United States 500 metric tons of highly enriched uranium that is being recovered from dismantling its nuclear weapons.[23]

The U.S. Central Intelligence Agency estimates that over 900,000 people in the former Soviet Union worked with security clearances in the nuclear weapons industry, including 2,000 with knowledge of nuclear weapons design and 3,000 to 5,000 working on uranium and plutonium production.[24] Mikhailov, once a nuclear weapons scientist himself, has stated that 10,000 to 15,000 people had access to "really secret information."[25] In response to Western fears that Russian nuclear scientists might sell their services abroad, he claimed in November 1991 that "not a single qualified nuclear scientist or nuclear-weapons designer ... has left the country. With confidence, I can say that my colleagues are patriots and won't leave the country to design nuclear weapons for anyone."[26]

Nevertheless, working conditions at once-privileged nuclear research centers have become so bad that scientists at Arzamas-16, who now mostly dismantle nuclear weapons, threatened to strike in July 1993 to protest their salaries and the fact that they had not been paid at all for two months.[27] In a grievance letter, they implied that "some experts might take up offers of lucrative jobs abroad if their demands are not met."[28] According to German Federal Intelligence, which claims to have extensive files on Russian nuclear specialists now working in other countries, many already have done so. These include fifty in Iraq (the largest contingent), forty in Israel, fourteen in Iran, four in India, two in Libya, and an unspecified (but presumably small) number in Algeria.[29]

The Russian Ministry of Atomic Energy (MINATOM) estimated that in 1992 it employed one million people, including 700,000 in 10 secret cities of approximately 70,000 people each. The ministry operates its

21. Cochran and Norris 1992, p. 19.

22. Bukharin 1993, pp. 2–5.

23. Portanskiy 1993.

24. Cochran and Norris 1992, p. 10.

25. As quoted in Khokhlov 1992.

26. As quoted in Goldanskii 1993, p. 25.

27. Bastable 1993, p. 9.

28. Bastable 1993, p. 9.

29. As cited in Goldanskii 1993, p. 26.

own health care system in these cities and, in exchange for the hardship of living in isolation, offers its employees a relatively high material standard of living. According to Alexei Yablokov, former official Adviser to the President of the Russian Federation on Ecology and Public Health, a total of 36 million workers and their dependents depend on the military-industrial complex for their economic survival. He calls the complex a "state within a state, a government within a government [that] does not sleep."[30] However, since the disintegration of the Soviet Union and the severe deterioration of the Russian economy, the standard of living of workers, including scientists and engineers, in the nuclear weapons complex has declined considerably.

As in the United States, weapons production was the overriding priority of the Soviet program. Nuclear authorities often callously disregarded public health and the environment. Measures to protect the environment, workers, and civilians were weak, at best, and demonstrably ineffective. For example, radioactive liquids were simply poured into canals, rivers, reservoirs, lakes, seas, and oceans for years. The normal operation of Soviet nuclear weapons facilities routinely released radioactive solids, liquids, and gases directly into the environment, contaminating air, soil, and ground and surface waters. In general, the environmental damage in the former Soviet Union is far greater than that in the United States, largely due to the greater quantity of radioactive wastes discharged directly into the environment.

The United States has offered Russia, Ukraine, Belarus, and Kazakhstan a variety of forms of financial and technical assistance to help with their disarmament commitments. Congress appropriated $400 million for this purpose in 1992 and another $400 million in 1993, with high priorities being to improve storage of fissile materials, to design new plutonium storage facilities, and to dismantle nuclear, chemical, and biological weapons. However, there has been criticism that most of the money remains unavailable, and that a large fraction is earmarked for spending in the United States.[31]

QUANTITY AND QUALITY OF DATA

Until recently, very little information was available in the West on the Soviet nuclear weapons industry. The most comprehensive compilation of data from primary sources on Soviet nuclear weapons and production facilities available from this period is *Soviet Nuclear Weapons*, the fourth volume of the *Nuclear Weapons Databook* series written by the Natural Resources Defense Council (NRDC) in 1989. The NRDC has issued periodic updates to this volume in the form of Nuclear

30. As cited in IPPNW 1992b, pp. 14–15.

31. Lockwood 1993, pp. 39–40.

Weapons Databook Working Papers. The latest revision consulted for this report, *Russian/Soviet Nuclear Warhead Production*, is dated September 1993. Another early and frequently cited source is Avraham Shifrin's *First Guidebook to Prisons and Concentration Camps of the Soviet Union*, first published in Russian in 1981.

More recent sources used to identify and characterize nuclear weapons facilities include *Plutonium: Deadly Gold of the Nuclear Age*, a global study of plutonium production, use, and disposal published in 1992 by the International Physicians for the Prevention of Nuclear War and the Institute for Energy and Environmental Research; trip reports written in 1991 and 1992 by IPPNW representatives, IEER staff, and other U.S. participants in various tours of Russian facilities; papers presented by Russian officials and activists at the 1992 Commonwealth of Independent States Nonproliferation Project Conference on "The Nonproliferation Predicament in the Former Soviet Union," sponsored by the Monterey Institute of International Studies; and annual directories and reports published by various U.S. and international agencies. For example, the U.S. Bureau of Mines, the U.S. Central Intelligence Agency, and the International Atomic Energy Agency have all produced reports on uranium mining.

Much information about specific sites came from personal contacts IPPNW and IEER representatives made with nationally and locally active environmental activists in Russia. These contacts also were very helpful in clarifying information from other sources. (Those who made important contributions are noted in the acknowledgments.)

In 1989, Russian newspapers and magazines began to publish many interviews and investigative articles about the health and environmental conditions around nuclear facilities. Articles came from IPPNW and IEER contacts in Russia, as well as from three sources that specialize in analyzing foreign media reports: the U.S. government's Foreign Broadcast Information Service (which, together with the Joint Publications Research Service, publishes complete English translations of media reports from around the world devoted to environmental issues), the independent JV Dialogue (a daily on-line news service based in Moscow that offers English abstracts of articles from over 200 Russian newspapers and magazines), and the Monterey Institute of International Studies' "Monitoring CIS Environmental Developments Project" (an on-line service that also offers English abstracts of Russian and Eastern European media reports on environmental issues in the Commonwealth of Independent States).

With the collapse of the Soviet Union, several scientists, physicians, military officers, and administrative officials associated with the nuclear weapons industry have spoken or written about their experiences. The Russian government has acknowledged nuclear accidents that had been previously denied, and it has begun opening some secret cities to Western visitors and impaneled scientific commissions to investigate and report publicly on the extent of radiological contamination around key nuclear weapons facilities. Several official government bodies have commissioned regional investigations. The most detailed is an official report on the radiological situation around Chelyabinsk-65 that President Mikhail Gorbachev commissioned.

While these sources have added greatly to outside knowledge of the Soviet nuclear industry, especially regarding the extent of environmental problems, little hard data have been released on public or occupational health conditions around facilities. Moreover, the quality of these data and the analyses that accompany them are highly variable. Both are affected to some degree by the nature, bias, and accuracy of their sources. As in the United States, the scope, perspective, and content of research presented by official and nuclear industry sources must be considered with care. In several instances (noted below), officials have issued sweeping and patently false assurances about the absence of *any* detectable health or environmental effects around nuclear weapons facilities.

"We have to realize that we have no right to believe [past data]," says Alexei Yablokov, the former director of the Institute of Developmental Biology in Moscow. "It was collected for political purposes, and often it does not tell us what we need to know."[32] In 1994 Yablokov served as the Russian Federation's official advisor to President Yeltsin on matters of ecology and public health. He is one of a growing number of Russian officials working to expose and correct the problems of the past.

Information and reports published by Russian nongovernmental sources, with no or only limited access to raw data on site operations, cannot paint a complete and accurate picture either. Even if unintended, some of their assumptions and results may be misleading, exaggerated, distorted, or misrepresented. For example, it is often difficult to verify assumptions and sift factual allegations from hearsay. Given the limited environmental and health monitoring equipment in place at these sites even now, the full scope of radioactive contamination caused over the years will likely never be known with confidence.[33] Not until June 1992, for example, did the Russian Federation approve and agree to fund a program to survey, locate, and map areas of radioactive contamination as a first step toward compiling priority lists of sites to be decontaminated.[34] Thus, the data presented here provide only a starting point for estimating the total impact at each

32. As quoted in Mervis 1992.

33. Cochran and Norris 1992; Saleska 1992; Bradley 1991.

34. Dunayeva 1992.

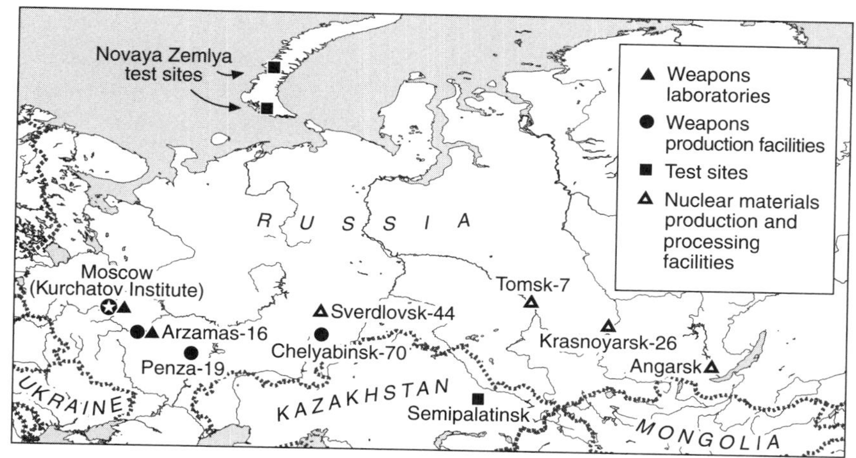

Figure 7.1 Major Russian nuclear weapons production and testing sites.

site. The data are often incomplete, reports rarely present methodologies, and many analyses are questionable and poorly documented.

NUCLEAR WEAPONS–RELATED ACTIVITIES AND FACILITIES

The names and locations of facilities in the territories of the former Soviet Union known to be related to the research, development, production, and testing of nuclear weapons and nuclear weapons materials are listed in table 7.1. The locations of major Russian nuclear production and testing sites are shown in figure 7.1. Figure 7.2 shows civilian nuclear power plants whose spent fuel was reprocessed to separate plutonium. While no evidence indicates that the Soviet nuclear weapons program used any of the plutonium separated from civilian power reactors, it is included for discussion here because of the health, environmental, and proliferation risks associated with the production and stockpiling of this element (see chapters 2 and 11).

Soviet sites for disposing of radioactive wastes are not shown. All military-related facilities dumped, stored, or released their radioactive wastes as close to the production site as possible—and usually within the boundaries of the site—as a matter of policy to minimize transportation costs.[35] There are other radioactive waste disposal areas, however. For instance, a 70-hectare site about 80 kilometers from Moscow in the Zagorsk region reportedly has concrete-lined underground burial chambers to dispose of radioactive waste from the Moscow area.[36]

35. Bolsunovsky 1992.

36. Dreckmann 1993.

Table 7.1 Location of Nuclear Weapons–Related Activities and Facilities in Russia and Other Territories of the Former Soviet Union

Secret Site Name	Official Facility Name	Geographic Location (in Russia unless noted otherwise)	Operating Status[1]	Affiliated Closed City	Source[2]
R&D Weapons Laboratories (only the largest are listed)					
Chelyabinsk-70	All Russia Scientific Research Institute of Technical Physics	20 km north of Kasli, Urals region	1 OP	Snezhinsk [Sunezhinsk?]	a
Arzamas-16	All Russia Scientific Research Institute of Experimental Physics	Sarova, Mordovian Republic	1 OP[3]	Kremlev [Sarov]	a
Lab No. 2	Kurchatov Institute of Atomic Energy (renamed the Russian Scientific Center (RSC))	Moscow	7 NO[3]		a
Uranium Mining[4]					
	Mine name(s) not available	Taboshar, Tadjikistan			c, k, n
	Mine name(s) not available	Naugarzan-Chigrik, northern Tadjikistan			k, n
	Mine name(s) not available[5]	Khudzhand [Leninabad], Tadjikistan			b
	Mine name(s) not available[5]	Sovetabad, Tadjikistan			b
	Mine name(s) not available[5]	Bekabad, Tadjikistan			b
	Mine name(s) not available[5]	Asht, Tadjikistan			b
	Mine name(s) not available[5]	Zeravshan, Tadjikistan			b
	Mine name(s) not available[5]	Adrasman, Tadjikistan			k
	Mine name(s) not available[5]	Rakhov, Carpathian region, Ukraine			b
	Mine name(s) not available[5]	Cholovka, Zhitomire region, Ukraine			b
	Mine name(s) not available[5]	Krasnoarmeysk, Chita region, Siberia			e

Zabaikalsky Mining and Refining Complex	Pervomaisky, Chita region, Siberia	NO	a
Mine name(s) not available[5]	Oimyakon, near Yakutsk, far eastern Siberia		b
Mine name(s) not available[5]	Achinsk, near Krasnoyarsk, Siberia		b
Mine name(s) not available	Vikhorevka, Siberia		n
Mine name(s) not available	Slyudyanka, Siberia		n
Mine name(s) not available	Aldan, Siberia		n
Mine name(s) not available[5]	Kyshtym, near Chelyabinsk, Urals region		b
Mine name(s) not available	Ozerny, southern Urals, near Kazakhstan border		f
Mine name(s) not available[5]	Omutninsk, Kirov region north of Moscow		b
Mine name(s) not available[5]	Tot'ma, Vologda region north of Moscow		b
Mine name(s) not available[5]	Cherepovets, Vologda region north of Moscow		b
Mine name(s) not available[5]	Vaigach Island, Barents Sea		b
Mine name(s) not available[5]	Cape Medvezhii, Novaya Zemlya Island, Barents Sea		b
Mine name(s) not available[5]	Shamor Bay, Primorsk region northwest of St. Petersburg		b
Mine name(s) not available[5]	Kavalerovo, Primorsk region northwest of St. Petersburg		b
Mine name(s) not available	Chupa District, northwest of St. Petersburg		n

Table 7.1 (continued)

Secret Site Name	Official Facility Name	Geographic Location (in Russia unless noted otherwise)	Operating Status[1]	Affiliated Closed City	Source[2]
	Mine name(s) not available	Lake Onega, northwest of St. Petersburg			n
	Mine name(s) not available	Lovozero Tundra, northwest of St. Petersburg			n
	Mine name(s) not available[5]	Groznyi, Chechen-Ingush region west of Caspian Sea			b
	Mine name(s) not available	Almalyk, Uzbekistan			k
	Mine name(s) not available	Naugarzan-Chigrik, southern Uzbekistan			k, n
	Mine name(s) not available[5]	Leninsk, Uzbekistan			b
	Mine name(s) not available[5]	Fergana, Uzbekistan			b
	Mine name(s) not available[5]	Margelan, Uzbekistan			b
	Mine name(s) not available[5]	Kokand, Uzbekistan			b
	Mine name(s) not available	Yangiabad, Usbekistan			c, g
	Mine name(s) not available	Uygursay, Uzbekistan			g
	Mine name(s) not available	Charkesar, Uzbekistan			k, n
	Mine name(s) not available	Uchkuduk, Uzbekistan			k, n
	Mine name(s) not available	Nurabad, Uzbekistan			k
	Mine name(s) not available	Zaf-Arabad, Uzbekistan			k
	Mine name(s) not available	Zarafshan, Uzbekistan			k
	Mine name(s) not available	Ak-Tyuz-Bordunsky, Kirghizstan			k, n
	Mine name(s) not available	Tyuya Myuyun, Kirghizstan			n
	Mine name(s) not available	Tonskiy Bay, Lake Issyk-Kul, Kirghizstan			c, k

	Mine name(s) not available	Kadzhi-Say, south shore, Lake Issyk-Kul		c, g, k
	Mine name(s) not available	Mayli-Say, Kirghizstan		g, k
	Mine name(s) not available	Min-Kush, Kirghizstan		c, g, k, n
	Mine name(s) not available	Granitogorsk, Kirghizstan		k, n
	Mine name(s) not available	Kyzyl-Dzhar, Kirghizstan		k, n
	Mine name(s) not available	Kara-Balta, Kirghizstan		k
	Mine name(s) not available	Sumsar, Kirghizstan		n
	Mine name(s) not available	Tassbulak, Kazakhstan		k
	Mine name(s) not available[5]	Aksu, Kazakhstan		b
	Mine name(s) not available	Aksuyek-Kiyakhty, Kazakhstan		n
	Mine name(s) not available[5]	Borovoe, Kazakhstan		b
	Mine name(s) not available[5]	Karagaily, Kazakhstan		b
	Mine name(s) not available[5]	Al'malyk, Kazakhstan		b
	Mine name(s) not available[5]	Rudnyi, Kazakhstan		b
	Mine name(s) not available	Chalgi, Kazakhstan		k
	Mine name(s) not available	Koktas, Kazakhstan		k
	Mine name(s) not available	Chavlisay-Krasnogorskiy-Yangiabad		n
	Mine name(s) not available	Sillamaë, Estonia		n
Uranium Milling (production of yellowcake, U_3O_8)				
	Chemical-Metallurgical Industrial Assn.	Sillamaë, Estonia	NO 1990	n
	Vostochny [East] Mining & Refining Complex[5]	Zheltyye Vody, Ukraine	OP	a

Table 7.1 (continued)

Secret Site Name	Official Facility Name	Geographic Location (in Russia unless noted otherwise)	Operating Status[1]	Affiliated Closed City	Source[2]
	Dneiper Basin Chemical Works	Dneprodzerzhinsk, Ukraine	NO		l
	Verkhnedneprovsky Mining & Chemical Works	Verkhnedneprovsky, Ukraine			k
	Tselinny Mining and Chemical Complex[5]	Stepnogorsk, Kazakhstan	OP		d, l
	Pricaspiysky [Caspian] Mining & Smelting Complex	Aktau [Shevchenko], western Kazakhstan	NO		d, l
Atomabad	Vostochny [East] Rare Metal Complex	Tchkalovsk, near Khadjent [Leninabad], Tadjikistan	NO 1991		k, l
	South Polymetal Complex [Production Assn.][5]	Bishkek [Frunze], Kirghizstan	NO 1991		l
	Priargunsky [Argun] Mining & Chemical Complex	Krasnokamensk, Siberia	OP		l
	Almaz Industrial Association[5]	Lermontov, Stavropol region north of Caucasus	OP		l
	Malyshevsk's Mining Utility	Malysheva, Sverdlovsk region (Urals)	OP		l
	Navoi Mining and Smelting Complex	Navoi, Uzbekistan	OP		a
	Facility name not available	Toytepa, Uzbekistan	NO		k
	Facility name not available	Uchkuduk, Uzbekistan	NO		k

Conversion of Yellowcake (U_3O_8) to Uranium Tetrafluoride (UF_4) and Uranium Hexafluoride (UF_6) (in preparation for enrichment)					
Angarsk	Electrolyzing Chemical Complex	Near Irkutsk, Siberia	OP		r
Sverdlovsk-44	Urals Electrochemical Plant	Yekaterinberg, Urals			r
Uranium Enrichment[6]					
Angarsk	Electrolyzing Chemical Complex	Near Irkutsk, Siberia	OP		r
Tomsk-7 [Seversk]	Siberian Chemical Complex	Near Tomsk, Siberia	OP	Seversk	r
Sverdlovsk-44	Urals Electrochemical Complex	Verkhniy-Neyvinskiy, near Yekaterinburg	OP	Novouralsk	r
Krasnoyarsk-45	Electrochemistry Plant	Near Kransk and Krasnoyarsk, Siberia	OP	Zelenogorsk	r
Uranium Fuel Fabrication (oxide fuel)					
Mashinostroitel'nii Zavod	Machine Building Industrial Association	Electrostal, Moscow region			r
Ust-K, PO Box 10	Ulbinskiy Metallurgical Industrial Association	Ust-Kamenogorsk, eastern Kazakhstan			r
	Chemical Concentrates Industrial Association	Novosibirsk, Siberia			r

Table 7.1 (continued)

Secret Site Name	Official Facility Name	Geographic Location (in Russia unless noted otherwise)	Operating Status[1]	Affiliated Closed City	Source[2]
Production of Heavy Water[7]					
	Agrokimservis Company	Dneprodzerzhinsk, Ukraine			k
	Soyuzagrochimimport	Kiev, Ukraine			k
	Facility name not available	Yavan, Tadjikistan			k
	Facility name not available	Kalininabad, Tadjikistan			k
Production of Tritium[8]					
Tritium is (was) produced in military reactors at Chelyabinsk-40 and Tomsk-7 (see below)					
Production of Plutonium from:					
Military Production Reactors					
Chelyabinsk-65	Mayak Chemical Complex[5]	15 km east of Kyshtym, Urals region	6 NO 1990	Chelyabinsk-65 [Ozersk]	a
Krasnoyarsk-26 [Atomgrad]	Mining-Chemical Complex[5]	Near Dodonovo and Krasnoyarsk, Siberia	2 NO 1992 1 OP	Zheleznogorsk	a
Tomsk-7	Siberian Atomic Power Station	Near Tomsk, Siberia	2 OP, 3 NO	Seversk	a

Civilian Power Reactors[9]					
	OP and NC are fast breeder reactors	Beloyarsk	1 OP, 1 NC (also 2 NO RBMKs)		j, m
	All fast breeder reactors	Chelyabinsk-40	3 NC		k
	All VVER-440 type light-water reactors	Kola	4 OP		j, m
	3 VVER-440 and 1 VVER-1000 type reactors	Novo-Voronezh	3 OP		l, m
	All VVER-440 type light-water reactors	Oktembryan, Armenia	2 NO 1988		j, k
	2 VVER-440 and 1 VVER-1000 type reactors	Rovno, Ukraine	3 OP, 1 NC		k
	Fast breeder reactor	Aktau [Schevchenko], Kazakhstan	1 OP		k
Chemical Separation of Plutonium (Reprocessing)					
Chelyabinsk-65	Mayak Chemical Complex[5,10]	Near Kyshtym, Urals region	1 OP, 1 NO	Chelyabinsk-65 [Ozersk]	a
Krasnoyarsk-26 [Atomgrad]	Mining-Chemical Complex[5,11]	Near Dodonovo and Krasnoyarsk, Siberia	1 OP, 1 NC	Zheleznogorsk	a
Tomsk-7 [Seversk]	Siberian Chemical Complex	Near Tomsk, Siberia	1 OP	Seversk	a

Table 7.1 (continued)

Secret Site Name	Official Facility Name	Geographic Location (in Russia unless noted otherwise)	Operating Status[1]	Affiliated Closed City	Source[2]
Fabrication of Plutonium Metal Components					
Pu-239 Pits and Triggers					
Tomsk-7 [Seversk]	Siberian Chemical Complex	Near Tomsk, Siberia	1 OP	Seversk	q
Pu-238 Radioisotope Thermal Generators					
Chelyabinsk-65	Mayak Chemical Complex[5,12]	Near Kyshtym, Urals region	1 OP	Chelyabinsk-65 [Ozersk]	q
Other Warhead Components					
Beryllium Mining[13]					
	Mine name(s) not available	Vitimskiy, Mama-Chuya region, Siberia			g
	Malyshev Beryllium Mines	Malysheva, Urals region			h
	Izumrud Mines	Near Asbest			g
	Mine name(s) not available	Northern Karelia			g
	Mine name(s) not available	Lugovskiy, Mama-Chuya region, Siberia			g
Beryllium Processing					
	Mine name(s) not available	Gorno Chuyiskiy, Mama-Chuya region, Siberia			g
Ust-K, PO Box 10	Ulbinskiy Metallurgical Industrial Association	Ust-Kamenogorsk, Kazakhstan	NO 1990		h

Electronic Components					
Penza-19	Facility name not available	550 km southeast of Moscow	OP	Zarchinuy	a
High-Explosives Testing, R&D					
Chelyabinsk-70	All Russia Scientific Research Institute of Technical Physics	Near Kasli, Urals region	OP	Snezhinsk	q
Lithium-6 and Lithium Deuteride Production[14]					
	Chemical Concentrates Industrial Association	Novosibirsk, Siberia			i
	All Russia Scientific Research Institute of Chemical Technology	Moscow			i
	Chemical-Metallurgical Factory	Krasnoyarsk, Siberia			i
Neutron Generators					
	All Russia Automatics Research Institute	Moscow			q
Assembly and Disassembly of Nuclear Weapons					
Arzamas-16	All Russia Scientific Research Institute of Experimental Physics	Sarova, Mordovia	NO	Kremlev [Sarov]	q
Penza-19	Facility name not available	Kuznetsk, 115 km east of Penza	OP	Zarchinuy	a
Sverdlovsk-45	Electrochemical Measurement Complex	Nizhnyaya Tura, Urals region	OP	Rusnoy	a

Table 7.1 (continued)

Secret Site Name	Official Facility Name	Geographic Location (in Russia unless noted otherwise)	Operating Status[1]	Affiliated Closed City	Source[2]
Zlatoust-36	Facility name not available	Yuryuzan, 85 km southwest of Zlatoust, Urals region	OP	Torifugornuy	a
Semipalatinsk-21	Facility name not available	Kurchatov, Kazakhstan	NO		k
Warehouse Storage of Nuclear Weapons					
Chelyabinsk-115	Facility name not available	Chelyabinsk region			p
Zlatoust-20	Facility name not available	Chelyabinsk region			p
Disassembly of Nuclear Warheads and Fissile Material Storage					
Sverdlovsk-45	Electrochemical Measurement Complex	Nizhnyaya Tura, Urals region	OP	Rusnoy	a
Krasnoyarsk-26 [Atomgrad]	Mining-Chemical Complex[5]	Near Dodonovo and Krasnoyarsk, Siberia	OP	Zheleznogorsk	o
Tomsk-7 [Seversk]	Siberian Chemical Complex	Near Tomsk, Siberia	OP	Seversk	o
Testing of Nuclear Devices					
"Peaceful" Nuclear Explosions			150+ sites		
Weapons Testing					
Novaya Zemlya	Central Test Site, includes Northern and Southern areas	2 islands above the Arctic Circle between the Berents and Kara Seas	2 NO 1992		a

Semipalatinsk-21	Semipalatinsk [Kazakh] Test Site, includes Shagan River, Degelen Mountain, and Konyastan test areas	Near Semipalatinsk, Kazakhstan	3 NO 1991		a

1. Status abbreviations (number of facilities is shown along with status; year numbers indicate shutdown dates; for multiple plants, shutdown date shown is for last closed plant):
NC = Not completed (construction suspended or abandoned)
NO = not operating and either certain or likely not to resume, with year stopped, if known
OP = operating at some capacity
UC = under construction

2. Sources:
a = Cochran and Norris 1993
b = Shifrin 1982
c = Morgachev 1992
d = Ruzicka 1992
e = Chemyakin 1991
f = Sanatin 1991
g = Shabad 1969
h = Levine 1991
i = Ray-Press Information Agency 1991
j = Lehman 1990
k = Potter 1993
l = Syzmanski 1992
m = Gosatomnzador, as cited in SocEco Agency 1992
n = U.S. CIA 1985
o = Mikhailov, as cited in Charles 1992
p = Fonotov 1993

Table 7.1 (continued)

q = Cochran and Norris 1992
r = Bukharin 1993

3. With research reactor.
4. Uranium mining is also likely in close proximity to the production centers listed below.
5. According to Shifrin (1982), facility was built and/or operated at least in part with forced labor.
6. All production of highly enriched uranium stopped in 1989; continuing operations produce only low-enriched uranium for nuclear power reactors.
7. Heavy water needed for lithium deuteride production and operation of heavy-water reactor at Chelyabinsk-65.
8. No information is available on tritium processing facilities; Cochran (1992, p. 64) assumes they are located at the same sites where tritum is produced.
9. Reactors whose fuel has not been reprocessed are excluded. An exception is sites with both VVER-440 and VVER-1000 reactors.
10. RT-1 plant separates plutonium from only civilian VVER-440 and naval reactor fuel; plutonium is stored for future possible use in breeder reactors.
11. RT-2, the uncompleted plant at Krasnoyarsk-26, is designed to separate plutonium from civilian VVER-1000 fuel.
12. Pu-238 is produced in a special isotope reactor and processed in a separate facility from Pu-239.
13. Beryllium is used in some neutron generators to boost the yield.
14. Lithium is mined and processed to extract lithium-6, which is used to make target rods for tritium production and combined with heavy water to make lithium deuteride, which is used inside thermonuclear warheads to produce more tritium.

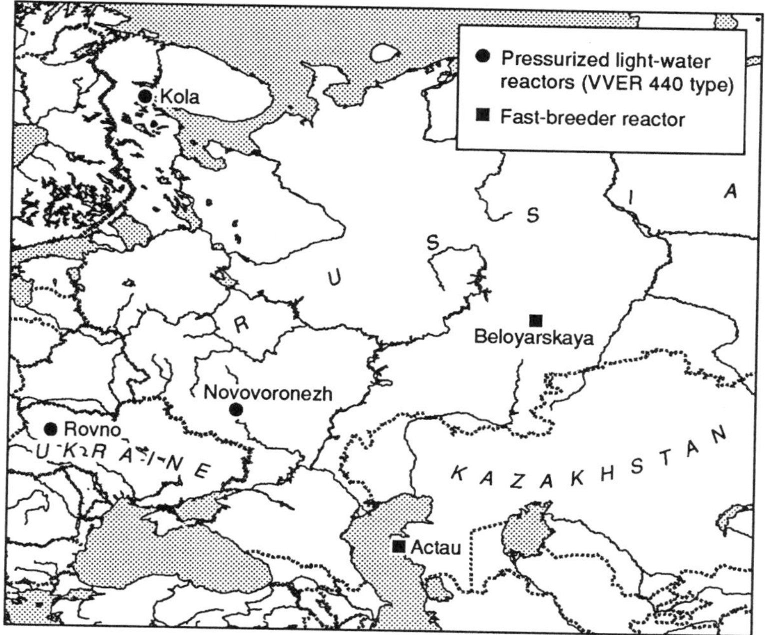

Figure 7.2 Civilian reactor sites in the countries of the former Soviet Union supplying spent fuel for reprocessing in Russia. Source: Cochran and Norris 1993, p. 163.

While we have not covered radioactive wastes from naval reactors in other chapters, we do so briefly in the context of this discussion of the former Soviet Union because of the extraordinary extent of Soviet dumping of radioactive wastes and the repeated official denials that such dumping was going on. In 1993, a commission led by Alexei Yablokov, environmental advisor to President Yeltsin, issued a detailed report on the subject.[37] The commission's report describes in detail intentional dumping of solid and liquid radioactive wastes at sea—largely in the northern seas, but also in the seas off the eastern coast of Russia and the Pacific and Atlantic Oceans. This included dumping of nuclear reactors with and without spent fuel aboard (6 with and 12 without, plus one spent fuel screen assembly) and accidental loss of five nuclear reactors at sea.[38] The upper bound estimate that the commission put on the dumping and accidental losses was 10 million curies (370,000 terabecquerels). The report accounts in detail for about 3 million curies (111,000 terabecquerels) of this total. More than 90 percent of it is due to the radioactivity in the spent fuel of dumped reactors and accidentally sunken submarines. The total solid

37. Government Commission on the Questions Related to the Dumping of Radioactive Waste at Sea 1993.

38. Government Commission on the Questions Related to the Dumping of Radioactive Waste at Sea 1993, tables 1 through 8.

and liquid radioactive wastes was about 57,700 curies (about 2,130 terabecquerels), dumped in the northern and far-eastern seas. In addition the commission reports that a few million curies consisted of "lost nuclear weapons, radionuclide sources, satellites, etc." and another few million curies (on the order of 100,000 terabecquerels) consisted of radioactive waste washed out from the Yenisey and Ob rivers into the Arctic Ocean.[39]

Given the excess of both nuclear warheads and nuclear wastes, the Ministry of Atomic Energy has suggested the use of underground nuclear explosions to get rid of high-level wastes. It has given one company exclusive rights to develop this and other commercial uses of nuclear explosives. The company, Chetek, was formed in 1990 as an international joint-stock venture by the Ministry of Atomic Power and Industry and Arzamas-16. Most of the staff at Arzamas-16 are assigned to Chetek, which has branches in eight cities promoting underground nuclear explosions as "the cheapest and safest method of eliminating highly toxic radioactive waste."[40] A branch in Hamburg, Germany, closed after receiving adverse publicity in local newspapers.[41]

Research and Development Laboratories

The military-industrial complex of the former Soviet Union includes over 450 research and development institutions and 50 major design bureaus.[42] In Russia, these bureaus now work under the direction of nine key military-related ministries, primarily the Ministry of Atomic Energy (MINATOM). The nuclear weapons ministry was first known as the Ministry of Medium Machine Building,[43] and then as the Ministry of Atomic Power and Industry (MAPI).

A Soviet nuclear weapons research facility, called "Laboratory No. 2," was set up in Moscow in 1942. It is now known as the Russian Scientific Center Kurchatov Institute or RSC Kurchatov Institute.[44] All 7 nuclear reactors there are now shut down.[45] The oldest, built in the 1940s, was shut in 1992.[46]

39. Government Commission on the Questions Related to the Dumping of Radioactive Waste at Sea 1993, appendix tables and table 8.

40. Kovalenko 1992a.

41. Kovalenko 1992a.

42. Cochran et al. 1989, p. 70.

43. Cochran et al. 1989, p. 72.

44. Cochran and Norris 1993, p. 8.

45. Lehman 1993.

46. Kolesnikov 1993.

In 1947, the Soviet Union transferred most nuclear weapons design work from Laboratory No. 2 to the newly created Design Bureau 11 (KB-11, or "Construction Bureau-11"). It is now known as the All Russia Scientific Research Institute of Experimental Physics and usually referred to by its code name, Arzamas-16.[47] This facility is in Sarovo, about 60 kilometers southwest of Arzamas.

The first Soviet nuclear bomb was designed and assembled at Arzamas-16, which continues to specialize in designing and fabricating nuclear warheads. It also appears to have been a large-scale warhead assembly site. One employee, quoted in a 1992 Moscow newspaper article, claims to have worked on assembling several thousand warheads over 14 years.[48]

The lab's talents are gradually being transferred to civilian projects. According to Atomic Energy Minister Viktor Mikhailov, military projects accounted for only 70 percent of the work at Arzamas-16 in 1992 and are expected to fall to 50 percent by 1995.[49] Like the RSC Kurchatov Institute, this facility houses several research reactors.[50]

The third major research facility involved in weapons design is the All Russia Scientific Institute of Technical Physics, started in 1955 as a backup to Arzamas-16. Unlike Arzamas-16, it employed few Jewish scientists. As a result, the two competing weapons labs were nicknamed "Israel" and "Egypt."[51] The institute and its secret closed city of Snezhinsk are surrounded by barbed wire and armed guards at a site known as Chelyabinsk-70, about 20 kilometers north of Kasli in the Ural Mountains. This facility also experiments with chemical high explosives. In 1988, it began conversion to nonweapons work, which constituted 50 percent of its research by 1992.[52]

In addition to these facilities, 17 other atomic research centers in the former Soviet Union house at least 30 experimental reactors. Most are located in Russia, including some at or near major nuclear weapons centers. But there are also research reactors and training centers in Ukraine, Kazakhstan, Georgia, Uzbekistan, Latvia, and Belarus.[53] At least 17 of 25 research reactors in the former Soviet Union use highly enriched uranium as their fuel source.[54]

47. Zaloga 1991, p. 178; Cochran and Norris 1993, pp. 29–30.

48. Filin 1992, p. 2.

49. Cherepanov 1993.

50. *Rossiyskaya Gazeta* 1992; Cochran and Norris 1993, pp. 29–30.

51. Zaloga 1991, p. 181.

52. Cochran and Norris 1992, pp. 15–16.

53. Potter and Cohen 1992.

54. Bukharin 1992, p. 5.

Little is known about the historical radioactive waste disposal practices of these research facilities or about their impact on human health and the environment. Gosatomnadzor, the State Committee on Atomic Safety, acknowledged in 1991 that one of the largest, the Scientific Research Institute of Atomic Reactors in Dimitrovgrad, injected 1.9 million cubic meters of radioactive liquid wastes with an activity of 88,000 curies (about 3,260 terabecquerels) into deep wells underground.[55]

At exchange rates prevailing in early 1993, the wages earned by scientists working in military facilities, although relatively high by Russian standards, amounted to just a few dollars per day. This does not begin to compare with what they could earn in many other countries, which has raised fears of nuclear proliferation if some scientists leave Russia for higher hard-currency wages elsewhere. To give the staff in these labs a reason for staying, the United States, Japan, and the European Community have approved $70 million in "conversion" aid. An international panel will select the proposals to be funded, with all the money going through the new International Center of Science and Technology in Moscow.[56]

Uranium Chemical Conversion and Enrichment Facilities

The chemical conversion of uranium from oxide or nitrate form to the uranium hexafluoride form needed for enrichment occurs at three locations. The two larger facilities are the Electrolyzing Chemical Complex in Angarsk, 30 kilometers northwest of Irkutsk near Lake Baikal, and a plant at the Tomsk-7 Siberian Chemical Complex. A smaller facility, called the Urals Electrochemical Plant, is located at Sverdlovsk-44, in Verkhniy-Neyvinskiy near Yekaterinburg (the Urals).

There are four uranium enrichment plants. Three of them are at Angarsk, Tomsk-7, and Sverdlovsk-44. The fourth plant is called Krasnoyarsk-45, on the Kan River about 90 kilometers east of the city of Krasnoyarsk.[57]

The Soviets first developed the gaseous diffusion enrichment method, with the first plant being built in 1949 at Sverdlovsk-44. A pilot scale gas centrifuge was first operated in 1957, and according to Oleg Bukharin, a Russian researcher at the Center for Arms Control, Energy, and Environmental Studies at the Moscow Institute of Physics and Technology, became "the backbone of the Soviet enrichment program in the 1970s."[58] Highly enriched uranium for nuclear weapons

55. As cited in SocEco Agency 1992.

56. Charles 1992.

57. Bukharin 1993, p. 5; and Bukharin paper, included in Cochran and Norris 1993, pp. 115–122.

58. Bukharin paper, included in Cochran and Norris 1993, p. 118.

was produced only at Sverdlovsk-44, which has a capacity of two to three million separative work units per year.[59]

All four of Russia's enrichment plants are operating at reduced capacity and only for the low enrichment of civilian reactor fuel. Their total estimated capacity is about 13 million separative work units per year. According to Bukharin, Russian enrichment requirements in the early 1990s were only 4.5 million separative work units, leaving 8.5 million units available for export. He estimated in January 1993 that Russia was exporting enrichment services at the rate of about 2 million separative work units per year.[60]

Uranium Fuel Fabrication Facilities

Three plants are centrally involved in converting uranium hexafluoride into uranium dioxide and fabricating it into fuel pellets and finally into fuel rods for commercial power reactors as well as naval propulsion and research reactors: the Ulbinsky Metallurgical plant in Ust-Kamenogorsk, Kazakhstan; the Machine Building Industrial Association in Electrostal near Moscow; and the Chemical Concentrates Industrial Association (Khimconcentrate Plant) in Novosibirsk, Siberia.[61] The last mentioned is a relatively small installation.

The Soviet Union fabricated highly enriched uranium metal into components for nuclear warheads at Tomsk-7, the Siberian Chemical Complex.[62] While we have no published information on where it machined depleted uranium metal into warhead casings, the facilities most likely to be involved in this type of processing, in addition to the three making fuel rods, are the two others listed in a Ministry of Atomic Power and Industry (MAPI) organizational chart under the Department of Nuclear and Structural Materials Metallurgy, Nuclear Fuel Fabrication. These are the Tchepetsk's Mechanical Works in Glazov and the Pridneprovskiy Chemical Industrial Association in Dneprodzerzhinsk.[63]

Plutonium and Tritium Production from Military Reactors

The Soviet Union built fifteen military reactors for the production of plutonium. Five graphite-moderated reactors (with a total capacity estimated by Cochran and Norris to be 6,565 megawatts thermal), now

59. Bukharin paper, included in Cochran and Norris 1993, p. 121.

60. Bukharin 1993, p. 5.

61. Bukharin 1993, p. 7.

62. Cochran and Norris 1993, p. 41

63. MAPI 1991. MAPI was renamed the Ministry of Atomic Energy (MINATOM) in 1992 after the breakup of the Soviet Union.

shut, and one light-water and one heavy-water reactor (each with a capacity of 1,000 megawatts thermal) are at the Mayak Chemical Complex at Chelyabinsk-65, near Kyshtym in the southeastern Urals. One of the light-water reactors is housed in a reactor building that held a heavy-water reactor for most of the Mayak complex's existence. The two 1,000 megawatt capacity reactors are used for producing tritium and other isotopes. Also shut down are two reactors that produced only plutonium at Krasnoyarsk-26, a completely underground "mining-chemical complex" on the Yenisey River in Siberia about 50 kilometers northeast of the city of Krasnoyarsk. Still operating at Krasnoyarsk-26 is a dual-use military reactor that produces plutonium as well as steam for district heating and some electricity.[64] The capacity of the Krasnoyarsk-26 reactors is not publicly known. Cochran and Norris assume them to be about the same size as the larger reactors at Chelyabinsk-65, which would give them each a capacity of about 2,000 megawatts thermal.[65]

The Siberian Atomic Power Station at Tomsk-7 houses five graphite-moderated reactors with a similarly estimated capacity of 2,000 megawatts thermal each. Four of the five were used to generate heat and electricity as well as plutonium; two are still operating.[66]

Some of the Tomsk-7 reactor capacity was no doubt used for tritium production, but no site-specific output data are available. Cochran and Norris estimate that the cumulative decay-corrected inventory of tritium as of the end of 1991 in Russia was 35.7 kilograms, which requires the same plant capacity as to produce 5.8 metric tons of plutonium.[67]

Plutonium Production from Civilian Power Reactors

The Soviet Union built four kinds of civilian power reactors by the end of 1991: twenty "RBMK" graphite channel reactors, eight "VVER-440" and sixteen "VVER-1000" pressurized light-water reactors, and two "BN" fast (breeder) reactors.[68] According to a 1992 report of Gosatomnadzor, Russia's equivalent of the U.S. Nuclear Regulatory Commission, reprocessing is technologically possible at this time only for

64. Cochran and Norris 1993, p. 45.

65. Cochran and Norris 1993, pp. 99–100.

66. Cochran and Norris 1993, pp. 87, 109; Cochran and Norris assume the capacity to be 2,000 megawatts thermal based on the capacity of a similar reactor at Hanford (the C-reactor).

67. Cochran and Norris 1992, p. 82.

68. IAEA 1992, pp. 20–22; Albright, Berkhout, and Walker 1993, p. 120, for information on BN reactors.

the spent fuel from VVER-440, BN, and naval reactors.[69] Russia explored the possibility of reprocessing civilian RBMK fuel, but later suspended these plans. The Ministry of Atomic Energy has announced its preferred option to bury this fuel under permafrost on the Arctic islands of Novaya Zemlya, after letting it cool in pools and dry casks for 50 to 70 years.[70] The spent fuel from civilian VVER-440 reactors goes to the RT-1 plant at Chelyabinsk-65 for reprocessing after being stored on-site in special pools for three years to allow short-lived radionuclides to decay.[71] As of 1991, the Chelyabinsk plant had received spent fuel from Hungary, Germany, and Czechoslovakia.[72] A new Russian law prohibits nuclear waste imports. This has caused shipments from Ukraine, Czechoslovakia, Germany, and Hungary to be held up, pending resolution of the issue of whether spent fuel is to be regarded as nuclear waste.[73] Total surplus plutonium in storage at the end of 1992 at Chelyabinsk-65 was estimated to be 25.4 metric tons and rising by about 1 metric ton per year.

Some spent fuel from VVER-1000 reactors is stored at Krasnoyarsk-26, where a reprocessing facility to handle it is under construction. MINATOM is aggressively promoting this plant, known as RT-2, although the project faces strong public opposition and funding problems. The spent fuel from RBMK reactors, for which reprocessing is not economical, is stored at reactors on-site.[74]

Facilities for the Chemical Separation of Plutonium and Uranium

Tomsk-7, Chelyabinsk-65, and Krasnoyarsk-26 operate chemical reprocessing plants to recover plutonium and uranium from reactor fuel. In 1976, Chelyabinsk-65 switched its RT-1 plant to processing only civilian VVER, research, and naval-propulsion reactor fuel. It sent to Tomsk-7 the irradiated fuel from its own plutonium production reactors until they were shut down in 1990. Chelyabinsk-65 received 190 metric tons of spent fuel in 1992, mostly from VVER-440 reactors in Russia (Novo-Voronezh and Kola) but also from naval and civilian reactors in Ukraine, Bulgaria, Hungary, and Czechoslovakia. Although Chelyabinsk-65 can process 400 metric tons of spent fuel per year, it

69. As cited in SocEco Agency 1992; existing reprocessing plants lack the technology needed to cut up the larger VVER-1000 fuel rods.

70. Hibbs 1993.

71. SocEco Agency 1992; Bradley 1991, p. 11.

72. Bradley 1991, p. 11.

73. Cochran and Norris 1993, p. 54.

74. SocEco Agency 1992; Miheev 1992.

has averaged only half this since 1976. Recently the rate of reprocessing has declined to only 120 metric tons per year.[75]

The underground plant at Krasnoyarsk-26 only reprocesses fuel from the site's own plutonium production reactors. A storage facility for the spent fuel from VVER-1000 reactors already going to Krasnoyarsk-26 opened in 1985 and has a capacity of 6,000 metric tons.[76] As of June 1992, only 402 metric tons from two VVER-1000 reactors in Russia (Tver—formerly Kalinin—and Balakov) had been collected.[77] By 1993, the total stockpile reached 1,100 metric tons. In 1992, delivery of spent fuel from VVER reactors outside Russia was blocked briefly due to a law rescinded in April 1993 prohibiting import of radioactive waste into Russia.[78]

Plutonium recovered from these reprocessing operations was then processed in other plants at the same sites. The processing involved blending freshly reprocessed plutonium with the plutonium recovered from retired warheads, which was contaminated to varying degrees with americium-241.[79]

Cochran and Norris estimate that the cumulative total quantity of plutonium produced from military sources ranges from 145 to 170 metric tons, with about 140 metric tons actually used in weapons at the time the stockpile peaked in 1986.[80] These estimates are roughly consistent with an estimate of plutonium production done by Frank von Hippel and his colleagues at Princeton University based on the amount of krypton-85 discharged to the atmosphere in chemical reprocessing.[81]

Production of Plutonium Components

Plutonium-239 Triggers (pits or cores) The only site known to manufacture plutonium pits is the Siberian Chemical Complex at Tomsk-7.[82] Given the importance of this step in the manufacture of nuclear weapons, the Soviet Union most probably maintained at least one other facility with this capability.

75. SocEco Agency 1992; Cochran and Norris 1993, pp. 54–57.

76. SocEco Agency 1992; Cochran and Norris 1992, p. 56.

77. SocEco Agency 1992.

78. Cochran and Norris 1993, pp. 101–102.

79. Cochran and Norris 1992, p. 19. Over time, as plutonium-241 undergoes beta decay to americium-241, the amount of americium-241 contaminating a given sample of plutonium increases. The half-life of plutonium-241 is 14.4 years.

80. Cochran and Norris 1993, p. 42.

81. Von Hippel, Albright, and Levi 1986, chapter 5, p. 6.

82. Cochran and Norris 1992, p. 48.

Plutonium-238 Radioisotope Thermal Generators Chelyabinsk-65 is the only site known to produce radioisotope thermal generators powered by plutonium-238. The plutonium-238 is made in the light-water reactor at the facility. There is also a separate reprocessing facility for separating special isotopes[83] like plutonium-238 for civilian as well as military purposes.

Production of Other Components

Beryllium The Soviet Union ceased all production of beryllium in September 1990 after an explosion at its only processing plant, the Ulbinsky Metallurgical Industrial Association in Ust-Kamenogorsk, Kazakhstan.[84] (This site also processes uranium.) Sparks from welding operations ignited an estimated 4,000 kilograms of beryllium that had accumulated in air ducts of the plant's ventilation system. The blast blew beryllium across the center of Ust-Kamenogorsk. According to a Soviet commission of inquiry, the accident was "the logical consequence" of years of mistakes, foul-ups, and neglect of elementary safety rules. Inside the fallout zone, where beryllium concentrations ranged from 60 to 890 times the permitted level, were the homes of 120,000 people, 42 kindergartens, 23 schools, and 5 colleges and universities.[85]

The public was not informed about the accident for two and a half hours and even then was told nothing about beryllium—only that there was no radiation danger. Of the 2,700 residents who were eventually screened for beryllium, 236 registered levels above normal, in the range of 1.5 to 5 times background. On the basis of these results, public officials canceled all further investigations.[86]

Large-scale mining of beryllium and lithium began together in the Mama-Chuya region of eastern Siberia, where three mining cities were built in 1952: at Lugovskiy on the Mama River, Gorno Chuyskiy on the Chuya River, and Vitimskiy on the Vitim River. Beryllium also was mined in northern Karelia[87] and the Urals, at the Malyshev mines (reportedly converting some capacity to civilian emerald mining),[88]

83. Cochran and Norris 1992, p. 23.

84. Levine 1991, p. 3.

85. Rich 1990.

86. Rich 1990.

87. Shabad 1969, pp. 65, 262.

88. Levine 1991, p. 3.

and the Izumrud mines near Asbest.[89] Until 1962, the Soviet Union also obtained beryllium from China.[90]

Chemical High Explosives It is not known where the Soviet Union produced conventional high explosives for its nuclear weapons programs. This production is probably the responsibility of the Ministry of Machine Building, which also makes fusing devices, solid propellants, and a variety of civilian products, including bicycles and refrigerators.[91] Research, development, and large-scale testing of the chemical explosives used in nuclear weapons are conducted at the All Russia Scientific Research Institute of Technical Physics.[92]

Depleted Uranium Metal for Warhead Casings There is no published information on where depleted uranium from uranium enrichment plants is converted to uranium metal and formed into warhead casings. The conversion to metal probably occurs at the enrichment plants at Angarsk, Tomsk-7, Sverdlovsk-44, and Krasnoyarsk-45, but the warhead casings may be produced at other locations such as Glazov.

Electronic Components The only facility known to manufacture electronic components for nuclear weapons is Penza-19 in Kuznetsk, about 550 kilometers southeast of Moscow.[93] One or more other sites may be engaged in manufacturing electronic components.

Heavy Water (deuterium) In the late 1940s, the Soviet Union developed electrolytic cells capable of producing deuterium from synthetic-ammonia plants. The necessary electrodes were produced at the Urals Chemical Machine Plant near Sverdlovsk and installed at four ammonia factories in the early 1950s.[94] While the location of these first deuterium factories is not known, two reportedly operate in Ukraine: the Agrochemical Service Company in Dneprodzerzhinsk and the Union Agrochemical Import plant in Kiev.[95] Given the strategic importance of this material—it was used in the military's heavy-water production reactor at Chelyabinsk-65 and for producing lithium deuteride—it is likely that at least one other production facility existed in

89. Shabad 1969, p. 65.

90. Lewis and Xue 1988, pp. 155, 269.

91. Cochran et al. 1989, p. 69.

92. Cochran and Norris 1992, p. 16.

93. Cochran and Norris 1992, p. 17.

94. Zaloga 1991, p. 180.

95. Potter and Cohen 1993, p. 85.

Russia. The only other places at which heavy water is reported to "perhaps" have been produced are in Yavan and Kalininabad, Tajikistan,[96] in the Pridneprovskiy Chemical Factory of the Ukraine, or in Armenia.[97]

Lithium-6 and Lithium Deuteride Some lithium is mined with beryllium in the Mama-Chuya region. The largest source of lithium is the Chita region of eastern Siberia, where lithium is often found with rubidium and cesium. Mining centers there include Klichka, Kadaya, Gornyy Zerentuy, and Aktau.[98] The Soviet Union also imported lithium from mines in China.[99]

Lithium is processed from ores at the Chemical Concentrates Industrial Association plant in Novosibirsk, Siberia. Casting and molding of lithium-6 and lithium deuteride take place at the Chemical and Metallurgical Factory in Krasnoyarsk, Siberia.[100]

Neutron Generators Very little information is available about the production of neutron generators. The All Russia Automatics Research Institute in Moscow manufactures commercial pulsed-neutron generators. According to Cochran and Norris, it probably also makes the neutron generators used in nuclear warheads.[101]

Assembly and Disassembly of Nuclear Weapons

Nuclear warheads and weapons were assembled and are now being disassembled at several locations. The first assembly site was at Arzamas-16.[102] Large-scale assembly and storage of nuclear warheads later shifted to the Electrochemical Measurement Complex at Sverdlovsk-45 (also called Rusnoy, population 54,700) in Nizhnaya Tura in the Urals, about 200 kilometers north of Yekaterinburg. Smaller assembly line production facilities are located at Penza-19 in Kuznetsk,[103] 115 kilometers east of Penza, and Zlatoust-36 near Yuryuzan,

96. Potter and Cohen 1993, p. 79.

97. Potter 1993, pp. 1, 89.

98. Shabad 1969, p. 265.

99. Lewis and Xue 1988, pp. 155, 269.

100. Ray-Press Information Agency 1991, p. 347.

101. Cochran and Norris 1992, p. 18.

102. The facility was known by several names, including: Obyekt No. 558, Volga Office Number (number unknown), Moscow Center 33, Military Installation "N," Kremlev City, and Khariton's Institute (informally). Cochran and Norris 1993, pp. 30–31.

103. Cochran and Norris 1993, p. 40.

85 kilometers southeast of Zlatoust in the Urals. The total dismantling capacity of these plants is 5,500 to 6,000 warheads per year. The most active dismantling site, according to the U.S. Central Intelligence Agency, is Zlatoust-36.[104] Estimates of the number of warheads actually dismantled per year range from 1,500[105] to 2,500.[106]

The final step of dismantling nuclear warheads takes place at Mayak and Tomsk-7. Current plans are to store the plutonium for at least 10 years until a final disposal method is adopted. Officials in the Ministry of Atomic Energy want to use the plutonium to make mixed-oxide fuel for civilian reactors, while weapons scientists at Chelyabinsk-70 want to store it indefinitely and Arzamas-16 scientists propose destroying the stockpile in underground nuclear explosions.[107]

DESCRIPTION OF KEY SITES

We will describe in more detail three sites in Russia involved in warhead and materials production on a large scale—Chelyabinsk-65, Krasnoyarsk-26, and Tomsk-7—in order to examine some environmental issues more closely. Data on facilities at these sites already provided above will not be repeated below, except as necessary to provide context.

Chelyabinsk-65 (Mayak)

"You could hardly call Mayak a monument to nature and humanity."
— Sergei Spitsin, public relations manager, Mayak Chemical Complex, 1990[108]

The Mayak [Beacon] Chemical Complex, commonly known by its code name of Chelyabinsk-65, is the oldest and largest of the former Soviet Union's three plutonium production centers. The highly contaminated site is about 70 kilometers north of the city of Chelyabinsk (population 1.1 million) on the eastern flank of the Ural Mountains, about 1,750 kilometers east of Moscow. It lies in a region of interconnecting lakes, marshes, waterways, and artificial reservoirs at the headwaters of the 240-kilometer-long Techa River (see figure 7.3).[109]

104. Cochran and Norris 1993, pp. 39–40.

105. Norris 1992, p. 27.

106. Berkhout et al. 1992, p. 30.

107. Cochran and Norris 1993, p. 44.

108. As quoted in MacLachlan 1990a.

109. Akleev et al. 1989.

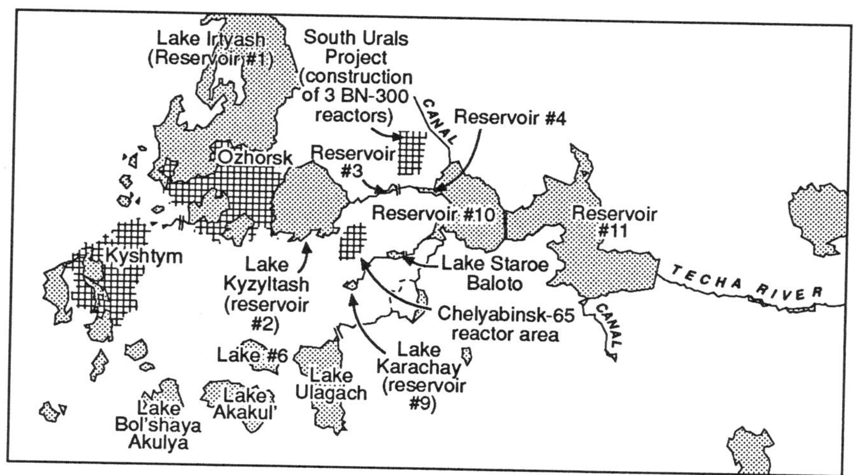

Figure 7.3 Reservoirs and lakes at Chelyabinsk-65. Source: Cochran and Norris 1993, pp. 170–174.

The heavily industrialized Chelyabinsk region is one of the most polluted in Russia.[110] The city of Chelyabinsk contains over 130 factories, including the largest zinc-processing plant and the second largest iron and steel complex in the Urals. Goskompriroda, the Soviet State Committee for Environmental Protection, put the city on its original 1989 list of the 68 most chemically polluted places in the Soviet Union.[111]

Most of the plutonium production facilities at Chelyabinsk-65 are concentrated in a 90-square-kilometer area on the southeast shore of Lake Kyzyltash (Reservoir No. 2) near Lake Irtysh, the source of the Techa River, and just north of the smaller Lake Karachay (Reservoir No. 9), which is not directly connected to the Techa River system. Both lakes were used to contain radioactive wastes from the Mayak Complex, as were all the other numbered reservoirs shown in figure 7.3. The artificial reservoirs (numbers 3, 4, 10, 11, and 17) along the Techa River are the result of dams built by the Mayak Complex in 1949, 1956, 1963, and 1964 to try to keep radioactivity from flowing downstream. The first dam, an earthen one, was constructed just below Staroye Boloto ("old swamp") to prevent this artificial lake—which was being filled with radioactive waste—from draining back into the Techa system through Reservoir No. 10.

The Techa River first flows into the Iset' River and then into the Tobol, the Irtysh, and finally the massive Ob' River. The Ob' flows on

110. Commission for Investigation of the Ecological Situation in the Chelyabinsk Region 1991, vol. II, section 1.1.

111. As cited in Monroe 1991, p. 543.

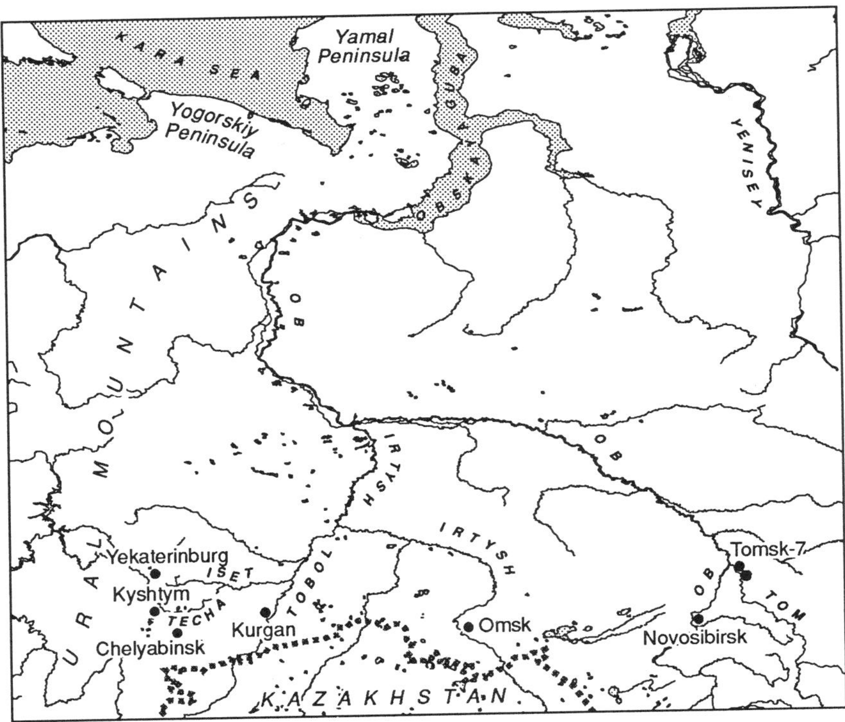

Figure 7.4 Western Siberia, showing Chelyabinsk, Tomsk, and Ob' River.

for another 1,500 kilometers before reaching its mouth at the base of the Kara Sea above the Arctic Circle (see figure 7.4)

About 10 kilometers northwest of the Mayak reactors, between lakes Kyzyltash and Irtysh, is the "closed" and secret military city of Chelyabinsk-65 (population 83,500), built to house workers from the complex. It is also known as Sorokovka ("Forties Town") and Ozyorsk, although none of these names appears on any public map. Before the official designations Chelyabinsk-65 or Mayak were known, people in the West often called it the Kyshtym Complex, after the city of Kyshtym (population 40,000), 15 kilometers to the west. The closest village to the reactor site was Metlino (population 1,200), just 7 kilometers downriver from the plant's main discharge point.[112] Metlino has been evacuated.

The Chelyabinsk-65 complex is enormous, with a militarily restricted area that covered as much as 2,700 square kilometers in the 1950s. By 1992, this had been cut back to 200 square kilometers, within which are five graphite-moderated, water-cooled plutonium production reactors and two special-isotope production reactors (one light- and one heavy-water). An earlier heavy-water reactor has been decommissioned. The

112. Cochran and Norris 1993, p. 65.

light-water reactor was built in the same reactor building. The isotopes produced include tritium for weapons as well as cobalt-60 for medical applications.

All the graphite-moderated reactors are shut down. Chelyabinsk-65 also contains an operating reprocessing plant and storage facility for plutonium, a separate reprocessing plant for special isotopes, a partially constructed mixed-oxide fuel fabrication plant for breeder reactors and four smaller mixed-oxide fuel facilities (two are still operating), the South Urals Atomic Energy Station (a new complex of three breeder reactors on which construction was barely begun before being suspended), and nuclear waste storage and treatment facilities, including about 60 high-level waste storage tanks and a vitrification plant for high-level liquid wastes.[113]

Chelyabinsk-65 also has an Institute of Biophysics set up to monitor the effects of the site's various discharges of radioactivity. These have had serious radiological consequences.[114] The site has plants under construction for solidifying medium-level liquid wastes and processing solid wastes, and plants for manufacturing defense industry equipment, machinery, and machine tools.[115] Little information about these plants is available.

The first graphite reactor at Chelyabinsk-65, known as the A-reactor, came on line in 1948 and shut down in 1987. The decommissioning plan for the reactor calls for removing all the control and operating systems, filling the empty reactor vessel with concrete, and letting its radioactivity decay for 20 to 30 years before deciding whether to bury the vessel on site or remove it. The second reactor operated from 1951 to 1989. The next three, AV-1, AV-2, and AV-3 were larger. The AV-1 reactor was shut in 1989 and the other two were shut in 1990.[116]

The light-water and the heavy-water reactor at Chelyabinsk-65 produced tritium, plutonium-238, and other special isotopes. The United States is purchasing at least 5 kilograms of plutonium-238 from this facility for $6 million and as much as 40 kilograms over the next 5 years for $57.3 million to power NASA space missions.[117]

The first chemical separation plant for reprocessing plutonium-239 from the graphite reactors used the process of coprecipitation of uranium and plutonium in the form of sodium uranyl acetate and sodium

113. Commission for Investigation of the Ecological Situation in the Chelyabinsk Region 1991, vol. II, section 2.1; Cochran and Norris 1993, pp. 48–59, 80–81, and 137.

114. These releases are discussed below.

115. Commission for Investigation of the Ecological Situation in the Chelyabinsk Region 1991, vol. II, section 2.1; also cited in Bolsunovsky 1992.

116. Cochran and Norris 1993, pp. 49–50.

117. Cochran and Norris 1993, p. 59.

plutonyl acetate from a nitric acid solution (mentioned in chapter 3). However, the process was inefficient, and much of the waste had to be processed again to further extract uranium and plutonium.[118] The final waste from this process also contained high concentrations of both sodium nitrate (over 100 grams per liter) and sodium acetate (60 to 80 grams per liter).

A second reprocessing facility known as DB came on line in 1956 and later used the same coprecipitation process. This plant was modified to handle spent fuel from the VVER-440 type of light water reactors. The modified plant, called RT-1, uses PUREX reprocessing technology (described in chapter 3). Large-scale industrial sterilization and food irradiation plants use the recovered cesium-137, while the strontium-90 is used as a power source in navigation gauges and other remote instruments.[119] Americium-241 and curium-244 are also sometimes extracted.

RT-1's capacity is rated at 400 metric tons of heavy metal per year.[120] In addition to reprocessing fuel from Soviet power reactors, the plant has also taken fuel from reactors of Soviet origin in Hungary, Germany, and the former Czechoslovakia.[121]

For each 100 metric tons of spent fuel processed, about 0.8 metric tons of reactor-grade plutonium were recovered per year.[122] This plutonium then was converted to oxide or metal form for further fabrication at a separate facility near Lake Karachay. These fabrication operations were suspended in 1987.[123] The recovered uranium is blended with higher-enriched uranium to produce new fuel for RBMK reactors.[124]

From 1976 until Mayak's last plutonium production reactor shut down in 1990, the spent fuel from these reactors went by rail to Tomsk-7 for reprocessing. The plutonium recovered from civilian fuel since 1976 has been stockpiled in a temporary surface storage facility on site,

118. Cochran and Norris 1993, p. 52.

119. Lehman 1990, p. 6.

120. Cochran and Norris 1993, p. 54.

121. Bradley 1991, p. 11; Monterey Institute of International Studies 1992. Although Article 50, Clause 3 of the Environmental Protection Law adopted by Russia in 1992 is supposed to prohibit the importation of "radioactive waste and materials for the purpose of storage and disposal," it has not stopped either the importation or the reprocessing of spent fuel from foreign nuclear-power plants. See Belyaninov 1993 for more on Russian nuclear imports. President Boris Yeltsin issued a special decree overturning this law in 1993 (Cochran and Norris 1993, p. 102).

122. Cochran and Norris 1993, p. 56.

123. Bradley 1991, p. 10.

124. Cochran and Norris 1993, p. 56.

to await eventual use in mixed-oxide fuel for fast breeder reactors.[125] None has yet been used.

A mixed-oxide fuel fabrication plant at Chelyabinsk-65 is almost 70 percent complete, but all work was suspended in 1987 due to strong public protests and a lack of funds. The German multinational corporation Siemens has indicated interest in completing this plant. It was designed to make mixed-oxide fuel rods for three liquid-metal fast breeder reactors planned at the South Urals Atomic Energy Station. Construction of these reactors, which began in 1984, also halted in 1987. Only a few support buildings and the concrete footings for two of the three reactors are finished. Russia is soliciting international aid to complete the project, especially from Japan, which shares Russia's interest in developing plutonium breeder reactors.

Whether to build the breeder is a subject of intense local debate. A 1990 Soviet Academy of Sciences report on the ecological situation around Mayak supported the breeder project,[126] and the Chelyabinsk regional parliament voted in favor of resuming construction in 1990 (after a favorable report on the project by a commission set up to study the ecological situation).[127] However, public outcry against this decision forced a referendum on the issue in March 1991.[128] Voters defeated the breeder proposal (about 76 percent opposed), as well as a question about importing more civilian spent fuel from abroad for plutonium reprocessing and waste vitrification (over 84 percent opposed).[129]

While the Ministry of Atomic Energy is researching the use of mixed-oxide fuels in its VVER reactors, the Mayak plant was only designed to produce such fuel for breeder reactors. In any case, Russia's two existing breeder reactors are using highly enriched uranium fuel and have only experimented with plutonium fuel.[130]

The South Urals breeder construction site at Chelyabinsk-65, about 10 kilometers from the production reactor area, is on the northwest shore of Reservoir No. 10. Part of the rationale for siting the reactors here is that they would draw cooling water out of the reservoir, hastening evaporation and thus helping prevent it from overflowing and reducing seepage into groundwater.

125. Commission for Investigation of the Ecological Situation in the Chelyabinsk Region 1991, section 2.1.1.

126. Bolshakov et al. 1991.

127. MacLachlan 1990.

128. Monroe 1991.

129. Mironova 1991; Monroe 1991.

130. Albright, Berkhout, and Walker 1993, p. 120.

A pilot radioactive-waste vitrification plant at Chelyabinsk-65 came on line briefly in 1986, after years of research and testing that began in 1967, but it was decommissioned in 1987.[131] A second melter, on line since June 1991, can solidify in phosphate glass the high-level liquid wastes from reprocessing both breeder and VVER reactor fuel. The process uses a single-stage ceramic melter to create glass blocks that are designed to remain intact for 300 years.

In July 1992, the production output of this plant was one metric ton of glass per day, with each metric ton containing an average of 148,000 curies (about 5,500 terabecquerels) of radioactivity.[132] About 680 metric tons of vitrified glass had been processed by December 1992, containing 90 to 100 million curies (3.33 million to 3.7 million terabecquerels).[133] The amount of radioactivity remaining in Mayak's high-level waste tanks is estimated at 400 million curies (about 15 million terabecquerels) and would take 10 years for this vitrification plant to process, according to the chief engineer, Evgeniy Dzekun.[134]

After vitrification, the glass blocks are stacked above ground in metal canisters (0.63 meters in diameter and 3.4 meters high) in a special building with forced-air cooling and an exhaust gas purification system. In 1991, Mayak officials projected that the site had enough canister storage space for another 10 to 12 years of production. The canisters are to be stored for 20 to 30 years, after which they will be transferred to a granite or salt formation for permanent burial.[135] No such site has been selected, however. According to Alexander Suslov, the chief engineer of the plant in April 1992, Germany has offered an 8-year contract to participate in the vitrification project, and several countries in the former Soviet Union and Eastern Europe have also asked to participate.[136]

Plutonium Production at Chelyabinsk-65 Cochran and Norris estimate that the five graphite reactors produced about 58.3 metric tons of plutonium-239 through 1990. The light-water and heavy-water reactors, if devoted principally to plutonium, could have produced another 14.7 metric tons, for a total of 73 metric tons.[137] It is not known how much, if any, of the light-water and heavy-water reactors' capacity was devoted to plutonium production. If devoted entirely to

131. Bradley 1991, p. 8.

132. Hibbs 1993a.

133. Hibbs 1993a; Bradley 1991, p. 8; Usoltsev 1992.

134. Von Hippel, Cochran, and Paine 1993, p. 7; Cochran and Norris 1993, p. 83.

135. Bradley 1991, p. 12; Hibbs 1993a.

136. As quoted in Tkachenko 1992.

137. Cochran and Norris 1993, pp. 150–151.

tritium production, they could have produced tritium at a peak rate of 7.3 kilograms per year.[138]

Not all the plutonium produced in these reactors was reprocessed at Chelyabinsk-65. As noted above, the Mayak complex started sending spent fuel from its graphite-moderated reactors to Tomsk-7 for reprocessing in 1976, when the former switched over to processing plutonium spent fuel from civilian, naval-propulsion, and other reactors (including two breeder reactors). Thus, in estimating the actual amount of plutonium reprocessed at Chelyabinsk-65, we deduct the 23.3 metric tons generated between 1976 and 1990 (inclusive) in the graphite-moderated reactors while adding the estimated 25.4 metric tons recovered from civilian and naval sources up to 1992, which gives a total of 60.4 to 75.1 metric tons.[139] Chelyabinsk-65 continues to reprocess this civilian fuel, generating over 28 million curies (about 1 million terabecquerels) of long-lived wastes per year from strontium-90 and cesium-137 alone.

Waste Output and Environmental Contamination at Chelyabinsk-65 Cochran and Norris estimate that Chelyabinsk-65's reprocessing operations generated approximately 430 million curies (about 16 million terabecquerels) of cesium-137 and 350 million curies (about 13 million terabecquerels) of strontium-90. Due to their high burn-up level (estimated at an average of 30,000 megawatt-days per metric ton), these fuels contain about 9.9 and 7.3 curies (366 and 270 gigabecquerels) of cesium-137 and strontium-90, respectively, per gram of plutonium, compared to 3.6 and 3.3 curies (133 and 122 gigabecquerels) per gram in the low burn-up fuel of the complex's own plutonium production reactors.[140]

No precise decay-corrected estimate is possible for the total activity remaining in the absence of the details of how much plutonium was processed each year. Cochran and Norris estimate a decay of almost 30 percent, with 560 million curies (about 21 million terabecquerels) of strontium-90 and cesium-137 remaining. The estimate above for the activity only from long-lived isotopes of cesium and strontium is half the total, as these isotopes are in equilibrium with their short-lived radioactive daughter products, yttrium-90 and barium-137m. The total remaining activity including these daughter products is therefore about 1.1 billion curies (about 41 million terabecquerels). In addition, there are relatively short-lived fission products, like cerium-144, since reprocessing is continuous. These mostly liquid reprocessing wastes

138. Estimated by Cochran and Norris 1993, p. 154.

139. Cochran and Norris 1993, pp. 51–53, 56–57, 150.

140. Cochran and Norris 1993, p. 64.

make up the vast majority of all the radioactivity in the waste generated by the site's various operations.

Mayak officials tried several methods for disposing of these wastes, first pouring them directly into the Techa River, then into open lakes, and finally storing them for decay in steel tanks before draining off some of them. All these techniques resulted in far-reaching ecological disasters that the Soviet government covered up for decades. Official silence ended in 1989, when Gospromatomnadzor, the Soviet State Committee for the Utilization of Atomic Energy, publicly confirmed the worst accident—a 1957 waste storage tank explosion known as the Kyshtym disaster—in a report to the International Atomic Energy Agency.[141] Since then, many official reports and personal accounts of this accident and other problems at the Mayak site have been presented in the press and at international meetings by scientists, physicians, and environmentalists.

Faced with a wide range of evidence and a growing awareness of the seriousness of the situation, the Soviet government under Gorbachev issued a decree in January 1991 establishing the Commission to Investigate the Ecological Situation in the Chelyabinsk Region. The three-volume report of the "Gorbachev Commission" is the most comprehensive review of the extent of radioactive waste production, disposal, and pollution around the Mayak site.[142] All the data on environmental contamination presented below comes from this report unless otherwise noted. Since the commission merely reviewed data collected and held in secret by official sources—including Mayak's own Experimental Research Station—its findings are tentative. Many of its sources may have underestimated the severity of the environmental contamination and health effects.

At Mayak, water for cooling the reactors was taken from Lake Kyzyltash and after cooling was discharged back into the lake. From 1948 until 1951, all of Mayak's liquid reprocessing wastes from its reactors were discharged directly into the Techa River, about six kilometers below its source.[143] Mayak continued to release intermediate- and low-level wastes into the river until 1956. The Gorbachev Commission reported that 78 million cubic meters of high-level wastes containing

141. These issues are discussed at length in IPPNW and IEER 1992, chapters 4 and 5; see also Medvedev 1990.

142. Commission for Investigation of the Ecological Situation in the Chelyabinsk Region 1991.

143. Cochran and Norris 1993, p. 65. The source of the Techa was originally Lake Kyzyltash. A series of dams and reservoirs were built on the Techa, along with two bypass canals. Now the Techa river actually consists of these two canals, which are filled with water from Lake Kyzyltash and city sewage. The canals meet downstream.

approximately 2.7 million curies (100,000 terabecquerels) of beta activity were poured into the river during this time, ending in 1956.[144]

Scientists from the Ministry of Health's Institute of Biophysics in Chelyabinsk reported that 99 percent of this activity accumulated in the first 35 kilometers downstream.[145] "We were counting on the river to dilute the concentration of radionuclides to a safe level," said the chief engineer at Mayak, A. Suslov, in 1991. "But we did not take into account, we simply did not know, that the radioactivity would be swallowed up by the bottom silt, that it would bind up and concentrate ... in the upper course of the river. The academicians of those times knew as much about the atom as ninth-graders do today."[146] Site officials acknowledge paying little attention to managing radioactive wastes in the early years. According to Nikipelov, during the first several years of reprocessing scientists had no practical experience or body of scientific inquiry to draw on to protect the health of the workers or the environment.[147]

A radiation survey of the area done in the summer of 1951 documented extensive radioactive contamination in the riverbed and its floodplain, with isotopes from the plant detected as far away as the Arctic Ocean.[148] The survey also found high levels of exposure and disease among the 124,000 people living along the Techa and Tobol rivers, especially 28,100 people in 38 villages along the Techa. None of them had been informed or warned about the dangers of using the water for drinking, bathing, or washing.[149] The greatest civilian exposure was in the village of Metlino, downstream from the release point, where the gamma dose at some points along the river bank was about 5 centigrays per hour and about 3.5 centigrays per hour near some households near the river. The background dose rate on streets and roads was about 10 to 15 microroentgens per hour (or about 0.10 to 0.15 micrograys per hour).[150]

The magnitude of the problems revealed by this 1951 report must have been severe, as authorities banned all public use of the river and fenced off much of the floodplain. Wells were dug in some villages to provide alternative sources of water, and a military guard enforced the restrictions. But officials never gave villagers any explanation for the

144. Commission for Investigation of the Ecological Situation in the Chelyabinsk Region 1991, vol. II, section 2.3.1.

145. Kossenko, Degteva, and Petrushova 1992; Cochran and Norris 1993, p. 66.

146. As quoted in Leonov 1991.

147. Nikipelov et al. 1989, p. 1.

148. As cited in Cochran and Norris 1993, pp. 65–66, and Soyfer et al. 1992.

149. Akleyev et al. 1989; Soyfer et al. 1992.

150. Cochran and Norris 1993, pp. 65–66.

prohibitions, which were widely ignored as the river was used not just for drinking but also for irrigating gardens, fishing, bathing, washing, and the watering of cattle and fowl.[151] Evacuation of villages began only in 1953 and took until 1960 to complete. The government eventually relocated 7,500 people from 22 villages.[152]

By September 1951, plant officials had built their first dam across the Techa and diverted the highest-level reprocessing wastes to nearby Lake Karachay.[153] However, medium- and low-level wastes continued to be dumped into the Techa until 1956 and 1964, respectively, while a series of dams and reservoirs was built along the river to prevent the wastes from flowing downstream.[154] According to the Gorbachev Commission report, liquid wastes in solution were defined as high-level if their radioactivity exceeded 1 curie (37 gigabecquerels) per liter.[155] Medium-level wastes ranged from 10^{-5} to 1 curie (370 kilobecquerels to 37 gigabecquerels) per liter; low-level wastes were anything below this. It is not known how closely this classification scheme was followed in the actual segregation of liquid wastes for disposal.[156] With the diversion of high-level liquid wastes to Lake Karachay, the total activity of discharges into the Techa fell in 1952 from the earlier average of about 4,000 curies (150 terabecquerels) per day to about 25 curies (about 930 gigabecquerels) (see figure 7.5). The rate declined further in the period 1953–1956, until the second dam across the Techa was completed in 1956.[157]

While most of the floodplain barricades erected in the 1950s are no longer in place and the level of radiation in the Techa has fallen greatly, the river still poses a health risk to those who live along it. The river now cascades from Lake Kyzyltash through reservoirs created by the four dams along the Techa (numbers 3, 4, 10, and 11). The combined surface area of these reservoirs is 84 square kilometers, and their reported volume is 380 million cubic meters.[158] The water levels in these reservoirs and in Lake Kyzyltash were close to the maximum in 1992, and the complex continues to dump 5.5 million cubic meters of residential sewage and low-level radioactive wastes into the system

151. Akleyev et al. 1989; Cochran and Norris 1993, p. 66.

152. Soyfer et al. 1992.

153. Bolsunovsky 1992, p. 9.

154. Cochran and Norris 1992, p. 33.

155. Commission for Investigation of the Ecological Situation in the Chelyabinsk Region 1991, vol. II, section 2.1.1.

156. Cochran and Norris 1993, p. 63.

157. Kossenko, Degteva, and Petrushova 1992; Soyfer et al. 1992. A third dam was built in 1957 and a fourth in 1964, creating two huge reservoirs along the river.

158. Cochran and Norris 1993, p. 68.

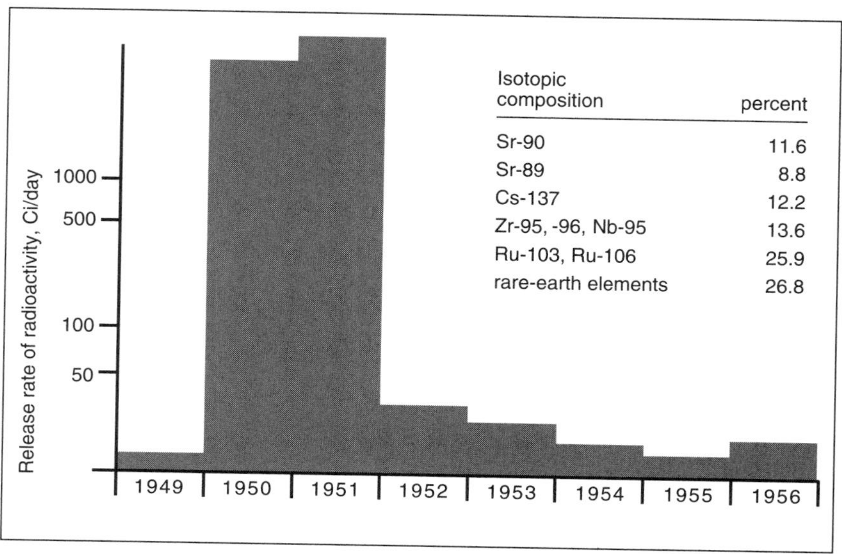

Figure 7.5 Discharge of radioactive wastes from Chelyabinsk-65 plutonium production to the Techa River (Chelyabinsk Region, Russia), in average curies per day, 1949-1956. Source: Kossenko, Degteva, and Petrushova 1992.

every year.[159] Channels built in 1963 and 1972 along the left and right banks, respectively, of the lakes and reservoirs divert the inflow of surface runoff, but these have not stabilized the rising water levels. Rather, they have become waste canals into which radioactive water from the reservoirs regularly seeps and spills as the absorptive capacity and integrity of the reservoir walls degrade.[160]

Now that the reactors are shut down and no longer draw water from the system, site officials are concerned that the reservoirs might overflow. The level of Reservoir No. 11, the last and largest, is just 20 centimeters below capacity.[161] This threat is even cited as a justification for completing the breeder reactors: their operation would take cooling water from No. 10, increasing its rate of evaporation.[162] Of course, the possibility also exists that one or more of the dams may fail before the reservoirs overflow, which would be catastrophic. All together, the reservoirs and lakes along the Techa contain over 122 million curies

159. Commission for Investigation of the Ecological Situation in the Chelyabinsk Region 1991, vol. II, section 2.1.7.

160. Commission for Investigation of the Ecological Situation in the Chelyabinsk Region 1991, vol. II, section 2.1.7; Galinka 1990.

161. Commission for Investigation of the Ecological Situation in the Chelyabinsk Region 1991, vol. II, section 2.1.7; Galinka 1990.

162. Cochran and Norris 1992, p. 48; Galinka 1990.

Table 7.2 Radioactive contamination in Chelyabinsk-65 Reservoirs

			Composition of Radionuclides							Accumulation, Ci		
			Concentration in Water, Ci/l					Ground Deposits, Ci/kg				
Reservoir Number	Area of the Reservoir (sq km)	Capacity of the Reservoir (million cubic m)	Sr-90	Cs-137	HTO	$\Sigma\alpha$	$\Sigma\beta$	Sr-90	Cs-137	In the Reservoir	In Ground Deposits	Overall
2	19	83	1.1×10^{-8}	4.5×10^{-8}	2.5×10^{-7}	?	—	1.3×10^{-6}	3×10^{-5}	2×10^3	18×10^3	20×10^3
3	0.5	0.75	1.6×10^{-6}	2.0×10^{-7}	1.4×10^{-6}	3×10^{-10}	—	1.4×10^{-4}	1×10^{-3}	2.6×10^3	15.4×10^3	18×10^3
4	1.3	4.1	1.7×10^{-7}	1.3×10^{-8}	5.2×10^{-7}	4.5×10^{-9}	—	4×10^{-6}	6×10^{-5}	1.7×10^3	4.2×10^3	6×10^3
6	3.6	17.5	3.7×10^{-10}	2×10^{-11}	1×10^{-8}	3.9×10^{-5}	—	3×10^{-7}	3.9×10^{-7}	2	300	300
9	0.25	0.4	1.7×10^{-3}	1.2×10^{-2}	5.3×10^{-5}	5.7×10^{-6}	1.9×10^{-2}	0.3	1.4	8.4×10^6	110×10^6	120×10^6
10	16.6	76	3.5×10^{-7}	8.6×10^{-9}	3.2×10^{-7}	1×10^{-11}	—	3.5×10^{-6}	1.5×10^{-1}	50×10^3	60×10^3	11×10^3
11	44	217	5.1×10^{-8}	2×10^{-11}	4.5×10^{-8}	2×10^{-12}	—	1.3×10^{-6}	1.3×10^{-7}	24×10^3	15×10^3	39×10^3
17	0.17	0.8	7×10^{-4}	4×10^{-6}	1×10^{-4}	1.2×10^{-3}	—	3.3×10^{-7}	3.3×10^{-2}	45×10^3	2×10^6	2×10^6

Note: HTO is water in which one hydrogen atom is radioactive tritium (T).

Source: Reproduced from Cochran and Norris 1993.

(about 4.5 million terabecquerels) of activity, most of it in reservoirs 9 and 17 (see table 7.2).[163]

More contaminated than all of the other reservoirs combined is Lake Karachay, also known as Reservoir No. 9. Since October 1961, a large quantity of highly radioactive wastes were dumped into Lake Karachay at a rate of about 1 million curies (37,000 terabecquerels) per year.[164] It is a small, shallow lake, less than 3 meters deep, with no surface outlet. Originally covering 45 hectares, its surface area in 1991 was about 20 hectares due to efforts to fill and cover the lake. There are plans to completely fill in the lake by 1995.[165]

Lake Karachay is probably the world's most radioactive open body of water, holding an estimated 120 million curies (4.44 million terabecquerels) of activity in 1990, according to scientists from the site.[166] Of this total, 98 million curies (about 3.63 million terabecquerels) are from cesium-137 and 20 million curies (0.74 million terabecquerels) are from strontium-90.[167] About 7 percent of this activity is in the water (0.019 curies or about 0.7 gigabecquerels per liter); 41 percent is adsorbed in the clay bed and 52 percent is in migratory sediment on the bottom.[168] (The next most radioactive lake at the site is Staroye Boloto, a 17-hectare swamp containing 2 million curies (74,000 terabecquerels) of beta-emitting radionuclides from medium-level liquid wastes and tritium condensate.)[169] Gamma radiation readings at the surface of Lake Karachay are 30 to 40 milligrays per hour, with readings along the shore in the range of 0.18 to 0.2 grays per hour due to the concentration of radioactivity in soil and sediment. The highest readings have been recorded near the plant's discharge pipe, where the dose is approximately 6 grays per hour—high enough to give an adult a lethal dose in less than one hour.[170]

In 1953, the construction of a tank storage facility near the reprocessing plant alleviated the dumping into Lake Karachay somewhat. Reprocessing solutions containing substantial residues of plutonium

163. Cochran and Norris 1992, p. 147; Bolsunovsky 1992.

164. Commission for Investigation of the Ecological Situation in the Chelyabinsk Region 1991, vol. II, section 2.1.5.

165. Cochran and Norris 1993, pp. 68–70.

166. Commission for Investigation of the Ecological Situation in the Chelyabinsk Region 1991, vol. II, section 2.1.5; also quoted in Cochran and Norris 1992, table 6.

167. Cochran and Norris 1993, pp. 69, 70.

168. Commission for Investigation of the Ecological Situation in the Chelyabinsk Region 1991, vol. II, section 2.1.5.

169. Commission for Investigation of the Ecological Situation in the Chelyabinsk Region 1991, vol. II, section 2.1.5; Cochran and Norris 1993, p. 72.

170. Cochran and Norris 1992, p. 37.

and uranium were held in these tanks for a year to decay before further processing.[171] The high-level wastes from this second round of processing also were kept in the tanks. A sludge containing most of the fission products, including strontium-90, settled out from this waste, while the supernatant fraction containing most of the cesium-137 was discharged into Lake Karachay.[172]

Radioactivity from the lake was detected in groundwater in the 1960s. Mayak officials started trying to eliminate the hazard posed by Lake Karachay in 1967. They began a multi-year effort to eliminate the lake entirely by filling it with rocks, dirt, and massive hollow concrete blocks. The backfill will then be capped with a layer of clay, while, beginning in 1994 or 1995, the lake water underneath is to be pumped out and treated. The supposedly volunteer dump truck drivers doing this dangerous work have just 12 minutes to enter the high-risk zone, unload, and leave before receiving their maximum allowable *annual* dose of radioactivity.[173] Between 1984 and 1991, 8,088 blocks and 736,600 cubic meters of rock and soil were dumped, reducing the surface area of the lake to 25 hectares and its volume to 400,000 cubic meters. (A five-year plan to similarly contain and fill in nearby Lake Staroye Boloto was under development in 1991.)[174] The backfilling operation was also meant to eliminate the potential hazard posed by floods and droughts, both of which are common in the area.

Some Mayak scientists are growing concerned that the now-contaminated topsoil cover may pose wind-erosion problems and that continuing to force down the waters of the lake may worsen groundwater contamination.[175] Radioactive groundwater flowing at the rate of 0.84 meters per day has spread 3 kilometers from the lake and been detected just 15 meters under the Mishelyak River.[176] Some 4 million cubic meters of groundwater (over 1 billion gallons) are estimated to be contaminated to a depth of 100 meters with about 6,000 curies (about 220 terabecquerels) of long-lived isotopes.[177]

A dry winter and dry early spring of 1967 led to the drying of Lake Karachay. Strong winds in April and May blew dust laden with 600 curies (22.2 terabecquerels) of radioactivity. Strontium-90 and cesium-

171. Nikipelov and Drozhko 1990.

172. Cochran and Norris 1993, p. 64.

173. Monroe 1991.

174. Commission for Investigation of the Ecological Situation in the Chelyabinsk Region 1991, vol. II, section 2.1.5; Cochran and Norris 1993, pp. 69–70.

175. Monroe 1991.

176. Commission for Investigation of the Ecological Situation in the Chelyabinsk Region 1991, vol. II, section 2.1.6; Bolsunovsky 1992, p. 8.

177. Bolsunovsky 1992, p. 10; Cochran and Norris 1993, p. 72.

137 were the two main long-lived fission products (three-fourths cesium-137 and the rest strontium-90), with 63 towns and villages with about 41,500 inhabitants and over 1,800 square kilometers affected. The area contaminated at levels greater than 0.1 curies (3.7 gigabecquerels) per square kilometer of strontium-90 and greater than 0.3 curies (11.1 gigabecquerels) per square kilometer of cesium-137 was in the range of 1,800 to 2,700 square kilometers. It overlapped partly with the larger area of contamination that had occurred a decade earlier as a result of the 1957 tank explosion described below.[178]

The 1957 Tank Explosion In the Chelyabinsk-65 site's worst accident, a high-level liquid-waste storage tank exploded on 29 September 1957. A few scientific reports and even a book about the accident's radiobiological effects on humans, animals, and the environment were published in 1974 for limited circulation within the government, but these sources refer only to an unspecified "industrial accident" and omit any information about its date or location.[179] Maps show neighboring towns and villages but identify them only by number; no geographical features are named. The Kyshtym accident, as it has come to be known, was not reported outside Russia until 1976, when the Soviet biologist and dissident Zhores Medvedev published an account in England.[180] The U.S. Central Intelligence Agency apparently knew of the accident by 1959 at the latest,[181] but Washington kept the information secret, apparently fearing the impact on the U.S. nuclear industry.[182]

Since the Soviet Union acknowledged the accident in 1989, some of the secret radiobiological reports by Soviet scientists and physicians have been released, and several detailed analyses have been published and presented at international conferences. Except where otherwise noted or when there are direct quotations, the account of the explosion that follows comes from the summary of Russian sources reported by IPPNW and IEER in *Plutonium: Deadly Gold of the Nuclear Age.* Most of the data on radioactive releases from the disaster come from official studies of the ecological situation around the Mayak site. These are comprehensively reviewed in the Gorbachev Commission report[183]

178. Commission for Investigation of the Ecological Situation in the Chelyabinsk Region 1991, vol. II, section 2.5; Cochran and Norris 1993, p. 71.

179. See Burnazyan 1991.

180. Medvedev 1976.

181. U.S. CIA 1959.

182. For further discussion see IPPNW and IEER 1992, pp. 81–84.

183. Commission for Investigation of the Ecological Situation in the Chelyabinsk Region 1991.

and also well summarized in another report by the USSR Academy of Sciences.[184]

The Kyshtym explosion occurred in the tank storage facility built in 1953 that, by 1957, consisted of 20 enormous stainless-steel tanks, each with a capacity of 300 cubic meters. They were located in a concrete canyon 8 meters underground. Water flowing in the space between the tanks and the concrete containment cooled the tanks. Among the numerous problems with the storage tanks was the loss of information as to what was going on in the system. Measuring instruments quickly deteriorated and were in unsatisfactory condition.[185] Further, due to the high radiation fields and design defects in the layout, the instruments were for all practical purposes inaccessible and could not be maintained.

The tanks initially contained liquid radioactive wastes, but these gradually began to dry out due to the heat generated by the radioactivity and the inadequacy of the cooling system. As the wastes began to dry and gases from their evaporation were vented, the tanks lightened and began to float in the cooling water around them. This put unanticipated stresses on the tanks' connecting pipes and seals. The tanks began to leak, contaminating the cooling water with their highly radioactive contents.

As this contamination increased, efforts to treat the water in the reprocessing plant required interrupting the cooling system, causing the wastes to overheat further. At the same time, a choice had to be made between safety and plutonium production, since the plant's chemical processing capacity could not simultaneously process cooling water and maximize plutonium production. Production came first, so the tanks were cooled only intermittently.

The tank that exploded in 1957 contained 70 to 80 metric tons of highly radioactive waste with a total radioactivity now estimated at 20 million curies (740,000 terabecquerels).[186] The force of the explosion has been estimated at between 70 and 100 metric tons of TNT.[187] Although a brief nuclear criticality cannot be ruled out as a factor in initiating the explosion, it is highly unlikely to have been the major source of explosive power. The explosion was mostly if not entirely chemical in nature, involving the high concentrations of sodium nitrate (up to 100 grams per liter) and sodium acetate (up to 80 grams per liter) in the accumulated wastes of the reprocessing process used at the time. It is possible that this mixture ignited spontaneously as it

184. Bolshakov et al. 1991.

185. Nikipelov and Drozhko 1990.

186. Nikipelov and Drozhko 1990; Commission for Investigation of the Ecological Situation in the Chelyabinsk Region 1991, vol. II, section 2.4.

187. Cochran and Norris 1993, p. 74.

Table 7.3 Characteristics of Radioactivity Released in the 1957 Accident

Radionuclide	Contribution to Total Activity of the Mixture (%)	Half-Life	Type of Radiation Emitted
^{89}Sr	traces	51 days	β, γ
$^{90}Sr + ^{90}Y$	5.4	28.6 years	β
$^{95}Zr + ^{95}Nb$	24.9	65 days	β, γ
$^{106}Ru + ^{106}Rh$	3.7	1 year	β, γ
^{137}Cs	0.036	30 years	β, γ
$^{144}Ce + ^{144}Pr$	66	284 days	β, γ
^{147}Pm	traces	2.6 years	β, γ
^{155}Eu	traces	5 years	β, γ
$^{239,240}Pu$	traces	—	α

Source: Nikipelov et al. 1989.

heated and dried out (dry mixtures of sodium nitrate-acetate begin to decompose spontaneously and violently at about 380°C); a spark or friction may have ignited it at a much lower temperature.

The explosion completely destroyed the tank and damaged two adjacent ones. All the waste was expelled from the tank in the explosion. About 90 percent of the waste fell back to the ground in the immediate vicinity. The remaining 2 million curies (74,000 terabecquerels) was carried off-site by a southwesterly wind in a massive plume that reached up to 1,000 meters high. This fallout plume contaminated an elongated area toward the northeast that is now known as the East Urals Radioactive Trace (see figure 7.6). About two-thirds of the radioactivity in the plume consisted of relatively short-lived cerium-144 and its daughter praseodymium-144; most of the rest was zirconium-95 and ruthenium-106, also relatively short-lived, and their daughter products. Strontium-90 and its daughter product yttrium-90, the main contaminants that remain today, were 5.4 percent of the fallout plume (see table 7.3). Strontium-90 contamination in the most contaminated fallout areas was as high as 10,000 curies (370 terabecquerels) per square kilometer.[188] The overall radioactive contamination was initially 50 times the strontium-90 levels; external gamma radiation in the most contaminated regions was about 0.6 centigrays per hour.[189]

Figure 7.6 shows a map of strontium-90 contamination of the region. The area with contamination levels greater than 0.1 curies (3.7 gigabecquerels) per square kilometer contains numerous rivers and lakes. It is an agricultural region growing grains and other foods as well as hay

188. Burnazyan 1990, p. II-6.

189. Commission for Investigation of the Ecological Situation in the Chelyabinsk Region 1991, vol. II, section 2.4.1.

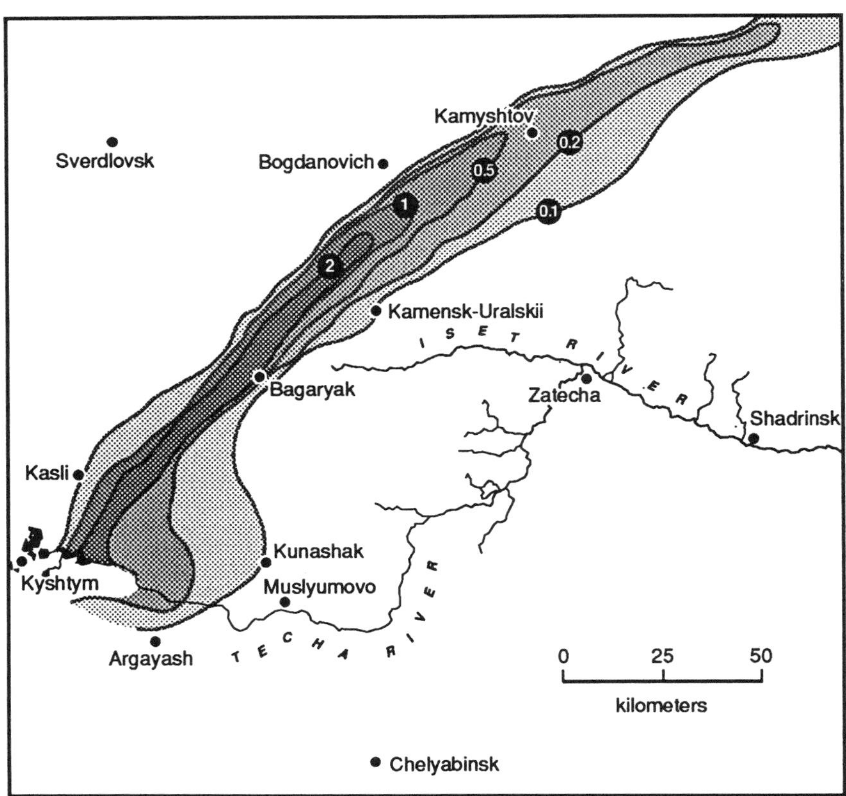

Figure 7.6 Radioactive fallout from the 1957 accident at Chelyabinsk-65. Contamination lines indicate the levels for strontium-90 in curies per square kilometer in 1957 and 1958, shortly after the Kyshtym accident. Source: Kossenko, Degteva, and Petrushova 1992.

for cattle. Since only the most highly contaminated region was initially evacuated, the authorities attempted to reduce exposures by placing restrictions on the consumption of contaminated food and on cultivation of the land. But in the first months after the accident, these measures were not very effective, according to a 1991 official account.[190]

After the evacuations, some restrictions continued on agricultural practices. Initially, about 20,000 hectares (200 square kilometers) of the most contaminated land were ploughed under to reduce wind dispersion of radioactivity. Agricultural restrictions were introduced in stages, and by 1959 applied to the entire evacuated area that was contaminated with more than 2 curies (74 gigabecquerels) of strontium-90 per square kilometer. By 1962 these restrictions were relaxed and production of various kinds was allowed on land contaminated up to 100 curies (3,700 gigabecquerels) of strontium-90 per square kilo-

190. Romanov et al. 1991.

meter.[191] As a result, the population consumed contaminated food. In particular, milk consumption "was responsible for up to 70–80% of the daily intake of ^{90}Sr [strontium-90] in the diet of the population."[192] Most of the lakes originally considered unsafe for fishing—in zones with contamination levels above 2 curies of strontium-90 (74 gigabecquerels) per square kilometer—were opened to unrestricted use in 1970.[193] By 1990, site scientists reported that more than 80 percent of the land within the official original trace had been returned to productive use.[194]

No attempt was made to clean up a 16,700-hectare zone in which contamination levels exceeded 100 curies (3,700 gigabecquerels) per square kilometer. The government at the time decided to set aside this area permanently for field research into the aftereffects of the accident on plants and animals. It is now known as the East Urals State Preserve.[195] Finally, within the most radioactive zone, almost all trees were damaged or killed.[196]

Official data show only a small amount of cesium-137 relative to strontium-90. Normally, these elements are present in about the same order of magnitude in irradiated reactor fuel rods and in high-level waste from reprocessing. The relatively small amount of cesium is due to the fact, noted above, that the supernatant fraction containing most of the cesium-137 was selectively removed from the waste and dumped into Lake Karachay. Therefore, there is much more cesium-137 than strontium-90 in Lake Karachay, as discussed above.

Estimates of the amount of plutonium released by the explosion depend on several unknown variables. The Gorbachev Commission reported that the explosion deposited plutonium in the area at densities of up to 1 curie (37 gigabecquerels) per square kilometer, which is not inconsistent with the above estimates.[197] As of 1993, there is no public estimate of the total amount of plutonium released.

Other Waste Disposal Practices After the 1957 explosion, high-level liquid wastes were again poured into Lake Karachay and Staroye Bo-

191. Romanov et al. 1991, p. 5; Nikipelov 1989.

192. Romanov et al. 1991, p. 4.

193. Romanov, Teplyakov, and Shilov 1990.

194. Romanov, Teplyakov, and Shilov 1990.

195. Romanov, Teplyakov, and Shilov 1990.

196. Medvedev 1979, pp. 110–111.

197. Commission for Investigation of the Ecological Situation in the Chelyabinsk Region 1991, vol. II, section 2.1.4. By comparison, deposition of plutonium from nuclear testing fallout in the north temperate latitudes is 1.4 millicuries (51.8 megabecquerels) per square kilometer, or about one seven-hundredth of this amount (Eisenbud 1987, p. 335).

loto until a new tank storage system was devised. High-level liquid wastes are now first evaporated and then mixed with barely soluble compounds, such as hydroxides and ferrocyanides. The concentrated waste is then stored in single-wall stainless-steel storage tanks, which are in turn "housed in metal-lined, reinforced-concrete canyons."[198]

Estimates of the amount of high-level waste stored at the site vary from a low of 546 million curies (about 20.2 million terabecquerels)[199] to a high of 976 million curies (about 36 million terabecquerels). The estimated volume of these wastes is 20,000 cubic meters, stored in 60 tanks, though one source has stated that there are as many as 99 high-level waste tanks in the Mayak facility.[200] Some of the waste has been vitrified, but most is still in liquid form.

Large quantities of solid radioactive wastes have also been generated. Mayak's operations contaminated about 525,000 metric tons of equipment and supplies from routine operations and accidents.[201] The total volume of solid waste buried on-site is over 725,000 cubic meters, and its activity is estimated at 12 million curies (444 million gigabecquerels). These wastes are buried in 227 locations, with a total area of over 21 hectares. Twenty-five of these sites are still active. Almost all of the radioactivity is contained in 41,300 cubic meters of high-activity waste. The 685,000 cubic meters of low- and medium-activity waste contain only 31,600 curies (about 1,170 terabecquerels) of the total activity.[202]

Radioactive solid wastes reportedly were segregated prior to burial by level of activity as well as by degree of combustibility, mercury concentration, and hazardous organic waste content. Most were not treated in any way to reduce their volume, whether via compaction, incineration, or melting. A solid waste treatment plant designed to do some of these things is now planned. The medium-level and low-level wastes were buried in clay-lined trenches, dug at least four meters above the water table and capped with 10 meters of clay and soil.[203] They are not actively monitored, and even the exact location of many is unknown.

High-level solid wastes were disposed of more carefully, in subsurface bunkers layered on all sides with clay soils, bituminous rock,

198. Cochran and Norris 1993, p. 80.

199. Commission for Investigation of the Ecological Situation in the Chelyabinsk Region 1991, vol. II, section 2.1.4.

200. Cochran and Norris 1993, pp. 80–81.

201. Cochran and Norris 1993, p. 83.

202. Commission for Investigation of the Ecological Situation in the Chelyabinsk Region 1991, vol. II, section 2.1.3.

203. Commission for Investigation of the Ecological Situation in the Chelyabinsk Region 1991, vol. II, section 2.1.3.

stainless steel, and concrete. The 24 high-level burial sites contain almost 12 million curies (444 million gigabecquerels). These storage bunkers have remote-control monitoring equipment that is wired to alarm systems.[204]

The Russian government took an important step toward addressing the legacy of all this radioactive waste in November 1990, when it declared the entire Chelyabinsk Oblast (region) an ecological disaster zone. After President Yeltsin visited the area to show his support for a massive cleanup program, the Council of Ministers appropriated 90 million rubles to the task. Unfortunately, the value of the sum has been reduced to insignificance due to the severe fall in the value of the ruble.

Environmental Contamination The many routine and accidental releases of radiation from the Mayak complex have caused background gamma radiation at the Chelyabinsk-65 site to increase to an average of between about 0.1 micrograys per hour and about 0.3 micrograys per hour, compared to the reported average around other sites of about 0.1 micrograys per hour.[205] An IEER scientist, Scott Saleska, measured spots along the shore of the Techa River in 1992 that had readings two or more orders of magnitude above this, including about 3 to 5 micrograys per hour at 78 kilometers from the plant in the village of Muslyumovo, and about 80 micrograys per hour under a bridge along the main highway near the breeder reactor construction site.[206]

Radioactive reprocessing wastes continue to be generated at the site at the rate of more than 91 million curies (about 3,370 million gigabecquerels) per year, including about one million curies (37 million gigabecquerels) of medium-level and low-level liquid wastes. These wastes are still being released into open reservoirs.[207]

Krasnoyarsk-26

Krasnoyarsk-26 is the code name of a once-secret, massive underground plutonium production complex on the Yenisey River in Siberia, about 40 to 50 kilometers northeast of the city of Krasnoyarsk (population 970,000) (see figure 7.1).[208] The plant's 11,000 workers and their

204. Commission for Investigation of the Ecological Situation in the Chelyabinsk Region 1991, vol. II, section 2.1.3.

205. Cochran and Norris 1993, p. 84.

206. Saleska 1992a.

207. Cochran and Norris 1993, p. 84.

208. IPPNW and IEER, p. 34; Miheev 1993.

families live in the closed city at Krasnoyarsk-26 known as Zheleznogorsk, which means "Iron Town" (population 96,500).[209]

Stalin ordered the building of Krasnoyarsk-26 in 1950 as a backup to Chelyabinsk-65.[210] Designed to withstand a direct hit from an atomic bomb, it is built entirely within a mountain (about 1 kilometer in by train from the surface of the mountain). The habitable rooms are about 80 to 200 meters from the surface, and are close to the top of the mountain. The reactors, research labs, one of two reprocessing plants, and other operations are scattered throughout a multilevel complex of huge interconnecting caverns and about 3,500 rooms.[211] Three tunnels connect the complex to the outside world: one for transportation, one for supplies, and one for fresh air, which is pumped in at the rate of 5.5 million cubic meters per hour.[212] More than 65,000 prisoners and 100,000 soldiers did the excavating.[213]

A "sanitary protection zone" around the mountain site encompasses 13,000 hectares of mostly forested land. No towns draw their drinking water from the Yenisey River in this area, and commercial fishing is prohibited. However, some 30,000 people live in the zone, in which farming and grazing are allowed. Meat and dairy cattle drink from the river.[214] People are prohibited from swimming, boating, or fishing in the river within five kilometers downstream from the plant's discharge pipes,[215] but a 1991 visitor to the site saw fishermen wading in the river less than one-half kilometer from the pipes.[216]

Atamanovo, the first village downstream from Krasnoyarsk-26, is six kilometers from the plant. Just beyond this is a Young Pioneers' summer camp: since 1940, 5,000 children a year have come here from Norilsk, a city near the mouth of the Yenisey, to swim, play, and fish in the river.[217] Several hundred thousand more people live along the 1,500 kilometers of the Yenisey River between its mouth at the Kara Sea in the Arctic and Krasnoyarsk-26.

Krasnoyarsk-26 is now officially called the Mining-Chemical Complex. It is referred to locally as the Mountain Chemical Complex or

209. Miheev 1993.

210. Cochran and Norris 1993, p. 96.

211. Cochran and Norris 1993, pp. 97–98. See also Spiridonov 1992.

212. Cochran and Norris 1993, p. 98.

213. Cochran and Norris 1993, p. 97.

214. Korogodin et al. 1990.

215. Zubov 1992.

216. Soler-Sala 1992, p. 97.

217. Korogodin et al. 1990.

Atomgrad—the Atomic City.[218] Krasnoyarsk-26 should not be confused with Krasnoyarsk-45, also near Krasnoyarsk, which houses an "electrochemistry" uranium-enrichment plant.[219]

Several other secret military complexes are in the heavily industrialized surrounding area, including plants for making ballistic missiles, space vehicles, and satellites, each with their own "closed" worker housing.[220] There are also 29 uranium-mining sites within 200 kilometers of the city of Krasnoyarsk.[221] The worst nonradioactive atmospheric pollution in the region apparently comes from an aluminum factory in the city of Krasnoyarsk that may be the largest plant of its kind in the world.[222]

At Krasnoyarsk-26 itself, three graphite-moderated reactors, each with a capacity of 2,000 megawatts thermal, produced plutonium but not tritium.[223] They came on line in 1958, 1961, and 1964.[224] Two of the three were shut down in 1992. They both relied on "once-through" cooling systems that drew cooling water from the Yenisey River and returned it there without any treatment, contaminating the river with radioactive wastes.[225]

The only reactor still operating also provides steam heat and/or electricity for the complex and the city of Krasnoyarsk-26.[226] It uses a dual-loop cooling system with a closed primary loop and a secondary loop that releases only heated water back to the Yenisey.[227]

In addition to the reactors, Krasnoyarsk-26 houses a reactor coolant water preparation plant, waste treatment and storage facilities, research laboratories, an engineering and repairs plant, 22 workshops, and two chemical reprocessing plants. The first reprocessing plant (of unknown name) was built in the underground complex in 1964 and is presumably still operating.[228] Construction of the second, "RT-2," was halted in 1989, after completion of its spent-fuel storage facility, due to

218. Miheev 1991; Cochran and Norris 1992, p. 53.

219. Miheev 1993.

220. Cochran and Norris 1992, p. 54; Litvinov 1991.

221. From a map provided by Miheev 1992b entitled "Areas with an increased radiation danger in the Krasnoyarsk Region, the Khakas Republic and the Tuvin Autonomous Republic" (primary source unknown).

222. Soler-Sala 1992, p. 95.

223. Cochran and Norris 1992, p. 55.

224. Bolsunovsky 1992, p. 19; Cochran and Norris 1992, p. 55.

225. Cochran and Norris 1992, pp. 54–55.

226. Bolsunovsky 1992, p. 19.

227. Cochran and Norris 1992, p. 55.

228. Cochran and Norris 1993, p. 97.

a lack of funding and overwhelming opposition from the public and local media.[229] More than 60,000 people signed a petition opposing the plant.[230]

RT-2 is one of two aboveground facilities at Krasnoyarsk-26; it is located just north of the underground reactors, overlooking the Yenisey River. Officials from the Mining-Chemical Complex, MINATOM, and the regional government still hope to complete RT-2. The other aboveground facility is a complex of 500 to 600 liquid-waste injection wells, known as Site 27, built to accommodate waste from RT-2. It is about 16 kilometers from the incomplete RT-2 on the other side of the river.[231]

In 1992, the Ministry of Atomic Energy won some new Russian funding to resume construction of RT-2.[232] South Korea, which reportedly is willing to pay the Mining-Chemical Complex one million dollars per metric ton of spent fuel accepted for treatment and disposal, is also providing assistance,[233] including an offer to pay part of the cost of finishing the reprocessing plant.[234] Ukraine, by comparison, paid just 960,000 rubles per metric ton in 1991 and only 94,000 rubles per metric ton in 1990.[235]

Whatever the fate of the RT-2 reprocessing plant, its storage facility for spent fuel from civilian VVER-1000 reactors has operated since 1985, accepting waste from seven nuclear power stations in the former Soviet Union.[236] Spent fuel is stored under three meters of water in guide-shelves, so that the fuel rods are in contact with the pool. The building in which they are stored is itself eight meters below ground. Concern has been expressed that the thin zirconium cladding on these VVER fuel rods may corrode and crumble in 25 to 30 years.[237]

In mid-1992, Gosatomnadzor reported that the Krasnoyarsk facility was only storing spent fuel from VVER-1000 reactors, but does not yet have the technology to reprocess it. The Russian government has not announced any long-term plan for disposing the wastes accumulating in this facility.[238] The main option being considered is reprocessing

229. Cochran and Norris 1992, pp. 54, 57; Miheev 1992b.

230. Cochran and Norris 1992, p. 56.

231. Cochran and Norris 1993, pp. 98, 103–104.

232. Cutter Information Corp. 1992.

233. Bolsunovsky 1992.

234. Miheev 1992b.

235. Tarasov 1992.

236. Bolsunovsky 1992, pp. 23–24; Miheev 1992b.

237. Solntsev 1992.

238. Bolsunovsky 1992, pp. 20–21.

the fuel to recover the plutonium and uranium it contains and then vitrifying the remaining wastes to solidify them in glass prior to deep geological burial.

It also is not clear how much of the remaining storage capacity will be made available to South Korea and other countries that have expressed an interest in paying the Krasnoyarsk-26 facility to accept their wastes.[239] The issue of accepting foreign waste, whether for reprocessing or disposal and at what price, remains one of intense local public and political debate. It also has important implications for nuclear proliferation.

The many published letters, statements, and articles on the subject reveal a wide range of opinion. Opponents of accepting foreign nuclear waste criticize the plan on ecological grounds, pointing out that an environmental review of the site's suitability only began nine years after construction.[240] Proponents cite the economic benefits.[241] The chief spokesperson for the Ministry of Atomic Energy, Sergei Yermakov, describes this waste as "a gold mine ... [containing] up to 90 percent uranium, 2 percent to 3 percent plutonium, and a number of useful isotopes."[242] Although Yermakov does not specify the sources of this foreign waste, his estimate of the amount of plutonium in civilian fuel seems high. Spent fuel from commercial reactors in the United States typically contains about 96 percent uranium, about 0.9 percent plutonium, and just over 3 percent fission products, such as cesium and strontium.[243] Yermakov also says that Russia is contractually bound to take back all the nuclear fuel it sells abroad. This means that Russia will take back the spent fuel from East European reactors, which are of Soviet origin.

From RT-2, a tunnel passes 50 meters under the Yenisey River and extends about two kilometers toward Site 27, the injection-well complex.[244] Radioactive liquid wastes from the reprocessing plant were to have been piped there for "disposal by injection" into a sandstone lens between two layers of clay at a depth of 700 meters.[245]

Krasnoyarsk officials are working to convert some of the facilities to other production. They plan to expand a pilot plant making such semiconducting materials as gallium arsenide.[246] They are also plan-

239. Miheev 1992b.

240. Cochran and Norris 1993, p. 102.

241. Melnik 1992.

242. Cutter Information Corp. 1992, p. 15.

243. LaMarsh 1983, p. 150.

244. Cochran and Norris 1992, p. 59.

245. Soltsnev 1992; Cochran and Norris 1993, pp. 103–104.

246. Cochran and Norris 1992, p. 59.

ning to convert much of the RT-2 reprocessing site into a full-scale plant for converting high-level liquid wastes into vitrified and ceramic forms.[247]

Production at Krasnoyarsk-26 Estimates of the cumulative amount of plutonium produced at Krasnoyarsk-26 depend on basic assumptions about the operating history of the three reactors, their capacity, and their power levels. The model used by Cochran and Norris assumes that the reactors each operated at a power level of 2,000 megawatts thermal from the year they started. This puts their cumulative estimate of plutonium production through 1992 at 44.7 metric tons.[248] Based on 3.6 and 3.3 curies (about 133 gigabecquerels and about 122 gigabecquerels), respectively, of cesium-137 and strontium-90 per gram of plutonium, the non-decay-corrected production of these fission products at Krasnoyarsk-26 would total about 169 million curies (6,253 million gigabecquerels) and 155 million curies (5,735 million gigabecquerels), respectively.[249] However, since the reprocessing plant started up in 1964, six years after reactor operation began, some reprocessing may have been done elsewhere.

Waste Output and Environmental Contamination at Krasnoyarsk-26 Krasnoyarsk-26 officials have said very little about their radioactive waste disposal practices and nothing about the annual or cumulative volume and activity of the radioactive wastes generated at the site.[250] On-site, the most dangerous wastes from both an environmental and public-health perspective are the highly radioactive liquids from plutonium reprocessing. According to Alexander Bolsunovsky, director of the Krasnoyarsk Ecological Center and co-chair of the Krasnoyarsk Regional Green Movement, these liquids were routinely disposed of underground, with some injected into wells to a depth of 270 meters and others poured into buried concrete tanks.[251] According to Bolsunovsky, the use of these injection wells is slated to continue until around the year 2000.[252] Pipes line the injection wells to prevent radioactive waste from penetrating into other layers.

The only estimates of the volume and activity of injected wastes come from Yuri Vishenevsky, chair of Gosatomnadzor, and Alexander

247. Revenko, as cited in Miheev 1992a.

248. Cochran and Norris 1993, p. 153.

249. The factors of 3.6 curies per gram and 3.3 curies per gram cesium-137 and strontium-90 are from Cochran and Norris 1993, p. 65, footnote 272.

250. Bolsunovsky 1992, p. 20.

251. Bolsunovsky 1992, pp. 20, 21; and as cited in IPPNW and IEER 1992, p. 67.

252. Bolsunovsky 1992, pp. 20–21.

Penyagin, former chair of the Supreme Soviet's Subcommittee on Atomic Power and Nuclear Ecology. According to a 1992 Gosatomnadzor report, 3.8 million cubic meters of highly radioactive liquids with a total activity of 660 million curies (24,420 million gigabecquerels) had been injected "into isolated layers of earth crust" since the practice began in 1967.[253] There is no information about how the reprocessing plant and other facilities disposed of high-level liquid wastes before 1967. Penyagin has also given a somewhat different estimate of waste injection amounting to 600 million curies (22,200 million gigabecquerels) injected into 2,500 boreholes. He also states that another 120 million curies (4,440 million gigabecquerels) of liquid wastes were poured into four artificial pond-sized reservoirs. In 1991, site officials decided to close these ponds and fill them in.[254]

On the subject of airborne emissions, all radioactive effluents are adequately controlled by a purification system that, according to Ministry of Health measurements in 1987, is 99.8 percent effective.[255] However, as Bolsunovsky points out, this does not include inert radioactive gases, which are released directly, without being held for decay.[256] Whatever the actual quantity of radioactive materials that escape the filters, they are quickly released to the outside by the complex's powerful ventilation system, which changes all the air within the mountain every 10 hours. The average concentration of cesium-137 in the top 5 centimeters of soil within a 12-kilometer radius of the plant is reported at 38 millicuries (1.41 gigabecquerels) per square kilometer, although 60 percent of this is ascribed to global fallout from nuclear weapons testing.[257]

As would be expected in a 30-year-old reprocessing plant, Bolsunovsky reports that surface spills of radioactive materials contaminate many of the rooms and much of the equipment at Krasnoyarsk-26.[258] No estimates are available of the total activity involved. Contaminated equipment at the site includes the shut-down reactors, which Russia plans to leave in place for 50 years, allowing some of their induced radioactivity to cool before interring them permanently in underground shafts.[259] A Krasnoyarsk newspaper reported in 1992 that the complex also stores contaminated cleanup equipment from the

253. As cited in SocEco Agency 1992 and Penyagin 1991a.

254. Penyagin 1991a.

255. Korogodin et al. 1990. Bolsunovsky states that some officials claim the purification efficiency is 99.9%. (Bolsunovsky 1992, p. 20.)

256. Bolsunovsky 1992, p. 20.

257. Bolsunovsky 1992, p. 20; Cochran and Norris 1993, p. 105.

258. Bolsunovsky 1992, p. 20.

259. Khots 1992.

Chernobyl accident.[260] Routinely contaminated solids—such as masks, gloves, and other disposable supplies and equipment—are buried in clay-lined trenches, covered with a special but unspecified material for biological protection and then with soil and clay.[261]

One far-reaching environmental legacy of Krasnoyarsk-26 is the pollution of the Yenisey River, into which two of the three reactors discharged their untreated cooling water contaminated with plutonium and fission and activation products. This contamination continued for 30 years because each reactor's cooling water could be shut down safely only after its fuel had been removed or cooled sufficiently. The river carried the radioactive effluents for hundreds of kilometers, and they accumulated along the way in sediment, fish, plants, and the floodplain.

Officials first acknowledged the extent of this pollution and its threat to human health and the environment in 1990. Under pressure from environmental groups in Krasnoyarsk, the Ministry of Natural Resources and the Soviet Council of Ministers established a commission of high-level scientists from many specialties to undertake a comprehensive assessment of radioactive pollution in the Krasnoyarsk region. The commission reviewed documents regarding the waste disposal practices and radioactive emissions of the Mining-Chemical Complex and examined gamma-radiation monitoring data from the plant, towns along the Yenisey River, and the city of Krasnoyarsk. The commission held meetings with plant officials as well as with grassroots groups and in communities. The brief final report—eleven pages with four small tables—provides useful information about the history of Krasnoyarsk-26 but only limited annually averaged data on radioactive contamination.[262]

The commission report presented airborne concentrations for seven specific radionuclides recorded inside the plant's sanitary zone and at communities upwind and downwind. While these are stated as annual averages, there is no information about the timing or methodology of the sampling. The annual average could range from a few samples a year to continuous monitoring; without knowing the range, it is not possible to evaluate the significance of the average. It is not clear whether the reported averages include background figures. Regardless, the highest reported readings (those taken inside the sanitary zone) are very low: four to seven orders of magnitude less than the Soviet legal standards for exposure of persons in Category B given in the same table. These standards, referenced only to Norms of Radiation Safety

260. As cited in Bolsunovsky 1992. Contaminated equipment from Chernobyl is also stored in Tomsk-7 and Chelyabinsk-65.

261. Revenko, as cited in Miheev 1992a.

262. Korogodin et al. 1990.

"NRB-76/87," are almost identical to the U.S. Nuclear Regulatory Commission's limits for the allowable concentrations in air (above background) for the same radionuclides. While the cited levels are low, they are plausible. However, as with other official records, and not only in the former Soviet Union, release data need to be independently confirmed by evaluation of primary plant data.

The commission's report also provides figures for levels of 27 radionuclides measured in water along the east side of the river, 95 kilometers downstream from the discharge point. Again, most of the reported readings are very low, ranging from 10^{-10} to 10^{-16} curies (3.7 to 3.7×10^{-6} becquerels) per liter, with most ranging from a few percent to a few thousandths of one percent of the allowable maximum limits. The highest detected activity was from sodium-24 at 33.2 percent of the allowable limits and manganese-56 at 5.5 percent of the limits. The lowest levels of contamination cited are so low that they would require extraordinary accuracy in measurement and laboratory analysis (see further discussion below).

Two official data collection expeditions quickly followed up on the work of the commission. The first was conducted jointly by scientists from the Institute of Applied Geophysics (part of the State Committee for Hydrometeorology) and the Krasnoyarsk Research Center (part of the Siberian Branch of the Russian Academy of Sciences). It focused on the 20 sections of the Yenisey River between Krasnoyarsk and Igarka (where the river meets the Kara Sea) that had the highest readings of gamma radiation in aerial surveys. Over 200 samples of water, sediment, fish, algae, and floodplain soil were collected from these sites in the summers of 1990 and 1991 and analyzed for their radioactive concentration and activity.[263] The Institute of Applied Geophysics (IAG) published the final report in 1991.[264] The second effort, known as the Noibinski Geological Expedition, was commissioned by the regional Krasnoyarsk government council in 1991 to study contamination along the river.[265]

Vladimir Miheev, cofounder and coordinator of the Krasnoyarsk Green Movement (KGM), reviewed the results of both studies in detail in the movement's newspaper.[266] The sampling data presented in his review are not always identical to those cited in the expedition reports, but the sampling locations match, and the data are all in the same range. The main difference between the data of the commission and the data (reviewed by Miheev) of the two expeditions that followed is that, in general, these latter sources all cite considerably higher readings for

263. Miheev 1991.

264. Ivanov, Ashanin, and Nosov 1991.

265. Miheev 1991; also cited in Soler-Sala 1992, p. 99.

266. Miheev 1991.

radioactive contamination. The KGM review, however, does not cite or comment on any of the commission's data, although it uses much of its background information.

None of the reports describe methodology or analytical techniques in any detail. However, the commission claims that its methodologies for analyzing radionuclide concentrations and migration were checked by the Ministry of Health and the Ministry of Atomic Energy.[267]

Table 7.4 shows the concentrations of some radionuclides in the Yenisey River and in the soil of the floodplain at various distances from the plant's discharge pipes and the total level of "gross beta" activity. The highest readings were found at the outflow of the plant's discharge pipes, which open underwater about 50 to 100 meters from the east bank of the river. Eleven radionuclides were detected at this point, with the highest single reading being sodium-24 at 10^{-7} curies per liter, probably from neutron activation of sodium in the reactor's once-through cooling water system. The furthest samples were taken 249 kilometers from Krasnoyarsk-26, near Strelka. At this distance, only sodium-24 (at 1.1×10^{-10} curies per liter) and chromium-51 (at 7.5×10^{-11} curies per liter) were detected. The sodium-24 level reported 50 to 100 km downstream is the same order of magnitude as that reported by the commission.

Gamma readings taken by the IAG expedition near the source of contamination (about a half meter below the surface) reached about 30 micrograys per hour.[268] The commission cites a much lower range of about 12 to 15 micrograys per hour,[269] while a 1992 official survey, by the Ministry of Natural Resource's Committee on Hydrometeorological and Ecological Monitoring, reports about 6 to 10 micrograys per hour.[270] Normal background levels of gamma radiation in uncontaminated regions are on the order of 0.1 micrograys per hour. The flow of the Yenisey at this point is enormous, about 2,760 cubic meters of water per second.[271]

Only the commission, in its aerial survey, measured gamma readings further downstream. It reported levels of 4 to 5 micrograys per hour at 85 kilometers from the site, 0.4 micrograys per hour at 95 kilometers, and 0.2 micrograys per hour at 103 kilometers where the Kan River joins the Yenisey. The report offered no explanation for the unexpectedly high level found at 85 kilometers or the order-of-magnitude drop in readings over the next 10 kilometers. It also found gamma radiation levels of up to 1 microgray per hour over previously flooded

267. Korogodin et al. 1990.

268. Miheev 1991.

269. Korogodin et al. 1990.

270. Zubov 1992.

271. Miheev 1992.

Table 7.4 Concentrations of Selected Radionuclides in Yenisey River Water and in Floodplain Soil

Sample Location	Distance from Plant (km)	Manganese-56 in water, pCi/liter (Bq/liter)	Sodium-24 in water, pCi/liter (Bq/liter)	Cobalt-60 in water, pCi/liter (Bq/liter)	Cesium-137 in water, pCi/liter (Bq/liter)	Cesium-137 in soil Ci/km^2 (gigaBq/km^2)	Strontium-90 in soil, Ci/km^2 (gigaBq/km^2)	Plutonium in soil, Ci/km^2 (gigaBq/km^2)	Beta in water, pCi/liter (Bq/liter)
Tartat Bay	–10	—	—	—	—	.08 (3)			—
Discharge point	0	40,000 (1,480)	100,000 (3,700)	8 (.3)	25 (.925)	NR			510,000 (18,870)
Near discharge	1	NR	NR	—			13.9 (514.3)	0.03 (1)	—
Atamonovo	6	3,000 (111)	21,000 (777)	—	5 (.185)	2.72 (101)		0.014 (.52)	47,000 (174)
Kononovo	25	330 (12.21)	3,100 (115)	—	—	1.57 (58)			4,300 (160)
Pavlovschina	57	—	930 (34.4)	—	—	NR			1,500 (56)
Strelka	249	—	110 (4.1)	—	—	1.7 (63)		0.031 (1.15)	—
Igarka	1,667	NR	NR	NR	NR	.07 (2.6)			—

Note: Negative distance relative to the plant means upstream; NR means not reported; soil data are averages for the location.

Sources Mikheev 1991.

islands in the river as far as 336 kilometers from the site. These are very high radiation levels to find so far from the plant.

Table 7.4 also gives the concentrations of various radionuclides in soil along the Yenisey flood plain. Contamination in the flood plain is very unevenly dispersed between the shoreline and the high-water mark of the last 30 years. Two flash floods, in 1966 and 1988, increased the river's flow to 21,000 cubic meters per second and stirred up and carried radioactive sediments far beyond the river's normal banks.[272]

The data clearly indicate higher concentrations close to the discharge pipe, decreasing with distance. The concentrations of sodium-24 are high in these data, conforming to those reported by the interministerial commission, discussed above. Elevated levels of strontium-90, cesium-137, and other fission and activation products are evident; the same is true of plutonium. The contamination extends hundreds of kilometers from the site despite the enormous flow of the river.

The KGM review also reports on radioactive contamination in algae and fish. The main gamma emitters in fish are zinc-65, in the range of 5 to 6 nanocuries (roughly 200 becquerels) per kilogram, and cesium-137, in the range of 0.3 to 0.9 nanocuries (about 11 to 33 becquerels) per kilogram. These isotopes are detectable even in fish caught over 600 kilometers from the site, evidence that migrating fish can accumulate and carry radioactivity for great distances. Both expeditions also reported finding unspecified but high levels of sodium-24 and phosphorous-32. However, the commission reports finding contaminated fish only as far as 350 kilometers from the site. The highest level of contamination (reported only by the IAG) was 150 nanocuries (5,550 becquerels) per kilogram of phosphorus-32 in a grayling caught 60 kilometers from the discharge site. (Graylings are a migrating species similar to salmon.) Both the KGM and IAG reported algae to be a high source of secondary contamination. According to the IAG, algae concentrates long-lived radionuclides accumulated from sediment by a factor of 1,000 to 6,000.

All these findings led to the filing in August 1992 of an unprecedented appeal for financial compensation signed by the leaders of the Krasnoyarsk Regional Environmental Movement, the Krasnoyarsk Survival Fund, and the Krasnoyarsk branch of IPPNW.[273] Their widely publicized claim, filed against Valery Lebedev, the director of the Mining-Chemical Complex, seeks decontamination of the Yenisey River and its flood lands. (Other environmentalists believe it may be unwise to stir up the contaminated sediments.) The environmental organizations estimate that such a cleanup would collect 70 million

272. Miheev 1992.

273. Tarasov 1992a; Miheev 1992a.

cubic meters of soil, which they recommend be disposed of at Krasnoyarsk-26. They also seek another 600 million rubles to provide 1.2 million people living in the region with personal radiation dosimeters and training in their use. When the director of the complex failed to respond to their appeal, they forwarded copies to the Ministry of Atomic Energy in Moscow and the Russian State Prosecutor's Office.[274] Only the ministry responded, denying that the site had released any plutonium and dismissing any cleanup as unnecessary.[275]

The total 6.73 billion rubles requested represents just over two percent of Russia's military spending in 1992.[276] The head of the Krasnoyarsk Survival Fund, R. Solntsev, said the groups "realize that the complex and the ministry will hardly have that kind of money, and we have no intention of making them pay up. A victory in court is what is important to us."[277]

The contamination of the Yenisey flood plain may be considerably aggravated if Russia proceeds with plans to build 10 hydroelectric dams along the river and its tributaries. The reservoirs created by just two of these proposed dams—at Sredneyeniseyskaya and Turukhanskaya—would flood contaminated land and at least seven radioactive waste dumps.[278]

Tomsk-7

The Siberian nuclear weapons production complex known as Tomsk-7 and its closed city of Seversk (population 107,700) are about 15 kilometers northwest of the provincial capital city of Tomsk (population 500,000) in southern Siberia, about 1,500 kilometers east of the Ural Mountains. The site is located along the Tom River, which flows into the Ob' River and then the Arctic Ocean. In 1949, Lavrenti Beria, head of the Soviet secret police, selected this isolated and remote region as ideal for nuclear materials production because it could be kept secret. The Tomsk-7 site includes the Siberian Chemical Complex, the Siberian Atomic Power Station, and waste management facilities. A storage facility for plutonium and highly enriched uranium recovered from nuclear weapons is in the planning stages.[279]

274. Miheev 1992a.

275. Bolsunovsky 1993.

276. *Izvestia* 1991.

277. As quoted in *Izvestia* 1991.

278. Nelyubin 1991.

279. Chelnokov 1992, p. 61; Cochran and Norris 1993, pp. 85–86. According to the IAEA report of the Tomsk-7 tank explosion in 1993, the population of the closed city, which it calls Tomsk-7 city, is about 200,000. González, Bennett, and Webb 1993.

Although the Atomic Power Station and entire site are usually referred to as Tomsk-7, each facility may have its own code number. For example, an ex-employee of the Tomsk complex refers to plutonium contamination at "Installations 1 and 10";[280] Cochran and Norris mention plutonium-blending plants numbered 5 and 25.[281] *The First Guidebook to Prisons and Concentration Camps of the Soviet Union* describes Tomsk-2 and Tomsk-3 as "underground military plants" at which "up to 5,000 prisoners are confined underground under strict regime and are rarely transferred to other facilities, except to special isolation prisons as an additional punishment."[282] (The book says nothing about Tomsk-7.)

The plants referred to had several purposes. Plant Number 1 was used for enriching fuel by uranium-235; Plant Number 5 contained reactors Number 1 through Number 3; Plant Number 10 produced uranium hexafluoride; Plant Number 15 was a radiochemical plant; and Plant Number 25 was a chemical metallurgical plant. In addition, Plant Number 45 (also called "Reactor Plant") contains reactors 4 and 5, which were still operating as of 1994. In light of the limited information about most facilities at the site, we continue the practice of using Tomsk-7 to refer to the entire complex.

All production facilities at Tomsk-7 and the city that houses the workers are ringed by four rows of barbed wire, a footprint-monitoring belt, and towers with armed guards.[283] Encircling the site is a vast "sanitary zone" declared in 1970 that encompasses 192 square kilometers inside a 68-kilometer perimeter.[284] Local villagers are not supposed to eat berries, mushrooms, plants, vegetables, or fish inside the zone—but they regularly do.[285] Others have been seen grazing and watering cattle, cutting hay along the canal banks, swimming, and boating inside the sanitary zone.[286] Beyond this, a "zone of observation" stretches 75 kilometers along the Tom and Ob' rivers, with an area of 1,560 square kilometers and a perimeter of 240 kilometers.[287]

No information is available on what radiation levels define these zones, although an official commission reported in 1990 that the "total internal and external radiation of the critical group of people in the zone of observation (taking into account global fallout) is 37 millirem

280. As cited in Bolsunovsky 1992, p. 18.

281. Cochran and Norris 1993, p. 91.

282. Shifrin 1982, p. 201.

283. Chelnokov 1992, p. 62.

284. Aleksrashin et al. 1990.

285. Bolsunovsky 1992, p. 15.

286. Aleksrashin et al. 1990.

287. Aleksrashin et al. 1990.

[370 microsieverts] per year."[288] This is about 50 percent more than the 250 microsieverts per year limit allowed by the U.S. EPA for public exposure to radiation from U.S. nuclear fuel cycle facilities, but less than the overall 1 millisievert per year limit from all nonmedical sources.

The Siberian Atomic Power Station houses five graphite-moderated reactors, three of which were permanently shut down between 1990 and 1992. The first reactor came on line in 1955, the second in 1958, the third in 1961, the fourth in 1964, and the fifth in 1965. The first reactor probably used a "once through" cooling system that took water directly from the Tom River and returned it, via a series of reservoirs and canals, to the river.[289] The other four used a closed cooling system with a secondary loop that generated electricity and heat for other Tomsk-7 facilities as well as for agricultural complexes and, according to official claims, 40 percent of the heat for the city of Tomsk.[290] All are assumed to have been upgraded from an initial capacity of 650 megawatts thermal to 2,000 megawatts thermal.[291]

The Siberian Chemical Complex can handle several stages of the nuclear fuel cycle, with facilities for the enrichment, processing, and fabrication of uranium, as well as for chemically separating (reprocessing) both uranium and plutonium from spent reactor fuel.[292] Facilities also exist for separating tritium, blending plutonium isotopes, and fabricating plutonium into triggers and other warhead components.[293] Tomsk-7 processed all its own plutonium and, as we have noted, the spent fuel from the military production reactors at Chelyabinsk-65, which it began receiving in 1976.

Russian officials claim that the centrifuge technologies developed at the Siberian Chemical Complex compete in cost and quality with those available outside the country.[294] Although it no longer produces highly enriched uranium for nuclear weapons, the plant still enriches uranium to low concentrations of uranium-235 for use in power reactors and reprocesses spent fuel from its own and other reactors. Several countries have contracts with Russia (including some signed originally with the Soviet Union) to obtain uranium-enrichment and plutonium-reprocessing services from Tomsk-7. For example, in 1991 the French

288. Aleksrashin et al. 1990.

289. Cochran and Norris 1993, p. 88; Bolsunovsky 1992, pp. 12–13.

290. Cochran and Norris 1993, pp. 86–88; Bolsunovsky 1992, p. 12.

291. Cochran and Norris 1993, p. 109.

292. Bolsunovsky 1992, p. 13.

293. Cochran and Norris 1992, pp. 21, 50; Carnegie Endowment for International Peace and Monterey Institute of International Studies 1994, p. 21.

294. Chelnokov 1992, p. 63; Bolsunovsky 1992, pp. 13–14.

reprocessing firm Cogéma negotiated a $50 million per year, 10-year deal that calls for Tomsk-7 to low-enrich up to 150 metric tons of mixed uranium oxides per year in 1992 and 1993 and up to 500 metric tons per year (in the form of uranium hexafluoride) from 1994 through 2000. Cogéma provides uranium from what it recycles out of reprocessing spent reactor fuel in France.[295] Although Cogéma has its own gaseous diffusion enrichment facilities, Cochran and Norris point out that it apparently prefers to send recovered uranium to Tomsk-7 to avoid contaminating its own diffusion plants with impurities, mostly uranium-232 and uranium-236, that come from using recycled uranium as feed stock.[296] Recycled uranium also carries along some fission products, notably technetium-99, so that enrichment processing facilities also become contaminated with them. The design of the Tomsk cascades allows uranium-232 to be isolated in a limited part of the production line and periodically cleaned out.[297]

MINATOM has a 10-year contract with South Korea requiring the complex to supply 260 to 400 metric tons of low-enriched uranium per year in exchange for electronic consumer goods in short supply, such as TVs, VCRs, and refrigerators.[298] Tomsk-7, Krasnoyarsk-26, and probably other closed cities as well received large shipments of consumer goods in 1991 as part of this agreement.[299]

In 1992, Tomsk-7 officials sought U.S. financial and technical assistance to begin constructing a "Fissile Materials Storage Facility,"[300] but it has faced public opposition within Russia. This was meant to be the sole secure facility in Russia for stockpiling plutonium and highly enriched uranium for the next 80 to 100 years.[301] Of the $400 million that the U.S. Congress appropriated in 1992 to assist Russia, Belarus, Ukraine, and Kazakhstan with disarmament-related expenses (along with another $400 million appropriated in 1993), $15 million is earmarked for the design of this facility.[302] Besides Tomsk-7, locations that have been considered for it include Krasnoyarsk-26, Chelyabinsk-

295. Cochran and Norris 1993, p. 94.

296. Cochran and Norris 1993, p. 95.

297. Popova 1993.

298. Levine 1991, p. 10; Bolsunovsky 1992, p. 13.

299. Bolsunovsky 1992, p. 13.

300. Berkhout et al. 1992, p. 30; Cochran and Norris 1993, p. 95.

301. Cochran and Norris 1993, p. 95.

302. Kolesnikov 1992; Lockwood 1993, p. 39. Other U.S. funds committed to Russia under the "Soviet Union Denuclearization Act" include up to $50 million for 10,000 containers in which to transport and store fissile material, up to $5 million for armored blankets to enhance the safety of warheads being transported, up to $20 million to improve the safety of rail cars for transporting this material, and up to $15 million for emergency response equipment and protective clothing in the event of a transportation accident.

65, and warhead-dismantling plants in Arzamas-16, Sverdlovsk-45, Penza-19, and Zlatoust-36.[303]

U.S. officials have expressed concern about the implications of the storage facility's enormous proposed size. Plans call for a 50,000-square-meter underground bunker designed to withstand tornadoes, earthquakes, fires, floods, airplane crashes, electricity blackouts, bombing, and terrorist attack—everything short of attack by nuclear weapons or conventional penetrating missiles.[304] The facility is to hold about 100,000 hermetically sealed containers, including a 40,000 container capacity in the first stage of construction, capable of storing either plutonium or highly enriched uranium from dismantled warheads. Each container will hold 4 to 5 kilograms of plutonium or 10 kilograms of highly enriched uranium.[305]

The planned capacity is enough, for example, to hold 250 metric tons of plutonium *and* 500 metric tons of uranium. Since Russia plans to sell the United States 500 metric tons of highly enriched uranium (out of an inventory estimated by Cochran and Norris to be 1,000 metric tons[306] and by Bukharin to be in the range of 520 to 920 metric tons)[307] over the next 20 years, this may be considerably more storage capacity than Russia needs.[308] The Russian inventory of highly enriched uranium may be larger. Atomic Energy Minister Mikhailov has stated it to be 1,250 tons.[309] (The United States plans to license a private consortium to dilute this highly enriched uranium so it can be used as fuel in commercial power reactors.)[310] Russia estimates that it will take about $30 million and eight years to construct enough capacity for the first 40,000 containers.[311] The merits of this large storage project are the subject of heated local debate, with 80 percent of the regional parliament voting against construction in October 1992.[312]

Production at Tomsk-7 Estimates of cumulative plutonium production at Tomsk-7 depend on basic assumptions about the operating history of the five reactors, their capacity, and power levels. The model

303. Lockwood 1993, p. 40.

304. Cochran and Norris 1993, p. 96.

305. Berkhout et al. 1992, pp. 29–30.

306. Cochran and Norris 1993, p. 122.

307. Bukharin 1993, p. 5.

308. Berkhout et al. 1992, p. 30.

309. National Academy of Sciences 1994, p. 131.

310. Bukharin 1993, p. 3.

311. Von Hippel, Cochran, and Paine 1993, p. 6.

312. Chernykh 1992.

used by Cochran and Norris gives a cumulative estimate through 1993 of 73.7 metric tons.[313] This estimate does not include plutonium received at Tomsk-7 from other sources, notably the 23.3 metric tons contained in military reactor fuel sent by Chelyabinsk-65 beginning in 1976, as discussed earlier.

Assuming that 3.3 curies of strontium-90 and 3.6 curies of cesium-137 were generated for each gram of plutonium recovered,[314] and that all the irradiated fuel generated on-site and received from Chelyabinsk-65 has been reprocessed, Tomsk-7 created approximately 320 million curies (about 11.84 million terabecquerels) of strontium-90 and about 349 million curies (about 12.91 million terabecquerels) of cesium-137. When we include the decay products of strontium-90 and cesium-137 (which are yttrium-90 and barium-137m, respectively) we get a total estimate for these long-lived radioactive nuclides and their decay products of about 1.3 billion curies (48 million terabecquerels). Since the half-life of both strontium-90 and cesium-137 is approximately 30 years, roughly 30 percent of the radioactivity of these two radionuclides has decayed away, assuming a production-weighted storage time of about 15 years. Thus approximately 900 million curies (33.3 million terabecquerels) of strontium-90, cesium-137 and their decay products remain as of the early 1990s.

These estimates indicate that high-level waste generation at Tomsk-7 was comparable to that at Chelyabinsk. Such an estimate is supported by Nikolai N. Guryev, a plant official who resigned in 1990 (see below).[315] In addition, Tomsk-7 undoubtedly created large volumes of other low-level and medium-level radioactive wastes from uranium conversion, enrichment, and processing. However, most of the radioactivity is contained in the high-level reprocessing wastes.

Waste Management at Tomsk-7 Information on high-level waste management at Tomsk-7 has been emerging slowly since about 1990. Alexander Penyagin, former chair of the Supreme Soviet's Subcommittee on Atomic Power and Nuclear Ecology, estimates that about one billion curies of low-level and high-level radioactive liquid wastes were injected underground.[316] He also believes that 100 million curies of medium-level and high-level liquid wastes, including neptunium, strontium-90, and cesium-137, were just poured into two aboveground reservoirs near Tomsk.[317] To lessen the risk of the

313. Cochran and Norris 1993, p. 109.

314. Cochran and Norris 1993, p. 64, footnote 268, suggest these figures for military production reactors, with burn-up levels of 500 Mwdt/metric ton.

315. As cited in Bolsunovsky 1992, p. 17.

316. Penyagin 1991a.

317. Penyagin 1991.

reservoirs overflowing or evaporating, Penyagin says, Tomsk-7 officials decided in 1991 to close and fill them.[318] Penyagin's total estimate of 1.1 billion curies of liquid wastes released from Tomsk-7 corresponds approximately with the estimates of cesium-137, strontium-90, and their radioactive progeny in liquid wastes made above.

In June 1992, Gosatomnadzor confirmed allegations of liquid wastes being injected underground. A report to the Russian deputy prime minister says that since pumping began in 1967, 36 million cubic meters of liquid wastes with a total activity of 1.06 billion curies have been injected "into isolated layers of earth crust."[319]

The most detailed critique of Tomsk-7 waste disposal practices comes from a statement issued by Nikolai N. Guryev upon his 1990 resignation from the Communist Party.[320] A retired engineer-physicist who worked on security at the complex from 1959 to 1988, Guryev claims that in 1982 he began trying, without success, to alert local and national party officials to the critical radiation situation. He paints a very different picture of radioactive-waste management procedures from that presented by site officials. To avoid censorship, Guryev uses the terms product A and product B when referring to the main long-lived radioactive substances processed at Tomsk-7, but, notes Bolsunovsky, "Every reader of the statement understands that A is plutonium and B is uranium."[321]

The impact of Guryev's claims on the Tomsk community can be judged from the response. On the one hand, others came forward with similar reports that local newspapers continued to publish.[322] In contrast, the directorate of the complex sued Guryev for libel.[323]

Guryev asserts that solutions of radioactive waste, including many contaminated with products A and B, were dumped on several thousand hectares of land and reservoirs. The discharges from the settling basin reservoirs are detected not just in the Tom River but also downstream in the Ob' River and as far away as the Arctic Ocean. Tens of thousands of metric tons of radioactive scrap metal and other solid wastes are buried "like cattle" under tens of hectares of land. He confirms that radioactive liquid wastes were injected underground, although he reports this was done at a depth of 180 to 210 meters, just above the 240-meter to 260-meter zone from which he says the city of Tomsk draws its water. Guryev gives the complex credit for trying to

318. Penyagin 1991a.

319. As cited in SocEco Agency 1992.

320. As cited in Bolsunovsky 1992, pp. 16–18.

321. Bolsunovsky 1992, p. 17.

322. See, for example, Chelnokov 1992.

323. Bolsunovsky 1992, p. 18.

reduce the release of radioactive substances into the atmosphere, estimating that from one complex chimney alone, releases fell from 1,500 kilograms per year of product B in the 1950s to 15 kilograms per year of product A and B combined in the 1980s.[324]

However, the complex is still plagued by what Guryev calls tens or hundreds of "microaccidents" annually that release fluorides of products A and B—most likely from processing and enriching uranium—that contaminate both air and soil. He claims that each kilogram of product released in these microaccidents contains 230 grams of hydrogen fluoride and 670 grams of products A and B.[325]

The Tomsk-7 Tank Explosion The only well-documented accident at Tomsk-7 occurred on 6 April 1993, in the No. 15 plutonium-reprocessing plant. There was an explosion in a 35-cubic-meter tank containing approximately 25 cubic meters of uranium, plutonium, and some radioactive fission products in a complex solution of nitric acid, paraffin and tributylphosphate. The solution needs stirring with compressed air to keep it well mixed, which is necessary to prevent an explosion. According to an investigation by the International Atomic Energy Agency, either the absence of or insufficient compressed air was responsible for the explosion, which "blew a hole in the roof and upper wall of the room, and a shockwave passed down a gallery of about 100 m[eters] length. When the wave reached the end of the gallery, it burst through the lateral brick wall. There was thus extensive damage to the physical plant. A small fire began on the roof, but this was quickly put out."[326]

Mixtures similar to the one that exploded are sometimes called "red-oils" in the United States. Four similar but less severe "red-oil" explosions involving processing tanks occurred in the United States. Three occurred during the 1950s: at the Savannah River Plant in South Carolina in 1953; at Hanford, Washington, in 1953; and at Oak Ridge, Tennessee, in 1959.[327] The *Energy Daily* recently also reported another red-oil explosion at Savannah River in 1975.[328]

Unlike their behavior after earlier accidents at Chernobyl and Chelyabinsk, Russian authorities informed the public and the press about this explosion within hours and invited a team from the International Atomic Energy Agency (IAEA) to inspect the damage. However, officials at Tomsk-7 and with the Atomic Energy Ministry in Moscow still

324. Bolsunovsky 1992, p. 17.

325. Bolsunovsky 1992, p. 17.

326. González, Bennett, and Webb 1993.

327. IPPNW and IEER 1992, pp. 57–58.

328. As cited in Cochran and Norris 1993, p. 89.

tried to downplay the seriousness of the accident, insisting at first that "the contamination zone is restricted to the plant's territory."[329]

The hole in the roof of the building and the collapsed wall allowed substantial releases of radioactivity. As with the 1957 Chelyabinsk-65 tank explosion, there were, broadly speaking, two zones of contamination. The first zone had contamination of about 1,500 square meters in the immediate vicinity of the plant, officially estimated to be 4 curies (148 gigabecquerels) of beta-gamma activity. The second zone included a far wider area, contaminated with about 40 curies (1,480 gigabecquerels) of radioactive materials (beta-gamma activity). This radioactivity was spread mainly through the 150-meter-high stack of the building.[330]

The International Atomic Energy Agency reported that there were 19 curies of plutonium, 3 curies of other alpha activity, and 537 curies of beta emitters in the tank at the time of the accident.[331] Since only 44 curies of beta-gamma emitters are reported as having been released, almost 500 curies presumably remained in the heavily damaged building as a cleanup problem.

Figure 7.7 of the Tomsk-7 explosion shows the map of external radiation readings produced by the International Atomic Energy Agency. It clearly shows that, contrary to early assurances by plant officials, substantial contamination occurred outside the plant's boundary. The isopleth of about 4 micrograys per hour (about 40 times natural background) appears to be less than a kilometer from the village of Georgievka (population 200), the closest and most seriously affected village. Yet officials waited more than a week before beginning a "precautionary" evacuation of children from Georgievka.[332] The delayed evacuation followed the pattern of previous nuclear accidents in the former Soviet Union.

A decontamination team in protective clothing worked in Georgievka for at least four days before the evacuation began. The unprotected adults and children of the village looked on and played while the team came for a few hours each day to take measurements and shovel contaminated soil and snow into plastic bags. The Russian Television Network filmed and broadcast this remarkable scene.[333]

Prevailing winds carried the radiation plume released by the explosion and fire away from the city of Tomsk and toward less-populated

329. Tomsk-7 Director Gennadiy Khandorin, as quoted in Illesh and Kostyukovskiy 1993.

330. González, Bennett, and Webb 1993.

331. González, Bennett, and Webb 1993.

332. WISE 1993b; Mackenzie and Bastable 1993.

333. As filmed by Pelt and Goryunov 1993.

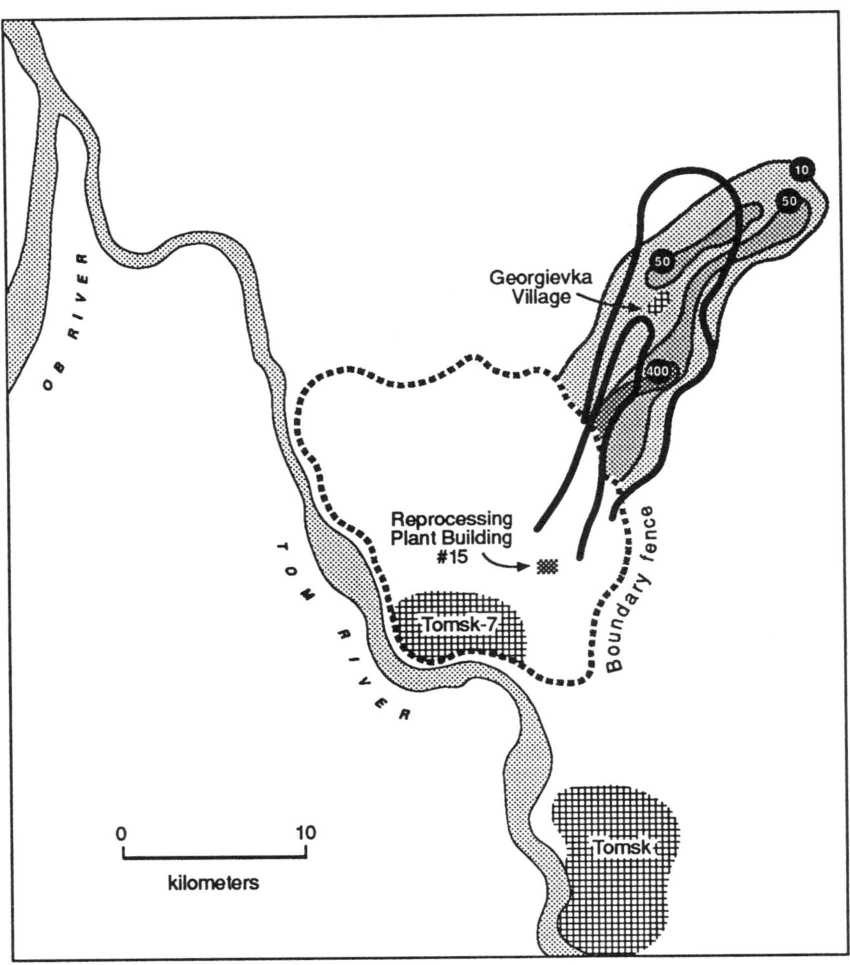

Figure 7.7 Rough diagram of the Tomsk area, showing dose rate contours from the 1993 explosion. Doses are indicated in μR/h. Heavy black lines represent aerial measurements. Source: Gonzalez, Bennett, and Webb 1993.

areas to the northeast.[334] The Ministry of Atomic Energy initially reported that less than 100 square kilometers were contaminated, but later MINATOM revised this to over 250 square kilometers in a narrow fallout trace about 37 kilometers long.[335] The ministry reportedly has paid local residents of this zone about three and a half dollars per person in compensation.[336]

The long-term distribution of the radionuclides is likely to shift. With the coming of spring and the melting of the region's snow cover,

334. Illesh and Kostyukovskiy 1993.

335. Morozova 1993; Cochran and Norris 1993, p. 90.

336. *New Scientist* 1993.

much of the fallout undoubtedly entered into streams and rivers.[337] In any case, longer-term cleanup efforts focused not on the environment but on the plant itself—on getting it back on line quickly to process the spent fuel of still-operating reactors.[338] Nevertheless, three weeks after the accident, Tomsk-7 director Gennadiy Khandorin admitted that radiation levels in the worst-damaged part of the plant (covering several hundred square meters) were still too high to begin reconstruction.[339]

Environmental Contamination Prior to 1989, there was essentially no public information about environmental contamination from Tomsk-7 operations. A considerable amount of information has emerged since that time. A variety of independent and official efforts have reported data on releases of radioactive materials, contamination of water, and radiation levels. This body of information is not internally consistent. A clearer, more consistent picture will only emerge after official plant documents are released for independent critical evaluation. We present a very preliminary and brief analysis of the data here.

Nuclear specialists from the Siberian Chemical Complex began providing some information to the public in late 1989. They did so under pressure from Ecology Initiative, a local environmental organization, and others in the Tomsk region concerned over rumors about the partial meltdown of a reactor at the site.[340] The officials went so far as to participate in an open roundtable discussion with Ecology Initiative at which they gave sweeping assurances that radioactivity detected beyond the site was below (unstated) background levels.

These comments and the refusal of site officials to provide firm answers to many other questions drew harsh criticism from Ecology Initiative coordinator Albina Biychaninova. In a widely publicized letter sent to high officials in Moscow in December 1989, she stressed the urgent need for an investigation into the facilities at Tomsk-7 and their waste disposal practices. Regardless of any rumors about accidents, she said, an investigation is warranted by "the increase in oncological [cancer] and haematological [blood] disease[s], and of leukemia, which are common for the inhabitants of Tomsk."[341]

337. There was snow on the ground at the time of the accident, and some cleanup consisted of shoveling it.

338. Yakushev 1993.

339. As cited in Yakushev 1993.

340. Biychaninova 1989. This source notes 1989 rumors that a runaway reactor incident at Tomsk-7 overheated, deformed, developed a fissure, released radioactive material into the atmosphere, and finally was enclosed in a hastily built sarcophagus. It attributes this information to "unofficial information from workers at the atomic plant." No other references to this alleged accident have been found.

341. Biychaninova 1989.

A reply from V. H. Dogudjiev of the USSR Council of Ministers, countersigned by the deputy ministers of health, atomic power, and meteorology, asserted that "according to all existing data the complex does not have any harmful effect on the health of the inhabitants of towns in this region." The reply denied the rumors of a runaway reactor, insisting that "the technical state of the nuclear reactors and their control systems cannot be cause for anxiety since they undergo systematic prophylactic repairs, permitting them to work with full safety."[342] The reply also cites unreferenced 1987 data showing gamma-radiation levels around Tomsk of about 0.09 to 0.13 micrograys per hour, in the range of natural background.

This series of discussions motivated several official and independent investigative efforts. While various reports arising from these investigations differ in some points, there is order-of-magnitude agreement on others. Readings in or near the canals and discharge points of the reactor cooling water (of the now-shut once-through cooling system reactors) are several micrograys per hour.[343] Radiation readings as high as about 5 micrograys per hour have also been recorded over the Tom River near the village of Chernilshikovo, and in a dramatic videotape, a reading of about 10 micrograys per hour was recorded over water, though the source does not provide the exact location.[344] Background radiation readings in the region are around 0.1 to 0.2 micrograys per hour.

The situation of contamination of water and soil along the Tom River appears to broadly resemble that along the Yenisey River downstream from Krasnoyarsk-26. There are indications of high levels of activation products in the cooling water, including activated alloying metals such as chromium-51, cobalt-58, and zinc-65.[345] These were discovered in canal sediments by French investigators from the independent laboratory CRII-RAD. According to Cochran and Norris these indicate corrosion of the reactors.[346] While the once-through-cooled reactors are now shut, there is still accumulated radioactivity in the ponds and reservoirs where cooling water and other wastes were discharged over the decades.

These accounts include varying estimates of the total amount and activity of wastes released from Tomsk-7. One Russian newspaper reports that the complex released more than eight million cubic meters of radioactive water into at least eight reservoirs; that it pumped 42,000

342. Dogudjiev 1990.

343. Bobrova 1990; Lariviere and Denis-Lampereur 1991, pp. 102–103.

344. Bobrova 1990.

345. Lariviere and Denis-Lampereur 1991, p. 103.

346. Cochran and Norris 1993, p. 94.

cubic meters of liquid radioactive wastes per day into the Tom River; and that it pumped another 175 cubic meters per hour (4,200 per 24-hour day) into underground zones 345 to 370 meters below the surface.[347] This source also mentions published statements of former employees who assert that not just radioactive wastes but also finished products—in the form of plutonium-239 and highly enriched uranium—were disposed of in reservoirs and burial grounds. According to a former senior foreman of the technical inspection section, site managers deliberately concealed these practices with the knowledge of the ministry responsible, the KGB, and town and regional committees of the Communist Party.[348] However, there is reason to be skeptical of this claim because such products are relatively valuable. This report may refer to discarded materials that were highly contaminated with plutonium or highly enriched uranium or to relatively high levels of unrecovered plutonium and uranium-235 left in reprocessing and enrichment wastes.

A second account, based on an official report of the Tomsk Oil and Gas Geology Office, claims that the underground waste reservoirs are actually sandy beds 220 to 360 meters underground, between layers of water-resistant clay, at a spot 10 to 12 kilometers from the Tom River.[349] The same source estimates that 127,000 metric tons of solid and 33 million cubic meters of liquid radioactive wastes of unknown concentration have been injected in these beds,[350] a volume that would take over 21 years to inject at the rate cited above.

Finally, there is also evidence of fish being contaminated, notably with phosphorous-32, an activation product. Since local fish consumption would tend to be a principal source of radiation dose, Bolsunovsky reports that officials have recommended storing fish for 143 days prior to consumption. He regards this as "impractical due to a shortage of food."[351]

There is some information available on current radioactive solid waste management. According to C. S. Andreev, head of a laboratory on the site, wastes are handled properly, and there are no serious environmental contamination problems other than those possibly arising from activation products.[352] However, journalistic accounts tend to dispute this.

347. Chelnokov 1992.

348. Chelnokov 1992.

349. As cited in WISE 1991; see also Kostyukovisky 1991.

350. Kostyukovisky 1991.

351. Bolsunovsky 1992, p. 15

352. Bolsunovsky 1992, p. 14.

HEALTH IMPACTS IN THE FORMER SOVIET UNION

In 1972, the economist Marshall Goldman warned that an environmental catastrophe loomed over the Soviet Union, caused by unmitigated contamination of air and water, but this assessment had a striking omission. He made only a few references to environmental contamination related to nuclear power and weapons, and one of these merely noted the positive impact of the 1963 Nuclear Test Ban Treaty and the generally low atmospheric exposures to the citizens of Moscow from strontium-90.[353]

Goldman's omissions were attributable to the lack of accessible information on deaths and diseases among plant workers and civilians exposed to and at risk from radiation from Soviet production of nuclear weapons. A closed administrative department in the Ministry of Health enforced secrecy around radiation health issues. The responsible investigators, as they now acknowledge, were required to collaborate in distorting and minimizing consequences, and medical researchers had to convert radiation-induced conditions into more benign-appearing diagnoses.[354] This occurred both despite and along with sometimes elaborate medical surveillance efforts following several notable releases.[355] Zhores Medvedev's reconstruction of the massive 1957 Kyshtym explosion only underlines the obstacles created by severe censorship and deliberate deception that would be encountered by a resourceful Soviet scientist working in the field.[356] The sophistication of Soviet physicians in establishing grades of lethality for radiation effects among firefighters at Chernobyl adds to this impression of an intricate secret world, featuring an extensive unpublished experience with highly exposed nuclear workers.

A substantial body of information on the Soviet nuclear program has been disseminated in the open literature since 1989. Unfortunately, it is not a straightforward archive. For one thing, methodology is poorly documented. Furthermore, the available information reflects the perspectives and limitations of analysts reliant on the methods and verities of an earlier generation of physicians and scientists. Its interpretation requires substantial clarification, background, and skepticism. The unfinished status of this record is expressed by A. V. Akleyev, a prominent Russian researcher who defines the Soviet experience as an invitation to international investigators to dig deeper.[357]

353. Goldman 1972, p. 128.

354. For instance, new diagnostic categories such as "chronic radiation sickness" were created. See Soyfer et al. 1992.

355. Penyagin 1991; *Los Angeles Times*, 3 September 1992.

356. Medvedev 1979.

357. Akleyev et al. 1992.

Perhaps this is sound advice, but the implications produce an unusual set of critical problems. The renewed analysis and subsequent publication of material more than 30 years after the major accidents in the south Urals links the reliability of information to the scientific integrity of the 1950s and 1960s. We know that in many cases this record was highly distorted. Moreover, as Medvedev shows in his account of Chernobyl, in the more recent open period reports from Soviet scientists—particularly on causes of the accident—were still flawed by calculated omission and misrepresentation, although the IAEA readily accepted them.[358]

The example of Chelyabinsk-65 is characteristic of the difficulties that can arise in interpreting Russian and Soviet data. When the circumstances of the 1957 accident were officially unveiled in 1989, reports included details on radionuclide dispersion, evacuation patterns, and dose reconstruction.[359] Although 1,054 people had received an effective dose equivalent of 0.52 sieverts and 1.5 sieverts to the gastrointestinal tract, neither short-term nor long-term surveillance were reported to have detected serious radiation-related effects.[360] The contrast of quite high individual and population exposures with little damage to human health that characterizes Soviet reports, particularly those associated with the weapons industry in the south Urals, is highly problematic. At such levels of radiation, effects in populations of over a thousand should be readily detectable. That they were not reported is reason for suspicion and extreme caution.

Health effects records were in all probability affected by official policies of putting nuclear weapon production priorities ahead of health considerations. The doses and health outcomes as presented here are official Soviet data, which are suspect for many reasons, external and internal. A basic external factor is that the data on health outcomes of people exposed to relatively high levels of radiation (tens of centisieverts to a few sieverts) are in clear conflict with well-established effects of such levels of radiation from medical practice and Hiroshima-Nagasaki survivor data. These Soviet health outcome data cannot be accepted at face value; rather they need to be carefully and independently reevaluated. Internal factors include cover-ups of accidents and the pressure to falsify health outcome records by not reporting radiation-related diseases. For example, physicians reportedly faced seven years of imprisonment if they explained to their patients that the reasons for medical intervention were linked to radiation.[361] Accordingly, references to "ABC disease" or "weakened vegetative

358. Medvedev 1990.

359. Nikipelov 1989.

360. Buldakov et al. 1989, table 11.

361. Hertsgaard 1992.

syndrome'' effectively blinded the exposed inhabitants to their actual clinical condition—and may have also blinded both examining physicians and record keepers. There is no simple way of knowing how this filtering of the kind and severity of diagnoses distorted the quality of data and the thoroughness of follow-up.

Because of the systematic problems mentioned above, Soviet dose and health data are not amenable to straightforward interpretation or to direct use in epidemiological studies or radiation risk calculations. Rather, we believe that the past and current health status of populations with high recorded doses should be carefully reevaluated. For populations with relatively low doses—a few tens of centisieverts lifetime exposure or less—careful reevaluation of the dose records, use of biological markers that may be able to independently verify dose, and assessment of current and past health status may yield valuable insights, especially in conjunction with health outcome data from Chernobyl. Such studies should properly be conducted within the context of treatment of affected populations so that they do not become ''guinea pigs'' for radiation risk studies, as has tended to happen in the past with certain populations.[362] Any successful outcome of such research depends critically on the Russian government's willingness to open up its raw data and records to independent investigators, much as the U.S. Department of Energy is now doing.

In fairness, we must also recognize that some problems faced by Soviet investigators were due to generic limitations in the general public health database available to them. For example, regional statistics for cancer deaths may not exist in a reliable form for the period under observation.[363] In addition, certain practices that would be considered unusual in the West were intrinsic to Soviet radiation surveillance. Thus, physical examinations of the civilians exposed near Chelyabinsk-65 focused on a somewhat unusual choice of intermediate symptoms and diseases, such as parasitic infections, nodular goiters, cervical erosions, and rapid heart rate.[364] Further, these assessments paid insufficient attention to long-term effects. The sanitary hygienic model for public-health investigation in the former Soviet Union seems to have been heavily weighted toward individual clinical examinations, emphasizing acute symptoms over subtle or longer term effects, particularly cancer. In certain respects, the medical assessment of both workers in nuclear weapons plants and people living near the facilities had more in common with the type of medical surveillance

362. The people of Rongelap in the Marshall Islands who were highly exposed to fallout from the 1 March 1954 test of a thermonuclear bomb had good reason to feel that they had been used as ''guinea pigs.'' See IPPNW and IEER 1991, chapter 5.

363. Yablokov, Vorobiev, and Pokrovski 1992.

364. Buldakov et al. 1989.

practiced in Western private industry, with its formalized yearly physical exams, than with the type of government-sponsored epidemiological investigations that have occurred in the United Kingdom and the United States. The latter would assume a more open and inquiring disposition toward endpoints, such as cancers, and toward cumulative doses. However, a primary interest in weapons production can compromise such studies, as controversies surrounding U.S. plant and resident populations exposed to radiation have shown.[365] Finally, as noted above, Soviet doctors were also forbidden to explicitly report radiation-related diseases, and this probably had the effect of encouraging misdiagnosis and falsification of data. We examine the data for Chelyabinsk-65 in some detail. There is as yet very little public health–related information on Tomsk-7 or Krasnoyarsk-26.

Hazards to Chelyabinsk-65 Workers

The oldest plutonium-production facility in the former Soviet Union, Chelyabinsk-65, is also the most studied in terms of exposure and health effects. For more than three decades, A. I. Burnazian of the Institute of Biophysics No. 1 directed an intensive sanitary inspection and medical observation unit at the plant.[366] As is the case in the United Kingdom and the United States, the plant monitored external exposures to the workers enlisted in producing nuclear weapons, and doses to workers exceeded most nonworker exposures. However, also as in the United States, there were problems with the reliability and universality of the use of film badges, and internal doses were infrequently assessed.

Begun in 1948, the medical registration of workers provides for direct comparison of chronic low-dose radiation effects at Chelyabinsk-65 with those at Hanford, Oak Ridge, and Sellafield, to the extent that data on health and exposures are valid. However, this intriguing potential for international comparison is belied by the very different and startling experience of the Soviet group. The reported exposures to Soviet workers differ from Western counterparts by orders of magnitude, with the cumulative mean dose exceeding the mean instantaneous whole-body dose to atomic bomb survivors.[367] Yet as the previous anecdote from Chelyabinsk-65 and much of the following information demonstrate, corresponding reported adverse health effects are low. Even the "low" doses are not low in the context of lifetime exposures of a few centisieverts or less which characterize the current low-dose controversies. Rather, as noted above, they fall into a

365. Geiger et al. 1992.

366. Nikipelov et al. 1990a, p. 373.

367. Kossenko, Degteva, and Petrushova 1992.

Table 7.5 Permissible and Measured Gamma-Radiation Exposures to Chelyabinsk-65 Workers, Selected Years

		Measured Exposures at A-Reactor		Measured Exposures at Reprocessing Plant	
Year	Permissible Exposure Limit (sieverts/year)	Percentage of Workers >0.25 Sieverts	Average Dose for All Workers (sieverts/year)	Percentage of Workers >0.25 Sieverts	Average Dose for All Workers (sieverts/year)
1949	0.3	90%	0.936	73%	0.48
1953	0.15	21%	0.196	49%	0.307
		>5 sieverts		>5 sieverts	
1961	0.05–0.12	4%	0.02	37%	0.11
1971	0.055	38%	0.013	5%	0.014

Sources: Energizdat 1981; Nikipelov, Lyzlov, and Koshurnikova 1990.

higher range where cancer risks have been reproducibly established from extensive observations on atomic bomb survivors and people who have received therapeutic radiation. This is an example of official Soviet data that are suspect.

The graphite-moderated, watercooled reactor at Chelyabinsk-65 (Mayak A) went on line in June 1948, and the first reprocessing plant (Mayak B) began operation in December. In August, the Ministry of Machine Building and a closed department of the Ministry of Health introduced the first regulatory standards for worker exposures to radiation.[368] Maximum daily occupational tolerances were set at 0.1 centisieverts per 6 hours (about 0.3 sieverts per year). In 1952, the daily and annual limits were halved to 0.05 centisieverts per 6 hours (about 0.15 sieverts per year).[369] Workers receiving that dose had to undergo medical surveillance and either temporary or permanent reassignment to an unexposed job.

Although these standards were comparable to those in the West, actual exposures at Chelyabinsk-65 were far higher, indicating lax enforcement. The situation is not unlike that in other Soviet workplaces, where limits on concentrations for inhaled toxins were stringent but compliance was limited. Recorded film-badge exposures for Chelyabinsk-65 workers frequently exceeded the upper limit of the standards (see table 7.5). In practice, significant allowances and contingencies were introduced, particularly for the high exposures encountered during repairs. In 1954, one-time emergency exposures of 1 sievert were permissible, with the mandatory and permanent removal of the worker from job-related radiation exposures afterwards. In 1954–55, to

368. Cochran and Norris 1993, pp. 62–63; Commission for Investigation of the Ecological Situation in the Chelyabinsk Region 1991, vol. II, sections 2.1, 2.1.3.

369. Cochran and Norris, 1993, p. 63.

accommodate actual exposures, new regulations specified mandatory work removal for six months after exposure to more than 0.45 sieverts per year or 0.75 sieverts over two years. Thus, the seeming loosening of standards between 1952 and 1954–55 in reality reflected a realism toward attainable radiation protection, replacing an acceptance of normative and unrealized exposure limits. In 1960, the standards of 0.1 centisieverts per week and 5 centisieverts per year were enacted, although the acceptable dose was extended to 12 centisieverts per year for workers older than 30. Finally, in 1970, a uniform standard of 5 centisieverts per year was enacted.[370]

Through Chelyabinsk-65's first decade, reported yearly average exposures to gamma radiation generally exceeded the standard. This was particularly true for the reprocessing plant. Badge monitoring for external radiation had a predicted error rate of 30 percent.[371] The reported reduction in the average dose received by overexposed workers in the early 1960s and again the mid to late 1960s may reflect tightened standards. However, it is not clear why the percentage of workers above the threshold varied from year to year in the context of declining average plant values or why so many workers were above threshold in 1971 (see tables 7.5 and 7.6). Moreover, despite the fact that these data are reported independently in the two Russian sources, at least some of them are in error. Specifically, the figures reported in table 7.6 for 1971 fail on the basis of arithmetic. The data show that 37.9 percent of the exposures were over 5 centisieverts while the reported average is 1.4 centisieverts.[372] This is arithmetically impossible, since the average on this basis alone ignoring all other exposures would be over 1.8 centisieverts. Further, if we average together the lowest dose possible in each of the reported ranges, we get an average for the group of about 2.8 centisieverts, or double the reported average.[373]

While it is possible that the wide year-to-year fluctuation in average yearly exposures to Chelyabinsk-65 personnel may have reflected an eccentric impact from maintenance and emergency problems, it is surprising that exposures to all the plant's workers clustered tightly around the average 1 to 2 centisieverts range after 1965, with so few values reported that exceeded 5 centisieverts, except in 1971, where the average is wrongly reported as low in line with other post-1965 years.

370. Doschenko 1991, p. 21; Nikipelov, Lizlov, and Koshurnikova 1990.

371. Nikipelov, Lizlov, and Koshurnikova 1990.

372. Nikipelov, Lizlov, and Koshurnikova 1990.

373. According to the data, 37.9 percent of the workers got more than 5 centisieverts, 36.5 percent got 2.5 to 5 centisieverts, and the rest got between 0 and 2.5 centisieverts. We assume in calculating the lowest possible average compatible with this data that 37.9 percent got exactly 5 centisieverts, 36.5 percent got exactly 2.5 centisieverts, and the rest got zero.

Table 7.6 Reported Exposures of Chelyabinsk-65 A-Reactor Workers to Radiation

Year	Percent over 5 Centisieverts	Plant Average per Worker, in Centisieverts
1961	3.7	2.0
1962	4.0	2.3
1963	5.9	2.4
1964	16.5	3.0
1965	26.4	4.0
1966	22.1	1.8
1967	13.1	1.4
1968	6.3	1.1
1969	4.3	1.0
1970	13.4	1.4
1971	37.9	1.4
1972	3.5	1.1
1973	9.7	1.0
1974	0.0	1.0
1975	0.0	1.0

Source: Cochran and Norris 1993.

In light of these arithmetic problems and the improbable clustering of reported average doses, the need is evident for further clarification of the methodology of data recording and of data accuracy before making definitive conclusions about dose, and therefore risk, to plant workers. The possibility of fabrication of data should be carefully evaluated.

Worker exposures are not uniform but rather depend on the type of job and affected areas. Accordingly, some Soviet workers suffered much higher exposures than apparent from even the inordinately high average worker values in the Soviet Union, particularly in the 1950s. At Oak Ridge National Laboratory in the United States, for example, 25.6 percent of studied workers had no demonstrable exposures,[374] 0.2 percent received lifetime doses in excess of 0.5 sieverts, and only 27 percent had reported cumulative external exposures of over 1 centisievert. At Chelyabinsk-65's reprocessing plant, over 44 percent of the workforce was exposed to more than 1 sievert in 1951 alone, and the average individual dose was 1.13 sieverts. About 1.8 percent had exposures over 4 sieverts in that year. At the A-reactor, about 32 percent of the workers were exposed to more than 1 sievert in 1949; the average dose was 0.936 sieverts.[375]

374. Wing et al. 1991, p. 267.

375. Nikipelov, Lyzlov, and Koshurnikova 1990.

Table 7.7 Exposures and Mortality among Nuclear Plant Workers

	United States		Chelyabinsk-65	
	Hanford[1]	Oak Ridge[1]	A-Reactor[2]	Pu-Reprocessing Facility[2]
Population size	23,704	6,332	1,795	3,291
Years of exposure	1944–81	1943–77	1948–68	1948–68
Average lifetime exposure (centisieverts)	3.23	2.09	101.4	167.2
Standard mortality ratio (all cancer)	.82 (.77–.88)	.82 (.70–.96)	.79	1.17
Leukemia deaths	701	28[3]	8	31[4]
Standard mortality ratio (leukemia)	.69 (.5–1.0)	1.63 (1.0–2.9)	1.13	2.90

Note: The Hanford and Oak Ridge deaths include all leukemias but no lymphomas. The Russian category of hemolymphoblastosis includes Hodgkin's disease and non-Hodgkin's lymphoma. At Chelyabinsk-65, 80 percent of hemolymphoblastosis cases were leukemias; 20 percent were lymphomas.

The cumulative exposure intervals between the U.S. groups (34 and 37 years) and the Russian (20 years) differ substantially. This is because the material was taken from published sources, using the best available data. The inclusion of more recent dose data would slightly depress even these modest differences. Also, the presentation of a standardized mortality ratio for all cancers is intended as an index of general health; many cancers are not radiation-sensitive or are relatively radiation-insensitive, so a general figure for all cancers is necessarily diluting.

1. From Gilbert et al. 1989.
2. From Koshurnikova et al. 1991. Data for two cohorts each of reactor workers and reprocessing plant workers have been combined.
3. ICD (204-207), leukemia excluding chronic lymphocytic leukemia. (ICD refers to a standard disease-classification system that is used in both the Soviet Union and the United States; see WHO 1980. Accordingly, compatible criteria should have been used in determining cancer deaths.)
4. ICD (200-208), hemolymphoblastosis.

Extensive long-term studies of production personnel began in 1943 for men at Oak Ridge[376] and in 1945 for men and women at the Hanford, Washington, facility.[377] This provides an obvious basis of comparison to Chelyabinsk-65 workers, whose first exposures occurred within five years of the U.S. experience. In fact, the hazardous working conditions prevailing at Chelyabinsk-65 are represented by mean annual exposures that are about 30 to 80 times greater than those experienced by U.S. workers (see table 7.7). Surprisingly, however, recorded cancer deaths and deaths from leukemia were only elevated among workers at the reprocessing facility. Similarly, for the A reactor,

376. Wing et al. 1991; Checkoway et al. 1985.

377. Gilbert, Petersen, and Buchanan 1989; Mancuso, Stewart, and Kneale 1977.

the mean total lifetime exposures among workers hired before 1959 exceeded 1 sievert, more than twice the exposure level at which the leukemia rate rose precipitously after Hiroshima and Nagasaki,[378] but reported mortality did not appear to differ from that of the U.S. workers.

The experience of Japanese atomic bomb survivors differed from the Russian workers in two critical ways. First, the Japanese were surveyed prospectively and followed with great care (after an initial lapse), whereas mortality among the Russian workers was evaluated long after the key exposures. Second, the Russian exposures were chronic, though there were also probably high one-time exposures, for instance during cleanup after the 1957 tank explosion. As noted in chapter 2, one current practice is to reduce cancer risk estimates by a factor of about 2 when exposures are continuous rather than acute. (This is called the dose rate effectiveness factor, or DREF.) The discrepancies between observed and expected mortality in the Soviet plant population exist even when calculated on the basis of a DREF of 2. This is remarkable because worker exposures were not uniform; rather, many high exposures appear to have been due to accidents, emergency repair, and maintenance activities. This means that worker doses were episodic, combining acute and chronic exposures. Thus, it is not evident that a DREF of 2 should be applied to these workers, or whether it would be more appropriate to use a DREF of 1. In any case the result of high episodic exposures would be to increase the expected number of fatal cancers. We have given a range of cancer fatality estimates based on risk factors of 0.04 fatal cancers per sievert and 0.08 fatal cancers per sievert. These correspond approximately to DREFs of 2 and 1 respectively, as discussed in chapter 2.

In comparing U.S. and Soviet data on leukemia deaths, it is important to emphasize differences in reporting and presentation. The totals in Soviet cancer records include chronic lymphocytic leukemia, a non-radiation-related leukemia. However, its inclusion (25 percent of deaths) does not alter the basic observation. Soviet investigators also tended to group lymphomas, multiple myeloma, and leukemia into the single category of hemolymphoblastosis, which may actually cause an appearance of elevated deaths, as multiple myeloma has sometimes been observed to be radiation sensitive.

More to the point, mortality was especially concentrated among workers employed at the reprocessing plant from 1948 to 1953, a period when exposures were especially high. According to one source, mean lifetime external penetrating exposures to this group were officially reported as 2.453 sieverts, compared with an exposure of 0.716 sieverts among workers hired between 1953 and 1958.[379] The stan-

378. National Research Council 1990, p. 242.

379. Baisogolov et al. 1991. Doses reported in Nikipelov, Lizlov, and Koshurnikova 1990 are considerably higher.

dardized mortality ratio for the former and more exposed group was 3.52 (95 percent confidence interval 1.54–8.09), compared with 1.67 (95 percent confidence interval 0.45–6.15) for workers hired later. This means that the more tenured group alone can account for the statistically significant excess of deaths.

Although recorded exposures to Chelyabinsk-65 workers are high compared with the U.S. experience, the reported number of total cancer deaths are far fewer than expected (although elevated in several highly exposed groups). This anomaly is particularly apparent when comparing leukemia risks. Shimizu et al.[380] calculated an absolute risk for leukemia of 2.9 (95 percent confidence interval 2.4–3.5) cases per 10,000 person-years-gray for atomic bomb survivors. This calculation was made based on follow-up over 31 years of a population containing people of all ages; the average period for observation of individuals was 27 years, because some individuals died before this 31-year period of observation ended. It estimates that if a population of 370 individuals, with the same proportion of children and adults as the atomic bomb survivors, received a dose of a gray of radiation per person and were followed for 31 years, some dying prior to the end of follow-up but most surviving, thereby contributing a cumulative total of 10,000 person-years of observation, about 2.9 excess leukemias would result.

At Hanford, cancer deaths appeared to increase with dose, but the database was too small to resolve the issue of doubling dose.[381] The Hanford data have generated substantial controversy over whether there was an effective increase of risk at higher doses, but the effect, if present, is subtle. There was no appreciable excess risk for the full population of plant workers; cancer and leukemia rates did not appear to vary significantly from the general population.[382] This also appeared to be the case at the A-reactor in Chelyabinsk-65. Only at the reprocessing plant and among those beginning work from 1948–53 was there an elevated absolute risk—0.9 cases of acute leukemia per 10,000 person-years-gray.[383] Again, the appearance of a lower general level of mortality in the plant populations probably reflects the overall superior health of a selected working population when compared to the general population, both employed and unemployed (the "healthy-worker effect").

The official data on the Chelyabinsk-65 workers suggest that the risk from chronic exposures, which sometimes exceeded low-dose criteria,

380. Shimizu et al. 1987, pp. 502–524.

381. Gofman 1979, p. 617.

382. Gilbert et al. 1989, p. 19.

383. Koshurnikova et al. 1991.

Table 7.8 Comparison of Leukemia Risk Studies

Characteristic	Techa River Cohort[1]	Plutonium Reprocessing Plant[2]	Atomic Bomb Survivors[3]
Study size	28,000	1,812	42,000
Person-years	422,000	61,649	1,134,000
Mean dose	0.40 gray	2.45 gray	0.24 gray
Type of irradiation	Chronic external, internal	Chronic external, internal	Instantaneous whole-body
Absolute risk (cases $\times 10^{-4}$ per person-year-gray)	0.68[4]	.80	2.94

1. Kossenko 1992; Koshurnikova et al. 1992. Resident data do not include plant workers. The Techa River cohort members were followed by the authors' indications from 1951 to 1981 to check for radiation-related diseases.
2. Radiochemical workers employed from 1949 to 1953; see Koshurnikova et al. 1992.
3. Darby 1986; Shimizu et al. 1987.
4. Absolute risk is median value using six comparison groups.

was considerably lower than the estimates derived from Hiroshima-Nagasaki data. Table 7.8 shows that the chronically exposed working and resident populations at Chelyabinsk-65 have risk profiles for leukemia that are similar to each other. (See also the later discussion on the consequences of the dumping of radioactive wastes in the Techa River.)

Acute Radiation Sickness Among Workers At Chelyabinsk-65, in the setting of heavy chronic exposures to radiation, lethal or near-lethal incidents occurred.[384] In the A reactor, a spontaneous chain reaction killed two people from exposures in excess of 30 sieverts; nonlethal acute radiation sickness hit 43 workers whose exposures were as high as 10 sieverts.[385] The documented survival of several Chelyabinsk-65 workers exposed to more than 4.5 sieverts was likely due to intensive medical attention. (Approximately 4 sieverts is lethal to half the exposed population in about 60 days, in the absence of medical care.)

Chronic Radiation Sickness Among Workers Chelyabinsk-65 workers, exposed to high yearly doses of external radiation and unknown levels of internal radiation from ingested and inhaled particulates, received periodic medical examinations. They were diagnosed with a constellation of symptoms that were called chronic radiation sickness (CRS), a term that has no parallel in the West.[386] In most cases, diag-

384. Doschenko 1991.

385. Guskova and Baisogolov 1985.

386. Soyfer et al. 1992.

noses were made after 1954–1955, often six to eight years after the highest exposure. About 1,500 Chelyabinsk-65 employees were diagnosed with CRS, including 5.8 percent of those who had worked at the A-reactor and 22.5 percent of those who had worked at the reprocessing plant.[387] These cases were heavily concentrated among those who had worked there the longest: 80 percent of the diagnoses were made among workers hired before 1953. No cases were reported among people hired after 1953 to work at the A-reactor or after 1958 to work at the reprocessing plant. Among workers diagnosed with CRS at the former, mean cumulative exposures were 2.64 sieverts, with a single-year maximum of 1.27 sieverts; for the latter, the respective exposures were 3.4 sieverts in a lifetime and 1.5 sieverts in a single year.

Much about the diagnosis of CRS is elusive. It was based on three main features: hematologic suppression, disruption of neurovascular function, so-called "hypotonic neurovascular dystonia," and micro-organic damage to the peripheral nervous system.[388] Only hematologic suppression (a low blood count) was based on a quantitative test, as the other diagnoses were based on the physical examination. Thus, except for a decrease in bone marrow myelocytes and a pancytopenia,[389] a CRS diagnosis seems to rely on a doctor's interpretation. The features of hypotonic neurovascular dystonia, said to affect half the people with CRS, included rapid heart rate and lowered blood pressure. Nerve damage was principally assessed by reflex tests rather than with more exacting studies. It is unclear how "micro-organic" damage would be recognized without biopsies and analyses of individual nerves. Suppressed immune function was assessed indirectly as an enhanced susceptibility to infectious diseases. This is a fragile marker in a region recognized for high levels of bacteriologic disease and low-quality medical care.[390]

In the long term, the health of CRS victims doesn't appear to have been affected significantly. Among 177 deaths occurring in Chelyabinsk-65 workers diagnosed with CRS, malignancies were not reported to have occurred more often than in people without CRS, although there were two cases of aplastic anemia, which could possibly be radiation-related.[391] Among survivors, all traces of abnormality were said to have disappeared within 30 years of diagnosis.[392]

387. Nikipelov, Lyzlov, and Koshurnikova 1990; Soyfer et al. 1992.

388. Soyfer et al. 1992; Kossenko 1992a.

389. Soyfer et al. 1992.

390. Feshbach and Friendly 1992, pp. 181–201.

391. Soyfer et al. 1992.

392. Kossenko 1992b.

The questions surrounding CRS are worth probing carefully because this diagnosis has not been reported elsewhere. There may be reasons for circumspection around the long-term outcome reports in a disease that lacked clear criteria and was so limited to a single historical episode. The possibility that a diagnosis of CRS was used to hide malignancies should be investigated.

Hazards to the Population from Chelyabinsk-65

The people living near Chelyabinsk-65 were the victims of three major releases of radioactivity from the complex. These were the intentional release of radioactive liquids into the Techa River from 1948 to 1951, the accidental release of radioactivity to the atmosphere caused by the 1957 explosion of a tank of high-level waste, and the uncontrolled dispersion of radioactive sediments from the shores of Lake Karachay during the drought years of 1967 and 1972.

The approximately 23 million curies (about 850 million gigabecquerels) released in these three episodes represents a fraction of the perhaps one billion curies (37 million terabecquerels) that Chelyabinsk-65 has introduced into the region's environment.[393] Therefore the three episodes do not cover all the risks to human health from operations. There were also large releases of radioactivity to the air due to routine operations. Airborne releases were unchecked before 1961–1963; plutonium monitoring commenced only in the 1970s.[394] According to Lehman, Chelyabinsk-65 officials have admitted "that 57 rem [0.57 sieverts] have now been received by the public" living around the facility since the 1957 accident.[395] We first consider population exposures from releases to the air from plant operations and then discuss exposures from each of the three incidents mentioned above.

Regional Population Exposures from Airborne Emissions M. M. Kossenko, director of the Clinical Department of the Urals Radiation Medicine Research Center, has attempted to reconstruct exposures from airborne emissions to population centers between 1948 and 1990 (see table 7.9). By her estimate, and assuming a hypothetical worst case of continuous exposure over the maximum of 42 years, the total dose to the more than 1.3 million residents was about 20,000 person-sieverts. This would have resulted in about 130 excess leukemia deaths over 42 years, based on the continuous lifetime exposure estimates

393. Penyagin 1991.

394. Commission for Investigation of the Ecological Situation in the Chelyabinsk Region 1991, vol. II, section 2.1.9.

395. Lehman 1992, p. 9.

Table 7.9 Mean Effective Equivalent Doses to Population Irradiated from Chelyabinsk-65 Airborne Emissions

Locality	Population	Individual Total Personal Dose (in centisieverts for maximum residences, 1948–90)	Total Population Dose (person-sieverts)
Chelyabinsk-65	83,000	10.3	8,550
Kasli	53,800	5.2	2,800
Vishnevogorsk	7,860	2.3	180
Kyshtym	50,200	2.1	1,050
Novogorny	6,800	14.7	1,000
Sarykulmyak	600	8.4	50
Bashakul	800	8.1	60
Khudaberdinsky	680	6.9	50
Argayash	44,900	4.4	1,980
Kunashak	4,800	3.2	150
Chelyabinsk	1,100,000	0.4	4,400
Total	1,353,440		20,270

Source: Kossenko 1992. Population doses rounded to the nearest 10 person-sieverts.

from BEIR V.[396] If the same excess risks had prevailed that were reported for the Chelyabinsk-65 workforce, there would have been a total of about 80 excess leukemia deaths. The total estimate for all fatal cancers (including leukemias) is from about 800 to 1,600 excess fatal cancers among the civilians of the region and from about 300 to 600 among workers.

Even these calculations may understate the hazards from internal doses. Kossenko states that more than 70 percent of long-term exposures since the 1970s have been from plutonium-239.[397] This point seems very conjectural, however, as the implied level of plutonium concentration far exceeds levels reported at other worldwide weapons production sites. However, we should note that the Gorbachev Commission on Chelyabinsk expressed the same opinion of the role of plutonium. It states that "within the past 20 years the structure of the mean annual effective equivalent dose is 70 percent due to Pu-239"

396. See National Research Council 1990, pp. 172–173. The risk factor used here is that for continuous exposure of 100,000 people to 0.01 sieverts per year for 47 years, resulting in 355 excess leukemia deaths. This yields a factor of 75.5 excess leukemias per 10,000 person-sieverts in a population observed for 47 years.

397. Kossenko 1992a.

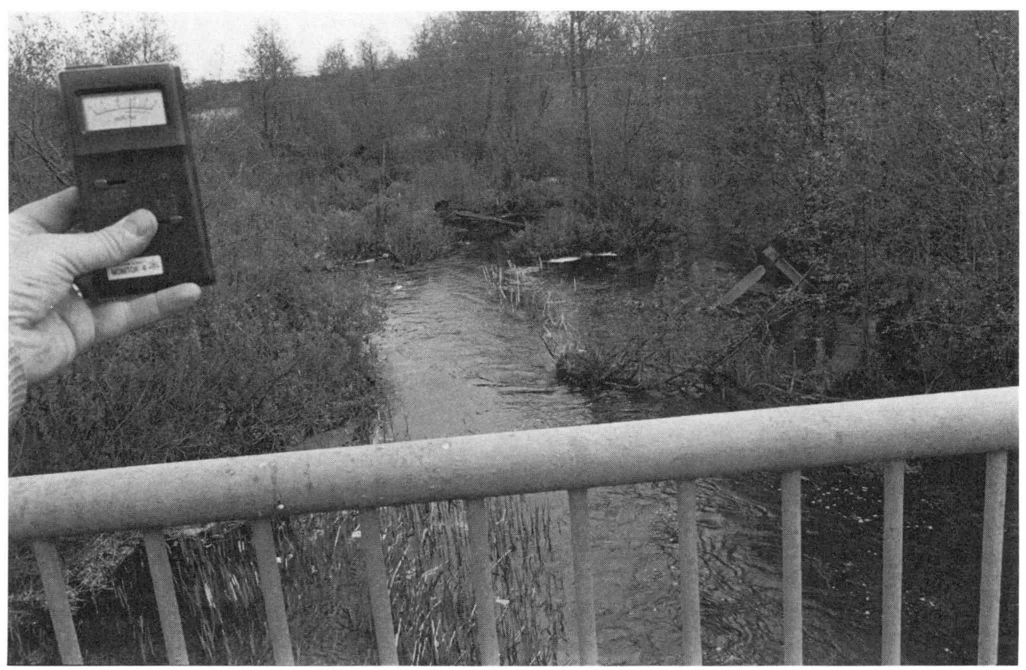

Figure 7.8 Measuring radiation at the Techa River. Photo by Robert del Tredici.

and calls for "a more detailed examination of the issue."[398] This situation needs careful evaluation. Doses from plutonium anything like the suggested levels would be most extraordinary, compared to doses resulting from environmental contamination due to plutonium in other nuclear weapons–producing countries.

Exposures due to Techa River Dumping Prolonged exposures resulted from Chelyabinsk-65's use of the Techa River as the primary disposal site for nuclear wastes. These occurred in the context of both intentional generalized contamination of surface and ground waters and from spills called "wild overflows."[399] Authorities eventually evacuated much, but not all, of the population living along the river in the vicinity of Chelyabinsk-65, but the potential health impacts from food and water contamination continue to this day.[400]

The 1991 Gorbachev Commission report includes an assessment of health effects based on extensive reconstruction of estimated doses.[401]

398. Commission for Investigation of the Ecological Situation in the Chelyabinsk Region 1991, vol. I, p. 45.

399. Soyfer et al. 1992.

400. Hertsgaard 1992.

401. Commission for Investigation of the Ecological Situation in the Chelyabinsk Region 1991.

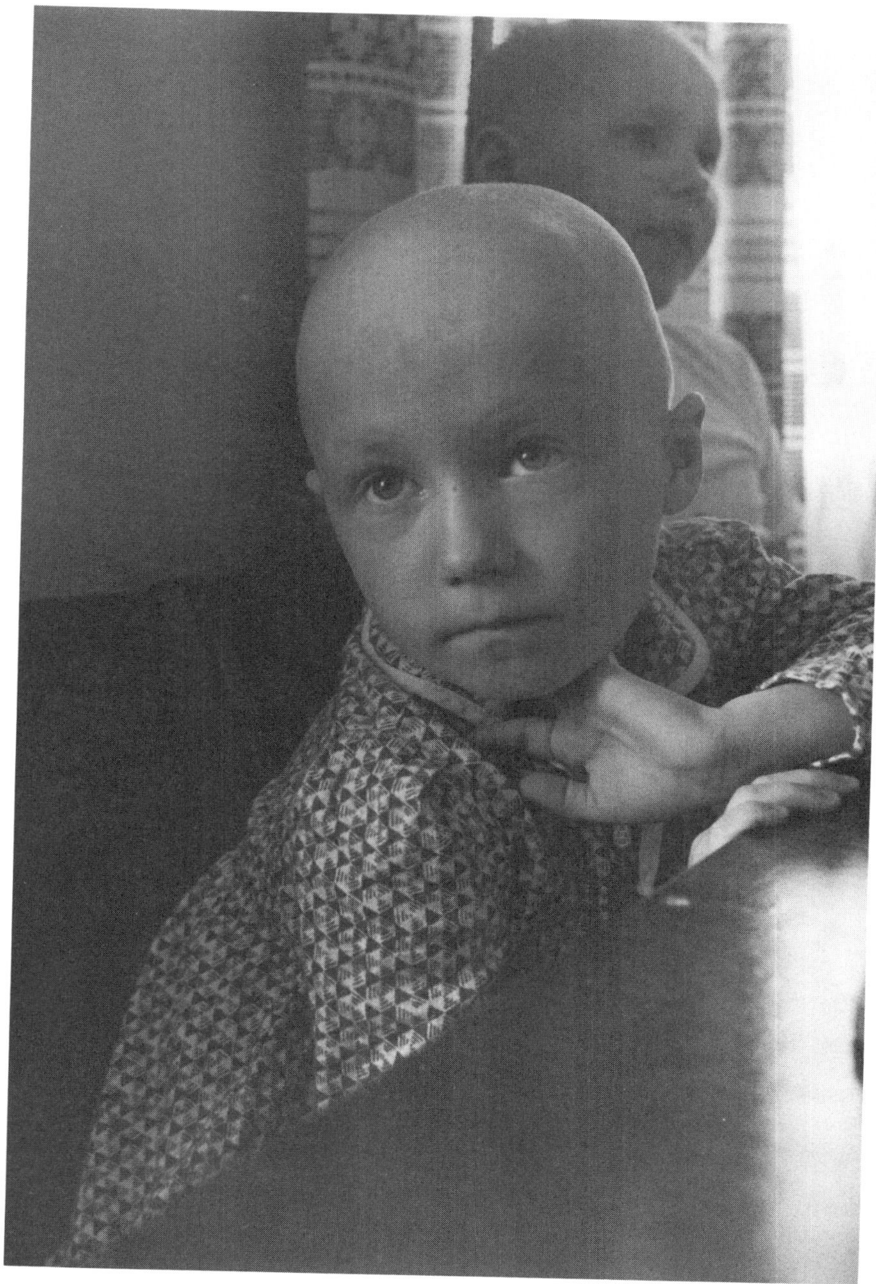

Figure 7.9 A leukemia patient at Chelyabinsk, 1991. Photo by Robert del Tredici.

Figure 7.10 Villagers from Muslyumovo on the Techa River, downstream from the Chelyabinsk-65 complex, 1992. Photo by Robert del Tredici.

The assessment is complicated by the many years between exposure and intervention and because people moved in and out of the region. Even the four-year interval for atomic bomb survivors posed difficulties, so the dose reconstruction task was especially formidable in the case of the South Urals. Moreover, during the early 1950s, radiation measurements were made only yearly throughout most of the Techa River Basin and only two to five times a year in the river's upper stretches.[402] This means that even careful individual and population dose reconstructions will likely have larger uncertainties than those developed for atomic bomb survivors.

There have been some quantitative internal dose measurements. For people living along the Techa River, the main radionuclide of concern is reported to have been strontium-90. Individual doses were estimated from 1960 by measurements of strontium-90 in teeth.[403] Whole body counting of strontium-90 in skeleton and cesium-137 in soft tissue

402. Kossenko, Degteva, and Petrushova 1992.

403. Kossenko, Degteva, and Petrushova 1992.

has occurred since 1974, with more than 12,000 measurements taken in all.[404] Dose reconstructions, including estimates of bone-marrow dose (which are pertinent to leukemia risk), have followed the recommended models of the International Committee on Radiological Protection.[405]

The long delay between exposure and measurement clouds any assessment of cancer risk, especially for leukemia. The registries have substantial omissions, most notably the 7,500 villagers evacuated because of their heavy exposures.[406] This crucial omission means that health effects have probably been underestimated. Russian investigators have had to extrapolate extensively because values are missing from the earliest and most crucial time periods and because surveillance was performed only on surviving people and without trying to include those who migrated or were relocated. Such problems may have biased dose estimates towards the low side.

In all, monitoring covered 28,000 people in 38 villages along the Techa River.[407] Some 7,500 individuals from 20 villages were evacuated.[408] Metlino, the most heavily exposed village, was not fully evacuated until 1956, and Muslyumovo, 78 kilometers downstream from Chelyabinsk-65 and also heavily exposed, was never evacuated.

People in both evacuated and nonevacuated villages were exposed to substantial radiation (see table 7.10). The 14,000-plus inhabitants of the 10 villages listed in the table were about half the total exposed population that was medically followed. Average individual whole-body-equivalent exposures in the most highly exposed villages of Metlino, Techa Brod, and Asanovo totaled about 1 sievert or more. The total population exposure in the 10 villages in table 7.10 was 4,180 person-sieverts. These data indicate an average individual exposure among the 14,362 highly exposed people of these 10 villages of about 0.3 sieverts. However, Kossenko, Degteva, and Petrushova report a larger average dose of 0.4 grays to the whole population of about 28,000.[409] They also report a larger maximum dose of three grays, more than twice that shown in table 7.10. This example illustrates the kind of discrepancies that exist in the data and that we have not been able to explain.

Kossenko and her colleagues report that 37 cases of leukemia were detected in the exposed population, about twice the rate expected in

404. Kossenko, Degteva, and Petrushova 1992.

405. ICRP 1989.

406. Soyfer et al. 1992, p. 14.

407. Kossenko, Degteva, and Petrushova 1992.

408. Bolsunovsky 1992, p. 11.

409. Kossenko, Degteva, and Petrushova 1992. Some exposures are assumed to be fission products; grays and sieverts are equivalent here (quality factor = 1).

Table 7.10 Exposure and Expected Leukemia Deaths Due to Techa River Contamination

Village	Population	Bone Marrow Dose (cSv)	Average Effective Dose Equivalent (cSv)	Total Effective Equivalent Population Dose (person-Sv)
Metlino	1,242	164	140	1,740
Techa Brod	75	127	119	90
Asanovo	892	127	100	890
Nadirovo	184	95	56	100
Muslyumovo	3,230	61	24	780
Brodokalmak	4,102	14	5.8	240
Russkaya Techa	1,472	22	8.2	120
N. Petropavlovskoye	919	28	10	90
Shutikha	1,109	8	5.6	60
Zatecha	1,135	17	6.6	70
Total	14,360			4,180

Sources: Soyfer et al. 1992; UNSCEAR 1988. Population doses have been rounded to the nearest 10 person-sieverts.

two control populations.[410] Thus an excess of about 20 leukemias appears to have been detected. This is about one-fourth the excess leukemias that would be expected in this population based on their dose estimates.

Other Health Effects on the Techa River Population V. N. Soyfer and colleagues have summarized the overall mortality among people living near the Techa River for selected villages and controls to determine whether deaths from cancer and from all causes combined were elevated.[411] However, the data are difficult to assess because the populations are not well characterized.

Concerns over health risks in the region around Chelyabinsk-65 have focused on leukemia, but the extensive surveillance effort directed toward people living near the Techa River actually focused more on other indicators of chronic disease. CRS, diagnosed among plant workers, was also recorded among Techa River residents for about a

410. Kossenko, Degteva, and Petrushova 1992.

411. Soyfer et al. 1992. The data appears to have a conversion error since elevated mortality appears at only the highest effective equivalent doses—140 sieverts and 52 sieverts—but the exposures are listed elsewhere as 1.4 sieverts and 0.52 sieverts. The first two numbers are probably 140 rem and 52 rem respectively, which would make them identical with 1.4 sieverts and 0.52 sieverts respectively.

decade following their most intensive exposures.[412] However, as noted above, the bases for this diagnosis remain elusive, particularly since no cases were recorded following the extensive exposures from the Chelyabinsk-65 1957 tank explosion.[413] Soyfer suggests that many presumed cases of diagnosed CRS may have been misdiagnosed infectious processes.[414] Nevertheless, the relatively high estimated bone-marrow doses to CRS patients demonstrating low blood counts—5.4 centisieverts to adults and 10 centisieverts to children—suggests they may have suffered from radiation-induced diseases.[415]

The time lag between event and recognition or reaction also affected diagnoses of CRS, which were made three to six years after the most intense exposures. As a result, the 935 identified cases of Techa River residents are probably a gross underestimate.[416] The association of CRS with exposures of over 0.5 sieverts would suggest that it afflicted more than half the residents in villages like Metlino.[417] The discussion of an apparent low prevalence of CRS is necessarily problematic, as it involves possible undercounting of a disease whose characterization is vague. Again, any potential use of CRS as a substitute diagnosis for cancers (including leukemia) needs to be investigated.

Table 7.8, discussed briefly above, compares leukemia risk studies of the Techa River cohort, the Chelyabinsk-65 reprocessing plant workers, and Hiroshima-Nagasaki atom bomb survivors. If adjusted for the fact that the Chelyabinsk-65 worker group included no children, the Russian risk profiles are virtually identical. This may mean that the same systematic bias(es) exist(s) in the database. If the source(s) of bias can be identified, it may be possible to use the data to draw less uncertain conclusions.

The comparison shows that Soviet reports of radiation leukemia deaths are internally consistent, but seriously at odds with much of the accepted literature on radiation hazards. It might be posited that the low life expectancy rates for the Soviet population in general and particularly for the Chelyabinsk region may have caused deaths from other causes before leukemia-induced mortality could occur. In non-irradiated populations, leukemia rates rise precipitously after the fifth decade.

Although childhood leukemia is relatively rare, children are particularly radiation-sensitive. After the atomic bomb exposures, these risks

412. Kossenko 1992b.

413. Soyfer et al. 1992.

414. Soyfer et al. 1992.

415. Kossenko 1992b.

416. Soyfer et al. 1992.

417. Bolsunovsky 1992, p. 14.

in young children were elevated by more than twentyfold, with a peak occurring five to seven years after exposure.[418] However, of 37 leukemia cases that were detected among the Chelyabinsk-65 residential population, only three occurred under age 20 and none under age 12. The apparent lack of radiosensitivity in children may be attributable to a systematic undercounting of childhood cases.

Substantial uncertainties in health and dose data make any general conclusions based on this data impossible, other than to say that the health data in particular are suspect. Better analysis can only be predicated on careful, independent analysis of the raw medical and dose data and on an evaluation of the health status of the surviving exposed populations.

The Chelyabinsk-65 1957 Tank Explosion The 1957 tank explosion at Chelyabinsk-65[419] was the most important airborne radiation contamination in the Soviet Union until the Chernobyl accident almost three decades later. According to one estimate, although the region is sparsely populated, the explosion exposed 20,000 people to over 5 centisieverts per year over the one to three years prior to evacuation.[420] Internal doses in the early 1990s remain in the range of 0.1 centisieverts per year due to residual strontium-90, which has a half-life of about 28 years. The short-lived radionuclides have decayed away.[421]

The explosion is often referred to as the radiation accident in the south Urals. Authorities handled it more attentively than the exposures linked to the Techa River releases, at least from the perspective of monitoring potential disease. Still, there was considerable risk from radiation doses due to delays in population evacuation and consumption of contaminated food.

The accident might seem to provide potentially valuable lessons on the long-term effects of radiation exposure. It was a single event with long-standing consequences from the prominence of strontium-90 in fallout.[422] By 1987 contamination concentration had decreased thirtyfold.[423] The unauthorized return of contaminated lands to grazing and dairy farming presented unwarranted health risks in the form of additional doses from contaminated food. Although the effectiveness of diluting measures has been documented and lauded by officials,[424]

418. Beebe, Kato, and Land 1977.

419. Medvedev 1979.

420. Akleyev 1993.

421. Akleyev and Shalaginov 1991.

422. Romanov et al. 1991.

423. Commission for Investigation of the Ecological Situation in the Chelyabinsk Region 1991, vol. II, section 2.4.2.

424. Romanov, Teplyakov, and Shilov 1990a.

two-thirds of contaminated acreage has gone back to general use even though strontium-90 levels average 43.4 nanocuries (about 1606 becquerels) per kilogram in soil. As early as 1961, lands contaminated with 25 curies (925 gigabecquerels) per square kilometer or less were returned to agriculture.[425] These are very high levels of contamination for a substance that mimics biologically essential calcium.

Delayed population monitoring and the complexity of dose reconstruction based on dietary estimates[426] have added to the controversies over the accident's health effects. In addition, several other major reservations pertaining to the data illustrate the problems involved in dose reconstruction, especially when a population is poorly characterized. There appears to be a lack of agreement about how many people were evacuated. Based on the same source data, the initial (within 7 to 10 days) evacuee population is variously presented as 1,150[427] 1,054,[428] and 600.[429] Given these uncertainties, serious reservations must be maintained toward population exposure and dose estimates, which are based on soil contamination and presumed duration of exposure.[430]

Of the approximately 10,700 people relocated, about 9,000 remained in their villages for about one to two years, receiving an added effective equivalent population dose of 600 person-sieverts (see table 7.11). Although medical surveillance reportedly began after the first year of exposure,[431] registration of the most exposed individuals did not formally begin until 1990. Therefore, information on the fate of the most highly exposed individuals is very limited.[432]

Soviet and Russian writers, describing the consequences of the 1957 explosion, point to the success of the surveillance and evacuation effort and the limited evidence of adverse health effects. For example, Buldakov et al. looked at cancer mortality, birth defects, birth rates, and nonmalignant disease among 1,154 highly exposed evacuees and almost 34,000 exposed but nonevacuated residents.[433] They found no adverse health effects. Buldakov has also claimed there was no risk

425. Commission for Investigation of the Ecological Situation in the Chelyabinsk Region 1991, vol. II, section 2.4.2; Romanov, Teplyakov, and Shilov 1990a.

426. Buldakov et al. 1989.

427. Soyfer et al. 1992, table 3.

428. Buldakov et al. 1989, table 11.

429. Nikipelov 1989, p. 3.

430. The uncertainties are exemplified by the controversy over plutonium releases. See Buldakov 1989, and below.

431. Buldakov et al. 1989.

432. Soyfer et al. 1992; Nikipelov et al. 1990a.

433. Buldakov et al., 1989.

Table 7.11 Dynamics of Population Evacuation and Dose to the Population Before Evacuation

Population Group and Size	Average Contamination Density (Ci/sq km of Sr-90)	Time to Evacuation, in Days	Average Dose Received up to Evacuation, in rems: External Exposure	Average Dose Received up to Evacuation, in rems: Effective Dose Equivalent	Population Dose Effective Dose Equivalent, in person-sieverts	Estimated Number of Fatal Cancers
1,150	500	7–10	17	52	600	24–48
280	65	250	14	44	120	5–10
2,000	18	250	3.9	12	240	10–19
4,200	8.9	330	1.9	5.6	240	10–19
3,100	3.3	670	.68	2.3	70	3–6
Total 10,730					1,270	51–102

Note: Values in the last column are calculated based on risk factors of 0.04 and 0.08 fatal cancers per person-sievert. Cancer mortality figures rounded to nearest whole number. Figures may not add to totals due to rounding.

Source: Buldakov et al. 1989.

from plutonium, citing a 100-percent-efficient plutonium extraction process at Chelyabinsk-65 and practically unmeasurable soil concentrations of 0.01 becquerels per 1 to 10 kilograms of soil, that is, 0.01 to 0.001 becquerels per kilogram of soil.[434]

Exception has been taken to both of these points: 100 percent efficiency for plutonium extraction would be technically exceedingly difficult and contradicts what is known about the Soviet, and even the U.S., plutonium extraction industry. Moreover, the measurement of such minute concentrations of radioactivity in soil would be not only difficult but highly implausible because it is far less than expected from nuclear test fallout alone![435] This claim is also flatly contradicted by the Gorbachev Commission finding that plutonium-239 contamination of 1 curie (37 gigabecquerels) per square kilometer exists in the most contaminated region around the accident site. This implies a plutonium concentration on the order of 1,000 becquerels per kilogram of soil, which is 100,000 to 1 million times larger than the range claimed by Buldakov.

There is also a curious absence of reports of CRS in the population exposed to the consequences of the 1957 explosion, even though CRS is presumably an exposure-related syndrome.[436] Similarly, the 1989 *Pravda* article that first disclosed the circumstances of the 1957 tank

434. Buldakov 1989.

435. IPPNW and IEER 1992, p. 86.

436. Soyfer et al. 1992.

Table 7.12 Radiation Doses and Leukemias from the 1957 Chelyabinsk-65 Tank Explosion

Group[1]	Bone Marrow Dose[2] (centigray)	Person-Years of Observation	Leukemia Cases	Leukemia Rate per 100,000 Person-Years (90% confidence interval)
I	20–25	22,627	3	13.3 (3.6–34.3)
II	3–12	112,900	9	8.0 (4.2–13.9)
III	0.3–3.0	7,940	23	6.4 (4.4–9.1)
IV	—	526,700	28	5.3 (3.8–7.3)

1. Group I consists of 1,054 evacuees relocated within 10 days from areas with an average density of contamination of over 400 curies per square kilometer of strontium-90. Group II consists of residents with a strontium-90 contamination of 1 to 4 curies per square kilometer of strontium-90. Group III consists of residents with a strontium-90 contamination of 0.1 to curie per square kilometer. Group IV are unexposed controls.
2. Kossenko, Degteva, and Petrushova 1991b, Table 6.

explosion to the Soviet public denied adverse health effects from radiation.[437] Instead, it attributed ill effects to disease-bearing ticks. Such connection of adverse health effects with other environmental factors has also included attributing excess cancer deaths to air pollution from metallurgical factories.[438] This claim has been challenged on the basis of a too-exact linear relationship between estimates of sulfur dioxide emissions and cancer fatalities.[439]

Table 7.11 shows dose data for various segments of the evacuated population. The total population dose to these 10,730 people was about 1,270 person-sieverts, or about 0.12 sieverts per individual on the average. Expected excess cancer deaths in this population would range from roughly 50 to 100, based on the risk coefficients discussed in chapter 2.

Table 7.12 summarizes available information on leukemia morbidity and mortality. Buldakov et al.'s finding of low level of health effects corresponds to the findings of Kossenko et al. Buldakov et al. (1989) give a substantially different estimate of bone marrow doses for the population evacuated in the first 10 days or so; we have found no explanation for the difference. The very different estimates of bone marrow dose among people evacuated immediately suggest significant problems with the methods of dose reconstruction. While the very different approaches to measurement of bone marrow dose may effectively prohibit comparisons between studies, both concur on the

437. Gubarev 1989.

438. Buldakov et al. 1989.

439. IPPNW and IEER 1992, p. 91.

absence of an observed leukemia risk to groups II and III. The apparent excesses in Group I are based on three cases, probably too few to infer much about risk. The absence of increased risk in Groups II and III again raises the question of possible under-reporting of leukemia diagnosis among people exposed to radiation from Chelyabinsk-65 operations.

Erosion-Related Releases from Lake Karachay Compared with the Techa River contamination and the 1957 tank explosion, the release of 600 curies from the shores of Lake Karachay in a wind storm during the drought of 1967 was less dangerous. Still, it exposed an estimated 41,500 people,[440] and, as discussed above, Lake Karachay was a particularly concentrated source for cesium-137, which was selectively dumped there. About 4,800 people received average external soil-related exposures estimated at 1.3 centisieverts, and people living further away had an average external dose of 0.7 centisieverts.[441]

No registration or medical evaluation was conducted on either affected group. High concentrations of cesium-137 and strontium-90 inevitably meant more risk from ingesting contaminated milk and other foods. About 40,000 people lived within the contaminated zone defined by levels of cesium-137 greater than 0.3 curies (11.1 gigabecquerels) per square kilometer and of strontium-90 greater than 0.1 curies (3.7 gigabecquerels) per square kilometer.[442]

Other Health-Related Concerns in the South Urals

Cleanup Crews According to one estimate, more than 20,000 soldiers, police, and emergency workers took part in an extensive cleanup following the 1957 tank explosion.[443] The 18 million curies released into the immediate environment of the tank likely caused large exposures to cleanup crews, but no data on exposures are public. Further, it will be difficult to determine the risks faced by these workers, who presumably were not subsequently followed unless they worked at Chelyabinsk-65 or lived near it.

Still, some speculative projections suggest potential tragedy. Given the 5.4 percent contribution of strontium-90 (plus yttrium-90) to the release, about one million curies of these two radionuclides alone would have been discharged locally. There would also have been very high levels of gamma radiation due to the short-lived radionuclides.

440. Nazarov 1992.

441. Cochran and Norris 1993, p. 71.

442. Commission for Investigation of the Ecological Situation in the Chelyabinsk Region 1991, vol. I, p. 12.

443. Soyfer et al. 1992.

Although anecdotal, one account suggests that 95 percent of the members of cleanup team No. 360 have already died, with high-level radiation a major cause of death.[444] However, no mention was made of the size of the team or the number of deaths. The exposures to the cleanup crews and resultant injuries remain unknown.

Fetal Injury and Population Genetic Effects The concern over genetic injury to people exposed at Chelyabinsk-65 is complicated by controversies over testing procedures and the fact that the exposures occurred so long ago. Cytogenetic testing for chromosomal damage has been conducted on people in the Techa River region in the last five years, but the results from bone marrow and peripheral-lymphocyte analyses have not been significant.[445] While several estimates based on extensive animal modeling[446] in mice and in monkeys have suggested a doubling of all genetic mutations at doses of about 1 gray, people appear to be more resistant than animals.[447] Fetal birth defects were reviewed in retrospect for intervals of both 2 years and 25 years after the incident. No significant effect was observed; the six recorded fetal deaths during 1958–1959 are too few to support any conclusions.[448]

Birth outcomes have also been analyzed retrospectively for 4,891 exposed women and 42,000 controls among Techa River residents.[449] The results are difficult to interpret because of differing birth control patterns between Russians and the many ethnic Tartars and Bashkirs living in the South Urals, the absence of information on spontaneous abortions, and an undeveloped surveillance system for congenital defects. For example, the study found that gonadal radiation doses of 3 to 126 centisieverts had no effect on birth outcomes.[450]

The total abortion rate was reported to be twice as high for the exposed population as the controls. Although no distinction was made by the authors of this report between miscarriages and voluntary abortions, it would appear unlikely that fear of radiation-induced birth defects would have played a role in increasing voluntary abortions since no one informed the women about their exposures. The suggestion thus remains that exposure may have played a role in increasing the rate of miscarriages for this population.

444. Sheleketov 1992.

445. Akleyev and Shalaginov 1991, pp. 46–51.

446. National Research Council 1990, pp. 68–71.

447. Russell 1977.

448. Krestinina, Kossenko, and Kostiuchenk 1991.

449. Kossenko 1991a.

450. Kossenko 1991a.

Chelyabinsk-65: Conclusions

Two features run through the severe contamination incidents due to Chelyabinsk-65. First, programs of medical surveillance were initiated, albeit well after the event, but systematic secrecy left exposed civilians ignorant of the motivations behind medical testing and evacuation. While several prominent Russian investigators have criticized journalistic sensationalism around these events[451] or documented a low level of resulting diseases,[452] they are contradicted by the realities that the cumulative population exposures to the Chelyabinsk-65 workforces vastly exceed those for any other recognized nuclear weapons plant population so far, and that the average individual doses for large groups of workers and some groups of residents exceed those recorded for either atomic bomb survivors (on average) or even many medically irradiated populations.

While a variety of serious inconsistencies and deficiencies in presented data have been cited, an even more fundamental question pertains to the enormous discrepancy between the South Urals reports on leukemia mortality and nearly 50 years of international observations on atomic bomb survivors and patients treated with therapeutic radiation. It is true that the nature of exposure experienced by atomic bomb survivors differs from that experienced by Western nuclear plant workers in that the former was single-dose and the latter involved comparatively low dose rates over long periods. However, ankylosing spondylitis (a disease of the spinal column) patients were irradiated on a regular intermittent basis,[453] which may be more comparable, and they still had a much higher excess risk of leukemia per dose. Among the Mayak A reactor worker cohort, no excess risk of leukemia was noted at lifetime exposures of 1 sievert. This is not credible. We believe the medical data in the Chelyabinsk-65 region need a thorough independent reevaluation based on original, raw data and documents before they can be used in any inferences regarding cancer risks.

The Effects of Krasnoyarsk-26 on Human Health

Cooling water from the Krasnoyarsk-26 nuclear reactors has been discharged into the Yenisey River for over 30 years. Even though commercial fishing, public drinking, and other agricultural uses are not practiced, activities such as grazing and recreational fishing continue on a personal level.[454]

451. Soyfer et al. 1992.

452. Nikipelov 1989.

453. Darby 1986.

454. Korogodin et al. 1990.

The irregular pattern of sediment distribution has created substantial hot spots. For example, contamination levels on Atamanovo are as high as 41 curies per square kilometer.[455] The flesh from Yenisey River fish is contaminated with phosphorus-32, sodium-24, zinc-65, cobalt-58, and cobalt-60.[456] Phosphorus-32 has been detected in fish as far as 300 to 500 kilometers from the waste sources. As many as 10 percent of the residents of Bolshoibalug and Atamanovo, two villages inside the Krasnoyarsk-26 sanitary zone, eat as much as 20 kilograms of Yenisey fish each year. Their yearly estimated bone marrow radiation dose is 3 to 6 millisieverts, compared with 1 to 1.5 millisieverts in the surrounding population.[457] This would be enough to raise the lifetime risk of death from leukemia between 10 and 25 percent.[458]

Radiation-Related Health Effects at Tomsk-7

The interministerial commission that studied the health and ecological impacts of Tomsk-7 in 1990 concluded that no significant adverse health effects were detected over 30 years of observation.[459] The commission also reported that internal radionuclide levels in residents of Somymus, a village inside the plant's sanitary zone, were not elevated.[460] The birth rates, death rates, and disease incidence of the workers and their families living at the site apparently differed little compared with the district population. However, the same problems with health statistics as noted above regarding Chelyabinsk-65 apply here.

Table 7.13 presents vital statistics for 1985 through 1989. Reported mortality at Tomsk-7 is slightly higher than mortality rates for the former Soviet Union—13.1 and 13.9 deaths per 1,000 in 1985 and 1989, respectively, compared with 10.6 and 10.0 per 1,000 for the nation, although the method of age adjustment is not stated.[461] The infant mortality rate at Tomsk-7, even when neonatal deaths are included, is clearly less than national norms, but regional variations in infant mortality are very large in Russia.[462] Although types of cancer were not specified, reported cancer rates at Tomsk-7 did not differ from the surrounding region.[463] Combined internal and external radiation doses

455. *Zelenyy Mir* 1992.

456. Korogodin et al. 1990.

457. Korogodin et al. 1990.

458. Based on National Research Council 1990 coefficients for leukemia risk.

459. Aleksrashin et al. 1990.

460. Aleksrashin et al. 1990.

461. Feshbach and Friendly 1992, p. 273; Alekrashin 1990.

462. Goskomstat 1990.

463. Aleksrashin et al. 1990.

Table 7.13 Vital Statistics for the Tomsk Region, 1985–1989

Health Outcome	Tomsk-7	Tomsk District
Birth rate (1,000)	15.1	16.1
Incidence of stillbirths (percent births)	0.6	1.3
Neonatal mortality (per 1,000 live births)	5.0	10.3
Infant mortality (per 1,000 live births)	10.1	17.3
Ischemic heart disease (rate per 1,000)	17.6	17.7
Myocardial infarction (rate per 1,000)	2.2	0.6
Mortality from disorders of blood and circulatory system (per 1,000)	3.0	4.8
Mortality from disorders of respiratory system (per 1,000)	0.11	0.5

Source: Adapted from Aleksrashin et al. 1990.

to occupants of the Tomsk-7 observation zone were reportedly stable between 1985 and 1990, with an average of 0.37 millisieverts per year.[464]

From the data we have, we cannot develop an overall perspective on the impact of Tomsk-7 operations on human health. Radioactive wastes from the complex have been buried in open areas accessible to grazing animals, and there are anecdotal reports of hospitalizations due to people eating animals contaminated with cesium-137.[465] Moreover, as in the Chelyabinsk region, restrictions on usage of the sanitary zone were often either evaded or unknown, "as established sizes and borders of zones have been set by outdated documents and do not comply with contemporary ... infrastructure and demography of the region."[466] These infractions include cattle grazing and watering along the Tom River; noncommercial fishing, boating, and bathing in the river; and harvesting of river-irrigated crops. Accordingly, humans take in phosphorus-32 by consumption of local fish and fresh milk, but the estimated dose for 1987–1989 was less than 7.5 microsieverts per year.[467] Cesium-137 and strontium-90 have been measured in the sanitary zone at less than 10^{-12} curies (0.037 becquerels) per liter. Estimated yearly intake of the two radionuclides for 1987–89 was 0.67 to 29.7 nanocuries (about 25 to 1,100 becquerels) per person for strontium-90 and 3.2 to 59.4 nanocuries (about 118 to 2,198 becquerels) per person for cesium-137. However, these fall within European standards.

464. Aleksrashin et al. 1990.

465. Dahlburg 1992, p. A3; Reicher and Suokko 1993.

466. Aleksrashin et al. 1990.

467. Aleksrashin et al. 1990.

8 The United Kingdom

David Sumner, Rebecca Johnson, and William Peden

British scientists played a significant role in the development of nuclear physics during the 1930s. In 1940, two Jewish emigré scientists from Germany, Rudolph Peierls and Otto Frisch, performed calculations showing that uranium-235 could be assembled into a critical mass small enough to make an atom bomb, and they wrote a memorandum on the subject.[1] The British government then set up a committee, code-named the MAUD Commitee, to oversee research on nuclear energy. Initial collaboration with the United States on developing nuclear weapons dates from that year. British scientists were ahead of their U.S. counterparts in crucial aspects of nuclear weapons work through June 1941, when they shared a report entitled *Report by MAUD Committee on the Use of Uranium for a Bomb.* The project was later code-named Tube Alloys.

By 1943, the U.S. Manhattan Project had surpassed British progress on developing an atomic bomb, and many Manhattan Project officials were no longer interested in joint research. Churchill negotiated hard to get British scientists a larger role in the Manhattan Project, leading to the 1943 Quebec Agreement. This agreement set up "arrangements ... to ensure full and effective collaboration between the two countries" on nuclear weapons, while at the same time Britain agreed that it "disclaims any interest" in gaining advantages in commercial applications of nuclear energy beyond those deemed fair by the U.S. President.[2] The next year, the Hyde Park Agreement seemed to ensure full collaboration after the war, but upon the death of Roosevelt, Churchill discovered this was not to be.

British scientists did participate in the Manhattan Project, and were among the observers who witnessed the awesome destruction of Hiroshima and Nagasaki by Little Boy and Fat Man. Yet in the immediate aftermath of World War II the United States sought to exclude its wartime ally, along with all other countries, from nuclear weapons

1. Rhodes 1986, pp. 322–325.

2. Quebec Agreement, as quoted by Sherwin 1987, p. 86.

technology. In August 1946, the U.S. Congress enacted the McMahon Act, outlawing the passing of classified atomic information to any country—including Britain.

However, after 1945, nuclear weapons appeared set to become the currency for international power. British politicians from all sides of the political spectrum did not wish to go "naked into the conference chamber."[3] And without nuclear weapons, it was thought, Britain would be ignored. Britain thus proceeded to embark on building a full nuclear weapons infrastructure, an effort that continues to the present.

Britain undertook a "flexible and open-ended" research program, notes Lorna Arnold, "'covering all uses of atomic energy'" and aimed at producing "fissile material in sufficient quantity to enable Britain to develop" atomic energy "'as circumstances may require.'"[4] Although to some people involved in the effort this did not necessarily mean bombs, according to John Simpson, a "definite decision had ... been taken by January 1946 to produce fissile material in the United Kingdom for use in a weapon development programme."[5]

Responsibility for designing, constructing, and operating Britain's nuclear reactors and the associated fuel production plants went to a new industrial organization that was part of the Ministry of Supply. The first task of this organization was to build a plant that would produce uranium metal from ore and make reactor fuel. The decision to build this plant (at Springfields, in Lancashire) was announced in March 1946.

Less than a year later, a top-secret committee of five ministers, handpicked by Prime Minister Clement Attlee, made the formal decision to build a British bomb. This decision was kept from Parliament until 1948. During the war, Winston Churchill had been a strong advocate of developing an atomic bomb, and he had kept British involvement secret. Upon his reelection as Prime Minister in 1951, he is said to have been astonished that the Attlee government had managed to spend almost £100 million on the nuclear program without informing Parliament. Churchill then proceeded to maintain the same veil of secrecy.[6]

From 1946 to 1954, the Ministry of Supply was responsible for nuclear activities. The Ministry's Division of Atomic Energy oversaw the construction of the Atomic Energy Research Establishment at Harwell, reactors at Calder Hall and Chapelcross, uranium factories at Springfields and Capenhurst, and reactors and a reprocessing plant at Windscale (renamed Sellafield in 1984). On 1 April 1950, the Ministry of

3. "You will send a British Foreign Secretary, whoever he may be, naked into the conference chamber." Aneurin Bevan, 3 October 1957 (speech at Labour Party Conference in Brighton).

4. Arnold 1987, p. 4.

5. Simpson 1986, p. 43.

6. Milliken 1986, p. 38.

Works took over a site near the village of Aldermaston, in Berkshire, for making atomic bombs. A nearby factory at Burghfield was also taken over and converted to nuclear weapons assembly. In 1958, a royal ordnance factory at Llanishen near Cardiff in Wales was converted to produce uranium shells and beryllium tampers for nuclear warheads. The nuclear weapons infrastructure was almost in place.

By April 1952, the first plutonium from Windscale had arrived at what was now called the Atomic Weapons Research Establishment (AWRE) at Aldermaston, ready to be turned into warheads in the new plutonium-processing facility. On 3 October 1952, Britain exploded its first atom bomb, at Monte Bello, Australia, and became the third nuclear weapons state.[7]

The British atomic explosion of 1952 was one of the factors that led to the reestablishment of Anglo-American nuclear relations in the 1950s. Other factors were the 1952 election of Dwight Eisenhower (who was sympathetic to the British desire for collaboration) as U.S. President, his administration's desire to equip NATO with tactical nuclear weapons, and the Soviet thermonuclear explosion in 1953. However, according to Lorna Arnold, "the real breakthrough did not come until 1958. Before then the Americans were not particularly impressed by what Britain had to offer as a partner.... [But] by 1958, the success of the British independent nuclear weapons programme had been demonstrated."[8] The foundation of renewed U.S.-British cooperation was laid in the same period.

Prime Minister MacMillan and President Eisenhower met in March and October of 1957 to discuss possibilities for increasing U.S.-U.K. cooperation on nuclear weapons. As a result, the Agreement for Cooperation on the Uses of Atomic Energy for Mutual Defense Purposes was passed by the U.S. Congress and Senate and signed by Eisenhower on 2 July 1958, going into effect on 4 August. The Agreement, with amendments, is still in force and is very broad. It not only covers production, design, and testing of nuclear weapons, but also training of personnel and plans for deployment and potential use of nuclear weapons.[9]

The British Atomic Energy Act of 1954 created the United Kingdom Atomic Energy Authority (UKAEA). The UKAEA was composed of three departments: research, weapons research, and production. The weapons research division operated somewhat independently as a result of a policy decision in 1953 to do this work at Aldermaston.[10] In 1971, British Nuclear Fuels Limited (BNFL)—a company with the

7. Greenpeace 1993, p. 5.

8. Arnold 1987, p. 7.

9. Norris, Burrows, and Fieldhouse 1994, p. 46.

10. Simpson 1986, p. 97.

government as sole shareholder—was formed to take over the Windscale reprocessing plant, the Capenhurst enrichment plants, and the Springfields nuclear fuel plant and to operate them on a commercial basis.

For the United Kingdom, the tie between civilian nuclear power and nuclear weapons came early on. The British nuclear establishment enthusiastically embraced the Atoms for Peace program, launched by President Eisenhower at the United Nations in 1953. For U.S. policymakers, Atoms for Peace meant capturing global markets for U.S. nuclear technology and controlling the global supply of nuclear fuels to prevent other countries from joining the nuclear club. For British leaders, Atoms for Peace meant gaining domestic support for a very expensive project that the public, still enduring food rationing, might not welcome. The ostensible promise of cheap, safe energy diverted attention from emerging questions about whether Britain needed a nuclear arsenal.

Britain's determination to produce its own bomb emerged in the context of the Cold War and the unstable international relations of the period following World War II. For instance, both the Polaris submarine program upgrade with Chevaline missiles and the acquisition of Trident missiles were justified using deterrence of the Soviet Union as the rationale. Although sections of the armed forces have been ambivalent about tactical nuclear weapons, there seems to have been a consensus of support for a British independent deterrent, the preferred public relations term for Britain's bomb. The concept of deterrence is at the heart of political arguments justifying Britain's nuclear weapons program: the United Kingdom must have nuclear weapons to ensure that they are never used. Yet deterrence has many contradictory elements, especially now that the perceived main threat—the Soviet Union—has been transformed. U.K. policy that the nuclear deterrent is needed only as a last resort does not exclude the use, or threat of use, of nuclear weapons, even in a nonnuclear conflict.

British production of nuclear weapons grew up to some extent along lines dictated by military planners. At a 1952 conference, the British Chiefs of Staff made a recommendation to the government that more reactors be built so that Britain could double its plutonium production. They argued that Britain should possess more nuclear weapons so it could attack strategic targets that the U.S. saw no direct national interest in attacking, and also to enable the use of smaller yield weapons in a tactical role during any ground engagement in Western Europe or at sea.[11] Four new nuclear reactors were constructed at Calder Hall (adjacent to Windscale) and another four at Chapelcross in Scotland. These reactors were designed to produce both plutonium and power.

11. Simpson 1986, p. 69.

In 1953, the Blue Danube, a crude 20-kiloton bomb deployed on bombers, came into service, to be replaced in 1958 by the lighter and more versatile Red Beard, yielding 5 to 20 kilotons.[12] By this time, Britain had begun experimenting with thermonuclear weapons, leading to the Yellow Sun series during the 1950s. In the early 1960s, Britain deployed the Yellow Sun Mark II version, with a yield "'in the megaton range.'"[13]

After Yellow Sun, the next stage of British nuclear weapons development was the Blue Steel air-to-surface missile, which probably used "a large thermonuclear device of approximately 1 megaton," apparently based on an American design. This weapon was stockpiled from 1962 through 1970.[14] In 1966, the Royal Air Force and the Royal Navy began replacing these nuclear weapons with the WE-177 free-fall and strike-depth bombs, 100 of which are still in service.[15]

Polaris, a submarine-launched ballistic missile, was stockpiled from 1968–1984. In 1982, under the Chevaline program, the Polaris warhead was upgraded. Trident II MIRVed missiles are currently under production for deployment on Vanguard submarines by 1995. It is expected that the British nuclear weapons stockpile will grow to almost 300 (including Trident II and WE 177 warheads) by the year 2000, from the 1994 level of 200. It peaked at 350 between 1975 and 1981. In addition, the U.S. supplied warheads for British forces for many years, starting at around 150 in 1958, rising to close to 400 during the mid-1970s, and dropping off during the 1980s to about 160, until they were phased out in 1992.[16]

QUALITY AND QUANTITY OF DATA

In the United States the Freedom of Information Act and numerous other protected freedoms (discussed in chapter 6) played a key role in revealing information about unsafe conditions, environmental contamination, and coverups in the U.S. nuclear weapons complex. Public pressure, following a series of disclosures about the environmental contamination and poor safety of nuclear weapons facilities, has forced the closure of essentially all U.S. facilities except laboratories and design centers, so far as new weapons production is concerned.

Britain's nuclear bomb production plants are similar in design and operation to U.S. facilities, but the different legal climate has kept the

12. Norris, Burrows, and Fieldhouse 1994, pp. 54–55, 64–65.

13. Norris, Burrows, and Fieldhouse 1994, pp. 55, 57, 64–65.

14. Norris, Burrows, and Fieldhouse 1994, p. 59.

15. Norris, Burrows, and Fieldhouse 1994, pp. 59, 64–65.

16. Norris, Burrows, and Fieldhouse 1994, p. 63.

British public in the dark about their problems. Even matters published in the newspapers in other countries can be and have been banned in Britain. The British bureaucracy's culture of secrecy, enshrined in the Official Secrets Act, is nowhere more stifling than in military matters. The Official Secrets Act is a kind of Anti–Freedom of Information Act. It makes whistleblowing virtually impossible, even for uncovering gross violations of safety regulations. Under the Official Secrets Act, not only is a government employee liable to prosecution for information he or she divulges, but any recipients, including journalists, are liable for publishing it. Another draconian provision makes it the duty of a citizen to reveal the source of information on issues covered by the Act. According to author Tony Bunyan, the body of law embodied in the Official Secrets Act gives the British Government a "formidable and all-embracing net" to prosecute civil servants, journalists, and ordinary citizens.[17]

Nevertheless, questions asked by persistent Members of Parliament, official enquiries into increases in cancers, safety problems, and other issues, and an increasingly vigilant public are gradually bringing some of the truth to light. This chapter presents some new and unpublished information, while bringing together details of incidents previously reported only in local press or internal plant and government publications. Relying on press accounts carries additional penalties of uncertainty. As noted, some data have been revealed through Parliamentary questions, but even that source has proven patchy and limited. Many Parliamentary questions relating to defense establishments are met with the stock response that "it would not be in the national interest to reveal" the information.[18] For years, data on discharges remained concealed on the grounds that it would give an enemy too much insight into the weapons. It was not until 1980 that data were released annually on discharges from Aldermaston and Burghfield. Parliamentary questions about the price, or price overruns, of particular weapons, facilities, or developments may be met with several different examples of stonewalling, such as "It is not our policy to disclose information of this nature" or "The information cannot be provided without incurring disproportionate cost."

Moreover, from the first years of Britain's nuclear weapons program until 1952, spending on nuclear weapons was concealed in other budgets. Even after the first nuclear weapons test in October 1952,

17. Bunyan 1977, p. 11.

18. Incidences are numerous; this exchange is representative: Mr. Llew Smith: To ask the Secretary of State for Defence under what heading in figure 12 in his statement on the Defence Estimates 1993, Cm 2270, his Department's protected expenditure on nuclear testing is included. Mr. Aitken: It would not be in the national interest to reveal the costs of the Defence nuclear program. For this reason such costs are not separately identified in the statement on the defence estimates. (*House of Commons Hansard*, 19 July 1993, p. 80.)

"creative accounting" persisted. In particular, research and development costs were hidden between the defense allocation and the growing nuclear energy industry, enabling the government to underrepresent the real costs of both efforts.

The Defence Select Committee, comprising members of all parties, has little power. Unlike comparable U.S. congressional committees, it cannot require civil servants to appear before it or to provide specific information. Thus, the Ministry of Defence reveals information to Members of Parliament only on the assumption that they will follow the "gentlemen and players rule"—in other words, that they will keep silent.[19] Consequently, MPs advocating a public right to know are kept in the dark.

Because of the considerable difficulties in obtaining information on Britain's nuclear weapons plants, this book can only scratch the surface of the problem. For the same reason, we have relied on journalistic reports to a larger extent than some other parts of this book. We recognize that a reliance on news accounts for descriptions of leaks and accidents prevents an underlying picture of the nature of the problems from emerging clearly. The accidents and routine releases of radioactivity and other toxic materials identified here are only a fraction of the story. The full extent of the contamination will not be known until the cloak of secrecy is removed from the nuclear weapons industry. Only the British government can remedy this problem by publishing all relevant historical data and documents.

NUCLEAR WEAPONS–RELATED ACTIVITIES AND FACILITIES

Eight plants have held key roles in the production of Britain's bomb (see table 8.1 and figure 8.1): BNFL Springfields, BNFL Capenhurst, BNFL Sellafield, BNFL Chapelcross, AWE Cardiff, AWE Aldermaston, AWE Burghfield, and AEA Technology Harwell. Of these, all but the three Atomic Weapons Establishments (AWE Aldermaston, AWE Burghfield, and AWE Cardiff) now have primarily civilian functions.[20] Still, in conception, production, and practice, the civilian nuclear program developed out of and is interdependent with the military program. The drive to acquire nuclear weapons for Britain probably greatly lessened the public scrutiny that would otherwise have accompanied the large financial commitment that the government made for nuclear power development.

The production process for Britain's nuclear weapons begins abroad with the mining and milling of uranium. Uranium (imported as yellow-

19. Coker 1988, pp. 17–18.

20. Other facilities, not discussed in this book, are involved in more peripheral processes and supply.

Table 8.1 Location of Nuclear Weapons–Related Activities and Facilities in the United Kingdom

Official Facility Name[1]	Location	Operating Status[2]	Comments	Source[3]
R&D Weapons Laboratories (only the largest are listed)				
Harwell site (Atomic Energy Research Establishment)	About 16 km south of Oxford	OC	Responsibility for operations transferred to AWE Aldermaston	a
AWE Aldermaston Warhead Research and Design	Near Reading in Berkshire	OP	Established 1950	a
Uranium Mining and Milling				
The UK imported yellowcake from abroad for both military and civilian uses (see chapter 5).				
Conversion of Yellowcake (U_3O_8) to Uranium Tetrafluoride (UF_4) and Uranium Hexafluoride (UF_6) (in preparation for uranium enrichment)				
Springfields	Near Salwick in Lancashire	OP	Operated by BNFL	b
Uranium Enrichment[4]				
UKAEA Capenhurst	Near Little Sutton, Cheshire	NO 1963	Gaseous diffusion	b
BNFL Capenhurst	Near Little Sutton, Cheshire	NO 1991	Gaseous diffusion	b
Uranium Processing and Fuel Fabrication				
AWE Cardiff	About 5 km north of Cardiff, Wales	OP	Fabricated tampers	a
Production of Tritium				
Chapelcross			Opened 1980	a

Plutonium Production from:				
Military Production Reactors				
Windscale/Sellafield		NO 1957	2 air-cooled, graphite-moderated reactors	a
Sellafield/Calder Hall		OP	4 gas-cooled, graphite-moderated reactors: Magnox	a
Chapelcross (Solway Firth)	Annan, east of Dumfries, Scotland	OP	4 Magnox reactors	a
Civilian Power Reactors				
Hunterton A	Scotland	NO 1990	Magnox: ceased power production 3/90	
Wylfa		OP	Magnox	
Trawfynydd	Snowdonia National Park, north Wales	NO 1993	Magnox	f
Berkeley		NO 1989	Magnox: ceased power production 3/89	
Oldbury		OP	Magnox	
Hinkley Point A		OP	Magnox	
Dungeness A		OP	Magnox	f
Bradwell		OP	Magnox	
Sizewell A		OP	Magnox	
Reprocessing Plants				
Dounreay		OP	Operated by AEA Technology	b, e

Table 8.1 (continued)

Official Facility Name[1]	Location	Operating Status[2]	Comments	Source[3]
Sellafield (Magnox)		OP	Operated by BNFL: Uranium metal feed	b, e
Sellafield (THORP)		OP	Operated by BNFL: oxide feed	b, e
Fabrication of Plutonium Metal Components				
Aldermaston (Buildings A45/A1)		OP	Fabricated pits	a
Aldermaston		OP	Fabricated components	a
Other Warhead Components				
Beryllium Processing				
AWE Cardiff		OP		a
Electronic Components				
AWE Burghfield	About 8 km southwest of Reading	OP		a
High-Explosives Testing, R&D				
Foulness Island	Thames Estuary, near Southend-on-sea	OP		a
Assembly and Disassembly of Nuclear Weapons				
AWE Burghfield		OP		

Testing of Nuclear Devices[5]			
Australia (12 AT)		1952–58	a, c, d
Pacific (9 AT)		1952–58	a, c, d
United States (24 UT)	Nevada Test Site	1962–92	a, c, d

1. Abbreviations:
AWE = Atomic Weapons Establishment
BNFL = British Nuclear Fuels, Limited
THORP = Thermal Oxide Reprocessing Plant
UKAEA = United Kingdom Atomic Energy Authority

2. Status abbreviations:
NO = not operating and either will not resume or unlikely to resume, with closure year indicated if known
OC = operating to fulfill primarily civilian functions
OP = operating at some capacity

3. Sources:
a = Norris, Burrows, and Fieldhouse 1994
b = Nuclear Engineering International 1991
c = Norris and Arkin 1994
d = IPPNW and IEER 1991
e = Albright, Berkhout, and Walker 1993
f = May 1989

4. Since 1963, all highly enriched uranium has been puchased or bartered from the United States.
5. AT = atmospheric or surface test; UT = underground or underwater test.

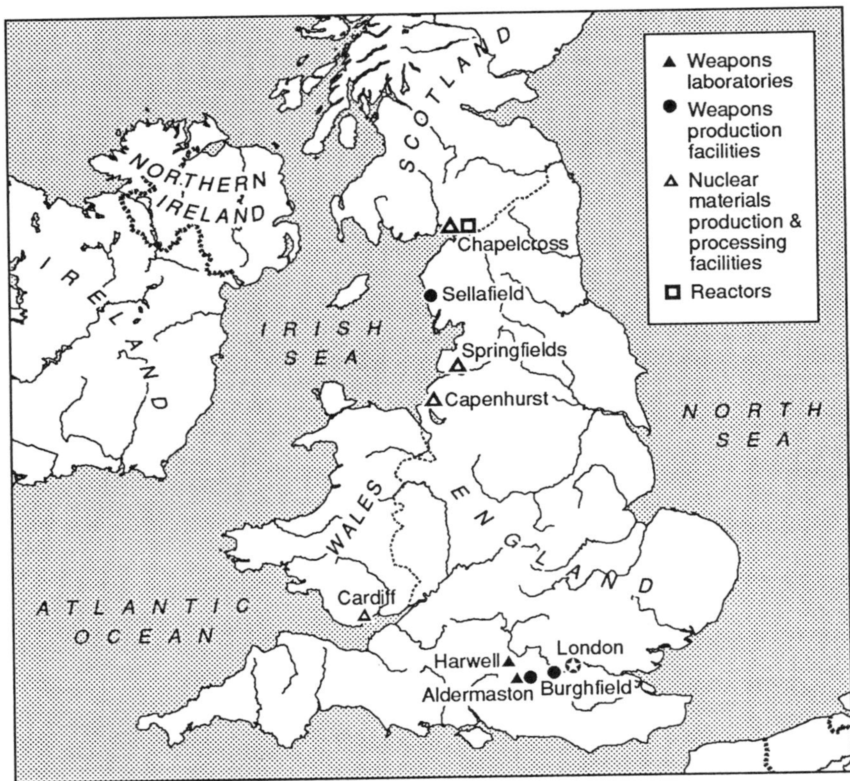

Figure 8.1 Nuclear weapons production sites in the United Kingdom.

cake) from some countries, such as Canada and Australia, can only be used for peaceful purposes. Regulations in those countries prevent Britain from using their uranium in weapons or as reactor fuel in submarines. Britain could not have legally manufactured new warheads and submarine reactors from safeguarded uranium. However, uranium from these sources does provide the raw material for the separated plutonium in Britain's civilian program; it is therefore potentially available for weapons purposes.

Britain's main source of military supply has been from sources without such peaceful end-use restrictions, notably South Africa or occupied Namibia. Britain has imported uranium from those sources, in particular from the Rössing mine in Namibia (formerly Southwest Africa), controlled by U.K.-based multinational Rio Tinto Zinc (RTZ). Rössing played a key economic role in South Africa's illegal occupation of Namibia, which ended with Namibian independence in 1990. British imports from occupied Namibia breached international law, including UN Decree No. 1 for the Protection of the Natural Resources of Namibia, and they defied UN Security Council resolutions and the International Court of Justice (see chapter 5). Without indigenous sources, Britain secured uranium for military use in part by breaking its international commitments—and keeping it secret.

The government only admitted these illegal imports in 1987 after detailed investigations by the Namibia Support Committee. Then-Energy Secretary Tony Benn finally revealed that the Labor Cabinet had secretly agreed to a 1,100-metric-ton contract from the Rössing mine in 1976. By 1987, Namibian and South African uranium accounted for around two-thirds of the initial processing work carried out at BNFL Springfields.[21]

Within Britain, the starting point for making nuclear weapons is BNFL Springfields, a 3,000-acre site next to the village of Salwick, between Preston and Blackpool in Lancashire. Springfields processes uranium ore concentrates into uranium metal fuel for Magnox reactors or converts them into uranium hexafluoride, which is either exported or sent to BNFL Capenhurst for enrichment. In the latter case, Capenhurst returns enriched uranium hexafluoride to Springfields, which fabricates it into fuel for advanced gas-cooled or light-water reactors.

BNFL Capenhurst, situated near the village of Little Sutton between Liverpool and Chester, enriches the uranium hexafluoride from Springfields. Begun in 1949, it was, writes Margaret Gowing, "designed above all against the clock, primarily to serve military purposes and secondarily to keep capital costs down."[22]

The UKAEA shares the Capenhurst site with BNFL; it pays BNFL for use of the site services (£1.5 million in 1988, the date of the last available data). As of 1988, the UKAEA employed about 600 people on laboratory research.

A gaseous diffusion plant to produce highly enriched uranium started operating at Capenhurst in 1953. The capacity of the plant has been reported as between 0.4 and 0.6 million separative work units per year.[23] Capenhurst ceased producing uranium-235 for military purposes in 1963. Evidently, the Ministry of Defence considered that its stockpile of highly enriched uranium, combined with imports from the United States, would ensure enough material for the weapons program.[24] However, the plant continued to operate until 1982, producing fuel for civilian nuclear power stations.

Since 1962, Britain has obtained all its enriched uranium from the United States, but this source has not always been deemed reliable. In January 1980, then-Secretary of State for Defence Francis Pym announced that the Ministry of Defence would build a new enrichment

21. Dropkin and Clark 1992, p. 8.

22. Gowing 1974, vol. 2, p. 440.

23. Bendict, Pigford, and Levi 1981, p. 816.

24. Albright, Berkhout, and Walker 1993, pp. 63–64.

plant at Capenhurst "to provide for the Royal Navy's long-term needs for fuel to be consumed in its nuclear propelled submarines."[25] According to Albright, Berkhout, and Walker, Britain "contracted for up to 100,000 SWU [separative work units] per year from 1981 until 1986. . . . We have no information about the amount of separative work actually used, or the purpose of this relatively large contract. But one objective was to obtain weapon-grade uranium for submarine reactors."[26]

The plant began operating in 1984 but may not have been used to produce highly enriched uranium. Nevertheless, the plant functioned as a part of the military system in that it produced uranium enriched to an intermediate level that was then further processed into highly enriched uranium in the United States. According to a June 1982 statement by John Nott, then a new Secretary of State for Defence:

> As a result of a careful review, it has been decided to modify the plan to obtain part of the high enriched uranium requirement for defense purposes by production of HEU at Capenhurst and part by procurement of HEU from the United States. The new proposal provides for the total requirement to be met by the enrichment of natural uranium to intermediate level at Capenhurst, the United States being restricted to final enrichment to the level required for defense purposes. As under the original plans, and to the same time scale, a new dedicated plant will be required at Capenhurst. It will be entirely separate from the facilities used for production of enriched uranium for civil purposes.[27]

In 1992, the government announced plans to transfer commercial uranium enrichment to URENCO (Capenhurst) Ltd. This company will be owned by International Nuclear Fuels Ltd., a subsidiary set up by BNFL to develop overseas business. In September 1992, the Capenhurst plant was split in two, with 600 workers transferring to URENCO and the remaining 550 continuing to dismantle and decommission the diffusion plant.

Plutonium for Britain's nuclear weapons is produced at the Sellafield (formerly called Windscale) reprocessing plant in West Cumbria. Run by BNFL, most of Sellafield's reprocessing serves the civilian nuclear program. There is also a reprocessing plant at Dounreay for recovering plutonium and uranium from fast breeder reactor fuels. Such reprocessing has separated a total of 3.08 metric tons of plutonium, mainly from mixed uranium-plutonium oxide (MOX) fuel.[28] We have not included this in net plutonium production figures, on the as-

25. *Hansard*, 15 January 1980, cols. 712–713.

26. Albright, Berkhout, and Walker 1993, pp. 64–65.

27. *House of Commons Hansard*, 23 June 1989, cols. 128–129.

28. Albright, Berkhout, and Walker 1993, pp. 95–96.

sumption that the amount of plutonium extracted is about equal to that in the fuel loaded in the breeder reactors.

BNFL also runs Chapelcross, just north of Annan, Dumfriesshire. The facility comprises four 50-megawatt reactors, with huge stacks towering over the surrounding farms. Originally designed for military production, Chapelcross produced weapons-grade plutonium for British nuclear weapons from 1957 to 1964. Modifications between 1961 and 1967 enabled the plant to take on civilian programs, primarily the production of electrical power, though some weapons-grade plutonium was produced after 1964.[29] In addition, Chapelcross began supplying tritium for nuclear weapon triggers in 1980.[30] This was because, under U.S. President Jimmy Carter's arms control and nonproliferation policies, the United States could no longer guarantee supplies of tritium.

In Llanishen, a northern suburb of Cardiff (population 405,900), a small factory with many chimneys sits next to a Safeway Supermarket and the suburban railway station of Birchgrove. Known locally as the Royal Ordnance Factory, Llanishen, its role in making nuclear weapons was formally acknowledged in 1987. At that time, along with ROF Burghfield, it was kept out of the government's privatization of Royal Ordnance factories. Llanishen took on the title Atomic Weapons Establishment, although there is some confusion over what the facility is now called, AWE Cardiff or AWE Llanishen; even official notices use both titles. Under either name, the plant makes beryllium and depleted uranium components for nuclear weapons. We use AWE Cardiff in this book to refer to this facility.

For 40 years, the Atomic Weapons Establishment (AWE—formerly the Atomic Weapons Research Establishment) at Aldermaston has been the central institution of the U.K. nuclear weapons infrastructure. Despite global nuclear arms reductions, Aldermaston still works at full stretch, producing components for Trident missiles. Aldermaston manufactures the plutonium fissile cores for nuclear warheads along with all the highly enriched uranium components.

Located southwest of Reading, Berkshire, AWE Burghfield (formerly known as the Royal Ordnance Factory) is the Ministry of Defence's factory for the final assembly and disassembly of Britain's nuclear bombs.[31] AWE Burghfield possibly also makes some nonfissile components of nuclear weapons, such as lithium and high-explosive components.

Formerly known as the Atomic Energy Research Establishment, the Harwell site, 12 miles south of Oxford, was at the forefront of early research into developing atomic energy for both military and civilian

29. Albright, Berkhout, and Walker 1993, p. 41.

30. Simpson 1986, p. 201.

31. Norris, Burrows, and Fieldhouse 1994, pp. 67–68, 73.

purposes.[32] Responsibility for research and development was gradually transferred to AWE Aldermaston throughout the 1950s.

DESCRIPTION OF KEY SITES

BNFL Springfields

The Springfields fuel fabrication plant, completed in 1948, was built to provide uranium metal fuel for the two Windscale reactors (known simply as the "Windscale piles"), plutonium piles, and later for the Calder Hall and Chapelcross reactors. The fabrication plant was closed in 1960 and replaced by another facility that continued to make the fuel assemblies for Calder Hall and Chapelcross and for a new generation of civilian Magnox reactors. In 1968, Springfields's plant for fabricating fuel for civilian advanced gas-cooled reactors also came on line.

At Springfields, yellowcake is dissolved in nitric acid to form soluble uranyl nitrate. This is further processed to produce uranium metal. Some solid waste is dissolved in water to form the primary waste stream, which is then discharged—untreated—by pipeline across farmland to the River Ribble, three kilometers south. A secondary waste stream arises from processing unfissioned uranium—called Magnox-depleted uranium (MDU)—that is recovered from spent fuel at Sellafield. Since the spent fuel has been irradiated in a nuclear reactor, MDU is contaminated with fission products. BNFL Springfields discharges some MDU into the River Ribble. The plant also produces uranium hexafluoride for feed into uranium enrichment plants.

For most local people, "the Salwick plant" (the common name for BNFL Springfields) is a chemical factory. Few people know that it is the domestic starting point for the entire British nuclear industry and the manufacture of Britain's nuclear weapons, or that thousands of truck shipments carry radioactive uranium oxide, uranium hexafluoride, and nuclear fuel rods to and from Salwick every year.

Discharges When Springfields opened, the plant flushed waste ore constituents and decay products of thorium and uranium down a 1.25-kilometer pipeline "directly from the refinery to the [River] Ribble, at a point a few hundred yards [meters] West of the mouth of Savick Brook." At that time, the Ribble was a heavily used, well-dredged, deep waterway. Because there has been no dredging for the last 12 years, the river has silted up considerably.[33] Since 1978, Springfields has discharged more beta-emitting radionuclides, excluding tritium, than

32. Norris, Burrows, and Fieldhouse 1994, pp. 22–23.

33. Brown 1993.

any other U.K. nuclear site except Sellafield. Since 1988, its beta discharges, excluding tritium, have been higher than those from Sellafield. Furthermore, Springfields's annual liquid beta discharges have consistently exceeded, by a significant margin, the combined annual alpha and beta discharges from all of the nuclear power stations owned by Nuclear Electric and Scottish Nuclear (the two U.K. nuclear power utilities) put together.[34]

In August 1991, Her Majesty's Inspectorate of Pollution and the Ministry of Agriculture, Fisheries, and Food (MAFF) changed the discharge authorization for Springfields. The new authorization reduced the maximum amount of radioactivity the plant can discharge in any one year. Despite this, BNFL has predicted that the actual amount of radioactivity discharged will increase over the next four years. BNFL has stated that "introduction of effluent reduction processes ... would result in costs which would be grossly disproportionate to the benefits and potentially financially crippling to Fuel Division. In the light of environmental impact, they are not justifiable."[35] In effect, BNFL argues that not treating Springfields's discharges to remove radioactivity is the "best practicable environmental option," given the cost of treatment, the benefit obtained, and the low risk of discharges. BNFL says the discharges contain thorium, uranium, and some radioactive daughters of uranium, all of which they say occur naturally.[36]

Despite such assurances, several incidents on local farms have been attributed to Springfields. In the mid-1970s, ewes in fields through which the discharge pipe runs suddenly began aborting their lambs and dying. BNFL denied any toxic or radioactive leak had occurred but eventually agreed to compensate the farmers.[37]

In August 1982, a mysterious illness on two farms adjacent to Springfields killed up to a dozen cattle and struck down many more. Farmers also claimed that pasture land had been scorched, apple trees had burned, paint on buildings facing the factory had blistered and flaked, and fish had vanished from ponds. Although Springfields is the only industrial plant for miles, BNFL denied responsibility. However, a MAFF investigator noted that "All the evidence does tend to point to the chimneys at Springfields because these farms are downwind."[38] In

34. Government Statistical Service 1991.

35. BNFL 1991.

36. *Guardian*, 6 March 1992.

37. Robert Chamley, Pear Tree Farm (personal interview by Rebecca Johnson, 26 February 1993).

38. *Lancashire Evening Post*, 12 October 1982, "A-Plant in Poison Alert"; *Lancashire Evening Post*, 13 October 1982, "'Acid rain' storm brews in Lancs"; *Lancashire Evening Post*, 4 September 1982, "Atom health scare—more information needed."

that case, chemical rather than radioactivity discharges from the plant are far more likely to be the cause.

In March 1984, an oil leak from a fractured valve on a storage tank "went unnoticed for days," contaminating a nearby farm.[39] In April 1985, several kilograms of uranium washed into a farm stream after being spilled from a uranium hexafluoride cylinder in the factory and entering the storm drain. BNFL dredged the stream bed and removed silt along a 200-meter stretch but did not inform the farmer why such action was taken. Subsequently, more dredging of Deepdale Brook was required "as a prudent measure."[40] There is something essentially incomplete or incorrect about this account, since uranium hexafluoride reacts vigorously with water to produce uranyl fluoride, which is soluble and would therefore not precipitate out into the silt. This is an example of the problems with environmental reporting in the absence of detailed official documentation and openness.

Accidents There have been reports of a number of accidents involving dangerous situations within the plant as well as releases of radioactivity to the environment. In the former category, there have been fires and explosions during the processing of uranium,[41] as well as spills of radioactive material.[42] As for releases off-site other than routine releases, there have been reports on a variety of these, including shipments of thousands of radioactive metal boxes containing radioactive dust to a scrap yard in Liverpool, and an accidental discharge of uranyl nitrate into the public sewer system in 1988.[43]

39. *Lancashire Evening Post*, 10 March 1984, "A-plant probe after fuel leak scare." BNFL said the leak should have been contained by a safety tank but somehow managed to seep through into local drains and onto Michael Tomlinson's Salwick Hall Farm.

40. *Lancashire Evening Post*, 13 May 1985, "Firm Acts on U leak: spillage into farm stream admitted by BNFL." See also *Lancashire Evening Post*, 25 May 1985, "BNFL leak: Don't panic say bosses."

41. *Blackpool Evening Gazette*, September 1987, "Fires bring probes at atom plant"; *Lancashire Evening Post*, 17 February 1987, "A-plant Uranium leak prompts inquiry"; *Lancashire Evening Post*, 18 February 1987, "All clear for A-plant after blast." Although it is not certain, the different description suggests this is not one of the two fires mentioned above; *Lancashire Evening Post*, 25 May 1988, "Nuclear disaster drill after accident."

42. *Lancashire Evening Post*, 12 June 1986, "Nuclear leak revealed"; *The Preston Citizen*, 18 December 1986, "Nuclear Safety Review: new check on transports at Springfields"; *Lancashire Evening Post*, 25 June 1991, "A-plant leak not picked up by alarm." According to BNFL, the leak was too small to be reportable to the Nuclear Installations Institute (NII).

43. *Lancashire Evening Post*, 17 January 1989, "Major leak at Lancs A-Plant"; *Lancashire Evening Post*, 18 January 1989, "MP to quiz minister on atom leak"; *Lancashire Evening Post*, 20 January 1989, "Radiation leak scare: call for urgent talks"; *Lancashire Evening Post*, 25 January 1989, "Nuclear leak was 'double' estimate," in which BNFL admitted that 78 kg of uranium had been lost, nearly twice their first estimate of 45 kg.

Figure 8.2 The Windscale/Sellafield complex. Photo © James Lerager.

Windscale/Sellafield

In 1939, a Royal Ordnance Factory to produce munitions was established on the coast of West Cumbria at a site known as Sellafield. In 1947, the Ministry of Supply acquired the facility, which it renamed Windscale, for constructing a reprocessing plant (B204), started up in 1951, and two graphite-moderated air-cooled nuclear reactors (the Windscale piles), which were started up in 1951 and shut in 1957. The B204 plant was shut down in 1964.[44] The tall chimneys of these reactors still dominate the skyline.

In 1953, Winston Churchill sanctioned the building of two more reactors to produce both plutonium for weapons and some power. These were carbon dioxide–cooled, graphite-moderated reactors, built at Calder Hall, adjacent to the Windscale site. These reactors went critical in 1956, and their fuel was reprocessed at Windscale the next year. However, at that time, the capacity of these two reactors and the original Windscale piles was deemed insufficient to meet Britain's projected plutonium demands.

The UKAEA, as the new controller of Windscale, started building two more reactors at Calder Hall and four at Chapelcross in Dumfriesshire, Scotland. The second pair of Calder Hall reactors was

44. Albright, Berkhout, and Walker 1993, pp. 41, 91.

completed in 1958. The first Chapelcross reactor achieved criticality later that year, and all four were operating by the end of 1959. Fuel from all these reactors went to Windscale for reprocessing. The military output of plutonium grew from about 50 kilograms a year to about 300 to 400 kilograms a year in the 1959–1964 period.[45] The Chapelcross and Calder Hall military nuclear reactors, operated by BNFL, at present produce electricity for civilian uses.

The Calder Hall and Chapelcross reactors were the first of the Magnox type: they use natural uranium as fuel, graphite as moderator, and carbon dioxide gas as coolant. The reactors take their name from the magnesium alloy in the fuel cladding. The choice of this design—dictated mainly by the requirement to produce both power and plutonium—has had far-reaching environmental and health consequences.

In 1964, the first Windscale reprocessing plant (B204) was replaced with a larger facility, B205, for reprocessing Magnox fuel not only from the military reactors but from the civilian nuclear reactors that were springing up across Great Britain. The rated capacity of the B204 plant had been 300 metric tons of fuel per year, which could be stretched to a maximum of 750 metric tons in the case of low burn-up fuel. The design capacity of the B205 plant is 1,500 metric tons of fuel per year.[46] Since the opening of the new plant, Sellafield has been primarily identified with commercial reprocessing, which BNFL presents as a method for managing nuclear waste and, increasingly, as a way to earn foreign currency.

In the mid-1960s, the UKAEA began to modify and refurbish the by-then disused B204 as a treatment plant in which the spent uranium oxide fuel from Britain's advanced gas-cooled reactors (more efficient carbon dioxide–cooled graphite-moderated reactors) and foreign light-water reactors would be dissolved in solvent in preparation for reprocessing in B205. This "Head End Plant" started up in 1969 and operated until mid-1972, when B205 was shut down for one year for repairs. According to May, "Unknown to the operators of the Head End Plant, previous batches of dissolved fuel had been depositing granules of insoluble radioactive fission products in the process vessel," which had heated up the floor of the vessel. Upon restart of the Head End Plant on 26 September 1973 the solvent that was poured into the process vessel "reacted violently on the hot surface and produced a steam explosion that sent a burst of radioactive gas out into the air of the plant."[47] In all, 34 workers were found to have skin and lung contamination, mostly from ruthenium-106. The Head End Plant was permanently closed in 1973.

45. Simpson 1986, p. 254.

46. Albright, Berkhout, and Walker 1993, pp. 90–91.

47. May 1989, p. 188.

In March 1977, the U.K. Environment Secretary announced the setting up of a public inquiry into BNFL's plans to build a new Thermal Oxide Reprocessing Plant (THORP) to reprocess spent nuclear fuel in oxide form, mainly from light-water reactors. The inquiry, which lasted for 100 days, found in favor of the proposed reprocessing expansion. Strong objections have been voiced against THORP on grounds of lack of economic justification, increased risks from routine discharges or accidents, as well as proliferation of nuclear weapons. The verdict of the inquiry was largely based on the financial advantages the plant will bring from processing Japanese fuels; indeed, Japan guaranteed a high proportion of the original cost of building the plant.[48]

THORP's construction took longer than anticipated and was concluded only in 1993. In the meantime, the financial advantages have become somewhat illusory, and concern about radioactive discharges has also increased. New objections, raised on these grounds, delayed the commissioning of the plant. It was commissioned in March 1994.

Total plutonium separation from the B205 plant in Britain through the end of 1990 is estimated at about 47.0 metric tons. About 4.2 metric tons of this was for foreign clients (2.6 for Italy and 1.6 for Japan), while the rest was domestic.[49] In addition, the B204 plant separated about 3.2 metric tons of plutonium from the Calderhall and Chapelcross reactors during the 1957–1964 period and 0.4 metric tons from the Windscale piles during 1951–1957.[50] This gives a cumulative plutonium production from 1951–1990 of 50.6 metric tons.

Discharges to the Environment Sellafield has discharged radioactive waste to the Irish Sea and to the atmosphere. Management of the radioactive waste arising from the weapons program was cloaked in secrecy, as is illustrated by Margaret Gowing's account of the suppression of information about the Windscale discharge pipe during its construction in 1950.[51] Moreover, Prime Minister Clement Attlee reportedly felt at the time that too much publicity about the pipe was undesirable because it would bring attention to the project. Minister of Supply Duncan Sandys concurred in this view, arguing that it "'would be better for us to keep [this] light under a bushel.'"[52]

48. U.K. Department of Environment 1993.

49. Albright, Berkhout, and Walker 1993, pp. 109, 111. This figure includes about 360 kilograms of plutonium from light-water reactors separated in B205 for which the modified B204 served as a "head-end" during 1969–1972.

50. Albright, Berkhout, and Walker 1993, pp. 41, 43.

51. Gowing 1974, vol. 2, pp. 106–108.

52. Berkhout 1991, p. 140.

Sandys also said that "'the only way to ensure absolute safety was to abandon the entire atomic energy project. If any sudden release of radiation should occur the sea was close at hand.'"[53] Indeed, from the beginning, it was assumed that a "dilute and disperse" approach to waste disposal would be adequate. In 1955, at a UN Conference on Peaceful Uses of Nuclear Energy, the international scientific community was first informed of an experiment in Cumbria that had taken the form of discharges of radioactive effluent by pipeline into the Irish Sea. These experimental discharges were intended to elucidate the fate in the environment of a cocktail of radioisotopes with widely differing chemical and biological properties.

During the 1960s and 1970s, discharges from Windscale rose steeply. Alpha discharges were principally from military plutonium-handling activities. From the mid-1960s, the discharges rose so much that in 1968 the UKAEA, the operator of the site at that time, sought an increase in authorized alpha discharges from the limit of 66.6 terabecquerels (1,800 curies) per year. The authorizing departments agreed in 1970 to increase the limit to 222 terabecquerels (6,000 curies) per year. However, at a 1977 Windscale Public Inquiry, it emerged that the Ministry of Agriculture, Fisheries, and Food (MAFF) had not started measuring transuranic elements—plutonium, for example—covered by the alpha-discharge limit until 1973.[54]

The contamination of the sea was partly due to increasing problems with Magnox reactor fuel processing and storage. It had been the practice at Sellafield to discharge the contaminated water from the pools holding Magnox fuel directly into the sea. Such a practice would cause contamination with any reactor or fuel design, but was specially damaging in the case of Magnox fuel because the fuel rods tend to corrode when stored in water. Contamination of the pools due to fission products leaking into them therefore contributed to the pollution of the sea when the pool water was discharged into it.[55]

These problems worsened in the early 1970s. Magnox reactors were discharging higher burn-up fuel (5,000 megawatt-days per metric ton instead of 3,000 megawatt-days per metric ton). Higher burn-up rates were causing operational problems in the reprocessing plant, leading to longer storage times for the fuel. This, of course, aggravated storage and corrosion problems. According to Berkhout, by 1973 these corrosion problems were causing a buildup of strontium-90 and cesium-137 in the pools used for storing Magnox fuel rods. Discharge of this pool water into the sea caused higher levels of pollution. Further, this buildup as well as other problems related to corrosion and processing

53. Berkhout 1991, p. 144; Gowing 1974, p. 110.

54. Isle of Man Local Government Board 1977, para. 12.15.

55. Berkhout 1991, pp. 148, 149.

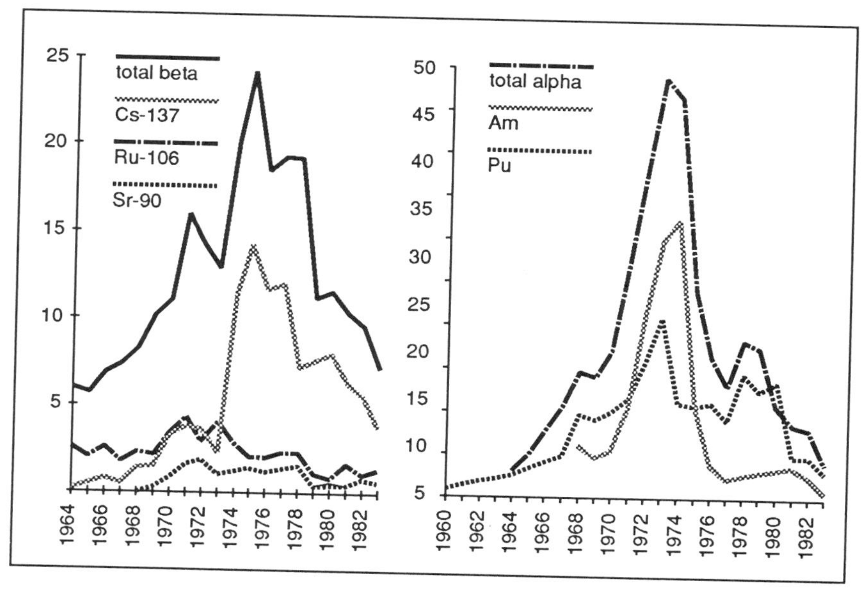

Figure 8.3 Annual radiation discharge from Windscale. Chart on left shows selected beta/gamma emitters, measured in 10^4 Ci; chart on right shows alpha emitters in 10^2 Ci. "Am" is americium-241, not including decay from Pu-241. "Pu" is plutonium-238, -240, and -241. Source: Jones and Southwood 1987.

of Magnox fuels caused higher radiation doses for operating and maintenance personnel, and could have resulted in the shutdown of Magnox fuel reprocessing if they had not been resolved.

Reprocessing was stopped in 1972 for repairs to the main plant, B205, and then again in 1973 because of the accident in the Head End Plant. It was resumed in 1974. But the corrosion and discharges of radioactivity continued.[56] Figure 8.3 shows the increasing level of discharge from 1960 onward. It is a curious fact that the authorized radioactivity discharge limits were raised as the discharges went up. From a report of discussions between MAFF and BNFL that was revealed at the 1977 inquiry, it emerged that MAFF had originally wanted discharges to sea of cesium-137 restricted to 370 terabecquerels (10,000 curies) per quarter, but that BNFL had held out for a limit of 555 terabecquerels (15,000 curies). Because of the additional costs that would be involved for BNFL, MAFF agreed to the higher figure. However, even this limit was unofficial and not translated into a revised authorization. At that time, BNFL was discharging nearly 1,480 terabecquerels (40,000 curies) per quarter.[57]

In his report on the Windscale Enquiry in 1977, the Hon. Mr. Justice Parker commented that:

56. Berkhout 1991, pp. 145, 148.

57. Isle of Man Local Government Board 1977, para. 12.16.

The process by which limits are fixed also requires improvement. At present, so far as aqueous discharges are concerned, limits are fixed by a process of negotiation in which it appeared to me that the public interest, although protected, was given insufficient emphasis. That interest requires that all discharges are kept as low as reasonably practicable and the authority should place the onus squarely on the operator to show that a discharge cannot practicably be avoided. There is, at present, a tendency either to ask the operator what he proposes to do and accept it if the result is within the levels regarded as permissible, or to suggest a limit, in which case the operator may accept it although he could at comparatively little cost have kept it to a lower level.[58]

The Inspector's Report on the Windscale Public Enquiry commented in relation to liquid discharges:

Discharge authorizations in the past have, in the case of aqueous discharges, not specified limits for each radionuclide. The authorization from Windscale in force since 1970 only specified limits for strontium-90 and ruthenium-106. Everything else was covered by a block limitation for alpha activity and beta activity. In the case of atmospheric discharges there is not even a block limitation. This is accepted as being unsatisfactory.[59]

These levels had arisen in part because the Windscale site had been a military one at the start and not subject to regulatory control. After the formation of BNFL in 1971, the site gradually came under control of the regulators, but full control was not achieved until 1984.

Although discharges had been substantially reduced by the early 1980s, they were still very high by prevailing industry standards. "One such discharge containing high levels of activity coincided with a well-publicized attempt by Greenpeace to block Sellafield's marine discharge outlet."[60]

Later discharges formed a slick just off Sellafield on 18 November 1983 that washed ashore. The surface dose on seaweed and flotsam was found to be so high that the Department of the Environment warned the public to avoid unnecessary use of the beaches along a 12-mile stretch of beach near the Sellafield plant. BNFL cleared debris from the beaches affected, but the government's advice to stay off them remained in place well into the following summer. Although at first a police investigation determined that BNFL had not contravened the "as low as reasonably achievable" (ALARA) condition for authorized discharge levels, BNFL was eventually convicted of failing to meet the ALARA criterion and fined "a nominal sum."[61]

58. Parker 1978, p. 59.

59. Parker 1978, p. 59.

60. Berkhout 1991, p. 171.

61. Berkhout 1991, p. 171.

Discharges from Sellafield are still much higher than from any other U.K. nuclear installation except Springfields, and about 20 times higher than the total activity discharged by all U.K. nuclear power stations. Sellafield is responsible for a major portion—90 percent in 1984, 60 percent in 1987—of the population dose attributable to radioactive wastes.[62]

Serious Accidents and Incidents From 1950 to 1986, over 250 safety-related incidents were recorded at the Sellafield site (including Calder Hall) and at Drigg (south of Sellafield) which is the national repository for all low-level waste produced in the United Kingdom. These have included two large accidents resulting in the shutdown of major facilities as well as many other serious problems. We first describe some of these other problems and then provide a brief account of the October 1957 reactor fire.

There have been large releases of radioactivity from Sellafield due to leaks. An inquiry by the Nuclear Installations Inspectorate (NII) published in 1980 found that the leak of radioactive water from silo building B38 (where Magnox fuel cladding was stored) may have been going on since 1972, and had resulted in 50,000 curies of radioactivity escaping into the environment. When the leak was discovered during excavations in 1976, 400 liters per day was seeping into the soil, and radioactivity levels were so high that work was suspended.[63] In another similar leak, 100,000 curies (3,700 terabecquerels) of radioactivity leaked into the soil over a period of three years or more from a disused building.[64]

From March to May 1992, BNFL detected radioactive single-celled coral-like organisms with high levels of radioactivity on the beach at Seascale, close to Sellafield, but was "at a loss" to explain this finding.[65] Numerous other discharges and emissions of radioactivity have been documented. We have already mentioned the 1973 accident that caused the shut down of the Head End facility. The most important accident was the reactor fire in 1957.

Reactor Fire in Windscale Pile No. 1 The Windscale piles were graphite-moderated, air-cooled reactors. Since graphite is carbon, it can catch fire if heated sufficiently. Such a source of heat exists in reactors of this design: it comes from the deformation of the graphite due to neutron bombardment, which causes energy storage. The

62. Hughes and Roberts 1984; Hughes, Shaw, and O'Riordan 1989.

63. Nuclear Installations Inspectorate 1980.

64. Nuclear Installations Inspectorate 1980.

65. BNFL 1992.

stored energy is called Wigner energy, after the physicist who described the phenomenon.

Wigner energy, to be controlled, must be released by an annealing process—else it can accumulate and be released suddenly. This can heat the graphite moderator enough to trigger a fire. The most probable cause of the accident appears to have been an improperly performed annealing procedure, which led to a fire breaking out on 10 October 1957 in Windscale Pile number 1. However, there is still some uncertainty as to the exact causes. The fire lasted for over 24 hours, until the morning of 11 October 1957. It was put out by dowsing the reactor core with water.[66]

Environmental Contamination Resulting from the Accident The first of the two major releases of radioactivity to the atmosphere as a result of the accident occurred early Thursday, 10 October, when the uranium was burning. The second big release was on the morning of 11 October "between 9 and 11 a.m., when the water was first put on the fire and a rush of steam carried radioactive particles and gases up the stack."[67]

An action level of 0.1 microcurie (3,700 becquerels) of iodine-131 per liter of milk was adopted. On 12 October, authorities stopped the distribution of milk originating from seventeen area farms.[68] However, just three days later, milk from a far wider area (200 square miles compared to the previous 80) was restricted, as a result of further milk sampling. Most restrictions on milk distribution were lifted on 4 November; the remainder of the controls were terminated on 23 November.[69] In all, authorities collected about two million liters of milk contaminated with iodine-131 and dumped it into the sea and local rivers.[70] This action is in contrast to the failure of the U.S. Atomic Energy Commission to collect contaminated milk downwind of Hanford (see chapter 6).

While the authorities denied large releases of radioactivity at the time,[71] this was not a correct portrayal of the situation. Various estimates have been made of fission product releases. Among the main ones, apart from the noble gases, were iodine-131 (16,200 to 27,000 curies or about 600 to 1,000 terabecquerels), tellurium-132 (12,000 to

66. Arnold 1992, pp. 42–52.

67. Arnold 1992, p. 54.

68. Arnold 1992, p. 58.

69. Arnold 1992, p. 61.

70. May 1989, p. 116.

71. May 1989, p. 116.

16,100 curies or 444 to 596 terabecquerels), and cesium-137 (600 to 1,230 curies, or about 22.2 to 45.5 terabecquerels).[72]

Chapelcross

"Britain's new atomic explosives factory—which produces electricity as a bonus—is being officially opened today," proclaimed Scotland's *Daily Record* on 2 May 1959. Under the headline "They'll carry H-bomb rods," the newspaper went on to explain how hand-picked lorry drivers would ferry plutonium, the "powerful atomic explosive," from Chapelcross to the "explosives refinery" at Windscale.

Following its 1976 decision to cease relying on U.S. tritium, the Ministry of Defence commissioned a new plant at Chapelcross for extracting tritium from irradiated lithium-6. The plant has operated since 1980. Tritium is made by bombarding lithium-6 with neutrons. This can be done in one of the Chapelcross reactors at the same time as it produces electricity. In the United States, the Savannah River tritium production reactors have produced tritium and weapons-grade plutonium at the same time; it is possible that the Chapelcross reactor can be used this way to produce about 173 grams of tritium and 38 kilograms of weapons-grade plutonium per year.[73] At present, the reactors are only producing electricity and tritium.

Discharges Periodically local newspapers have reported findings of radioactivity in local waterways, including the Solway Firth, once famous for Solway cockles. The findings would generally be described as "harmless" by BNFL spokespeople or blamed on Sellafield or Chernobyl.[74] Chapelcross discharges liquid waste to the Solway Firth under authorization by the Scottish Office. However, the discharges from Sellafield are numerically much greater, and are transported by prevailing currents to the north shores of the Solway Firth. As a result, concentrations of artificial radionuclides in the Chapelcross vicinity are mostly blamed on Sellafield discharges.

Accidents The most serious known accident occurred in May 1967. Newspaper reports refer to an inspection camera becoming detached

72. Arnold 1992, p. 185.

73. Simpson 1986, pp. 201, 202.

74. See, for example, *Annandale Observer*, 21 August 1992, "Radioactive particles found on Solway foreshore," which describes five "particles" found within 10 meters of the Chapelcross discharge point near Seafield, Annan. It was subsequently reported in the *Chapelcross Newsletter* that strainers had been fitted to the pipeline to reduce the possibility of contaminated limescale from the discharge pipe being deposited in the Solway Firth.

from its bracket and falling into the reactor. Under the heading "Chapelcross Reactor shut down," the articles also speak of a leak in a metal container of uranium fuel in the Number Two Reactor, causing a fire in which some of the metal melted. BNFL denied the fire could have resulted in a meltdown.[75] But to repair the damage, three scientists were kitted up in pressurized suits and gas masks in what was headlined "Engineer for 3 min dive into reactor" and "A-scientists were 25 mins in reactors."[76]

Concerns have been expressed about the dangers of low-flying RAF jets breaching the exclusion zone around Chapelcross. Chapelcross site superintendent Peter Jenkinson said that since January 1989, "official complaints had been made 8 times regarding incidents involving 13 aircraft either to the RAF or MoD," and one complaint regarding civilian aircraft. "I think the pilots are going so fast they don't realize it takes only 30 seconds to fly through the zone," he said.[77] Given the size of the four Chapelcross reactor stacks, such comment scarcely gives reassurance.

Radioactive particles were found on a site road in June 1986. The probable cause was given as transport of equipment between reactors.[78] In 1988, an empty nuclear flask carrier was involved in a crash on public roads.[79] A transport flask from Sellafield arrived at Chapelcross with its gas venting valves open.[80]

AWE Cardiff (also known as AWE Llanishen)

AWE Cardiff, which started out as an engineering establishment for small field ordnance, was established in 1939 on a 15-acre site. In 1958, special facilities were built to handle depleted uranium, and the beryllium shop followed two years later. AWE Cardiff uses beryllium and depleted uranium to manufacture crucial components of British nuclear weapons including depleted uranium tampers, beryllium reflectors, and depleted uranium blankets (see chapter 3).

Discharges In 1983, it was revealed that 50 metric tons of natural or depleted uranium are held at AWE Cardiff and that the factory was authorized to discharge radioactivity into the local environment,

75. *Annandale Observer*, 19 May 1967, "Chapelcross Reactor Shut Down."

76. *Cumberland News*, 26 May 1967.

77. *Annandale Observer*, 9 October 1992, "Atom Plant Low Flying Claims."

78. *Annandale Observer*, 13 June 1986, "Leak Probe at County Atom Plant."

79. *Annandale Observer*, 19 August 1988, "Nuclear Carrier in Crash."

80. *Chapelcross Newsletter*, 7 October 1992.

despite the proximity of the city of Cardiff. The main emissions are voided through a 26-meter stack. Adjacent office buildings tower over this and AWE Cardiff's other chimneys. In 1991, local Members of Parliament and city councilors condemned a decision of the Secretary of State for the Environment to exclude, on grounds of national security, "pollution events" at Atomic Weapons Establishments from the public registers set up under the new Environmental Protection Act.[81]

Accidents One incident in 1985 is recorded under "dangerous occurrences." Between 1970 and 1982, there were 13 "reportable incidents" involving major injury but no fatalities. There were also 6 additional "dangerous occurrences" and a total of 434 incidents involving more than three days absence during that time. Records are not held prior to 1970 for Royal Ordnance Factory Cardiff.

The Report of the Chief Executive, South Glamorgan County Council Public Protection Committee (which is responsible for Cardiff), 8 June 1987, stated that "a fire at the factory some years ago had identified beryllium being stored." This appears to relate to an incident, possibly in 1981 or 1985, in which South Glamorgan fire officers required radiation checks after being called to deal with a fire at the Royal Ordnance Factory. The 8 June 1987 report also raised concern that if there was a fire at Llanishen, the council's Fire and Rescue Service would not be called until ten minutes had elapsed, and that there was no participation in emergency exercises at the factory, unlike other major industries in the county. A subsequent memo of the South Glamorgan County Council of 24 July 1987 states that, following discussions with the Royal Ordnance Factory, it had been agreed that any incident not dealt with (extinguished) within one minute of its outbreak would result in an immediate call to the local authority fire brigade, and that any mechanical faults or alarms caused by internal detection equipment would result in an immediate call.

A major fire and accident at the beryllium processing plant at Ust Kamenogorsk in Kazakhstan (in the former Soviet Union) caused a flurry of concern about safety procedures and emergency planning at AWE Cardiff. The parliamentary reply from Kenneth Carlisle (Parliamentary Undersecretary for Defence Procurement) was that the Ust Kamenogorsk incident involved a production process in which beryllium metal is obtained from ores and that "such processes are not carried out in any U.K. facilities operated by, or contracted to, the Ministry of Defence."[82]

81. *South Wales Echo*, 24 June 1991, p. 8, "Anger Over N-Factory 'Cover Up.'"

82. *House of Commons Hansard*, 22 October 1990.

Such reassurance failed to satisfy local residents. Though AWE Cardiff may not use the same production process, there are enough dangers attached to the beryllium handling it does carry out that councilors asked for a meeting to discuss safety arrangements. Eventually the Ministry of Defence allowed the Third Officer of the County Fire and Rescue Service to brief councilors. However, the meeting began with a warning from a council solicitor that he would be prevented by the Official Secrets Act from telling them what he had seen or giving details of the fire service's obligation there.[83]

AWE Aldermaston

In 1950 the major effort of British nuclear weapons development was concentrated at a large military site in Berkshire called Aldermaston. In 1952, Aldermaston began processing plutonium for Britain's first nuclear test, code-named Hurricane, which took place on 3 October of that year. Since that time, AWE Aldermaston, whose high security A Area has come to be known as "The Citadel," has functioned as the pivotal establishment in British weapons manufacturing.[84] This role continues to the present, with a new facility to produce Trident missile components completed in late 1991.[85]

Despite the official secrecy which hampers access to information on this and other British nuclear facilities, AWE Aldermaston's safety record has on more than one occasion been the center of widespread public concern. In 1978, a series of revelations carried in national newspapers told of buildings being closed and boycotted by workers after tests showed that a number of workers had received plutonium overdoses. By August 1978, the laundry (where one worker was found to have twice the permissible level of plutonium in her lungs) as well as waste management and other facilities had been shut down because the staff refused to work there. That month the Ministry of Defence ordered an inquiry into health and safety at the Establishment, and closed all plutonium-handling facilities. The inquiry was headed by Sir Edward Pochin, a radiologist for the National Radiological Protection Board based at the Atomic Energy Research Establishment at Harwell.[86]

Pochin's most severe criticism was reserved for the standards of two main buildings handling nuclear waste. According to a 1993 Green-

83. *South Wales Echo*, 11 October 1990, p. 15, "New law would have barred city atom plant."

84. Greenpeace 1993, pp. 1, 5.

85. Norris, Burrows, and Fieldhouse 1994, p. 120.

86. Greenpeace 1993, pp. 27–29.

peace report entitled *Aldermaston: Inside the Citadel*, "these buildings and even the ground between them had been contaminated by radioactivity from past operations. They were badly designed and badly run ... [featuring] ventilation systems which blew contaminated air into workers' faces."[87] Pochin concluded that new buildings to handle these waste processes were needed as soon as possible. According to the Greenpeace report, by 1993 the key recommendation of the Pochin report to replace facilities found to have major safety problems with new ones had not been implemented, despite large construction programs for new facilities that were begun or completed since that time.[88] No further government-sponsored investigations into health and safety at Aldermaston have been made public.

At least eight workers with plutonium overdoses subsequently died of illnesses suspected of being linked to the workers' exposure and to the unsafe conditions at Aldermaston described by the Pochin Inquiry. There are nearly 70 claims for compensation filed with the Ministry of Defence by AWE workers and their families who believe they are suffering from radiation-linked illnesses.[89] *Inside the Citadel* also chronicles accidents and hazardous incidents since 1955, which include:[90]

- 22 fires involving radioactive materials in top security "A Area";
- 5 serious explosions that caused fatalities and wrecked buildings and equipment;
- 2 incidents involving lithium (highly caustic but not radioactive);
- 4 fires involving beryllium, exposure to which can cause lung disease;
- 2 accidents involving radioactive tritium;
- 9 electrical fires in areas subject to explosion;
- a serious leak of radioactive waste from a pipe running through a nearby village;
- a leak of radioactive tritium gas into the atmosphere that was equivalent to a quarter of Aldermaston's annually permitted discharge.

The most recent accident to be made public at Aldermaston occurred in December 1992. The incident resulted from the release of plutonium from a damaged storage container and led to the contamination of five workers. According to official statements the interior of the building where the accident took place was contaminated. The incident occurred while a routine accounting audit of plutonium holdings was

87. Greenpeace 1993, p. 30.

88. Greenpeace 1993, p. 32.

89. Greenpeace 1993, pp. 13–14.

90. Greenpeace 1993, pp. 3, 17–26.

being undertaken. The base was closed for two hours after the accident, and workers were told to remain inside the buildings with windows and doors closed. No warning was given to people living and working around the Aldermaston site, which is surrounded by residential areas and a business park.[91]

In April 1993 one business adjacent to the base, Blue Circle Cement, announced that it was pulling out of the area after contamination was found on its land. This was followed by an admission from the Aldermaston management, two months later, that radioactive water had been leaking from a discharge pipeline into the Berkshire countryside for years. The 40-year-old underground pipe running to the Thames had developed leaks in four places, allowing radioactive waste to seep into the soil.[92]

AWE Burghfield

Located to the southwest of Reading, Berkshire just five miles from Aldermaston, AWE Burghfield (formerly known as Royal Ordnance Factory, Burghfield) is the factory used by the Ministry of Defence for the final assembly and disassembly of Britains's nuclear bombs. It is also used to manufacture nonfissile components of the weapon. The factory is described in the 1981 Ministry of Defence List of Assessed Contractors, an Open Government Document, as DEF-STAN 05-24, which means that it satisfies MoD's inspection requirements for industry. It can therefore handle a variety of nuclear and nonnuclear components, such as highly enriched uranium and electronic parts.[93]

AWE Burghfield was constructed in 1941 and converted to nuclear weapons assembly in the 1950s. It covers 265 acres and was removed from Ordnance Survey maps as of 1976. The factory is estimated to employ around 600 people.[94]

A series of parliamentary inquiries and a search of the Berkshire Fire and Rescue Service records revealed a certain amount of information regarding some accidents. Typical reports involved fires with nonradioactive hazardous materials such as solvents or electrical fires. There are also reports of "major injuries" to workers on a number of occasions. However, we have no detailed information on the events and therefore cannot evaluate them.

91. "Checks follow plutonium leak," *The Independent*, 10 December 1992.

92. Letter to the editor, *Reading Evening Post*, from D. J. Hawkings, Chief Administrative Officer, Atomic Weapons Establishment, Aldermaston, 8 June 1993.

93. Ministry of Defence 1981.

94. Norris, Burrows, and Fieldhouse 1994, p. 73.

AEA Technology, Harwell

AEA Technology, Harwell (formerly the Atomic Energy Research Establishment) was founded in 1946 at a site on the Berkshire Downs near Oxford. In the early days of British nuclear weapons development Harwell was at the forefront of research into the bomb and the development of atomic energy for military and civilian purposes. Responsibility for research and development was gradually transferred to the Atomic Weapons Establishments.[95] Harwell at present carries out some research into the metallurgy of plutonium, among other materials, for the Ministry of Defence. It also receives low-level nuclear waste for incineration as well as toxic waste for storage from AWE Aldermaston.[96]

Some accidents at Harwell are also reported in the Berkshire Fire and Rescue Service records. These reports document spills and equipment and worker contamination with radioactive materials, including worker inhalation of plutonium.[97] Another example involved contamination of a roadway. Radioactive contamination of a site roadway and verges was discovered near the motor transport garage adjacent to building B462 during a routine survey on 18 July 1990. On 22 July, a further survey was conducted and over 100 spots of contamination were discovered, the main constituents of which were cesium-137, antimony-125, and cobalt-60. The contamination was caused by a release of radioactivity into the air from an unfiltered ventilation extract system on the decontamination unit situated adjacent to the garage. According to the Health and Safety Executive, neither the workforce nor the public was endangered.[98]

HEALTH AND ENVIRONMENTAL EFFECTS OF THE BRITISH NUCLEAR WEAPONS COMPLEX

The jury is still out on the question of whether discharges from Britain's nuclear establishments have resulted in health effects in surrounding communities. The most serious accident was the 1957 Windscale reactor fire. As described below, the next most pronounced effect observed so far—an increase in childhood cancer (mainly leukemia) around Sellafield—has defied any clear explanation. This is a very important problem that must be understood, as it may have considerable implications for the estimation of risks from environmental radioactivity.

95. Norris, Burrows, and Fieldhouse 1994, pp. 23, 66, 68.

96. UK Nirex Report 1991; Ministry of Defence 1989.

97. UKAEA 1987, chapter 2.

98. Health and Safety Executive 1990.

The Effects on British Workers in the Nuclear Weapons Complex

There have been three separate studies of health effects among Britain's nuclear industry workers, covering the UKAEA,[99] Sellafield,[100] and AWE.[101] These have all been carried out by the Department of Epidemiology, London School of Hygiene and Tropical Medicine. Clearly all the workers in the AWE study are connected with nuclear weapons production or development, but for the other two studies it is difficult to separate workers engaged in military activities from those involved with the civilian nuclear power program (see table 8.2). The main findings of these three studies are summarized below.

UKAEA Mortality from four cancers (testicular, leukemia, thyroid, and non-Hodgkin's lymphoma) was above the national average; however, none of the increases were significant at the 5 percent level ($p > 0.05$). Prostate cancer was the only condition in which the death rate clearly increased with exposure. The excess mortality was in younger men and concentrated in the small group of workers who were both monitored for tritium and had accumulated exposures exceeding 50 millisieverts. Tritium is not known to be concentrated in the prostate gland, so it may be a surrogate for something else that causes the cancer, although no clear occupational causes of prostate cancer are known. A recent case-control study has shown that "the risk of prostatic cancer was significantly increased in men who were internally contaminated with or who worked in environments potentially contaminated by tritium, chromium-51, iron-59, cobalt-60, or zinc-65."[102]

Sellafield A 1973 study by Smith and Douglas found statistically significant increases in several cancers, but these were only for "ill-defined and secondary sites" (i.e., metastases are found but no primary cancer is identified).[103] In internal comparisons across workers categorized by exposure levels, there was also an association between accumulated radiation dose and death rates from bladder cancer, multiple myeloma, leukemia, and all lymphatic and hematopoietic neoplasms. These were not statistically significant when exposure up to the time of death or up to two years before death was

99. Beral et al. 1985.

100. Smith and Douglas 1986.

101. Beral et al. 1988.

102. Rooney et al. 1993.

103. Smith and Douglas 1986.

Table 8.2 Studies of Health Effects of Nuclear Weapons Production on British Workers

Study	Number of Workers	Number of Deaths	Average Period of Follow-up	Average Radiation Dose
UKAEA	39,546	3,373	16 years	32 millisieverts
Sellafield	14,327	2,277	22 years	124 millisieverts
AWE	22,552	3,115	19 years	7.8 millisieverts

Sources: Beral et al. 1985; Smith and Douglas 1986; Beral et al. 1988.

considered. However, when the exposure used was that recorded up to 15 years before death (to allow for the fact that radiation-induced cancers take years to appear) these associations (except for leukemia) became significant.

AWE Aldermaston The only significant difference in death rates between radiation workers and other workers was in prostate cancer and cancers of "ill-defined and secondary sites." In workers who had been monitored for possible internal exposure, mortality from malignant neoplasms as a whole was not increased. However, if one accounts for a 10-year latency period between radiation exposure and onset of cancer, death rates from prostate and renal cancers were generally more than twice the national average, the excesses arising in a small group of workers monitored for exposure to multiple radionuclides. Though mortality from lung cancer in workers monitored for exposure to plutonium was below the national average, it was some two-thirds higher than for other radiation workers, the excess being of borderline statistical significance. Mortality from malignant neoplasms as a whole showed a weak and nonsignificant increasing trend with total dose from external radiation. When exposures up to 10 years before death were used, the trend became stronger and significant, the estimated increase in relative risk per centisievert being 0.076 (95 percent confidence interval 0.004 to 0.153). This trend was confined almost entirely to workers also monitored for exposure to radionuclides. The main contributions came from lung and prostate cancers.

The risk estimates relating site-specific cancer mortality to radiation dose for all three studies have large uncertainties; on the one hand, the upper confidence limit indicates a risk several times higher than the current ICRP risk estimate; in the other direction, the lower confidence limit is compatible with a conclusion that there is no excess risk at all. The most important finding is that for prostate cancer, which "suggest[s] a specific occupational hazard in a small group of workers in the nuclear industry who had comparatively high exposures to external radiation and who were also monitored for internal exposure to multiple radionuclides."[104]

Two other important points should also be noted. First, the studies were based on cancer *deaths*, not incidence. Moreover, the follow-up periods were comparatively short. Experience from the survivors of Hiroshima and Nagasaki has shown that radiation-induced cancers can appear 30 or more years after exposure. There is an indication, at least in the Sellafield and AWE studies, that longer follow-up may reveal more. If the doses received in the last 10 to 15 years were ignored, the relationship between cancer mortality and dose became stronger. This suggests that the excess cancers are related to radiation.

To overcome the problems of relatively small sample size, it has been argued that studies should be pooled. Although larger samples in general will increase the validity of the findings, this may be at the expense of masking or diluting local factors, such as prostate cancer and tritium at Winfrith (UKAEA).

In Britain, a National Registry for Radiation Workers set up in 1976 published its first report in 1992.[105] This report studies deaths from cancer in 95,217 workers employed by British Nuclear Fuels, the Ministry of Defence, the UKAEA, and Nuclear Electric. The average lifetime radiation dose was 3.36 centisieverts. This study shows a clear "healthy worker effect"—that is, worker cancer rates were generally lower than the national average, but, as remarked above, what is important is whether cancer rates increase as radiation dose increases. In fact, there was a positive trend with dose for all cancers combined, but it was not statistically significant. However, there was a significant association between leukemia and radiation dose. Central estimates of lifetime risk derived from these data were 0.1 total cancer per sievert (90 percent confidence interval less than 0 to 0.24) and 0.0076 leukemia per sievert (90 percent confidence interval 0.0007 to 0.024). This is the first study over such a large worker population to show a relationship between radiation exposure and leukemia at low doses, and it makes it difficult to continue to claim that low doses may not be harmful.

Effects of the Nuclear Weapons Complex on Public Health: The Seascale Cluster

It is a complex and difficult procedure to obtain the radiation dose to an individual member of the public from figures of total activity discharged. Presented as averages over the whole of the United Kingdom, the doses appear to be very small. However, an average can conceal a

104. Beral et al. 1988.

105. Kendall et al. 1992a.

Table 8.3 Critical Group Doses, 1976–1990

Year	1976	1977	1978	1979	1980	1981	1982	1983	1984	1985	1986	1988	1989	1990
Dose (millisieverts)	2.20	1.55	1.30	1.05	1.95	3.45	2.70	2.25	0.84	0.73	0.34	0.34	0.40	0.30

Source: Directorate of Fisheries Research, MAFF 1992.

wide variation. To estimate maximum possible values, the doses to *critical groups* are evaluated. A critical group is defined as a small homogeneous group of individuals who, due to their habits or way of life, represent the most highly exposed individuals in the population. In the Sellafield area, the critical group is now said to consist of members of the local fishing community who consume significant quantities of locally caught fish and shellfish. This group is contaminated with a number of radionuclides; plutonium and americium are particularly important from the point of view of potential toxicity (see table 8.3).

At its peak in the early 1980s, the dose to this critical group was over 3 millisieverts, a level that would be considered unacceptable now. In 1991, the National Radiological Protection Board recommended an upper value of constraint on effective dose for members of the public of 0.3 millisieverts per year from a single source.[106] This exposure would correspond to an average annual fatal cancer risk of about 1 in 100,000. The attributable lifetime probability of fatal cancer to age 75 works out to about 1 in 700. According to the NRPB, this is about the upper limit of acceptable risk. Despite this, the board recently modified its advice to recommend that the constraint apply to a single *new* source and not include the contribution from radioactivity already in the environment, effectively loosening the standard.[107]

There is widespread environmental contamination from Sellafield, and, to a considerably lesser extent, from other nuclear installations. But what health effects does this contamination cause? If the NRPB's estimates are correct, the average dose to an individual in Britain from radioactive discharges is only a very small fraction of natural background radiation and would not be expected to have any noticeable health effects. Even the doses to critical groups, although higher, should only cause a small increase in the already large "natural" risk of cancer.

It is worth mentioning at this point that in the 1970s some scientists had warned of possible future health problems resulting from the Sellafield discharges. In 1975, V. T. Bowen, a leading marine geochemist at the Woods Hole Oceanographic Institute in the United

106. NRPB 1991.

107. NRPB 1993.

States and a specialist in radioecology of fallout plutonium, warned the British government of potential problems from the high levels of discharge of radioactivity. During the 1977 Windscale Inquiry epidemiologist Edward Radford (at one time chair of the BEIR Committee) reinforced concerns regarding the effects of increased plutonium contamination on public health.[108] Nevertheless, until recently, the official opinion was that there would not be any observable health effects around nuclear installations.

In 1983, this changed after a Yorkshire Television film claimed that the village of Seascale, near Sellafield, had a much higher incidence of childhood leukemia than would be expected from national rates. Prime Minister Margaret Thatcher, usually a staunch supporter of the nuclear industry, was concerned enough to set up a Committee of Enquiry, chaired by Sir Douglas Black. Subsequently, on Black's recommendation, the government set up the Committee on the Medical Aspects of Radiation in the Environment (COMARE).

The Black Committee asked the NRPB to estimate the probable radiation doses to children in Seascale from the discharges. These calculations showed that the radiation doses likely to have been received from the discharges were much too low to have caused the incidence of leukemia.[109] However, another possibility was that the environmental or metabolic models were incorrect or incomplete in some way. Perhaps other exposure pathways had not been allowed for. A related possibility, which could make particular isotopes very important, is that radioactivity reaches critical cells, such as the blood-forming stem cells in the bone marrow, by unexpected routes that depend on how the radioactive material is chemically combined. The children may also have been indirectly affected by a parent's exposure to radiation at work.

The finding of the Seascale cluster, as it came to be known, stimulated other research into incidences of leukemia or other cancers around nuclear installations. Paula Cook-Mozaffari and co-workers have reported a slight but significant increase in leukemia, especially lymphoid leukemia, in people under the age of 25 in the vicinity of 15 nuclear installations of various types in England and Wales.[110] Of most interest here is the increased incidence of childhood leukemia in the vicinity of the plants at Aldermaston and Burghfield, which are nearby each other.[111] Although in terms of relative increase in incidence this is not as marked as the Seascale cluster—a twofold

108. Taylor 1987, pp. 27–28.

109. Stather, Wrixon, and Simmonds 1984.

110. Cook-Mozaffari et al. 1989.

111. Roman et al. 1987.

increase rather than a tenfold one—it is numerically more important because of the higher density of population in the Reading area. The Aldermaston and Burghfield cluster is difficult to explain on the basis of exposure to environmental radioactivity, as there is a very large discrepancy between the estimated doses and the observed leukemia excess. However, there has been very little environmental monitoring around Aldermaston, and the true extent of contamination by radionuclides such as plutonium is not known. In a way this is the reverse of the problem at Chelyabinsk-65. In Britain public health monitoring has resulted in data that allow for detection of increases in leukemia, but the data on releases of radioactivity, and possibly other non-radioactive carcinogens, are not good enough to enable firm conclusions as to doses.

Several follow up studies of the Seascale leukemias have been carried out on the recommendations of the Black Committee. The most important, a case-control study by Martin Gardner and colleagues, appeared in 1990.[112] The Gardner team concluded that an association existed between pre-conception radiation dose to a father and leukemia risk in his children. They estimated that a dose of 10 centisieverts or more to a father was associated with a sixfold increase of leukemia risk. This attracted some criticism: no such increased risk has been observed in the children of fathers exposed to radiation at Hiroshima and Nagasaki. However, these fathers were exposed to a single intense dose of radiation, the consequences of which may be quite different from the chronic exposures sustained by the Seascale fathers. Internal doses—from plutonium, for example—may be significant for many of the Sellafield workers; these were not included in the doses used by Gardner.

Gardner's hypothesis has received some qualified support from a similar case-control study of the leukemias around Aldermaston and Burghfield.[113] An investigation by the Health and Safety Executive into child cancer in the children of Sellafield workers concluded that "there was a clear distinction between the risk of leukemia and NHL [non-Hodgkin's lymphoma] for the children of Sellafield workers resident in Seascale when the child was born, compared with those resident elsewhere."[114] The rate of leukemia and non-Hodgkin's lymphoma was about fourteen times the national average for the Seascale children born to Sellafield fathers, and about twice the national average for the children of Sellafield fathers resident in locations other than Seascale.

However, it has been recently claimed by Leo Kinlen that the increase in childhood leukemia around Sellafield is not confined to

112. Gardner et al. 1990.

113. Roman et al. 1993.

114. Health and Safety Executive 1993, para. 143.

children born in Seascale.[115] Kinlen's hypothesis is that population mixing results in the spread of viral infections and that childhood leukemia is a rare consequence of such an infection. This hypothesis appears to be supported by data on the incidence of leukemia in England, Wales, and Scotland.[116] But the increases observed in these circumstances were transient, and in magnitude typically a factor of two or so. The increase around Sellafield is by an order of magnitude or so and is a continuing excess.[117] R. Doll and his colleagues have also disputed Gardener's hypothesis, saying that it was not consistent with current knowledge of radiation genetics or heritability of leukemia. They believe that this children's leukemia cluster is a chance event.[118]

In addition to leukemia there have been claims of an increased incidence of retinoblastoma in children born to mothers who have lived in Seascale.[119] The present situation is summarized in a recent letter from the COMARE Committee to the Pollution Inspectorate:

> The Committee consider that the cause of the excess rate of cancer in the 0–24 year old age range in the village of Seascale is currently unknown. There are a number of possible causes which may have led to this excess. There is insufficient evidence to point to any one particular explanation and a combination of factors may be involved. As exposure to radiation is one of these factors, the possibility cannot be excluded that unidentified pathways or mechanisms involving environmental radiation are implicated.[120]

A study carried out at the West Cumberland Hospital in Whitehaven concluded that "there is no good evidence to suggest a deleterious effect from Sellafield on local pregnancies but that miscarriage needs further investigation."[121]

Nonradiation Hazards to Workers

Radiation is not by any means the only hazard to which workers in the nuclear industry are exposed. For example, Sellafield uses about 3,000 chemicals, but little information is available on exposure of people to chemicals. Such information is not quantified to the same extent as radiation exposure.[122]

115. Kinlen 1993.

116. Kinlen 1990.

117. Draper et al. 1993.

118. Doll, Evans, and Darby 1994.

119. Morris et al. 1993.

120. HMIP and MAFF 1993.

121. Jones and Wheater 1989.

122. NRPB 1990, p. 2.

Beryllium is used at AWE Cardiff in the manufacture of nuclear weapons. Inhalation of beryllium can result in acute beryllium disease (chemical pneumonitis) or more commonly chronic beryllium disease (granulomatous pneumonitis)[123] (see chapter 4).

Sometime in 1963, a machinist at AWE Cardiff cut his finger on a grinding wheel contaminated with beryllium oxide, causing a small wound, which would not heal. Fifteen months later, his finger was amputated. Six months after that, ulcerative nodules had to be removed from the man's arm. Five years later, it was discovered that similar nodules had grown in his lung as a result of airborne beryllium contamination. Discovery of this case led to the setting up of a special nationwide register on the toxic effects of beryllium.[124]

According to a 1984 Ministry of Defence letter, AWE Cardiff "fulfills all the requirements of the Radioactive Substances Act and other relevant legislation and discharges into the atmosphere and sewers are maintained in accordance with levels set by the Department of Environment. Monitoring is undertaken on a regular and frequent basis by the factory personnel and outside inspectors."[125] The letter goes on to say that "there is no independent U.K. medical evidence to support a statement that beryllium is carcinogenic to man and there are no such cases reported in the United Kingdom."

Other Data on Health Effects around Nuclear Installations

In addition to the published literature, there are a large number of anecdotal reports on the health of workers at nuclear installations in Britain and members of the public living near them. There is some evidence of an increase in some adult cancers, particularly multiple myeloma.[126] As cancer in children occurs relatively rarely, it may be that any increase is more evident than a comparable increase in adult cancer.

According to a recent report by the Department of Public Health Medicine in Dumfries, the incidence of acute leukemia in Dumfries and Galloway is about 50 percent higher than expected.[127] It is not clear whether this relates in any way to discharges from Chapelcross or Sellafield. Chapelcross discharges three times as much tritium to the

123. Jones Williams 1988.

124. *New Statesman*, 25 March 1983, and Jones Williams, Lawrie, and Davies 1967. In a letter to the editor of *Western Mail*, 22 April 1983, Alex Farrow claims that approximately one-quarter of the register worked at Llanishen.

125. Letter from Ministry of Defense to Alex Farrow, 11 April 1984.

126. McSorley 1990, p. 126.

127. Maclean, Breen, and Chalmers 1992, p. 73.

Table 8.4 Estimates of Cancer Cases from the 1957 Fire in Windscale Pile Number 1

Study	Estimated Maximum Number of Nonfatal Cancer Cases (UK)	Estimated Maximum Number of Cancer Deaths (UK)
November 1957 (Cmnd 302)	0	0
December 1960 (Cmnd 1225)	0	0
July 1981 (Taylor)	248	10–20
1982 (Crick and Linsley)	237	20
September 1983 (Crick and Linsley)	?	35
1988 (Clarke)	90 (plus 10 hereditary defects)[1]	100[1]

1. Clarke considered these unlikely upper bound estimates.

Source: Arnold 1992, appendix X.

atmosphere as does Sellafield. Although conventional radiation protection has generally regarded tritium as not particularly hazardous, it has recently been suggested that organically bound tritium may be more dangerous than previously thought.[128]

Various estimates of fatal cancer and non-fatal cancer cases from the 1957 Windscale reactor fire are shown in table 8.4.

128. Fairlie 1992.

9 France

Albert Donnay and Martin Kuster

French involvement in nuclear physics dates from the end of the nineteenth century, when Henri Becquerel discovered X rays emitted from a uranium salt, uranyl potassium sulfate.[1] In the early 1900s, Pierre and Marie Curie discovered other radioactive elements, and the Radium Institute they founded in Paris became a world-renowned center for nuclear research. Their daughter, Irène Joliot-Curie, and her husband Frédéric discovered induced radioactivity at the institute in 1933, winning the Nobel Prize in 1935. Irène, on her own, discovered spontaneous fission in 1939.[2] All these advances were major steps in the chain of events leading to the global pursuit of nuclear weapons.

Many French scientists were already conducting research related to nuclear fission before World War II.[3] After the German occupation of much of France in 1940, many fled to Britain, Canada, and the United States, with some continuing to work on nuclear research for the Manhattan Project. However, French scientists were excluded from the project's inner scientific circles: Britain and the United States, which controlled the project, feared the left-wing sympathies of some scientists.

Under the occupation in France, German troops seized the laboratories of Frédéric Joliot-Curie and other nuclear physicists who stayed behind. The scientists were pressured to reveal the results of their research and turn over their stocks of uranium and heavy water. Joliot hid what he knew, agreeing only to work on fundamental (nonmilitary) research, and secretly supported the French Resistance.[4]

In October 1945, just two months after the bombing of Hiroshima, the French provisional government, which was headed by Charles de

1. Rhodes 1988, pp. 41–42.

2. Weart 1979, pp. 45–46, 64.

3. See Weart 1979, pp. 107–138.

4. Rhodes 1988, pp. 41–42.

Gaulle, created the Atomic Energy Commission (its French acronym is CEA), giving a high priority to nuclear research.[5] By March 1946, CEA research facilities had been established in the old Châtillon Fort in Hauts-de-Seine just outside Paris (known as Fontenay-aux-Roses) and in an explosives factory at Le Bouchet in Vert-le-Petit, Essonne.[6] The CEA's mission was broadly defined: "To pursue scientific research and technologies with the aim of using atomic energy in the various fields of science, industry, and the national defense."[7]

The scientists involved in these early years were divided as to the military extent of this mission. At the time, the Communist Party, which had been part of the leadership of the anti-Nazi struggle, was a very popular political party in France. Many Communists, including Frédéric Joliot-Curie, flatly refused to work on nuclear weapons that might be used against the Soviet Union.[8] France's government expelled Joliot-Curie from the CEA in April 1950.[9]

During the 1950s the CEA became more and more linked to the military under the influence of increasingly nationalistic Gaullist politicians like Pierre Guillaumat (from 1951 to 1958 the administrator of the CEA), military officials in the CEA led by General Albert Buchalet, and scientists like Bertrand Goldschmidt.[10] The latter was part of a group known as "the Canadians"—including Lew Kowarski and Jules Guéron—who had worked on plutonium separation and the large-scale production of other materials for nuclear weapons in Montreal with British and Canadian researchers during World War II.

Even after the expulsion of Joliot-Curie and other Communist Party members, many CEA scientists and technicians felt strongly that they should not be involved in manufacturing nuclear weapons but only in civilian work. In the early 1950s, 665 CEA employees signed a petition to this effect. This act and other internal opposition led advocates of military applications to pursue their research as covertly as possible, even hiding it from CEA colleagues.

The first CEA research center devoted exclusively to military applications was built near Le Bouchet at Bruyères-le-Châtel, Essonne, in 1956. It is known as Le Bouchet III or BIII. The clandestine period of nuclear weapons development lasted until 1958, during which time the

5. Norris, Burrows, and Fieldhouse 1994, p. 182

6. Barrillot and Davis 1994, pp. 21, 47.

7. From CEA founding ordinance, as quoted by Mongin 1990.

8. Weart 1979, pp. 249–259.

9. Weart 1979, pp. 259–261.

10. Weart 1979, pp. 249–267; Mongin 1990.

CEA researched and developed plans not only for making nuclear weapons but also for submarines, aircraft, and other delivery systems. The government financed the work during this period secretly, without even the knowledge or consent of Parliament.

During 1945 to 1958, the CEA:

• started up France's first atomic pile, the Zoé reactor, at Fontenay-aux-Roses on 15 December 1948, producing tiny quantities of plutonium that was extracted a year later at the Le Bouchet facility;[11]

• discovered the first uranium deposit in France in the Limousin region in 1948;

• got Parliament to adopt a five-year plan in 1952, directing the CEA to expand from purely scientific research to building an industrial-scale facility (at Marcoule, in Gard) for producing plutonium (although still ostensibly for civilian purposes at that stage);[12]

• created a joint "Bureau of General Studies" attached to the CEA in 1955 to direct research and development on nuclear weapons;[13]

• built a pilot reprocessing plant at Fontenay-aux-Roses and commissioned it in 1954;[14]

• started up France's first graphite-gas reactor for producing plutonium at the Marcoule Complex in 1956;

• started up France's second uranium conversion plant in Malvési in 1959;[15]

• established a nuclear test site near Reggane, Algeria, in the Sahara desert in 1956;

• started up a plutonium separation plant at Marcoule in 1958.[16]

The CEA expanded rapidly in these early years. From 1952 to 1960, the agency doubled its personnel and budget every two years.[17] It established many research centers devoted to different aspects of nuclear weapons production, including facilities for study of plutonium production and chemical reprocessing, reactor research, and weapons design. The main design center is the Limeil-Valenton Research center in Val-de-Marne just outside of Paris, and the fabrication and assembly

11. Barrillot and Davis 1994, pp. 21, 47.

12. Mongin 1990.

13. Mongin 1990.

14. Albright, Berkhout, and Walker 1993, p. 97.

15. Barrillot and Davis 1994, p. 31.

16. Barrillot and Davis 1994, p. 129.

17. Weart 1979, p. 266.

of nuclear weapons is carried out at the Valduc Research Center located in the Côte d'Or.[18] In addition to these research facilities, the CEA controls dozens of corporations through the Compagnie Générale des Matéhres Nucléaires (Cogéma). The CEA owns 99.7 percent of Cogéma, which in turn controls numerous subsidiaries involved in all aspects of the nuclear fuel cycle and nuclear engineering, from uranium mining and enrichment to fuel production, reprocessing, and nuclear waste treatment.

The military role of the CEA was officially announced on 11 April 1958, a few weeks before de Gaulle returned to power. On 22 July of the same year, de Gaulle confirmed this decision to openly pursue nuclear weapons with the creation of the Department of Military Applications. Several reasons were behind the decision. The country was still smarting from the 1954 defeat of its conventionally armed forces at Dien Bien Phu, Vietnam. The balance of power and influence in the Atlantic alliance and globally were clearly shifting to countries with nuclear weapons. France and Britain had not received U.S. support in October 1956, when they attacked Egypt over control of the Suez Canal. This lack of U.S. backing had emboldened the Soviet Union to issue a nuclear ("rocket weapons") threat.[19] De Gaulle clearly felt that France, faced with U.S. and British nuclear superiority, needed its own nuclear force to remain a "major power." The traditional hostility between France and Britain may have played a role as well.[20]

Although the creation of the Department of Military Applications within the CEA in 1958 enabled politicians to make a formal distinction between civilian and military nuclear programs, this distinction is largely bureaucratic. The programs tend to merge in many areas, with civilian and military research projects pursued at the same facilities by scientists whose roles are not differentiated. Nuclear materials and wastes produced at these facilities also are largely commingled, unlike the case in the United States (for the most part). Through the Department of Military Applications, the CEA continues to manage the development and production of nuclear weapons, closely overseen by several commissions, committees, and officials of the Ministry of Defense.

Ties between the CEA and the military were strengthened in 1958 by the promotion of CEA director Pierre Guillaumat to Armed Forces Minister. Guillaumat was promoted again in 1960, to Minister of State for Atomic Affairs, and just seven days after this promotion, on 13 February 1960, he presided over the explosion of France's first nuclear

18. Norris, Burrows, and Fieldhouse 1994, p. 199.

19. Kaplan 1981, pp. 154–155.

20. Mongin 1990.

bomb, code-named Blue Jerboa (Gerboise Bleue), at the Algerian test site.

Within an hour of this 60- to 70-kiloton blast (about four times as powerful as the one that destroyed Hiroshima), Guillaumat received a message of congratulations from de Gaulle: "Hurrah for France! As of this morning, she is stronger and prouder. From the bottom of my heart, thanks to you and those who have helped her achieve this magnificent success."[21] This official statement clearly expresses the nationalist character of the French government's strong commitment to nuclear weapons.

Defending France's independent *force de frappe* in 1966, de Gaulle said that nuclear weapons liberated France from the "yoke of double hegemony agreed on by the two rivals. Given that the United States and the Soviet Union did not destroy their weapons, it was necessary for us to break the spell. We did it, for our own reasons and by our own means."[22]

France's effort to build an independent arsenal was not entirely without foreign assistance. A 1959 agreement with the United States gave France access to military-grade (highly enriched) uranium-235 for its naval reactors, which it could not produce itself on an industrial scale until the Pierrelatte enrichment plant opened in 1967. Under a 1961 secret agreement, the two countries began to share information on nuclear-weapons technologies.[23] French-U.S. cooperation continued even after France's 1966 withdrawal from NATO's military command structure.

Not all of France's nuclear foreign relations have been mutual. France exploited its colonies in Gabon and Niger for uranium deposits, without royalties, and the people of colonial Algeria had no say in France's use of their territory for testing nuclear weapons. Even after Algeria gained its independence in 1962, France took four years to move its nuclear testing operations to the atolls of Mururoa and Fangataufa in its colonial Pacific territory of Polynesia, whose people were also given no say in France's nuclear testing decisions.

France exploded its first thermonuclear bomb at Mururoa in 1968.[24] France refused to sign the Partial Test Ban Treaty of 1963, by which the United States, the Soviet Union, and Britain agreed to stop atmospheric testing, but France did stop its atmospheric testing in 1974. As France built up its arsenal in the 1970s and 1980s, it followed the U.S. model of deploying its nuclear weapons in a strategic triad, with silo-based

21. De Gaulle, as quoted in Mongin 1990; translation by IEER.

22. From press conference, 28 October 1966, as quoted in Mongin 1990; translation by IEER.

23. *Médecine et Guerre Nucléaire* 1989.

24. IPPNW and IEER 1991, pp. 133–137.

ballistic missiles on land, submarine-launched ballistic missiles at sea, and gravity bombs aboard long-range aircraft. By 1992, France had an estimated 538 nuclear weapons, more than half of them deployed on nuclear-powered submarines. Although its arsenal is much smaller than those of Russia and the United States, France still has roughly twice as many nuclear weapons as Britain.[25] It has conducted 210 nuclear tests,[26] about one-fifth as many as the United States, though its arsenal size (as of the early 1990s) was only one-thirtieth that of the United States.

Civilian nuclear power also grew dramatically in the 1970s and 1980s, as 55 nuclear power plants came on line across France, all run by Electricité de France, the country's national utility.[27] A decision to go all out in developing civilian nuclear power and plutonium breeder reactors was made around the time of the dramatic increase in oil prices that accompanied the 1973 Middle East crisis. The breeder reactor program has also produced plutonium for military purposes (see below).

France's network of mostly light-water reactors now provides about three-fourths of France's electricity.[28] CEA's Cogéma subsidiary reprocesses the spent fuel from these reactors, along with spent fuel imported from other countries in Europe and Japan, to recover plutonium and uranium. France's reprocessing contracts require the suppliers of spent fuel to take back all their plutonium as well as all the nuclear wastes generated by reprocessing.

In 1966, Cogéma built a civilian-run plant for this purpose, at La Hague on the coast of the Cotentin Peninsula. The plutonium recovered at La Hague, intended originally to support a new generation of breeder reactors, has little current market since the world's only commercial-scale breeder reactor is France's Superphénix. Superphénix has been only recently restarted after being shut down due to severe technical problems, but it wll not operate as a breeder of plutonium. Its demonstration breeder reactor of 560 megawatts thermal, Phénix, built in 1973, was shut down for most of the period between since September 1990 and the end of 1993.[29] France has used some of this plutonium to make mixed-oxide fuel, which can fuel some of its existing commercial reactors, but this is more expensive than using natural uranium alone.[30]

25. Norris, Burrows, and Fieldhouse 1994, p. 214.

26. Norris, Burrows, and Fieldhouse 1994, pp. 65, 214.

27. EDF 1989.

28. EDF 1989.

29. Barrillot and Davis 1994, p. 117.

30. Berkhout et al. 1992.

A shipment of 1.5 metric tons of plutonium across the oceans from France's international contracts was sent from the port of Cherbourg on 7 November 1992, reaching Japan on 5 January 1993. Despite a storm of international protest, France plans many more shipments to Japan and other countries in the coming years. However, the protests appear to have significantly slowed the schedule of shipments, especially as Japan is facing a surplus of plutonium in its civilian program.[31]

With the collapse of the Soviet Union, France has begun to make some unilateral cuts in both its nuclear and conventional arsenals. For the first time, spending on nuclear weapons fell in 1993.[32] Nuclear cutbacks announced in 1992 included taking out of service the "Pluton" ground-launched nuclear missile regiments and reducing from six to four the number of new strategic nuclear submarines to be built; the four were to be ready by 2005. The government also said it would stockpile the new Hadès short-range nuclear missile instead of deploying it. In June 1992, President François Mitterand announced that the Hadès program would stop "immediately and definitively."[33] The place of these missiles in the French nuclear force after the end of the Cold War remains unclear, according to NRDC researchers Norris et al.[34]

France supports U.S. and Russian nuclear disarmament initiatives, but insists that its participation in negotiations to reduce its own weapons meet the conditions it set forth in 1983.[35] France's general stance, like China's, is that the Russian and U.S. stockpiles of nuclear weapons should be cut back to levels comparable to its own before it can participate in disarmament talks. Nevertheless, France has begun to support multilateral initiatives aimed at slowing nuclear proliferation. In 1992, France ratified both the Nuclear Non-Proliferation Treaty and the Latin American Nuclear Free Zone "Treaty of Tlatelolco." The latter commits signatory countries within the zone to using nuclear materials and installations only for peaceful purposes, and it requires France and other nuclear weapons countries outside the zone to refrain from introducing or using nuclear weapons in the region. The Treaty of Tlatelolco covers only French "territory" within Latin America, which includes about 94,000 square kilometers of French Guyana, Martinique, and Guadeloupe.[36]

31. New York Times, 22 February 1994.

32. *Médecine et Guerre Nucléaire* 1993; *Médecine et Guerre Nucléaire* 1993b.

33. *Médecine et Guerre Nucléaire* 1992c.

34. Norris, Burrows, and Fieldhouse 1994, p. 274.

35. *Médecine et Guerre Nucléaire* 1991a.

36. *Médecine et Guerre Nucléaire* 1993a.

Figure 9.1 The *Rainbow Warrior* in Auckland harbor, 1985. Greenpeace photo.

In response to the moratorium on nuclear testing initiated by Russia in 1991 and joined by the United States, France in 1992 "suspended for this year" all its nuclear testing in the Pacific. Foreign Minister Roland Dumas described the superpowers' moratorium as "a good thing," adding that "we're studying the possibility" of joining them.[37] The moratorium continues as of the summer of 1994.

France is gradually moving toward a nuclear position in Europe coordinated with the United States and Great Britain, although it still wants to keep its nuclear weapons outside NATO. Minister of Defense Pierre Joxe reported in 1992 that high-level discussions were underway with Germany regarding "conditions under which our nuclear forces might be combined."[38] Such discussions may be a prelude to a European policy. If achieved, this would mean, in effect, that Western Europe would go from having two nuclear powers (France and the United Kingdom) to twelve—all the members of the European Community, including Germany. Prospects for such a development are uncertain, given the various political and financial crises afflicting EC members.

Nuclear weapons programs make up an estimated 25 percent of France's 1993 defense budget of about 200 billion francs. The cumu-

37. As quoted in *Médecine et Guerre Nucléaire* 1993b.

38. As quoted in *Médecine et Guerre Nucléaire* 1992b.

lative cost of the French nuclear arsenal since 1950 is estimated at 860 billion francs or roughly $170 billion.[39]

Along with the United States and Great Britain, France has agreed to provide Russia with financial and technical assistance in dismantling nuclear weapons. In 1993 France allocated 50 million francs (about $10 million) to bring Russian scientists to work in French laboratories.[40]

NUCLEAR WEAPONS–RELATED ACTIVITIES AND FACILITIES

France has dozens of facilities related to the research, development, and production of nuclear weapons and nuclear weapons materials (see table 9.1 and figure 9.2). The nuclear test sites developed in the Pacific and Algeria are shown in figures 9.3 and 9.4. France also recovers plutonium from the spent fuel of its civilian nuclear power plants, as well as spent fuel sent from other countries in Europe and from Japan (see table 9.2). While ostensibly for civilian purposes, this plutonium has potential military applications in nuclear and radiological weapons.

Research and Development Laboratories

France has a widely dispersed network of research and development laboratories, many of which also take part in the production, assembly, and maintenance of nuclear warheads and their components. Research centers in the CEA Division of Military Applications conduct nuclear weapons–related work. Further, most production sites have some research facilities. The centers are listed in table 9.1.

The extensive involvement of the civilian CEA in weapons-related research dates from its creation in 1945 as the sole agency responsible for nuclear programs. In its early years, it focused almost entirely on building reactors to produce plutonium for nuclear weapons. In 1954, the government established a secret CEA department (the nondescriptly named Office of General Studies, later renamed the Office of New Technology and finally, and more frankly, the Military Applications Division) to pursue military applications for nuclear technology.[41] This secret department built France's first production reactors, plutonium reprocessing plants, and nuclear weapons.

The CEA began its nuclear research program in March 1946 with the establishment of its first Nuclear Research Center (CEN) at Fontenay-aux-Roses, the old Châtillon Fort in the Hauts-de-Seine region south-

39. *Médecine et Guerre Nucléaire* 1993. We use an approximate exchange rate of five francs to one U.S. dollar throughout this book.

40. *Médecine et Guerre Nucléaire* 1993b.

41. Norris, Burrows, and Fieldhouse 1994, p. 183.

Table 9.1 Location of Nuclear Weapons–Related Activities and Facilities in France

Official Facility Name	Location	Operating Status[1]	Operating Agency of Company[2]	Source[3]
R&D Weapons Laboratories				
CEA Research Centers				
Bruyères-le-Châtel Research Center (B III)	Near Arpajon, Essonne		CEA-DAM	d
Limell-Valenton Research Center (CELV)	Val-de-Marne, 20 km from Paris		CEA-DAM	d
Vaujours Research Center (CEV)	Vaujours, Seine-St.-Denis		CEA-DAM	d
Vaujours Research Center Annex	Moronvillers, Marne		CEA-DAM	d
Valduc Research Center, includes two research reactors	Lamargelle, St. Seine L'Abbaye, Côte-d'Or	2 RR	CEA-DAM	d
Aquitaine Scientific and Technical Research Center (CESTA)	Between Arcachon and Bordeaux, Gironde		CEA-DAM	d
Le Ripault Research Center	Monts, Indre-et-Loire		CEA-DAM	d
CEA Nuclear Research Centers (CEN)				
CEN Fontenay-aux-Roses	Chatillon Fort, Hauts-de-Seine		CEA	d
CEN Grenoble	Grenoble, Isère	3 RR	CEA	d
CEN Saclay	Gif-sur-Yvette, Essonne	2 RR	CEA	d
CEN Cadarache	St.-Paul-les-Durance, Bouches-du-Rhône	13 RR	CEA	d
CEN Valrho	Pierrelatte, Drôme, and Bagnols-sur-Cèze, Gard		CEA	d

DGA Research Centers				
Central Technical Establishment for Weaponry including the Center for Defense Analysis	Montrouge Fort, Arcueil, Val-de-Marne		DGA-DRET	d
Gramat Research Center	Gramat, Lot		DGA-DRET	d
Le Bouchet (Essonne) Research Center (CEB), includes Center for Nuclear, Biological and Chemical Defense	Vert-le-Petit, Essonne		DGA-DRET	d
Le Bouchet Research Center Annex	Odeillo, Pyrénées Orientales		DGA-DRET	d
Center of Defense Analysis	Arcueil, Val-de Marne		DGA-DRET	d
Uranium Mining				
Aveyron Region				
Bertholène (underground and surface)	Bertholène, Aveyron		TCMF	d
Cantal Region				
St.-Pierre-du-Cantal	St.-Pierre-du-Cantal, Cantal	NO	TCMF	g
Corrèze Region				
La Besse (underground)	Auriac and St.-Julien-aux-Bois-Auriac, Corrèze		SMUC	d
Creuse Region				
Hyverneresse (underground)	Croze and Gioux, Creuse		CFM	d
Coussat (strip mine)	Bonnat, Creuse		CFM	d

Table 9.1 (continued)

Official Facility Name	Location	Operating Status[1]	Operating Agency of Company[2]	Source[3]
Haut-Rhin Region				
Vosges	St.-Hyppolyte, Haut-Rhin	NO		g
Haute-Vienne Region				
Brugeaud	Bessines, Haute-Vienne	NO	SIMO	g
Puy Teigneux	Bessines, Haute-Vienne		COGEMA	b
La Traverse	Bessines, Haute-Vienne		COGEMA	b
Sagnes-Sud	Bessines, Haute-Vienne		COGEMA	b
Puy de l'Âge	Bessines, Haute-Vienne		COGEMA	b
Les Gorces	St.-Léger-la-Montagne, Haute-Vienne		COGEMA	b
Bellezane (underground and surface)	Bessines-sur-Gartempe and Bersac-sur-Rivalier, Haute-Vienne		COGEMA	d
Le Bernardan (underground and surface)	Jouac, Haute-Vienne		TCMF	d
Le Fraisse (underground)	Razès, Haute-Vienne		COGEMA	d
Silord (underground)	Razès, Haute-Vienne		COGEMA	d
Fanay (underground)	St.-Sylvestre, Haute-Vienne		COGEMA	d
St.-Sylvestre (underground)	St.-Sylvestre, Haute-Vienne		COGEMA	d
Margnac (underground)	Compreignac, Haute-Vienne		COGEMA	d
Pény (underground)	Compreignac, Haute-Vienne		COGEMA	d
Piegut (underground)	Cromac, Haute-Vienne		TCMF	d

Venachat (surface)	Compreignac, Haute-Vienne		COGEMA	d
La Vauzelle (surface)	Compreignac, Haute-Vienne		COGEMA	d
Les Loges (surface)	St.-Léger-Magnazeix, Haute-Vienne		TCMF	d
Hérault Region				
Le Mas laveyre (underground)	Le Bosc, Hérault		COGEMA	d
Le Mas d'Alary-Treviels (surface)	Soumont, Hérault		COGEMA	d
Loire Region				
Bois Noirs-Limouzat-Forez (surface & underground)	St.-Priest-la-Prugne, Loire	NO 1980	COGEMA	g
Loire-Atlantique Region				
Le Chardon (underground)	Gorges, Loire-Atlantique	NO 1991	COGEMA	d
L'Ecarpière (underground and surface)	Gétigné, Loire-Atlantique		COGEMA	d
Tesson-la-Garenne (surface)	Guerande, Loire-Atlantique		COGEMA	d
Piriac (underground)	Piriac, Loire-Atlantique		COGEMA	d
Lozère Region				
Le Cellier-Villeret (surface)	St.-Jean-de-la-Fouillouse, Lozère	NO 1988	MOKTA	d
Pierres Plantées (underground and surface)	Grandrieu, Lozère	NO 1987	MOKTA	d
Maine-et-Loire Region				
Basses Boissière Baconnière Bastille (surface)	Montigné-sur-Moine, Torfou, and Roussay, Maine-et-Loire		COGEMA	d

Table 9.1 (continued)

Official Facility Name	Location	Operating Status[1]	Operating Agency of Company[2]	Source[3]
Saône-et-Loire Region				
Grury (underground and surface)	Issy I'Evêque, Saône-et-Loire		COGEMA	d
Bauzot (underground and surface)	Issy I'Evêque, Saône-et-Loire	NO 1985	COGEMA	g
Vendée Region				
La Commanderie (underground)	Treize Vents and Le Temple, Vendée		COGEMA	d
Le Cottereau (surface)	Les Epesses, Vendée		COGEMA	d
Poitou-la-Gabrielle (surface)	Mortagne-sur-Sèvre, Vendée		COGEMA	d
Uranium Milling and Production (production of yellowcake from uranium ore)				
Bessines	Bessines-sur-Gartempe, Haute-Vienne		COGEMA/SIMO	d
Mailhac-sur-Benaize	Jouac, Haute-Vienne		TCMF	d
St.-Martin-du-Bosc/Lodève	Le Bosc, Hérault		SIMO	d
L'Ecarpière	Gétigné, Loire-Atlantique		SIMO	d
Bertholène (pretreatment plant only)	Bertholène, Aveyron		TCMF	d
Le Cellier	St.-Jean-de-la-Fouillouse, Lozère		CFM	d
Cherbois	Cherbols, Haute-Vienne		94.2% TCMF	e
Lodève	Hérault		SIMO	d
Gueugnon	Saône-et-Loire	NO 1980	COGEMA	e
Inguiniel, La Calardieu	Inguiniel, La Calardieu	NO 1980	MOKTA	e

St.-Pierre-du-Cantal	St.-Pierre-du-Cantal, Cantal	NO 1984	TCMF/SCUMRA	e
St.-Priest-la-Prugne	St.-Priest-la-Prugne, Loire	NO 1980	COGEMA	e
Rophin	Lachaux, Puys-de-Dôme	NO	COGEMA	g
Conversion of Yellowcake (U_3O_8) to Uranium Tetrafluoride (UF_4) and Uranium Metal (for use as fuel in gas-graphite reactors)				
Le Bouchet Research Center (CEB)	Vert-le-Petit, Essonne	NO 1971	CEA	f
Malvesi Plant (stopped uranium metal production in 1991)	Narbonne, Aude		Comhurex/CEA	d
Uranium Enrichment				
Low Plant (enriches to 2%)	Pierrelatte, Drôme	NO 1982[4]	COGEMA	d
Middle Plant (enriches to 6%)	Pierrelatte, Drôme	SP 1995	COGEMA	d
High Plant (enriches to 25%)	Pierrelatte, Drôme	SP 1995	COGEMA	d
Very High Plant (enriches to 90+%)	Pierrelatte, Drôme	SP 1995	COGEMA	d
Chemex pilot plant PP35 at CEN Valrho	Pierrelatte, Drôme	NO	CEA	d
Chemex pilot plant PL 81 at CEN Grenoble	Grenoble, Isère	NO 1988	CEA	d
Atomic Vapor Laser Isotope Separation (SILVA) Pilot Plant A-2 at CEN Saclay	Gif-sur-Yvette, Essonne		CEA	d
SILVA Pilot Plant at CEN Valrho	Pierrelatte, Drôme		CEA	d
Uranium Fuel Fabrication				
Fuel for Production Reactors				
SICN Plant	Annecy, Haute-Savoie	NO	COGEMA	f
Le Bouchet	Vert-le-Petit, Essonne	NO 1982	DGA-DRE	d

Table 9.1 (continued)

Official Facility Name	Location	Operating Status[1]	Operating Agency of Company[2]	Source[3]
Fuel and Blankets for Breeders				
Workshop for Treatment of Enriched Uranium at CEN Cadarache	St.-Paul les-Durance, Bouches-du-Rhône		CEA	d
SICN Plant	Veurey-Voroize, Isère		SICN	d
Fuel for Light Water Reactors[5]				
Franco-Belge de Fabrication de Combustibles (plant name unknown)	Romans-sur-Isère, Drôme		FBFC	d
COGEMA Framatome Combustibles Plant	Pierrelatte, Drôme		FBFC	d
CEN Cadarache	St. Paul lex Durance, Bouches-du-Rhône		CEA	d
CEN Saclay	Gif-sur-Yvette, Essonne		CEA	d
Enriched Uranium Metal and Oxide				
Valduc	Is-sur-Tile, Cote d'Or		CEA-DAM	f
Production of Tritium				
Célestin I Reactor (heavy water)	Marcoule, Gard		COGEMA	d
Célestin II Reactor (heavy water)	Marcoule, Gard		COGEMA	d
Tritium Separation				
Workshop for the Extraction of Tritium	Marcoule, Gard		COGEMA	d

Plutonium Production from:				
Military Production Reactors				
EL 1 pile (Zoé) at CEN Fontenay-aux-Roses	Chatillon Fort, Hauts-de-Seine	NO 1975	CEA	d
EL 2 pile at CEN Saclay	Gif-sur-Yvette, Essonne	NO 1965	CEA	d
G1 reactor (graphite-gas) (also used for nuclear power)	Marcoule, Gard	NO 1968	CEA	d
G2 reactor (graphite-gas) (also used for nuclear power)	Marcoule, Gard	NO 1980	CEA/EDF	d
G3 reactor (graphite-gas) (also used for nuclear power)	Marcoule, Gard	NO 1984	CEA/EDF	d
Célestin I and II reactors (heavy water)	Marcoule, Gard		CEA	d
Research Reactors (not known if plutonium reprocessed)				
EL 4 heavy water reactor	Brennilis, Finistère	NO 1985	CEA	d
Breeder Reactors				
Rapsodie reactor at CEN Cadarache	St.-Paul-les-Durance, Bouches-du-Rhône	NO 1983	CEA	d
Phénix breeder (liquid metal) reactor	Bagnols-sur-Cèze, Gard	NO 19??	EDF & CEA	d
Super-Phénix breeder (liquid metal) reactor	Creys-Malville, Isère	NO 1992	51% EDF	d
Civilian Power Reactors[6]				
Chinon A1 graphite-gas reactor	Chinon, Indre-et-Loire	NO 1973	EDF	d
Chinon A3 graphite-gas reactor	Chinon, Indre-et-Loire	NO 1990	EDF	d
St.-Laurent-des-Eaux A2 and A1 reactors	St.-Laurent-des-Eaux, Loir-et-Cher		EDF	d

Table 9.1 (continued)

Official Facility Name	Location	Operating Status[1]	Operating Agency of Company[2]	Source[3]
Chemical Separation of Plutonium Reprocessing				
Le Bouchet Research Center (CEB)	Vert-le-Petit, Essonne	NO 1950	CEA	b
Chatillon Pilot Reprocessing Plant	Chatillon Fort, Hauts-de-Seine	NO 1958	CEA	d
Cyrano Laboratory at CEN Fontenay-aux-Roses	Chatillon Fort, Hauts-de-Seine		CEA	d
UP1	Marcoule, Gard		COGEMA	
UP2-800 (expansion of UP2)[7]	La Hague, Manche		COGEMA	d
UP3 (for oxide fuel of light water reactors)	La Hague, Manche		COGEMA	d
Breeder Oxides Treatment Plant (TOR) at CEN Valrho (for metal and oxide fuels of breeder reactors, including Phénix and Super-Phénix)	Bagnols-sur-Ceze, Gard		COGEMA	d
Fabrication of Plutonium Components				
MOX Reactor Fuel				
Complex for the Fabrication of Fuel Elements Containing Plutonium at CEN Cadarache	St.-Paul-les-Durance, Bouches-du-Rhône		CEA	d
Melox Plant (MOX fuel for civilian reactors)	Marcoule, Gard		COGEMA	d
Pu-239 Pits and Triggers				
Valduc Research Center	Is-sur-Tille, Côte-d'Or		CEA-DAM	f

Other Warhead Components				
Electronic Components				
CEN Limeil-Valenton	Villeneuve-St.-Georges, Val-de-Marne		Sodern	b
Bruyères-le-Châtel Research Center (B III)	Near Arpajon, Essonne		CEA-DAM	f
High-Explosives Production				
Le Ripault Research Center	Monts, Indre-et-Loire		CEA-DAM	d
High-Explosives Testing, R&D				
Vaujours-Moronvilliers Center	Moronvilliers, Marne		CEA-DAM	b
Lithium-6 Production				
Miramas Plant	Miramas, Bouches-du-Rhône		COGEMA	d
Testing of Nuclear Devices				
"Peaceful" Nuclear Explosions				
None				
Weapons Testing, All Outside France				
Centre Saharien des Expérimentations Mmilitaires	Near Reggane, 150 km south of Adar in Sahara Desert, Algeria	NO 1961	CEA	a, c
Centre Saharien des Expérimentations Militaires	Hoggar Massif, 560 km southeast of Reggane, Algeria	NO 1966	CEA	c
DIRCEN[2] Pacific Test Center	Test sites in Moruroa and Fangataufa, affiliated bases in Tahiti and Hao, administrative headquarters in Villacoublay, Yvelines		CEA-DAM	d

Table 9.1 (continued)

Official Facility Name	Location	Operating Status[1]	Operating Agency of Company[2]	Source[3]
Spent Fuel Storage				
CEN2 Cadarache (dry vault for HWR and submarine fuel—180 metric tons heavy metal total)	St.-Paul-les-Durance, Bouches-du-Rhône		CEA	e
La Hague HAO/N (pools for LWR fuel—250 + 1000 metric tons heavy metal)	La Hague, Manche		COGEMA	e
Marcoule (pool for GGR fuel—800 metric tons heavy metal)	Marcoule, Gard		COGEMA	e

1. Status abbreviations:

NO = not operating and either certain or likely not to resume, with year stopped, if known

RR = research reactors

SP = stop production (year, if known or anticipated)

2. Abbreviations:

CEA-DAM = Commission d'energie atomique (Atomic Energy Commission), Direction des applications militaires (Military Applications Directorate)

CEN = Centre d'études nucléaires (Nuclear Research Center, CEA-administered)

CFM = Compagnie française de Mokta, a wholly owned subsidiary of COGEMA

COGEMA = Compagnie générale des matières nucléaires, a wholly owned subsidiary of CEA-Industrie, which is in turn 99.7% owned by CEA

Comhurex = Société pour la conversion de l'uranium en métal et en Hexafluorure: company created in 1971 to merge the uranium conversion operations of the Malvési and Pierrelatte plants; wholly owned by COGEMA

DIRCEN = Direction des centres d'experimentation nucléaire (Management of Nuclear Experimentation Centers)

DGA-DRET = Délégation général pour l'armament (General Armaments Commission), Direction des recherches, études et techniques (Research, Study, and Technical Management)

EDF = Electricité de France (Electricity of France), the national electric utility company

FBFC = Franco-Belge de fabrication de combustibles (Franco-Belgian Fuel Fabrication), 50% owned by Uranium Pechiney 25% by COGEMA and 25% Framatome.

SCUMRA = Société centrale de l'uranium, des minerais et métaux radio-actifs

SICN = Société industrielle des combustibles nucléaires, a wholly owned subsidiary of COGEMA

SIMO = Société industrielle des minerals de l'Ouest, a wholly owned subsidiary of COGEMA

SIMURA = Société industrielle et minière de l'uranium

Sodern = La Société anonyme d'études et de réalisations nucléaires (Nuclear Research and Production Company, Ltd.), an independent French company created at the initiative of the DGA and CEA-DAM

SMUC = Société des mines du centre, 33.32% owned by COGEMA, with 33.31% each owned by CFM and Smac Acieroid

TCMF = Total compagnie miniere France

3. Source:

a = Mongin 1990

b = Barrillot 1991

c = May 1989

d = Davis 1988

e = Nuclear Engineering International Publications 1991

f = Barrillot and Davis 1994

g = *Gazette Nucléaire* 1992a

4. Low-enrichment operations subcontracted to EURODIF in Tricastin beginning in 1982.
5. See below, under "Frabication of Plutonium Components," for facilities that make MOX fuel containing both uranium and plutonium.
6. Includes only those from which weapons-grade plutonium was reprocessed; note that France recycles civilian-grade plutonium from all its power reactors.
7. UP2 reprocessed metal fuel from gas-graphite reactors and MOX fuel from breeders until 1987; this is now done only by UP1 in Marcoule.

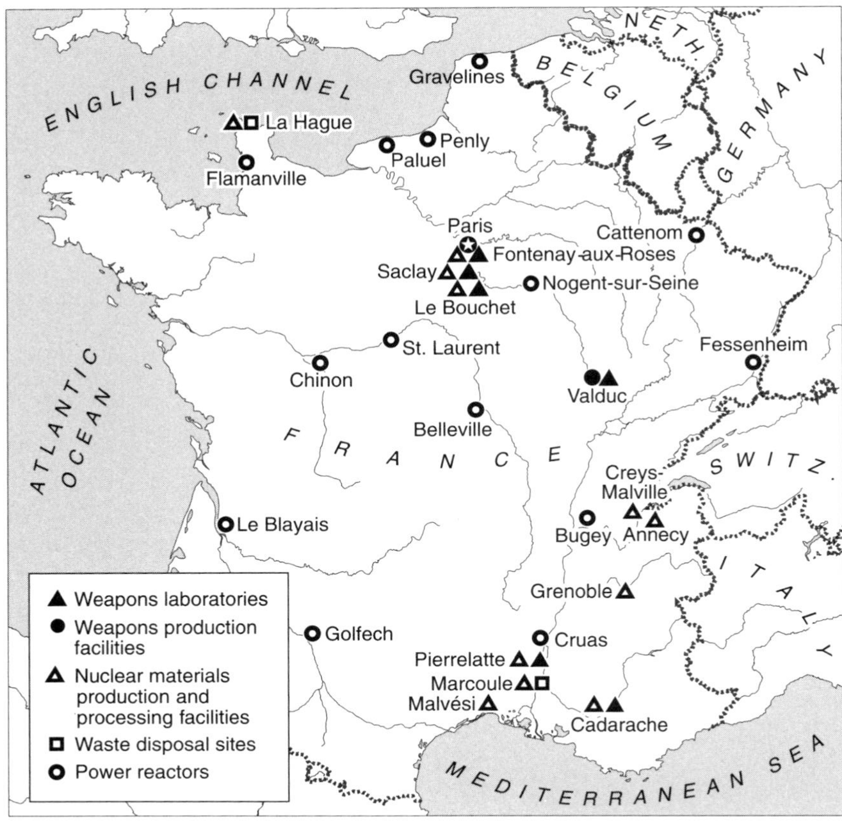

Figure 9.2 French nuclear weapons production and nuclear reactor sites.

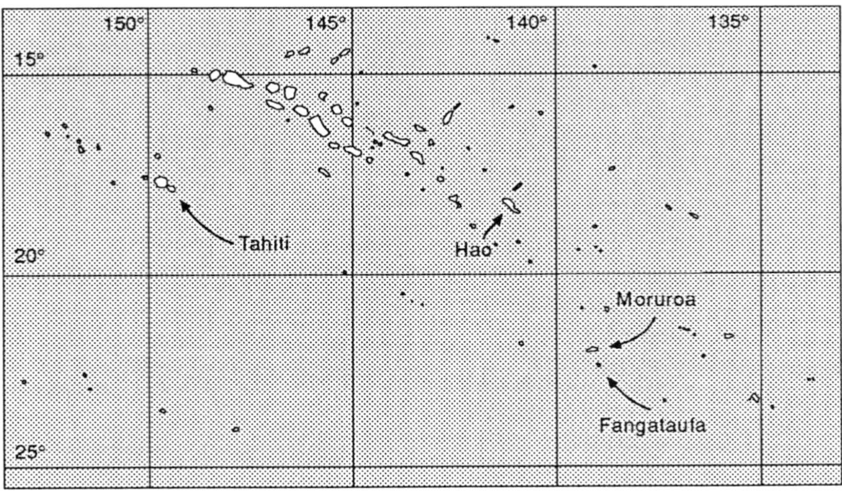

Figure 9.3 French nuclear weapons testing sites in the Pacific Ocean. Tahiti and Hao were staging areas for tests on Moruroa and Fangalaula.

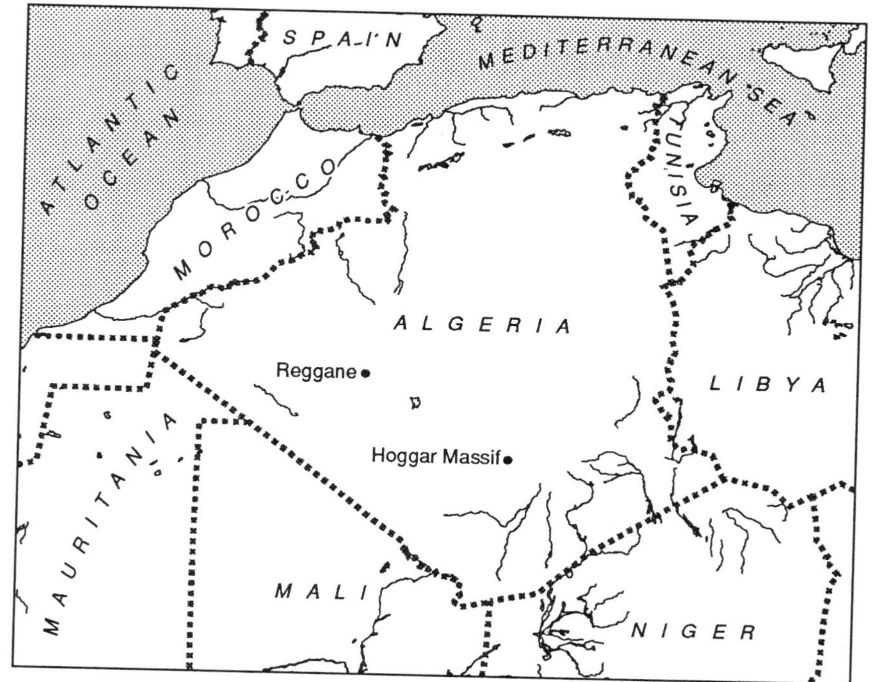

Figure 9.4 French nuclear weapons testing sites in Algeria.

west of Paris. France's first experimental atomic pile, named Zoé, went critical here in December 1948.[42] The Fontenay-aux-Roses Center also built and operated France's pilot plutonium-reprocessing plant, which was commissioned in 1954. The decommissioning of the plant began in 1959 and lasted until 1963.[43] The center also performs research on the physical and chemical properties of plutonium and on reprocessing techniques.

Four other nuclear research centers under CEA control are CEN Cadarache at St.-Paul-les-Durance in Bouches-du-Rhône, CEN Grenoble in Isère, CEN Saclay in Essonne, and CEN Vallée du Rhône (also known as Valrho), with facilities in Pierrelatte, Drôme, and Bagnols-sur-Cèze, Gard.[44] They do both civilian and military-related work.

The Saclay center operates two reactors and was the site of EL2, the CEA's second atomic pile. It studies irradiated fuel, researches laser isotope separation and particle accelerators, produces radioisotopes for commercial and medical use, and maintains a storage facility for solid radioactive wastes.[45] The Valrho center in Bagnols-sur-Cèze (part of

42. Barrillot and Davis 1994, p. 21.

43. Barrillot and Davis 1994, pp. 49–50.

44. Davis 1988, p. 44.

45. Davis 1988, p. 140.

Marcoule) houses the Phénix breeder reactor and a pilot reprocessing plant for the spent fuel of breeders.[46] The research center at Limeil-Valenton, near Paris, is France's main nuclear weapons design center and developed the trigger used in France's first nuclear test. The Vaujours center, including facilities for high-pressure shock studies, was established in 1955. The center specializes in the study of explosions and explosives, especially those used to trigger fission reactions inside warheads. Perhaps, the CEA's most diverse lab is Bruyères-le-Châtel Center. Since opening in 1957, it has engaged in a wide variety of research programs related to nuclear weapons in the fields of plutonium and uranium metallurgy, chemistry, electronics (especially the miniaturization and hardening of nuclear weapons components), nuclear physics, seismology, and instrumentation for nuclear weapons testing.[47,48] It provides other CEA labs with nuclear materials and also collects and repackages all their manufacturing and machining wastes. These wastes, mostly metal fragments and turnings originating from metallurgical operations on uranium and plutonium, are remelted and reused.

The waste-processing facilities at Valduc treat and package most of the radioactive wastes produced by the Department of Military Applications during all stages of the fabrication of nuclear weapons. It then turns these over to the National Agency for the Management of Radioactive Wastes (ANDRA).[49] The waste that does not meet ANDRA's criteria remains on the site. Valduc's facilities include a concrete casing plant, an incinerator being built for alpha-emitting wastes such as plutonium that will have a capacity of 7 kilograms per hour, a treatment plant for wastes contaminated with tritium, and an automated installation for dismantling contaminated equipment and recycling the plutonium from nuclear warheads. Significant quantities of tritium are handled at Valduc. The CEA's long-term plans for waste containment include developing a specialized center at Valduc for storing solid wastes contaminated with tritium.[50]

The Ripault Research Center specializes in manufacturing chemical-explosive detonators for nuclear weapons. Founded in 1962, this center now covers over 100 hectares.[51] It also shares responsibility with the

46. Davis 1988, p. 76.

47. Norris, Burrows, and Fieldhouse 1994, p. 200.

48. Barrillot and Davis 1994, pp. 197–201.

49. Davis 1988, p. 95.

50. Barrillot and Davis 1994, pp. 235–244.

51. Norris, Burrows, and Fieldhouse 1994, p. 199.

Aquitaine Center for the maintenance of nuclear warheads.[52] The Aquitaine Center (CESTA) was built in 1965 and occupies 700 hectares.[53] It adjoins a deforested firing range covering another 900 hectares. It is described as a pyrotechnic research and testing center responsible for hardening nuclear warheads against electromagnetic pulses and other shocks.[54] It develops warhead prototypes and plans for their testing and manufacture for weapons systems.

Uranium Chemical Conversion[55]

In 1948, a plant to make uranium metal was brought on line at Le Bouchet in Vert-le-Petit. The process converts U_3O_8 to uranium tetrafluoride and then reduces uranium tetrafluoride to metal. The plant used calcium until 1964 and magnesium thereafter (see chapter 3). The plant also prepared the natural uranium fuel used in the second French atomic pile, the EL2 in Saclay. From 1948 until its closure in 1971, the Le Bouchet plant produced a total of at least 2,600 metric tons of uranium metal. Decommissioning of the plant began in 1972.[56]

Planning began in 1957 for a second, larger facility that could process 1,000 metric tons of uranium per year either into uranium metal or uranium tetrafluoride. The metal was for the Marcoule G1, G2, and G3 reactors; the tetrafluoride was the feed material for producing uranium hexafluoride for France's military uranium enrichment program. Built in 1959 in Malvési, near Narbonne, which is directly north of the eastern portion of the Pyrénées, this plant was operated initially by the Société de Raffinage de l'Uranium (Uranium Refining Co.), then in 1971 by Comurhex, which in turn became part of Cogéma in 1992. Malvési processes both yellowcake and the decontaminated uranyl nitrate from the UP1 reprocessing plant at Marcoule. Modifications to the plant over the years have increased the plant's capacity for uranium tetrafluoride production until it reached 11,000 metric tons per year in 1987.[57] However, by this date most uranium was for civilian power reactors.

Malvési operated one facility until 1991 using magnesium for reducing uranium tetrafluoride to metal, and produced a total of 23,000 metric tons of uranium metal between 1959 and 1990. Until 1967

52. Davis 1988, p. 92.

53. Norris, Burrows, and Fieldhouse 1994, p. 199.

54. Davis 1988, p. 92.

55. Barrillot and Davis 1994, pp. 22–36, unless otherwise cited.

56. Barrillot and Davis 1994, p. 23.

57. Barrillot and Davis 1994, pp. 31–35.

Malvési used calcium to reduce uranium tetrafluoride to metal, after which it switched to using magnesium.[58] It also has a large storage facility on site for its surplus yellowcake, with a capacity of 25,000 metric tons. In 1984 this facility contained 11,000 metric tons.[59]

In 1963, a fluorination plant was commissioned in Pierrelatte to convert uranium tetrafluoride into uranium hexafluoride. This is the form needed for uranium enrichment, which is also done at Pierrelatte. Since 1981 the plant has had a rated capacity of 12,000 metric tons of uranium hexafluoride per year. The site houses a plant for producing fluorine (with a capacity of 3,000 metric tons per year), which is needed for uranium hexafluoride production.[60]

In 1971, the activities of the Malvési plant (centered around the production of uranium metal and tetrafluoride) and Pierrelatte (converting uranium tetrafluoride into hexafluoride) were merged in a new company, Comurhex. Its responsibility was converting yellowcake into both metal and hexafluoride. The CEA managed all industrial operations at Pierrelatte, though the plant is partly privately owned. In 1976, the CEA turned over management to Cogéma. Since 1992, when the industrial firm Péchiney withdrew from its involvement at Pierrelatte, Cogéma has owned all of Comurhex which became a subsidiary of Cogéma.[61]

Uranium Enrichment, Processing, and Fabrication

Under a French-U.S. agreement signed in 1959, France obtained 440 kilograms of highly enriched uranium from the United States for its land-based naval reactor prototype for use until its own uranium enrichment plants opened at Pierrelatte in the mid-1960s. Four sections of a gaseous diffusion plant were built there to span the full range of enrichment needs: from low (2 percent uranium-235) to very high (over 90 percent, for both nuclear warheads and naval propulsion reactors). In 1967 these plants produced France's first highly enriched uranium, which was used in the first French thermonuclear weapons tested the next summer and in the country's first nuclear-powered submarine, *Le Redoubtable*, launched in 1969.[62]

Both the low-level and the medium-level enrichment plants at Pierrelatte were closed in 1982 and their activities subcontracted to Eurodif, an internationally owned civilian gaseous diffusion plant in Tricastin.

58. Barrillot and Davis 1994, p. 34.

59. Davis 1988, p. 66.

60. Barrillot and Davis 1994, p. 269.

61. Barrillot and Davis 1994, pp. 35–36, 269.

62. Norris, Burrows, and Fieldhouse 1994, pp. 201–202, 245.

This is France's newest enrichment facility and is devoted to low enrichment of uranium for civilian purposes. Highly enriched uranium production for France's military program continues at Pierrelatte.[63]

In the north of the Pierrelatte site are the pilot enrichment facilities and laboratories at which the French gaseous diffusion technology was developed. A chemical laboratory studied fluoride products and a chemical isotopic separation process (Chemex) it developed in 1968. The Chemex process is no longer under study and work on this process was ended in 1988. France is focussing its efforts on the development of atomic vapor laser isotope separation at Pierrelatte, CEN Saclay, and other locations.[64]

To increase the efficiency of the Cogéma enrichment operations at Pierrelatte, Comurhex built a pilot plant there in 1977 to convert uranyl nitrate recovered from reprocessing into uranium hexafluoride so that the uranium they contain can be reenriched and used for fuel. It has a capacity of 400 metric tons of uranium per year.[65] Since 1984, Cogéma also has removed fluorine in the form of hydrofluoric acid from the depleted uranium hexafluoride produced at Pierrelatte. It apparently resells the hydrofluoric acid into the civilian economy, though the acid is slightly contaminated with uranium.[66]

Cogéma also maintains a storage site for up to 2,000 metric tons of enriched uranium hexafluoride (up to 5% enrichment) at the Miramas facility in Bouches-du-Rhône. Storage space at this site not holding uranium hexafluoride is used for natural and depleted uranium.[67]

Enriched uranium destined for nuclear weapons goes to the Centre de Valduc near Dijon for machining and fabrication.[68]

Plutonium and Tritium Production from Military Reactors

France has operated seven military reactors for plutonium production. France's first atomic pile, the EL1 or Zoé reactor, was also its first source of plutonium. This heavy-water–moderated pile of uranium oxide fuel started operation in December 1948. It later used uranium metal fuel.[69]

63. Barrillot and Davis 1994, pp. 263–265.

64. Barrillot and Davis 1994, pp. 256–257; Davis 1988, pp. 68–69.

65. Davis 1988, p. 69.

66. Barrillot and Davis 1994, pp. 270–271, 292.

67. Davis 1988, p. 67.

68. Barrillot and Davis 1994, p. 221.

69. Barrillot and Davis 1994, p. 48.

France's second atomic pile, the EL2, was built at the CEA Saclay Nuclear Research Center in Gif-sur-Yvette, Essonne. It was of a heavy-water–moderated, gas-cooled design. The 2.8 megawatt-thermal EL2 served as a research reactor and a pilot plant for testing methods of plutonium production.[70]

Based on this experience, the CEA moved quickly to build three larger metal-fueled plutonium production reactors at a massive new complex in Marcoule, Gard, along the Rhône River. Designated G1, G2, and G3, these gas-cooled graphite-moderated reactors operated from 1956 to 1968, 1958 to 1980, and 1959 to 1984, respectively. The rating of the G1 reactor is 38 megawatts; G2 and G3 are rated at 150 megawatts thermal. They produced an estimated 2.5 to 2.8 metric tons of plutonium altogether. Although designed primarily for plutonium production, they all were eventually connected to the grid of Electricité de France (EdF) and generated some electricity. Their graphite-gas design also became the model for EdF's first civilian power reactors. The G1 produced only about 100 kilograms of the total plutonium, while the G2 and G3 reactors produced the rest.[71]

To obtain tritium for thermonuclear weapons, the CEA built and still operates two heavy-water reactors at Marcoule, Célestin I and II. Rated at 190 megawatts thermal each, they came on line in 1967 and 1968, respectively. The Célestin reactors have produced 0.5 to 1.5 metric tons of plutonium. The total plutonium production in France from all military sources through 1991 is 3 to 4.3 metric tons. In addition, France's military stockpile contains 0.9 to 1.6 metric tons of plutonium from the Rapsodie and Phénix breeder reactors and 1.1 to 2.3 metric tons from the Chinon reactors 1, 2, and 3 and initial fuel loadings of some other civilian Magnox reactors. After accounting for losses, mainly through nuclear weapons testing, the total stockpile of military plutonium is estimated at about 6 metric tons.[72]

Plutonium Production from Dual-Use Civilian Power Reactors

According to Albright, Berkhout, and Walker, "No clear distinction is made in France between military and civil[ian] nuclear materials. Indeed the French have been open about their use in weapons of plutonium produced in power reactors operated by Electricité de France."[73] France has used its graphite-moderated gas-cooled reactors and the two plutonium-breeder reactors, Rapsodie and Phénix, for the dual purposes of electricity and plutonium production. As noted

70. Barrillot and Davis 1994, pp. 70–71.

71. Albright, Berkhout, and Walker 1993, p. 44.

72. Albright, Berkhout, and Walker 1993, pp. 44–45.

73. Albright, Berkhout, and Walker 1993, pp. 96–97.

above, plutonium from these has gone into the weapons program. (Fuel fabrication for the gas-graphite reactors is done at Annecy in a plant owned by Cogéma.)[74]

France also has an extensive program for reprocessing plutonium from commercial spent fuel for its own breeder reactor program and for other countries, notably Germany and Japan but also Switzerland, Belgium, and the Netherlands.[75]

Chemical Separation of Plutonium

As noted above, France began reprocessing in 1949, when it separated the first milligrams of plutonium from the spent fuel of the Zoé reactor. A pilot plant was commissioned in 1954, followed by France's first full-scale reprocessing plant, the UP1, which was completed at Marcoule in 1958. Like the pilot plant, the full-scale plant was built by Saint-Gobain; it was designed to extract plutonium from the spent metal fuel of the new G reactors being constructed on the same site. The UP1 also later recovered plutonium from the blankets of the Phénix breeder reactor. It also processes spent fuel from the Célestin I and II reactors at Marcoule and various CEA research reactors. Its capacity is estimated at 400 metric tons per year.[76]

From 1974 onward the UP1 in Marcoule also reprocessed spent fuel from Magnox reactors in France and Spain (the plutonium from these reactors belongs to France), including the now shut Chinon 1, 2, and 3 reactors, the shut units at Saint-Laurent-des-Eaux (No. 1 and 2) and Bugey (No. 1).[77] Much of this was for military purposes.

In 1966 at La Hague, Cogéma built a second plutonium-reprocessing plant, UP2, originally only for civilian Magnox fuel. Its original capacity was 400 metric tons per year of heavy metal. It initially reprocessed Magnox reactor fuel. To accommodate France's growing dependence on light-water reactors (with oxide fuels) and its foreign contracts, the UP2 plant was modified to allow reprocessing of oxide fuels as well.[78] The UP2 plant was expanded to handle 800 metric tons per year of heavy metal; its operation was authorized in May 1994 and it is now called UP2-800.

Of the total of 45.2 to 46.5 metric tons of plutonium separated in France through 1990 (see table 9.2), 18.7 metric tons have been for for-

74. Davis 1988, p. 70.

75. Albright, Berkhout, and Walker 1993, p. 99.

76. Albright, Berkhout, and Walker 1993, table 6.2 and pp. 96–103.

77. Albright, Berkhout, and Walker 1993, p. 98; Norris, Burrows, and Fieldhouse 1994, p. 202.

78. Albright, Berkhout, and Walker 1993, table 6.2.

Table 9.2 Plutonium Production in France through 1990

Reprocessing Center	Production (metric tons)[1]	Source (reactor type)[2]	Notes
Marcoule (UP1 plant)	7.8	Civilian Magnox	Some other fuel also
	2.5–2.8	G1, G2, G3 Magnox	
	0.5–1.5	Célestin 1 and 2 HWR	
Subtotal	10.8–12.1		
La Hague (UP2 + UP3 plants)[3]	1.17	LWR	For Belgium
	14.58	LWR	For (West) Germany
	1.17	LWR	For Japan
	0.67	LWR	For Netherlands
	1.11	LWR	For Switzerland
	7.0	LWR	Internal French production
	8.7	Magnox	Includes plutonium belonging to France produced in the Spanish Vandellos reactors
Subtotal	34.4		
Total	45.2–46.5		

1. Plutonium recovered from breeder reactor fuel is *not* included, under the assumption that the plutonium fuel loaded into the breeder reactors has approximately equaled the plutonium present in the discharged spent fuel.
2. LWR = light-water reactor; HWR = heavy-water reactor.
3. UP3 reprocessed only LWR fuel, producing 1.7 metric tons of plutonium in all. The rest of the La Hague reprocessing up to 1990 was in the UP2 plant.

Source: Albright, Berkhout, and Walker 1993, tables 3.8, 6.7, and 6.8; and Frans Berkhout, personal communication, 29 September 1993.

eign clients, of which 78 percent has been for Germany.[79] From 5.0 to 8.2 metric tons (including losses) have been for the military program.[80] To accommodate growing sales of reprocessing services to foreign clients, Cogéma built another reprocessing plant for oxide fuels of light-water reactors at La Hague, UP3, which was started up in 1990.[81] Like UP2-800, UP3 has a capacity of 800 metric tons of heavy metal per year.

The Cyrano laboratory at Fontenay-aux-Roses began separating plutonium on a small scale in 1968 from the fuel of the Rapsodie and Phénix breeder reactors, but the total recovered is estimated at only 15

79. Albright, Berkhout, and Walker 1993, p. 111.

80. Albright, Berkhout, and Walker 1993, table 3.8.

81. Albright, Berkhout, and Walker 1993, table 6.2.

to 20 kilograms. A larger treatment workshop (AT1), now shut, came on line at La Hague in 1969. It also reprocessed Rapsodie and Phénix fuel. Reprocessing of breeder fuel began at Marcoule in 1974, with the modification of CEN Valrho's Pilot Workshop Service (SAP) plant. A pilot oxide-processing system (TOP) was added to enable it to process the oxide fuels of breeder reactors. This pilot oxide-processing system (SAP/TOP) separated plutonium from the spent fuel of the Rapsodie reactor as well as German experimental reactors MZFR and KNK III until 1976. This plant was closed in 1983 to be renovated and expanded. It was renamed the Breeder Oxides Treatment Facility (TOR) and reopened in 1988 with a capacity of 6 metric tons of spent fuel per year.[82]

Note that in the plutonium production table (see table 9.2) we do not include plutonium separated from breeder reactor fuel. This is because plutonium is also loaded into these reactors and the net plutonium production from the French breeder program, if any, has been small. Through 1990, an estimated 300 kilograms of plutonium of breeder origin was recovered at AT1, 3.6 metric tons at the Marcoule facilities, and 2.9 metric tons in UP2 at La Hague.[83]

Tritium Separation

Cogéma began operating a tritium extraction plant at Marcoule in November 1967.[84] It extracts tritium from the lithium-6 alloy targets that are irradiated in the Célestin I and II reactors. No information is available on its capacity or output.[85]

Production of Plutonium Components

The plutonium-239 pits (or cores) of nuclear weapons are machined and fabricated at the Valduc Research Center, which is the only known fissile component production site.[86]

Production of Other Components

• *Beryllium:* Although France used graphite as the moderator for its plutonium production reactors, it also began manufacturing beryllium

82. Albright, Berkhout, and Walker 1993, pp. 100–102.

83. Albright, Berkhout, and Walker 1993, table 6.7.

84. Barrillot and Davis 1994, p. 124.

85. Albright, Berkhout, and Walker 1993, p. 44.

86. Barrillot and Davis 1994, p. 222.

in 1965 to support the beryllium fusion research conducted at the Péchiney Metallurgical Research Center in Chambéry.[87]

• *Chemical high explosives:* The Vaujours-Moronvilliers Research Center is responsible for developing the chemical high explosives used in French nuclear weapons. The chemical explosives, perhaps manufactured outside the CEA, are assembled into nuclear detonators by the CEA's Le Ripault Research Center in Monts, Indre et Loire.[88]

• *Electronic components:* Most miniaturized electronics for warheads are developed and manufactured at the Bruyères-le-Châtel Center near Arpajon, about 35 kilometers south of Paris.[89]

• *Lithium-6 and lithium deuteride:* The Cogéma plant in Miramas provides the lithium-6 needed for tritium production at Marcoule. Since 1976, this plant also has produced boron for various purposes.[90]

• *Neutron generators:* The French company Sodern has produced deuterium-tritium neutron sources for triggering nuclear weapons since 1960 at its plant in Limeil-Valenton.[91]

Assembly and Maintenance of Nuclear Weapons

The Department for the Production of Nuclear Assemblies fabricates nuclear warheads partly at the CEA's Valduc Research Center in Lamaragelle near St. Seine L'Abbaye, Côte-d'Or and partly at other locations.[92] The Valduc Center shares responsibility for nuclear warheads with the Aquitaine Scientific and Technical Research Center (CESTA) near Bordeaux. The main purpose of CESTA is "to militarize nuclear warheads."[93]

Waste Disposal

Many aspects of military and civilian radioactive waste disposal in France—from selecting, constructing, and operating sites to the rules governing waste classification, packaging, and transportation—are the responsibility of ANDRA, the French acronym for the National Agency for the Management of Radioactive Wastes, established within the

87. Barrillot 1991, part 2, 2.43.

88. Davis 1988, p. 92.

89. Barrillot and Davis 1994, p. 200.

90. Barrillot and Davis 1994, p. 345.

91. Barrillot 1991, part 2, section 5.21.

92. Barrillot and Davis 1994, pp. 221–222.

93. Norris, Burrows, and Fieldhouse 1994, p. 199.

CEA in 1979.[94] The CEA's Valduc Research Center, Bruyères-le-Châtel Research Center, and other centers, such as Marcoule and Pierrelatte, treat and package radioactive wastes from military sources before sending them to ANDRA.

France categorizes its nuclear wastes according to specific activity—that is, according to radioactivity per unit weight and, to some extent, according to longevity. This is in general accord with European practice and in contrast to the haphazard scheme of waste classification in the United States, which defines wastes according to the processes that generate them.[95] France has three categories of waste, with Category A being the least radioactive and Categories B and C having progressively higher specific activity. Both categories B and C are slated for repository disposal. Category C corresponds most closely with the U.S. high-level waste category.

ANDRA's first storage facility, the Centre de Stockage de la Manche (La Manche Storage Center), was built in 1969 by Infratome. ANDRA took over the facility from this private company in 1979.[96] The 12-hectare site is on the eastern edge of the La Hague complex. Its 400,000-cubic-meter capacity was three-fourths filled by 1987.[97] ANDRA closed the site, now full, in 1992.

La Manche accepted for shallow burial and surface storage all "Category A" radioactive wastes, both military and civilian, that were not incinerated elsewhere. ANDRA defines Category A wastes as those of weak to moderate total activity contaminated by beta- or gamma-emitting isotopes with half-lives of less than 30 years. France's civilian and military programs generate about 30,000 cubic meters of these low-level wastes every year.[98] By the year 2000, ANDRA projects that 800,000 cubic meters of Category A wastes will have accumulated in France, making up almost 95 percent of the total volume but just 1 percent of the activity.[99]

Category A wastes can contain up to 0.1 millicuries (3.7 megabecquerels) of alpha emitters per kilogram, although the average alpha activity at a Category A site after 300 years is not to exceed 0.01 millicuries (0.37 megabecquerels) per kilogram.[100] Until 1979, La Manche accepted wastes with alpha radioactivity as high as 10 curies (370

94. Davis 1988, p. 95.

95. Makhijani and Saleska 1992.

96. ANDRA 1988, p. 9; Davis 1992, p. 3.

97. Davis 1988, p. 97.

98. Behar 1992.

99. ANDRA 1988, p. 5.

100. CRII-Rad 1992a; Davis 1988, p. 96.

gigabecquerels) per cubic meter. La Manche either buried wastes in concrete monoliths equipped with drainage channels or placed them on aboveground concrete slabs, covered with clay, soil, and grass. Some older wastes were buried in unlined trenches that had only plastic covers.[101] ANDRA says it will monitor surface and ground waters at the site for 300 years, then release the area for other uses without restriction.[102]

Category A solid wastes are usually compacted before shipment to reduce their volume, while liquids are treated to obtain concentrated sludges that can be solidified in cement, bitumen, or resins. Facilities for concentrating sludges were first built at Marcoule in 1966 and also operate at various CEA Nuclear Research Centers. As of 1973, there were incinerators for Category A waste at La Hague and Marcoule as well as CEN research centers in Grenoble, Strasbourg, Cadarache, and Fontenay-aux-Roses.[103]

ANDRA opened a new 112-hectare site for surface storage of Category A wastes in 1992. Located near Soulaines-Dhuys in l'Aube, the site has a nominal capacity of more than one million cubic meters, which is expected to serve for 30 years.[104] As at La Manche, wastes go into concrete bunkers, with cement poured like mortar around the waste packages.[105] These are covered with soil and landscaped.

Also stored "temporarily" at La Manche, pending permanent deep underground disposal in a waste depository, are Category B wastes, defined as those with long half-lives but containing only weakly to moderately radioactive alpha-emitting isotopes in concentrations greater than 0.1 millicuries (3.7 megabecquerels) per kilogram. France generates 4,700 cubic meters of Category B wastes per year; ANDRA projects a total of 45,000 cubic meters to be in storage by the year 2000.[106] Incinerators for Category B wastes operate at several facilities, including Marcoule, Pierrelatte and the CEA's research center in Cadarache.[107] Another incinerator is under construction at Valduc.

After reprocessing, the leftover high-level liquid wastes (defined as Category C) are stored in double-walled, stainless steel, refrigerated tanks at Marcoule and La Hague until they can be vitrified. While France generates just 126 cubic meters of Category C wastes each

101. Davis 1992, pp. 3–4.

102. ANDRA 1988, p. 8.

103. Davis 1988, pp. 96–98.

104. CRII-Rad 1992a; Davis 1988, p. 98.

105. Davis 1992, p. 6.

106. ANDRA 1988, p. 5; Behar 1992.

107. Davis 1988, p. 99.

year (about 0.5 percent of the country's radioactive waste by volume), these contribute 98 percent of the total activity.[108] By the year 2000, France is projected to have 3,000 cubic meters of vitrified Category C wastes.[109]

France's first pilot plant for vitrifying Category C wastes operated probably at Fontenay-aux-Roses before being transferred to Marcoule. Known as Gulliver, its dates of operations are not well known but closure is believed to have been in the mid-1960s.[110] A pilot plant for vitrifying Category C wastes, Piver 1, operated in Marcoule between 1969 and 1980. Between 1969 and 1973 it vitrified about 5 million curies (185,000 terabecquerels) of fission products, mostly derived from the reprocessing of graphite-gas reactor fuel.[111] A period of research and development followed. The vitrification plant resumed activity between 1979 and 1980 to demonstrate the feasibility of vitrifying plutonium produced in Phénix. Piver 1 was dismantled between 1982 and 1991, and was replaced by Piver 2.[112] Cogéma also built the world's first commercial-scale vitrification plant (AVM) for high-level liquid wastes, which opened at Marcoule in 1978.[113]

ANDRA is looking for a suitable site for deep underground disposal of the Category B wastes accumulating at La Manche and the Category C reprocessing wastes at La Hague and Marcoule. In addition to examining sites in France, the CEA is taking part in an international research study of the possibilities of sub-seabed disposal.[114] Opposition, especially in farming regions, halted the search for a land-based site for several years. The search for a repository started again in 1993.

The plutonium recovered from French and other civilian spent fuel, which Cogéma regards as a resource, is stored at La Hague. This stockpile contains about 40 metric tons of plutonium. A much smaller quantity of plutonium-contaminated wastes are stored at the Cadarache site, which develops naval nuclear reactors. This center also stores plutonium wastes from the Bruyères-le-Châtel Research Center.[115]

108. *Médecine et Guerre Nucléaire* 1992d.

109. ANDRA 1988, p. 5.

110. Barrillot and Davis 1994, pp. 164–165.

111. IPPNW and IEER 1992, p. 70.

112. Barrillot and Davis 1994, pp. 165–166, 170.

113. Barrillot and Davis 1994, p. 168.

114. Davis 1988, p. 103.

115. Barrillot and Davis 1994, p. 207.

MAJOR SITES

La Hague

The La Hague complex is about 25 kilometers west of Cherbourg and 5 kilometers south of Cap de la Hague. The 300-hectare site is on the west coast of the Cotentin Peninsula, which juts into the English Channel (called La Manche in France). The CEA chose this area because it is thinly populated, exposed to high winds, and close to strong currents—all important for reducing the local impact of radioactive releases into the air and sea.[116] It is the principal reprocessing center in France. We have already described plutonium separation activities above, and so we will focus on waste management and environment activities in this section.

On the eastern edge of the La Hague site is ANDRA's now closed La Manche Waste Storage Center, which accepted Category A and B wastes from all over France. The rest of the La Hague site, inside a double-fenced enclosure, houses Cogéma's UP2 (now UP2-800), UP3, and AT1 reprocessing plants; its international stockpile of civilian plutonium; and facilities for the temporary storage of some nuclear wastes (including 10,000 metric tons of spent fuel, reprocessing liquids, and Category B wastes). The incineration of Category A and B wastes, the vitrification of liquid (reprocessing) wastes, and the treatment of other solid and liquid wastes, which are packed in concrete or bitumen, are also carried out at that site.

A series of specialized processing "workshops" or "processing cells" are associated with each step in reprocessing, from removing cladding from incoming spent fuel and separating the uranium and plutonium to treating outgoing wastes. The waste treatment operations include removing and concentrating radionuclides from high-volume liquid wastes. The concentrated sludges that result had been, until recently, treated and solidified in bitumen. La Hague is now phasing out the bitumen process.

Waste Output and Environmental Contamination at La Hague The spent fuel from civilian light-water reactors (LWRs) contains much more radioactivity from fission products than the spent fuel of Magnox reactors. This is due to the much higher "burn-up" of LWR fuel. Although LWR spent fuel lies in pools for at least three years before reprocessing to allow short-lived radioactivity to decay and thermal heat to cool, the high-level liquid wastes that remain after reprocessing still contain about 1.7 million curies (62,900 terabecquerels) of radioactivity per cubic meter. This is much greater than

116. CEA brochure, as cited in Davis 1992, p. 12.

the average concentration of U.S. high-level wastes. The radioactivity of these LWR reprocessing wastes falls by more than half in the five years after discharge from the reprocessing plant, however, leaving long-lived activity of about 0.8 million curies (29,600 terabecquerels) per cubic meter. These reprocessing wastes—approximately 1 billion curies (about 37 million terabecquerels) total—are stored in liquid form in electrically cooled, double-walled, stainless steel tanks until they can be vitrified in glass.[117] The first vitrification plant came on line in 1989 and the second in 1993.

La Hague reprocessing operations routinely release radioactivity into the environment through a 285-meter chimney and a polyethylene pipe that carries liquid wastes across the site above ground and then, underwater, out into the English Channel. The aboveground section of the pipe suffered 39 ruptures in just one year (1976–1977). These leaks contaminated the beach and the site's freshwater reserves.[118]

In 1990 and 1991, independent researchers with ACRO (l'Association pour le Contrôle de la Radioactivité dans l'Ouest) found short-lived ruthenium-106 and rhodium-106 in sediment samples taken from the small Sainte-Hélène stream on the Hauts-Marais plateau on the east side of the Cogéma site.[119] The stream's source is actually within the boundaries of La Hague, from which it flows just 4 kilometers to the sea, through the villages of Digulleville and Omonville-la-Petite. ACRO also detected cesium-137 levels of 2.2 becquerels (about 59 picocuries) per gram (dry) in February 1991 and 3.1 becquerels per gram (dry) (about 84 picocuries) in March 1991.[120]

These findings, confirmed by other independent laboratories and reported to the Special and Permanent Commission on Information about La Hague, forced Cogéma to undertake its own comprehensive sampling. In 1992, Cogéma reported finding the source of the contamination: an abandoned pipe within the double-fenced enclosure of the neighboring reprocessing center. This canal, made of fibro-cement, carried runoff waters from the northeast part of the site toward the Ste. Hélène River. These waters seeped through the particularly porous pipe, contaminating soil along the length of the pipe as well as the Ste. Hélène River into which it flowed. Cogéma has said it will repair the pipe, remove contaminated soil, and build a new controlled drainage system.[121]

In 1988, the latest year for which data are available, Cogéma reported liquid discharges as follows: about 67,500 curies (2500 tera-

117. IPPNW and IEER 1992, pp. 68–69.

118. Davis 1992, p. 12.

119. Davis 1992, p. 12; ACRO 1992.

120. *Médicine et Guerre Nucléaire 1991.*

121. *Gazette Nucléaire* 1992b.

becquerels) of tritium (about 6.8 percent of the limit); about 15,500 curies (575 terabecquerels) of other beta emitters (34.5 percent of the limit); about 1,300 curies (48 terabecquerels) of strontium-90 and cesium-137 (21.6 percent of the limit); and 10 curies (370 gigabecquerels) of alpha emitters (21.9 percent of the limit).[122]

Emissions from La Hague were not subject to any fixed limits until 1980. Since limits were adopted, Cogéma claims to have never exceeded them. Cogéma's own data on the volume and activity of La Hague's radioactive emissions to the sea from 1966 through 1987, published by the Special and Permanent Commission on Information about La Hague, support this claim.[123] However, this official commission, whose composition was defined by the prime minister, does not conduct any inspections and has no budget for independent studies. It receives and examines reports sent by official bodies like the Ministry of Industry, EdF, and Cogéma. It has published graphs showing the volume and radioactivity of La Hague's emissions to the sea over this 11-year period for tritium, strontium-90 and cesium-137, total beta and gamma, total alpha, and plutonium. (Tritium measurements began in 1970.)

Only in 1971 did La Hague exceed the current limits, and then only for strontium-90 and cesium-137. About 6,750 curies (250 terabecquerels) of cesium-137 and strontium-90 were released in 1971, about 13 percent more than the 1981 standard allows. The commission report offers no explanation for this peak. The data also show an unusual peak in alpha emissions reported for 1974, more than half of which was due to plutonium.[124]

In 1988, the latest year for which data are available, Cogéma reports discharges to air as follows: about 729,000 curies (27,000 terabecquerels) of krypton-85 (5.6 percent of the allowable limit); about 570 curies (21,000 gigabecquerels) of tritium (about 10 percent of the limit); about 0.6 curies (22 gigabecquerels) of halogens (one-third of the limit); and 6.8 millicuries (25 megabecquerels) of aerosols (less than 0.04 percent of the limit). According to Cogéma, La Hague has never exceeded its airborne release limits, and only in 1984 did emissions even reach 20 percent of the allowable maximum (in the case of halogens).[125]

The most radioactive solid wastes in "temporary storage" at La Hague as of 31 December 1991—aside from spent fuel awaiting reprocessing—were 1,653 metric tons of spent-fuel casings (cladding) and end caps stripped from spent fuel, averaging 10,000 curies (370 terabecquerels) per metric ton, and the 8,500 cubic meters of sludges

122. CGT 1989.

123. *Gazette Nucléaire* 1988.

124. *Gazette Nucléaire* 1988.

125. CGT 1989.

precipitated out of reprocessing and other liquid wastes, averaging 500 curies (18.5 terabecquerels) per cubic meter.[126] Most are stored underwater in a concrete silo meant for high–specific activity waste. In 1991, this silo contained 1,075 cubic meters of spent fuel casings and end cap scraps made of zircalloy and steel (the result of cutting open 2,500 metric tons of spent fuel); 3,500 contaminated aluminum covers from the sliding rack assembly used to move spent fuel; about 1 metric ton of highly radioactive fine-particle wastes collected from the chopping and dissolution steps of reprocessing; and 100 cubic meters of crushed filtration resins. Only filter resins and clarification residues have been added to the silo.[127] Cogéma hopes to treat all these wastes in a continuous production line process by the late 1990s.[128]

The PLH pool contains cladding wastes from 450 metric tons of spent fuel, while the newer S1 pool contains cladding wastes from 350 metric tons of spent fuel. At capacity, S1 is designed to accommodate cladding wastes from 2,000 metric tons of spent fuel.[129] The activity of these different wastes at the time of their creation varies considerably (see table 9.3).

Cogéma has promised the Service Central de Sûreté des Installations Nucléaires that it will repackage sludges and magnesium hulls for long-term storage by 1998. It has promised to describe new installations for the rest of the waste and procedures to prepare these wastes.[130] Cogéma plans to transfer all the cladding wastes in the HAO silo to the S1 pool pending final disposal. Three options have been proposed for the final disposition of the cladding wastes and end caps: permanent encasement in cement, fusion into ingots in a cold crucible, and compaction of the cladding after separation and decontamination of the end caps. A final choice is not expected until 1994, with treatment of all the pool and silo wastes expected to take until at least 2010.[131]

Sludges now in silos will be recovered and bituminized, while the radioactive water will be passed over beds of crushed ion-exchange resins. Finally, the filter residues and contaminated resins stored in silos will be removed and treated at the same time as the resins from the water filtration unit. Cogéma proposes to mix these wastes in cement before burial.[132]

126. ANDRA 1993.

127. *Gazette Nucléaire* 1991g.

128. Aycoberry 1990.

129. *Gazette Nucléaire* 1991g.

130. As cited in Davis 1992, p. 12.

131. *Gazette Nucléaire* 1991g.

132. Aycoberry 1990.

Table 9.3 Activity of Wastes in High-Activity Oxide Silos and Pools

Waste Type	Beta/Gamma[1]	Alpha[1]
Zircalloy cladding	888 TBq per cubic meter (24,000 curies per cubic meter)	17.76 TBq per cubic meter (480 curies per cubic meter)
End caps	11.1 TBq per cap (300 curies per cap)	222 GBq per cap (6 curies per cap)
Crushed ion-exchange resins	37 TBq per cubic meter (1,000 curies per cubic meter)	740 GBq per cubic meter (20 curies per cubic meter)
Filter residues (mostly ruthenium)	37 TBq per kilogram (1,000 curies per kilogram)	Negligible
Aluminium covers	Negligible	
Silo water	370 GBq per cubic meter (10 curies per cubic meter)	1.85 GBq per cubic meter (0.05 curies per cubic meter)

1. Tbq = trillions of becquerels, Gbq = billions of becquerels.

Source: *Gazette Nucléaire* 1991g.

The La Hague complex has suffered at least two serious power outages. Electricity for cooling is necessary for maintaining the integrity of its high-level waste tanks; a prolonged cooling failure could result in their contents overheating and possibly exploding.[133] Power failures between three and twelve hours could result in the release of fission products to the environment. However, in all instances, restoration of power has prevented any serious accident.

The most serious incident occurred on 15 April 1980 when La Hague lost all electric power from EdF. When the power came back on, the main transformer at the site caught fire. The fire disabled the emergency generators, which were located in the same building, and forced a shutdown of the entire complex. It took officials 45 minutes to bring in replacement generators from Cherbourg, two hours to put out the fire, and 24 hours to restore ventilation and negative pressure to other buildings at the site. Negative pressure ensures that contaminated air does not flow out via cracks and other leakage-prone areas. Exhaust fans blow air out of filtered stacks designed to trap most radioactive particles in this exhaust.[134]

Full or partial power outages at La Hague have affected the SPF2 waste-storage facility for fission product solutions (on 8 August 1989); the UP2 plant (on 10 September 1990, 14 March 1989.[135]

At least two fires have occurred, including one in the AT1 reprocessing plant on 20 June 1991, while the facility was being dismantled.

133. Davis 1992, p. 11.

134. Davis 1992, pp. 10–11.

135. Davis 1992, p. 11.

The other came at 4 a.m. on 6 January 1981 in a solid-waste silo containing graphite hulls and magnesium end pieces removed from spent fuel as part of reprocessing.[136] The Confédération Française Démocratique du Travail (CFDT), the union representing La Hague workers, has in the past pointed out several design flaws making for hazardous conditions and difficult routine maintenance.[137]

Marcoule

The Marcoule complex is on the west bank of the Rhône River in southern France, just south of the small town of Bagnols-sur-Cèze, Gard. The site is 20 kilometers northwest of Avignon and about 100 kilometers from the mouth of the Rhône at the Mediterranean Sea. Marcoule actually is two distinct facilities: Cogéma's Marcoule Establishment, with about 2,250 employees, and the CEA's Vallée du Rhône (or Valrho) Nuclear Research Center, with about 1,000.[138]

Construction at Marcoule began in 1954. Under the control of Cogéma are the G1, G2, and G3 plutonium production reactors (started up in 1956, 1958, and 1959 and shut down in 1968, 1980, and 1984, respectively; they are being dismantled); the Célestin I and II tritium reactors (modified to also produce plutonium); the UP1 reprocessing plant; the Melox plant for the production of mixed-oxide fuel for civilian (pressurized-water) power reactors; and numerous waste storage and treatment facilities.[139] The latter include facilities for the interim storage of nuclear wastes (spent fuel, reprocessing liquids, and Category B wastes not sent to La Manche); for compacting, solidifying, and incinerating wastes; and for vitrifying Category C wastes. Liquid wastes not vitrified or incinerated are concentrated via evaporation, coated in bitumen, and encased in concrete.[140]

Marcoule also has the Phénix breeder reactor and the pilot plant for reprocessing breeder fuel (now known as the Breeder Oxides Treatment Facility). We have already discussed plutonium production at Marcoule, and so will focus on waste management and environmental contamination in this section.

Waste Output and Environmental Contamination at Marcoule Based on the above estimates of separated plutonium, excluding breeder-recycled plutonium, reprocessing operations generated about 130

136. Davis 1992, p. 11.

137. As cited in Davis 1992, p. 11.

138. CFDT 1984, p. 9.

139. Norris, Burrows and Fieldhouse 1994, p. 202.

140. Barrillot and Davis 1994, Section on Centre de Marcoule, pp. 91–188.

million curies (4.81 million terabecquerels) of combined strontium-90 and cesium-137 wastes through 1990 (figures are not decay corrected). The fission products are stored in liquid form in electrically cooled, double-walled, stainless steel tanks until vitrified in glass. Marcoule's pilot vitrification plant, Piver 1, operated from 1969 to 1980 and vitrified 5 million curies (185,000 terabecquerels) in 12 metric tons of glass, for an average of 0.42 million curies (15,540 terabecquerels) per metric ton.[141] Its commercial-scale vitrification plant began operation in 1978, and through the end of 1985 it had produced about 400 metric tons of glass, which is stored on-site in more than 1,100 air-cooled containers.[142]

Sludges precipitated out from UP1's effluent treatment facility are packaged in drums with bitumen for final disposal.[143] By 1983, 20 percent of these drums had been sent to ANDRA's waste facility at La Manche; the rest may be stored at Marcoule.[144] Other UP1 radioactive wastes include the spent fuel cladding that is stripped off prior to reprocessing and the noble gases (primarily krypton-85 and tritium) released during reprocessing.

Organic wastes from the Melox plant are scheduled to be burned in a new incinerator with a rated capacity of 25 kilograms per hour.[145] There have been three incinerators at Marcoule *M*. One for low-level waste, one for wastes high in alpha activity, and a pilot plant.[146] According to workers at the site, none ever functioned well enough to get beyond the testing stage. All have experienced problems with clogging or corrosion of their filters due to the acidity of the wastes being burned.[147]

The CFDT's Syndicat National de l'Energie Atomique, which represents workers at Marcoule, has been one of the strongest critics of radioactive waste management practices at the site. In 1980, it characterized conditions in the UP1 plant, where cladding is removed from irradiated fuel, as "an indescribable mess. Each available square meter was a place for storing contaminated and radioactive materials, sometimes up to 500 millirem per hour [5 millisieverts], including in the passageways."[148]

141. IPPNW and IEER 1992, p. 70.

142. Davis 1988, pp. 101–102.

143. Barrillot and Davis 1994, p. 171.

144. Davis 1988, p. 5.

145. Davis 1992, p. 10.

146. Barrillot and Davis 1994, p. 149–154.

147. Davis 1992, p. 10.

148. Barrillot and Davis 1994, p. 110.

A report issued by the CFDT and published in 1984 by Gazette Nucléaire provided detailed criticisms of the Marcoule operations. It claimed that the advanced age, outmoded design, overloaded capacity, and contaminated conditions of the reprocessing, waste treatment, and waste storage facilities were responsible for numerous problems that threatened worker safety and the environment. The CFDT report gives an example of a fire on 22 March 1983 in the decladding facility where the magnesium casing around spent fuel is removed. The fire started in accumulated magnesium wastes and led to an explosion that completely disabled the facility. Other reports of this accident say the fire and explosion dispersed approximately 30 kilograms of uranium that had accumulated in the production line throughout the basement and several halls of the building. Surface contamination was on the order of 0.05 milligray per hour (5 millirad per hour).[149]

Little has been published about radioactivity in the environment around Marcoule. A study in 1990 by the Institut de Biogéochimie Marine in Montrouge on the origins of artificial radionuclides discharged into the Mediterranean Sea from the Rhône River found that Marcoule was responsible for most of the radioactivity in the river:

> The total activity supplied by the river averages 70 gigabecquerels [about 1.9 curies] per day and is one order of magnitude less than the effluent discharges from [the] La Hague reprocessing plant. However, because of a lower degree of dispersion and dilution in a relatively small water body compared to the situation at La Hague [where wastes are discharged directly into the sea], the artificial radionuclide concentrations of gamma emitters and plutonium isotopes are higher at the mouth of the river: 1,200 Bq [about 32.4 nanocuries]/kg in suspended sediments and 0.05 to 0.08 Bq [about 1.35 to 2.2 picocuries]/l in waters. Overall particulate activities are the highest that have been measured so far in French estuarine systems.[150]

While there are several nuclear facilities along the Rhône upstream from Marcoule—including the Super-Phénix breeder in Creys-Malville; civilian power reactors in Bugey, Cruas-Meysse, and Saint-Alban; and the Pierrelatte and Tricastin uranium-enrichment complexes—Marcoule is the only significant source of plutonium emissions.[151]

At the request of Avignon authorities, the independent laboratory CRII-Rad (Commission de Recherche et d'Information Indipendantes sur la Radioactivité) examined soil, vegetation, and river-sediment samples taken near Marcoule in 1990. It found plutonium-239 and plutonium-240 at levels of about 0.4 picocuries (0.015 becquerels) per gram of dry sediment taken from the Rhône just downstream of

149. *Gazette Nucléaire* 1983; CFDT 1984, pp. 8–9; Barrillot and Davis 1994, pp. 110–111.

150. Martin and Thomas 1990.

151. CRII-Rad 1992, p. 1.

Marcoule.[152] This is somewhat above fallout levels due to atmospheric testing.[153]

The regulatory limits set for Marcoule allow total alpha releases into the Rhône of up to about 150 gigabecquerels (about 4 curies) per year.[154] The alpha limit is the strictest; other permitted releases into the Rhône include up to 2,500 terabecquerels (about 67,500 curies) per year of tritium and 6 terabecquerels (about 160 curies) per year each of strontium-90 and cesium-137.[155]

Found upstream from Marcoule, as expected, was evidence of contamination from activation products released by nuclear power plants, including cobalt-58, cobalt-60, and magnesium-54.[156] A much wider variety of radionuclides appeared downstream in aquatic plants, including the same activation products as well as fission products (such as rhodium-106, iodine-129 and -131, and cesium-134 and -137) and transuranic isotopes (americium-241 and plutonium-238, -239, and -240).[157] All these are characteristic of reprocessing wastes. Only cesium-134 and cesium-137 were found in fish downstream.

Cesium-137, iodine-129, and americium-241, a decay product of plutonium-241, were found in samples of soil, thyme, and moss around Marcoule, with the highest levels in areas immediately downwind of the plant.[158] These radionuclides are characteristic of releases from Marcoule.

Decommissioning activities at Marcoule also have generated solid wastes. The plant's oldest plutonium-purification workshops, known as Room 82 and Room 100 and shut down in 1963 and 1973, respectively, were completely dismantled in 1985.[159] The operation took four years of planning. It required the removal of 1,000 cubic meters of solid wastes, all contaminated with alpha-emitting radionuclides (principally plutonium). The wastes included everything inside the four-story building, from steel beams and tanks to conveyor belts and glove boxes. Almost 300 containers of waste were transferred to ANDRA; 200 barrels of the most heavily contaminated plutonium wastes remained at Marcoule. Also recovered and recycled were 150 metric

152. CRII-Rad 1990, p. 16.

153. Based on Eisenbud 1987, p. 335. An areal density of 1.4 millicuries per square kilometer has been converted to an approximate density per gram of soil, assuming the plutonium extends at least one centimeter deep into the soil.

154. CRII-Rad 1992, p. 1.

155. CRII-Rad 1992, p. 2.

156. CRII-Rad 1992.

157. CRII-Rad 1992.

158. CRII-Rad 1992.

159. Dubois 1988.

tons of lead and 20,000 metric tons of barium sulfate. To complete the decommissioning of larger facilities at Marcoule, Cogéma announced in 1992 that it was building a waste treatment facility specifically for processing mildly contaminated steel; some 4,000 metric tons from decommissioning the G1 and G3 reactors are to be treated.[160]

While the total radioactivity of wastes is not known, according to the CFDT of Marcoule the number of barrels containing sludges in bitumen amounted to 58,000 in the spring of 1993. The number of barrels is expected to grow to 65,000 by the year 2000. At the moment they are covered with dirt overgrown with grass; they are too radioactive for surface storage. Surface storage would imply repackaging, which would entail the creation of more radioactive wastes.[161]

ENVIRONMENTAL PROBLEMS REPORTED AT OTHER NUCLEAR FACILITIES

In Saint-Aubin, Essonne, a waste storage site known as l'Orme des Merisiers was used by the CEA's nearby Saclay Nuclear Research Center, as well as by the CEA's Fontenay-aux-Roses Center and Le Bouchet. The site is contaminated from radioactive solid wastes and sludges stored there from 1961 to 1972.[162] The wastes—1,484 concrete blocks—were stored in the open, above ground, on a concrete slab. Exposure to the elements created cracks and fissures in 243 blocks, allowing water to penetrate in and radioactivity to leak out.[163]

The CEA claims to have cleaned up all traces of radioactivity from the site in 1974, after it repaired the leaking canisters and transferred most of the stockpile to La Manche. However, tests by CRII-Rad in 1990 revealed about 58 picocuries (about 2.15 becquerels) per gram of plutonium-239/240.[164] This is hundreds of times greater than plutonium from fallout due to atmospheric testing.[165] The CEA insists that this is far below maximum allowable limits.[166]

After 1976, the Saclay center sent its radioactive wastes to a municipal dump in Bailleau-Armenonville, Eure-et-Loire, including 1,720 metric tons in 1988 alone.[167] Some 1,800 cubic meters of dry sludge

160. *Médecine et Guerre Nucléaire* 1992c.

161. Personal communication with CFDT as cited by Barrillot and Davis 1994, p. 176.

162. CRII-Rad 1990; CRII-Rad 1992a; see also *Gazette Nucléaire* 1991c for a detailed account of the history of the Saint-Aubin site.

163. *Gazette Nucléaire* 1991b.

164. CRII-Rad 1990, p. 16.

165. Based on Eisenbud 1987, p. 335.

166. CRII-Rad 1990.

167. CRII-Rad 1992a.

originally from l'Orme des Merisiers were received in August 1989. The total activity from its cesium-137, plutonium-230, and cobalt-60 content is 1 millicurie (37 megabecquerels).

At the Le Bouchet Plant in Itteville, where the CEA made uranium fuel for production reactors from 1946 to 1971, radioactive wastes are stored in two large open areas. A 5,000-square-meter decantation basin holds about 15,000 metric tons of sludges contaminated with about 15 curies (555 gigabecquerels) of radium-226. The average alpha activity in this basin, counting radium-226 and its four alpha-emitting daughters, with which it is in equilibrium, is about 55.8 millicuries (2,065 megabecquerels) per dry metric ton or about 28 millicuries (1,038 megabecquerels) per wet ton. Adjacent to this is a 3,500-square-meter "hydroxides park" that holds 2,000 metric tons of hydroxides left over from uranium treatment and 2,500 metric tons of soil contaminated with about 5 curies (185 gigabecquerels) of radium-226. Its average alpha activity, similarly accounted, is 21 millicuries (780 megabecquerels) per dry metric ton or 10.5 millicuries (390 megabecquerels) per wet ton.[168]

Although the concentration of alpha-emitters in both the decantation basin and the hydroxides park greatly exceeds France's average allowable limit of 10 millicuries (370 megabecquerels) per metric ton for surface storage facilities, the CEA has refused to remove these wastes from the site. Under strong pressure from environmental groups and local authorities, it did agree in 1991 to cover the wastes with an impermeable clay cover that it claims will reduce radon emissions one hundred-fold.[169]

Le Bouchet also shipped much of its uranium-processing wastes to abandoned uranium mines for disposal, including Brugeaud, Bauzot, and Lavaugrasse, as well as La Crouzille and Montbouchet.[170] About 2,750 metric tons of lightly contaminated soil and rubble containing 0.2 curies (7.4 gigabecquerels) from Le Bouchet were used to make an embankment along Autoroute A87 between Chilly-Mazarin and Longjumeau.[171] The site's most radioactive wastes, some 2,135 metric tons of hydroxides and sulfates containing 400 curies (14.8 terabecquerels) of radium, went to the La Manche facility at La Hague.[172]

The CEA Research Center at Bruyères-le-Châtel, Essonne, intentionally released 1 gram of tritium, about 10,000 curies (370 terabecquer-

168. *Gazette Nucléaire* 1991d; 1992c.

169. *Gazette Nucléaire* 1991e; 1992a; 1992.

170. *Gazette Nucléaire* 1992a.

171. *Gazette Nucléaire* 1992a.

172. *Gazette Nucléaire* 1992a.

els), into the air in September 1986.[173] It did so to study tritium's dispersion in the environment, an ostensibly civilian experiment directed by the European Community's fusion research program. The center also released about 5680 curies (210 terabecquerels) accidentally on 28 April 1988.[174]

The CEA's Cadarache Nuclear Research Center maintains a stockpile of thorium nitrates in an abandoned hangar. It holds an estimated 2,245 metric tons of thorium in 21,700 drums with a total activity estimated at about 1940 curies (71.8 terabecquerels).[175]

The Malvési uranium conversion plant operated by Comurhex in Narbonne sent 176,000 drums of partially decontaminated uranate wastes—about 4,226 metric tons with a total activity of about 5.94 curies (220 gigabecquerels)—to the Margnac and Fanay mines in Haute Vienne between 1975 and 1989.[176] Since 1989, these wastes have been stored on site. Raffinate effluent from the plant is decanted in seven settling basins that hold a total of 200,000 metric tons of solids and 70,000 cubic meters of liquids, containing about 275 metric tons of uranium.[177] Comurhex also maintains a storage depot for industrial wastes, holding another 20,000 cubic meters of waste, including some low-level radioactive waste.[178]

The uranium enrichment facilities at Pierrelatte, Drôme, have generated about 7,000 metric tons of depleted uranium hexafluoride.[179] From 1968 to 1971, about 1,770 cubic meters of these and other assorted wastes from Pierrelatte were disposed in the Brugeaud mine. The dismantling of older chemical treatment facilities at the site has generated 2,118 drums of waste sent to La Manche, while 400 metric tons of steel and 40 metric tons of mercury were recovered and sold. Another 45 metric tons of mercury-contaminated wastes that ANDRA refuses to accept remain at the site, along with 240 cubic meters of rubble.[180]

HEALTH IMPACT OF THE NUCLEAR WEAPONS COMPLEX

France's extreme secrecy about its nuclear weapons program is most evident with regard to information regarding health effects. Almost no

173. CRII-Rad 1989.

174. CRII-Rad 1989.

175. *Gazette Nucléaire* 1992a.

176. *Gazette Nucléaire* 1992a.

177. *Gazette Nucléaire* 1992a.

178. *Gazette Nucléaire* 1992a.

179. CRII-Rad 1987.

180. *Gazette Nucléaire* 1992a.

information is publicly available with which to directly assess health effects.[181] On the other hand, French officials tend to publicly downplay the health hazards of ionizing radiation. This is perhaps a reflection of France's sensitivity over its nuclear power industry, which provides about 75 percent of the country's electricity, as well as over its nuclear weapons program.

Major inconsistencies are apparent in the reporting of the little data that exists. For example, two official groups—EdF and the Service Central de Protection contre les Rayonnements Ionisants (SCPRI) or Central Service for Protection from Ionizing Radiation—have produced estimates of collective radiation doses to nuclear industry workers that differ significantly, even though the data applies to the same population during the same years. For 1976 to 1981, SCPRI reported mean collective doses that were 1.72 times lower than the mean collective doses estimated by EdF. The existing data cannot explain this difference.[182]

Health Effects in Workers

EdF and CEA monitor their workers for external radiation exposures. In nuclear power plants between 1964 and 1974, the average individual external dose for these workers rose steadily to exceed 5 millisieverts per year by the end of 1974, compared with the current ICRP recommended limits of 20 millisieverts per year. J. C. Zerbib, a health physics expert who works for the CEA and is a member of the CFDT labor federation, attributes this apparent rise to the greater releases of radioactivity that are associated with the aging of nuclear plants.

Outside contract employees, working only temporarily in a nuclear plant, are a special group of exposed workers. Neither the CEA nor EdF record their individual exposures to radiation. In contrast to Germany, "exposure passports" that register the time and place of a worker's exposure to ionizing radiation have not yet been introduced in France. Nevertheless, these contract workers may be receiving an increasing share of the total burden of radioactivity exposure from French nuclear power plants—approximately 75 percent in 1989—mostly in the course of doing overhauls.[183] These data covering civilian facilities are relevant here because French nuclear weapons contain plutonium produced in civilian power reactors operated by EdF. It is not possible to segregate data for that portion of the civilian nuclear power industry that was devoted to military purposes.

181. Zerbib 1983.

182. Zerbib and Forest 1991.

183. Zerbib 1983.

At the reprocessing plant in La Hague, the mean annual radiation dose per worker remained below 5 millisieverts from 1968 to 1977.[184] Data on exposure are not available publicly for the Marcoule reprocessing plant, where most of the work is military-related.

Just how many workers received high amounts of ionizing radiation is difficult to ascertain. SCPRI, in a 1991 publication, chose to use 15 milligrays as a cutoff and reported that 5 percent of all exposed workers received doses over that amount for 1984–1985. SCPRI has published no additional data for later years. However, the CFDT, using data generated by Lefaure and Lochard,[185] estimates that for 1988, 45 percent of all exposed workers had exposures greater than 15 millisieverts, and 27 percent received doses over 20 millisieverts.[186] No information is available with which to independently determine if these estimates actually represent an increase in radiation exposure during the 1980s. Without centralized databases, a uniform measuring and reporting system, or exposure passports available to independent researchers, it is impossible to evaluate the accuracy of these estimates.

One source, albeit an indirect one, regarding the possible scope of radiation-induced illnesses among exposed workers is France's social security system.[187] From 1962 to 1985, it accepted about 18 disability claims per year as induced by occupational exposures to radiation. The review process, from filing to acceptance, takes about a year. The accepted claims comprised only 0.42 percent of all claims. Of the total number of people who died while their claims to the social security system were being reviewed, 12.3 percent filed a claim for radiation-induced illness.[188] However, we cannot evaluate the merit of these claims without dose and health data.

Clearly, many potential biases make social security data a seriously flawed source for estimating the number of radiation-induced illnesses. These biases could influence the numbers upward or downward. For example, workers have an economic incentive to claim that their disease relates to radiation. On the other hand, these statistics do not reveal diseases occurring after retirement, reducing the number of cancers that are likely to be counted.[189] Unfortunately, no information is available with which to assess the criteria used to determine whether diseases were radiation-related as claimed. Although the French government acknowledges that radiation-induced disease occurs, the rate

184. LeFaure and Lochard 1990.

185. LeFaure and Lochard 1990.

186. Zerbib 1983.

187. Zerbib 1988.

188. Zerbib 1985.

189. Zerbib 1985.

at which this occurs is in dispute. Further investigation into the French social security system database may be warranted.

Health Effects in the Population at Large

There is a total absence in the published literature of occupational or community epidemiology studies of France's nuclear weapons industry that use established techniques to follow well-defined populations over periods of time. Instead, only a handful of studies have appeared utilizing the relatively crude method of comparing mortality rates between geographically defined areas (ecological studies).

In 1983, CEA researchers M. Dousset and H. Jammet published the first ecological study focusing on the La Hague reprocessing plant.[190] They compared specific mortality rates in the Department of La Manche with those for France as a whole. It was not possible to study cancer incidence (rather than mortality rates) because only five of France's ninety-five departments had cancer registries at that time.

Dousset and Jammet found a trend of increasing overall age-adjusted cancer mortality rates for men over the observation period (1962–1982) in both La Manche and the country as a whole. Between 1975 and 1982, the absolute number of cancer deaths ranged between 200 and 400 per year for La Manche. The main diagnoses were tumors of the digestive tract, a finding the authors speculated related to the frequent use of alcohol. The standardized mortality ratio comparing La Manche with the general population demonstrated an annual increased risk ranging from 1.02 to 1.2 during the same period. In contrast, the cancer mortality rate for women decreased in La Manche and France as a whole between 1962 and 1982, with women in La Manche slightly less likely than other French women to get cancer. A number of methodological problems plagued this study, such as incomplete details on the causes of death on many of the death certificates studied. In addition, the authors stratified the population only by age and sex, omitting more detailed analyses that could control for differences in socioeconomic status, alcohol ingestion, and other potential confounders.

In a subsequent, more detailed study, Dousset analyzed the cancer mortality of Beaumont-Hague, the canton within La Manche containing La Hague, compared with cancer mortality for the whole department.[191] Between 1970 and 1982, Beaumont-Hague did not seem to experience a greater rate of either cancer overall or cancers of specific organs. However, the total population of Beaumont-Hague is small (less than 8,000), so the study could not accurately reveal anything less than a dramatic effect. Moreover, the first significant radioactive re-

190. Dousset and Jammet 1983.

191. Dousset 1989.

leases began in 1968, so the observation period was brief in view of the long latency period for most types of cancer that are related to ionizing radiation.

J. F. Viel and S. T. Richardson studied childhood leukemia mortality from 1968 to 1978 and from 1979 to 1986 in three circles around the La Hague plant, using death certificate data collected by the Institute National de la Santé et de la Recherche Médical (INSERM—National Institute for Health and Medical Research).[192] The standardized mortality ratio for all age groups, periods, and areas showed no significant effects. Nor was any significant trend found in the death experience comparing the two periods of time. Again, however, the small population studied gave the study very little ability to reveal significant effects. Furthermore, as the authors noted, no cancer registry was available to standardize data collection or to record cancer incidence as opposed to mortality.

C. Hill and A. Laplanche, focusing on six nuclear sites, have analyzed mortality data on diseases known to be radiation-induced (brain tumors, lung cancers, lymphomas, and leukemias) among people aged 0 to 24.[193] The observation period was from 1968 to 1987. The methods used were comparable to British studies of mortality rates around the Sellafield reprocessing plant (see chapter 8). The authors analyzed data from concentric circles around each site. Every commune (local administrative district) in these circles was matched with a similar commune in the same department outside the defined area. The study population for each year of observation was extrapolated from the censuses of 1968, 1975, and 1982.

In the group defined as exposed—people living within the circles around the sites—only the standardized mortality ratio for Hodgkin's disease was significantly elevated ($SMR = 197$, $p\text{-value} < 0.02$). The standardized mortality ratio for leukemia showed no significant effect, in contrast to the excess of leukemia observed around British nuclear sites. In view of the fact that some significant statistical associations are bound to appear by chance in a study examining multiple outcomes, the authors suggested that the association with Hodgkin's disease may fit in this category.

SUMMARY

Overall, the few epidemiological studies that exist concerning French nuclear installations do not suggest a significant increase in cancer mortality that can be readily associated with the installations themselves. However, a sanguine conclusion regarding the health effects of

192. Viel and Richardson 1990.

193. Hill and Laplanche 1990.

the French nuclear industry is premature. The methods employed in these few studies were less than optimal. They were largely ecological, rather than based on actual follow-up of defined populations and exposure measurements. It was not possible to ascertain cancer incidence in the absence of cancer registries, while the accuracy of death certificate data is questionable.

Moreover, most of the French plants are of relatively recent origin. The long latency of many radiation-induced cancers makes the period of epidemiological follow-up too short to detect a rise in cancer rates at this time. With regard to leukemia, which has the shortest latency period, improvements in medical technology and public health have reduced mortality, making studies of morbidity (and the establishment of cancer registries) especially important.

Occupational studies, which carry the theoretical advantage of having data on individual exposure doses, suffer in France from the problem of conflicting information, as demonstrated by apparent discrepancies between the dosimetric measurements reported by government agencies and those reported by unions.[194]

Finally, a point made throughout this book has particular relevance to France. The validity and credibility of all studies on the environmental and health effects of nuclear weapons production depends on lifting the shroud of secrecy that has enveloped the industry since its inception.

194. Zerbib 1979; Zerbib and Forest 1991.

10 China

Alexandra Brooks and Howard Hu

We have discussed many times the use of the atomic bomb, tactically.
—General Omar Bradley, chair, U.S. Joint Chiefs of Staff[1]

If we are not to be bullied in the present-day world, we cannot do without the bomb.
—Mao Zedong[2]

China exploded its first nuclear bomb on 16 October 1964, joining the United States, the Soviet Union, Great Britain, and France as a nuclear weapons state. This event introduced new complexities into the division of the world among nuclear states, which had been, at its core, a U.S.-Soviet confrontation. By 1964, the Sino-Soviet split had become a long-term reality of global politics. The successful Chinese nuclear test meant that there were now three different nuclear confrontations instead of one: U.S.-Soviet, U.S.-Chinese, and Soviet-Chinese, assuming that the British and French arsenals were strategically oriented in the same way as that of the United States.

Arguably, China's nuclear weapons program grew directly out of the superpower confrontation. One effect of the U.S. military policy of "containment" was to spur China to develop nuclear weapons, since explicit threats of the first use of nuclear weapons against both China and the Soviet Union were part of that policy.[3] Moreover, the United States tended to view the Soviet Union and China as a Communist monolith.

Against the backdrop of the Korean War and discussions about rolling back Communism, President Harry Truman announced on 30 November 1950 that he was considering a nuclear bombardment of China. The President said that the United States would use every

1. As quoted in Ege and Makhijani 1982, p. 11.

2. As quoted in Lewis and Xue 1988, p. 107.

3. For examples of first-use threats, see Ellsberg 1986, p. 40.

weapon in its arsenal to counter what he called "Chinese aggression" in Korea.[4]

Such statements were issued not solely for public effect. Consider the words of a 11 January 1951 U.S. National Security Council document on the military situation in Korea. It implies using the existence of nuclear weapons as a political tool, and perhaps actually using them in battle:

> The free nations are on the defensive because they are fighting the war ... on the basis which most favors the Soviets.... We are attempting to match men for men and tanks for tanks, instead of fighting most effectively with those elements of military supremacy we now have in the Far East.... The free nations do not in political discussion bring up their prime power advantage, the atomic bomb and the capacity to deliver it. That advantage now gives possible superiority of power in the free world.[5]

In 1951 the United States deployed nuclear-armed B-29 bombers to Guam "for possible use against targets in China."[6] Two years later, testifying in secret to the Senate Foreign Relations Committee, General Omar Bradley, chair of the Joint Chiefs of Staff, confirmed that the United States "would have to consider very seriously the use of the A-bomb" as soon as a "suitable target" in Korea was found.[7] In early 1954, General Curtis LeMay, commander of the Strategic Air Command, suggested that given the lack of "suitable strategic air targets in Korea ... I would drop a few bombs in proper places in China, Manchuria, and Southeastern Russia."[8]

Apparently the commander in chief agreed. "In order to compel the Chinese Communists to accede to an armistice, it was obvious that if we were to go over to a major offensive the war would have to be expanded outside of Korea," Eisenhower wrote in his memoirs. "Finally, to keep the attack from becoming overly costly, it was clear that we would have to use atomic weapons.... We dropped the word, discreetly, of our intention."[9]

A 1953 National Security Council document, "U.S. Policy Towards Communist China," urged the use of both overt and covert means to hinder Sino-Soviet relations.[10] If there were to be a war with China, the United States would consider "employing all available weapons"

4. Ege and Makhijani 1982, p. 11.

5. NSC 1951, p. 11.

6. Fieldhouse 1991, p. 1.

7. As quoted in Ege and Makhijani 1982, p. 11.

8. As quoted in Lewis and Xue 1988, p. 18.

9. As quoted in Coates 1986, p. 14.

10. Lewis and Xue 1988, p. 18.

against the Chinese air force and its facilities, which "might absorb a considerable proportion of the U.S. atomic stockpile."[11]

The United States threatened to attack China with nuclear weapons on other occasions as well. In 1958, the United States deployed the first atom-armed fleet in world history—the Seventh Fleet—in the Western Pacific, as well as placing nuclear-capable howitzers on Quemoy Island to support the Taiwan government in its conflict with China.[12] The United States also deployed nuclear-capable weapons systems on Taiwan. Surface-to-air Nike Hercules missiles became operational in Taiwan in 1958, and the "Improved Nike Hercules" was placed in operation in December 1962.[13]

The United States was not the only country to provide China with a rationale for developing its own nuclear weapons. The Soviet Union, at different times, provided both aid and a second threat, stimulating China's development of nuclear weapons in two ways.

Five years after China and the Soviet Union signed a 30-year Treaty of Friendship, Alliance, and Mutual Assistance in 1950,[14] Chinese leaders asked for Soviet assistance in developing nuclear weapons. Two years later, on 15 October 1957, the two countries signed the secret Sino-Soviet New Defense Technical Accord, stipulating that the Soviets would provide China with "a sample of an atomic bomb and technical data concerning its manufacture."[15] The Soviet Union also agreed to aid China in such technologies as rocketry and aviation[16] and to supply blueprints and a working prototype for an atom bomb and missiles.[17]

However, deep policy differences and mistrust on both sides led the Soviets to terminate the nuclear assistance agreement on 20 June 1959,[18] and on 24 August 1960, Soviet aid to the Chinese nuclear program ended. The Soviet Union gradually pulled its 233 advisors in the nuclear-industrial program out of China. In addition, it never supplied 40 percent of the agreed-upon equipment and materials.[19]

John Lewis and Xue Litai cite the Soviet view that the Sino-Soviet split arose in part from issues related to the actual sharing of advanced

11. NSC 166/1, 6. November 1953, as quoted in Lewis and Xue 1988, p. 19.

12. Ege and Makhijani 1982, p. 12; Ellsberg 1986, p. 40.

13. Fieldhouse 1991a, p. 39.

14. Fieldhouse 1991a, p. 39.

15. Coates 1986, p. 18.

16. Coates 1986, p. 53.

17. Fieldhouse 1991a, p. 40.

18. Coates 1986, pp. 18–19.

19. Lewis and Xue 1988, p. 72.

weapons.[20] Soviet Premier Nikita Khrushchev also believed the Chinese were not reliable and that the alliance was one-sided and dangerous.[21] Lewis and Xue note further differences between Khrushchev and Mao Zedong over the implications of nuclear weapons and the proper stance to take toward the U.S. nuclear threat.[22] People's Liberation Army Marshal Nie Rongzhen has provided the Chinese view of why the Soviet Union withdrew its aid. According to Nie, coordinator of scientific and technical developments throughout China between 1956 and 1966, the "Soviets wanted to leave China at a stage where we could only replicate what they made, keeping us dependent and several steps behind their own development."[23]

Whatever the causes, border tensions between China and the Soviet Union grew in the late 1950s, as manifested in increasingly frequent incidents. Also, the Soviet Union sided with India in a border dispute that led to the 1962 Sino-Indian War.[24] The Soviet Union and China had entered into their own cold war. Beginning in 1965, soon after China exploded its first nuclear weapon, the Soviet Union gradually increased the numbers and readiness of its troops on the Chinese border, in addition to equipping its forces in the Far East with surface-to-surface nuclear-tipped rockets.[25]

China also felt threatened by the doctrine of "limited sovereignty" that the Soviets had used to justify the 1968 invasion of Czechoslovakia.[26] Moreover, by 1968, the Soviet Union was holding major troop maneuvers in Mongolia. Border incidents and intense fighting between Soviet and Chinese troops took place throughout the spring and summer of 1969;[27] some of these occurred in the Xinjiang-Kazakhstan region near China's nuclear testing site at Lop Nur.[28] During this conflict, "China ordered its nuclear weapons retargeted on the Soviet Union," a condition that continued at least until 1991, according to Richard Fieldhouse, a researcher of Chinese nuclear weapons history.[29] Chinese scholar Shen Dingli also notes that as of 1990 Chinese nuclear

20. Lewis and Xue 1988, p. 63.

21. Lewis and Xue 1988, p. 63.

22. Lewis and Xue 1988, p. 60.

23. As quoted in Coates 1986, p. 54.

24. Lewis and Xue 1988, pp. 71–72.

25. Kaplan 1981, p. 271.

26. Coates 1986, pp. 22–23.

27. Kaplan 1981, pp. 271–283.

28. Kaplan 1981, p. 280.

29. Fieldhouse 1991a, p. 40.

weapons were still aimed at the Soviet Union.[30] (As is the case for the other nuclear weapons states, the current targets of Chinese nuclear weapons are not clear.)

Escalating pressure on China to settle their border conflict, in August 1969 the Soviet Union "allowed it to be known that they were contemplating a nuclear strike against China."[31] In addition, says Stephen Kaplan, Soviet leaders "sounded out their Warsaw Pact allies on the possibility of a nuclear strike."[32] By 1976, the Soviet Union had 40 armed divisions near the Sino-Soviet border, equipped with nuclear-tipped missiles.[33] China also began concentrating on developing medium-range missiles capable of reaching Moscow and most other large Soviet cities.

History of China's Nuclear Weapons Program

Chinese research in nuclear science and technology dates from 1949, when the Chinese Academy of Sciences established the Institute of Modern Physics under the direction of physicist Qian Sanqiang, who went on to lead China's nuclear weapons development.[34]

Shortly after the Communist Party of China came to power in 1949, the new government gave Qian foreign currency with which to buy China's first nuclear instruments when he attended a peace conference in Europe. According to Lewis and Xue, Nobel laureates Frédéric and Irène Joliot-Curie, with whom Qian had studied in wartime Paris, helped arrange the purchases in England and France. It is not clear if he bought the instruments with the intention of developing nuclear weapons, a civilian nuclear power program, or both. However, Lewis and Xue say that when the Joliot-Curies gave a Chinese radiochemist 10 grams of radium salts, they accompanied it with a message to Mao Zedong: "Please tell Chairman Mao Zedong ... you should own the atomic bomb. The atomic bomb is not so terrifying."[35]

By the mid-1950s, China's Institute of Atomic Energy, set up in 1950 and headed by Qian, had developed a heavy-water nuclear reactor.[36] The first experimental reactor was constructed in Beijing, and a gaseous diffusion uranium enrichment plant was built in Lanzhou.

30. Kaplan 1981, pp. 288–289; Shen 1990, p. 4.

31. Coates 1986, p. 24.

32. Kaplan 1981, pp. 280–281.

33. Kaplan 1981, p. 287.

34. Xinhua 1989.

35. Lewis and Xue 1988, p. 36.

36. Xinhua 1989.

On 15 January 1955, the Chinese government formally decided to make nuclear weapons and enlist the assistance of the Soviet Union.[37] Full-fledged development of nuclear weapons began,[38] with the Soviet Union agreeing to help China locate uranium reserves if China would sell it uranium. China established a Bureau of Uranium Geology in 1955.[39] In 1956, it created the Second Ministry of Machine Building, later renamed the Ministry of the Nuclear Industry and now known the Ministry of Energy.[40] In August 1956, the Soviet government agreed to help build China's nuclear industries and research facilities.[41]

In all, the Chinese and Soviets signed six agreements between 1955 and 1958 regarding nuclear science, industry, and weapons.[42] Pursuant to the 1957 secret accord, the Soviets provided stocks of uranium hexafluoride. They also helped China create an aircraft industry and a nuclear weapons infrastructure, providing equipment, plans, and training. By that time, "sustained Soviet backing seemed assured," according to Lewis and Xue.[43] Over 1,000 Chinese scientists and technicians received training in the Soviet Union at the Joint Institute for Nuclear Research in Dubna and other locations.[44]

By 1958, China had selected sites for the core of its nuclear weapons–related facilities. The Hengyang Uranium Hydrometallurgy Plant No. 414 was to produce uranium oxides from uranium ore. The Baotou Nuclear Fuel Element Plant was for production of uranium tetrafluoride, fuel rods for research and production reactors, and lithium-6 and deuteride for hydrogen bombs. Gansu Province was selected as the location for the Lanzhou Uranium Enrichment Plant. The Jiuquan Integrated Atomic Energy Enterprise was to produce and separate plutonium. And the Northwest Nuclear Weapons Development Base Area in Haiyan, Qinghai, was to handle research and design.[45]

China began exploring for uranium on a large scale in the summer of 1958, sending out thousands of peasants in many areas with pocket

37. Lewis and Xue 1988, pp. 38–39.

38. Xinhua 1989.

39. Lewis 1986, p. 24.

40. Burke 1988, p. 317; Fieldhouse 1991, p. 63.

41. Lewis and Xue 1988, pp. 61–62.

42. Lewis and Xue 1988, p. 41.

43. Lewis and Xue 1988, p. 115.

44. Bromley and Perrolle 1980, p. 33. Many Chinese scientists were also educated in the United States and Germany. The chief engineer of the Jiuquan Atomic Energy Complex, Jiang Shengjie, trained at Columbia University in New York.

45. Fieldhouse 1991, p. 63.

beta-gamma radiometers—and minimal training—to help search for deposits.[46] These efforts resulted in China's first uranium mines: in Chenxian, Hengshan Dapu, and Shangrao.[47] (See chapter 5 for a discussion of China's uranium mines and mills.)

After Soviet assistance ended in 1960, China adopted a policy of self-reliance. Although Chinese leaders discussed the possibility of slowing down work on modern weapons, the view that the country had to rid itself of "imperialist bullying" prevailed. "At least then we could effectively counterattack if China were subject to imperialist nuclear attack," wrote People's Liberation Army Marshal Nie.[48]

After the Soviet pullout delayed completion of the Lanzhou uranium enrichment plant, China made uranium enrichment the highest priority. In order to focus its resources, China temporarily suspended work on plutonium production.[49] China produced its own uranium hexafluoride at Tuoli and later at the Jiuquan complex. Uranium ore was converted to uranium oxide and uranium tetrafluoride at Baotou, Hengyang, and Tongxian.[50] In November 1963, a plant at the Jiuquan Complex produced its first satisfactory uranium hexafluoride. In January 1964, the Lanzhou plant began producing highly enriched uranium.[51]

China made the nuclear components for its first atomic bomb at the Jiuquan Complex in April 1964.[52] Preparations continued over the next several months until 16 October, when China exploded an atomic bomb in the Xinjiang desert 150 kilometers northwest of the Lop Nur marshes, near the Huangyanggou oasis.[53] The 20-kiloton bomb used the implosion design fueled by enriched uranium.[54]

After the success of its enrichment effort, China returned to its efforts on plutonium, completing a production reactor and a reprocessing plant at Jiuquan in 1967 and 1970, respectively. The plutonium production reactors at Jiuquan and Guangyuan are China's principal plutonium sources for nuclear weapons.[55]

46. U.S. Bureau of Mines 1959, p. 1151; Lewis and Xue 1988, pp. 87–88.

47. Fieldhouse 1991, p. 63.

48. Coates 1986, p. 56.

49. Fieldhouse 1991a, p. 40.

50. Fieldhouse 1991, pp. 53, 54, 58.

51. Fieldhouse 1991, p. 66.

52. Fieldhouse 1991, p. 66.

53. *Dangdai Zhongguo* Series Editorial Committee 1988, p. 39; Fieldhouse 1991, p. 12; Lewis and Xue 1988, p. 1.

54. Lewis and Xue 1988, p. xviii.

55. Fieldhouse 1991, p. 12.

On 14 May 1965, China exploded its second nuclear bomb, a 40-kiloton uranium-235 fission weapon dropped from an airplane. This event, says Richard Fieldhouse, marked the "beginning of a deliverable arsenal of [nuclear] weapons."[56] China moved quickly into thermonuclear designs, and on 17 June 1967, it tested a hydrogen bomb.[57]

China carried out its first underground nuclear test in September 1969. However, after political purges of scientists in 1968 and 1969,[58] testing stopped due to disruptions arising from the Cultural Revolution.[59] (The U.S. Bureau of Mines, which keeps track of uranium production, also reported in 1968 that "nuclear development in mainland China has temporarily slowed down by military intervention in scientific affairs to which scientists objected.")[60] China conducted its next underground nuclear test in 1975, followed by another in 1976.[61] While China's last atmospheric test took place in 1980, it was six years before the country announced that it would no longer carry out such tests.[62] China's most recent test was on 5 October 1993.

Besides building nuclear weapons, China prepared for the possibility of a nuclear war by building a vast tunnel system of bomb shelters in Beijing and other major cities during the early 1970s.[63] Coates visited the Beijing shelters and explains that they demonstrate China's great fear of nuclear attack: "It is quite apparent that no one would contemplate such a vast diversion of resources unless there were reason to think it necessary."[64]

China has never signed the 1963 Partial Test Ban Treaty. However, in March of 1992, the Chinese government formally became a party to the Nuclear Non-Proliferation Treaty.[65] Economic and political realities confronting China at the end of the Cold War probably played a strong role in this decision as nonproliferation issues came to play a larger role in U.S. foreign policy. Thus, the United States put more pressure on China and France, the two nuclear powers that had not signed the NPT in 1970, to do so. During the Cold War, the highest U.S. priority

56. Fieldhouse 1991, p. 67.

57. *Dangdai Zhongguo* Series Editorial Committee 1988, p. 40.

58. U.S. Bureau of Mines 1969, p. 208.

59. The Cultural Revolution appears to have disrupted the nuclear weapons program between 1966. and 1973. See *Dangdai Zhongguo* Series Editorial Committee 1988.

60. U.S. Bureau of Mines 1968, p. 196.

61. *Dangdai Zhongguo* Series Editorial Committee 1988, p. 46.

62. Fieldhouse 1991, p. 74.

63. Coates 1986, p. 26.

64. Coates 1986, p. 35.

65. Norris and Arkin 1993b.

had been confronting the Soviet Union (for instance, in Afghanistan), while competing nonproliferation goals (for instance, relative to Pakistan) were relegated to a lower priority. Further, after the Cold War, the United States no longer needed a "China card" to play against the Soviets, weakening China's leverage relative to the United States.

Estimates of the size of China's nuclear arsenal vary. The Washington-based Center for Defense Information puts the nuclear stockpile at about 500 nuclear warheads, with a total "explosive force equal to 50,000 Hiroshima bombs"[66] which amounts to 750 megatons of TNT. One estimate puts the number at about 1,245 fission and fusion weapons, ranging from 2 kilotons to 5 megatons, based on estimates of China's production of nuclear weapons materials and its organization of forces for waging a battle.[67] Researchers Norris and Arkin estimate 450 nuclear warheads,[68] while Shen states that China has 276 to 398 warheads, with a total force of 400 to 500 megatons.[69]

Since China began its nuclear weapons program, it has produced an estimated 23.5 metric tons of weapons-grade uranium-235 and plutonium-239.[70] Albright, Berkhout, and Walker give a range of 1 to 4 metric tons for the Chinese inventory of weapons-grade plutonium.[71]

QUANTITY AND QUALITY OF DATA

Except for the locations of uranium mines, data presented here primarily come from *China Builds the Bomb* by John Lewis and Xue Litai, the most authoritative historical work on China's nuclear weapons program. The data Lewis and Xue provide go up to 1988. Additional data come from Richard Fieldhouse of the Natural Resources Defense Council, which has published "Working Papers" and other materials on the nuclear weapons production of various countries. However, Fieldhouse also relied largely on the work of Lewis and Xue.[72] Some information comes from the Foreign Broadcast Information Service of the U.S. National Technical Information Service; its *JPRS Reports* translates some Chinese news articles. Data on uranium mining also come from the NUEXCO Trading Corporation, the Uranium Institute of London, and the U.S. Bureau of Mines. Finally, some information

66. CDI 1993, p. 2.

67. Hahn 1987, pp. 12–13.

68. Norris and Arkin 1993, p. 48.

69. Shen 1990, pp. 41–43.

70. Shen 1990, pp. 2–3.

71. Albright, Berkhout, and Walker 1993, p. 46.

72. Fieldhouse 1991, 1991a.

comes from a report by Shen Dingli, produced for Princeton University's Center for Energy and Environmental Studies.

Lewis and Xue use a variety of sources, including interviews with Chinese specialists. There is very little definitive official information on health and environmental effects of nuclear weapons production and testing, although some articles and interviews by officials have been cited in books and in the press. A principal official source is a book edited by the China Today [Dangdai Zhongguo] Series Editorial Committee, portions of which were translated into English in 1988 by the Foreign Broadcast Information Service of the U.S. State Department. This source provides many specifics about the historical development of the nuclear weapons production program. It makes brief references to some problems and has hints of others, but there are no details that allow for even a preliminary environmental assessment. Overall, the approximate locations and start-up dates of major facilities are now known with considerable confidence; the dates of China's nuclear tests are known (except perhaps for smaller tests that may still be secret); but little else can be said definitively.

LOCATION OF NUCLEAR WEAPONS–RELATED ACTIVITIES AND FACILITIES

As of the end of the 1980s, China's nuclear industry comprised 27 research institutes, 100 production facilities, and 18 universities and technical schools (see table 10.1 and figure 10.1). It employed 300,000 people, including 70,000 scientific workers.[73] The China National Nuclear Industrial Corporation, which heads China's civilian nuclear program, says the industry develops both military and civilian products.[74] Companies involved in building nuclear power plants include the China Nuclear Industry Corporation, the Shanghai Nuclear Power Office, and the Qinshan Nuclear Power Company. Also, Shanghai Machine Tool Plant No. 1 produces essential reactor components.[75] The civilian side of the nuclear industry has been expanding in recent years.

The Chinese government gave names to its nuclear weapons facilities and bureaus similar to those used by the Soviets, in order to hide their mission. For example, the early name of the agency responsible for these facilities was the Second Ministry of Machine Building, very similar to its Soviet counterpart, the Ministry of Medium Machine Building. China built almost all its original nuclear facilities according to a Soviet plan. The Second Ministry of Machine Building/Ministry of

73. Xinhua 1989, p. 1.

74. Xinhua 1989a.

75. Shanghai City Service 1989.

Table 10.1 Location of Nuclear Weapons–Related Activities and Facilities in China

Official Facility Name	Location	Operating Status[1]	Source[2]
R&D Weapons Laboratories			
Beijing Nuclear Weapons Research Institute	Western suburb of Beijing	NO (1958–62)	b, c, a, l
Harbin Military Engineering Institute	Harbin		b
Physics and Chemistry Engineering Academy	Tianjin		b
Institute of Atomic Energy 401 [Institute of Modern Physics, 601]	Tuoli, 35 km south of Beijing	OP, S = 1949	c, a, b, l
Research Unit on Underground Nuclear Testing Phenomena, Nuclear Weapons Research Base Area			b
Northwest Nuclear Technology Institute	Near Malan, Lop Nur Test Site		c, a, l
Northwest Nuclear Weapons Research and Design Academy, Ninth Academy	Haiyan, Qinghai Province	1962	a, b, l
Beijing Nuclear Engineering Research and Design Academy	Beijing		a, b
Isotope Applications Research Office, Shanghai Atomic and Nuclear Institute	Shanghai		a, l
Uranium Mining and Metallurgical Processing Institute—Tongxian Institute [Sixth Institute, renamed Fifth Institute]	Tongxian, a few miles east of Beijing		b, l
NA	Mianyang, Sichuan Province		e
Hengyang Institute of Uranium Mining and Metallurgy	Hengyang, Hunan Province		b
Hengyang Uranium Mining and Metallurgy Design and Research Academy	Hengyang, Hunan Province		b, l
Uranium Mining			
Chenxian Uranium Mine	Chenxian County, Hunan Province	S = 1960	b, l
Dapu Uranium Mine	Hengshan County, Hunan Province		b, l
Shangrao Uranium Mine	Shangrao County, Jiangxi		b, l
26 major mines of unknown name	Jiangxi and Guangdong provinces		b
Mashan	Guangxi Province		f

Table 10.1 (continued)

Official Facility Name	Location	Operating Status[1]	Source[2]
Lianxian Mine	Lianxian, Guangdong Province		b, l
Nanyang Mine	Nanyang		f
Jiuquan	Jiuquan, Subei		f
Jiuzhan	Jiuzhan		f
Xia-Chuan	Xia-Chuan		f
Linxian Uranium Mine	Hunan Province	S = 1962	a, l
Uranium Milling and Production (production of yellowcake from uranium ore)			
Fuzhou Mining and Metallugical Complex	Jiangxi at Fuzhou		a, l
Guangdong Mining and Metallurgical Complex	Guangdong Province		b, l
Liangshanguan	Liaoning Province		h
Chengxian	Chenxian, Hunan		b
Hengyang Uranium Hydro-metallurgy Plant Plant 272 [Plant 414]	Hengyang, Hunan Province	OP	g, b, l
Shao Kuan Mill	Jiangxi Province		h
Hengjian Mill	Zhejiang Province		h
Zhuzhou Mill	Guangdong Province		h
Conversion of Yellowcake to Uranium Oxide (in preparation for conversion to uranium tetraflouride)			
Uranium Oxide Production Plant [Plant 2], Fifth [Sixth] Institute	Tongxian, a few miles east of Beijing	S = 1961	b, l
Hengyang Uranium Hydro-metallurgy Plant Plant 272 [Plant 414]	Hunan Province	OP	g, b, l
Conversion of Uranium Oxide to Uranium Tetrafluoride (in preparation for conversion to uranium hexafluoride)			
Uranium Tetrafluoride Workshop, Plant 202 [Nuclear Fuel Component Plant]	Suburbs of Baotou, in Nei Mongol		a, b, l
Plant 812 [Nuclear Fuel Component Plant]	Yibin, Sichuan Province		a
Conversion of Uranium Tetrafluoride to Uranium Hexafluoride (in preparation for enrichment)			
Uranium Hexafluoride Plant, Jiuquan Atomic Energy Complex [Plant 404]	Subei Mongolian Autonomous County, Jiuquan Prefecture, Gansu Province, south of Gobi	S = 1963	a, b, f, l

Table 10.1 (continued)

Official Facility Name	Location	Operating Status[1]	Source[2]
	Desert and east of Lop Nur		
Research Department 615B, Institute of Atomic Energy 401 [Institute of Modern Physics 601]	Tuoli, just south of Beijing	S = 1960	b, l
Uranium Enrichment			
Lanzhou Gaseous Diffusion Enrichment Plant [Fifteenth Bureau, Plant 504]	15 miles north of Lanzhou, on the bank of the Yellow River	OP, S = 1964	a, b, f, l
Plant 812 [Nuclear Fuel Component Plant]	Yibin, Sichuan Province		a, e, l
Helan Shan I (centrifuge enrichment)	Ningxia		i
Research Department 615A, Institute of Atomic Energy 401 [Institute of Modern Physics 601]	Tuoli, just south of Beijing	S = 1960	b, a
Uranium Processing and Fuel Fabrication			
Plant 202 [Nuclear Fuel Component Plant]	Suburbs of Baotou, in Nei Mongol	S = 1964	c, a, e, f
Uranium Mining and Metallurgical Institute, Sixth [Fifth] Institute	Tongxian, a few miles east of Beijing		a, b
Plant 812 [Nuclear Fuel Component Plant]	Yibin, Sichuan Province		a, e
Nuclear Fuel Processing Plant, Jiuquan Atomic Energy Complex [Plant 404]	Subei Mongolian Autonomous County, Jiuquan Prefecture, Gansu Province, south of Gobi Desert and east of Lop Nur		a, b, l
Production of Heavy Water			
Ministry of Chemical Industry Plant	Location unknown		b
Production of Tritium			
Plant 202 [Nuclear Fuel Component Plant]	Baotou	S = 1968	a, b, c
Production of Plutonium from Military Production Reactors			
Jiuquan Atomic Energy Complex [Plant 404]	Subei Mongolian Autonomous County, Jiuquan Prefecture, Gansu Province, south of Gobi Desert and east of Lop Nur	S = 1966	a, b, l

Table 10.1 (continued)

Official Facility Name	Location	Operating Status[1]	Source[2]
Plant 821	Guangyuan, Sichuan Province	OP	a
Institute of Atomic Energy 401 [Institute of Modern Physics 601]	Tuoli, 35 miles south of Beijing		a
Chemical Separation of Plutonium (plutonium reprocessing)			
Plant 821	Guangyuan, Sichuan Province		a
Plutonium Processing Plant, Jiuquan Atomic Energy Complex [Plant 404]	Subei Mongolian Autonomous County, Jiuquan Prefecture, Gansu Province, south of Gobi Desert ad east of Lop Nur		a, b
Fabrication of Plutonium Metal Components			
Pu-239 Pits and Triggers			
Nuclear Component Manufacturing Plant, Jiuquan Atomic Energy Complex [Plant 404]	Subei Mongolian Autonomous County, Jiuquan Prefecture, Gansu Province, south of Gobi Desert and east of Lop Nur		c, a, b
Pu-238 Radioisotope Thermal Generators			
Nuclear Component Manufacturing Plant	Subei, Jiuquan Prefecture		b
Other Warhead Components			
Beryllium Mining		NA	
Beryllium Processing		NA	
Electronic Components		NA	
High-Explosives Testing, R&D			
Explosives Test Range of the Engineering Corps	Near Donghuayuan, Huailai County, just northwest of Beijing		b
Lanzhou Chemical Physics Institute	Lanzhou, Gansu Province		b
Northwest Nuclear Weapons Development Base Area, Ninth Academy	Haiyan, Qinghai Province	S = 1964	

Table 10.1 (continued)

Official Facility Name	Location	Operating Status[1]	Source[2]
Lithium-6 and Lithium Deuteride Production			
Lithium-6 Deuteride Workshop, Plant 202 [Nuclear Fuel Component Plant]	Suburbs of Baotou, in Nei Mongol	S = 1964	a, b, e
Plant 812 [Nuclear Fuel Component Plant]	Yibin, Sichuan Province		a
Neutron Generators		NA	
Assembly of Nuclear Weapons			
Assembly Workshop, Jiuquan Complex [Plant 404]	Subei Mongolian Autonomous County, Jiuquan Prefecture, Gansu Province, south of Gobi Desert and east of Lop Nur		k, f
Haiyan Nuclear Plant at Northwest Nuclear Weapons Research and Design Academy	Haiyan, Qinghai Province	OP	f, i, j
[Name unknown]	Guangyuan, Sichuan Province	OP	k
[Name unknown]	Harbin, Heilongjiang		a, i
Testing of Nuclear Devices			
"Peaceful" Nuclear Explosions			
None			
Weapons Testing			
Lop Nur Nuclear Weapons Test Base	Headquarters city for test site in Xinjiang Uygur Autonomous Region, 300 miles south of Urumqi; site traverses the boundaries of Turpan Prefecture and Bayingolin Mongol Autonomous Prefecture	OP	a, b, d

1. Status abbreviations:
NA = no information available
NO = not operating and either certain or likely not to resume, with year stopped, if known
OP = operating at some capacity
S = startup date

2. Sources:
a = Fieldhouse 1991
b = Lewis and Xue 1988
c = Dangdai Zhongguo 1988

Table 10.1 (continued)

d = Hughes 1992
e = Fieldhouse 1991a
f = Arkin and Fieldhouse 1985
g = Nuclear Engineering International Publications 1991
h = NUEXCO 1991
i = Hahn 1967
j = Shen 1990
k = Norris and Arkin 1993
l = Norris, Burrows, and Fieldhouse 1994

the Nuclear Industry/Ministry of Energy has controlled every facet of the production of nuclear weapons. Today, the Ministry of Energy plays a major role in determining and assigning development priorities in both military and civilian nuclear programs.[76]

Research and Development Laboratories

In the mid-1950s, China established nuclear science research institutions in or near Beijing, Xi'an, Harbin, and Tianjin. Today, the Chinese Academy of Sciences, under the State Science and Technology Commission, supervises research on nuclear weapons. The academy directs the Institute of Atomic Energy and the Institute of Mechanics and controls about 110 others.[77]

The Chinese Academy of Atomic Energy Science was organized in 1950 by the Chinese Academy of Sciences.[78] Located at Tuoli, about 50 kilometers south of Beijing, it conducts research on nuclear physics. It has had four names. From 1950 until 6 October 1953, it was known as the Institute of Modern Physics (first code named 601 and later 401). From 6 October 1953 until 1 July 1958, it was the Institute of Physics. Until 1984, it was the Institute of Atomic Energy, and at the end of 1984, it acquired its current name.[79]

China has two main nuclear weapons design labs. The Northwest Nuclear Weapons Research and Design Academy, also known as the Ninth Academy, is situated east of Lake Qinghai, in Qinghai Province.[80] This center was once referred to by U.S. Government intelligence as "Koko Nor," which formerly was the name of the nearby

76. Burke 1988, p. 317.

77. Burke 1988, pp. 320–321.

78. Lewis and Xue 1988, pp. 43, 261, note 32.

79. Lewis and Xue 1988, p. 261, note 32.

80. Fieldhouse 1991, p. 57.

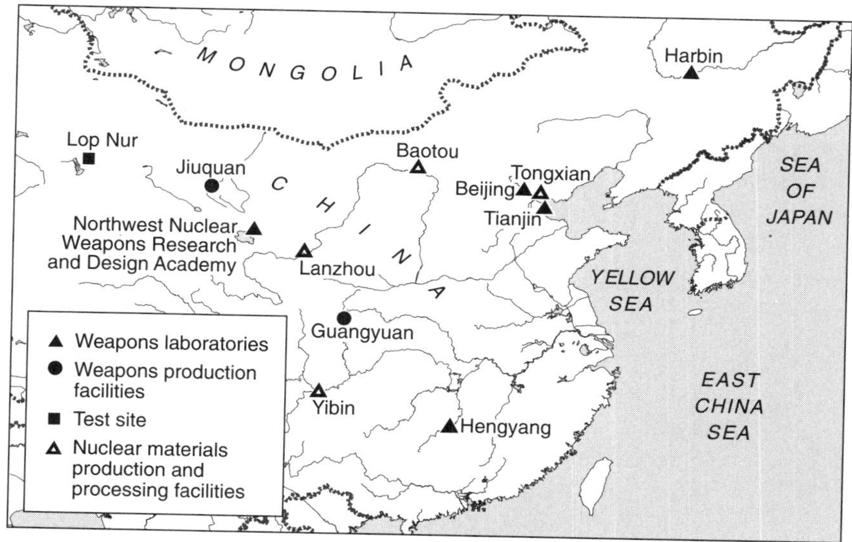

Figure 10.1 Major nuclear weapons testing and production sites in China.

lake.[81] The other main weapons research and design center, in Mianyang, Sichuan Province, dates from the 1960s.[82] Mianyang is now China's primary research and design center for nuclear weapons, with most work having been moved from Haiyan.[83]

The Beijing Nuclear Weapons Research Institute operated from 1958 to 1962.[84] It conducted the first research on the atomic bomb.[85] Located in Beijing's western suburbs, it operated as a transitional facility until the Ninth Academy opened in 1962. The institute conducted research in "theoretical physics, explosion physics, neutron physics, metal physics, projectile ballistics and other areas."[86]

The Beijing Nuclear Engineering Research and Design Academy worked with the Chinese Academy of Atomic Energy Science on research and development related to uranium hexafluoride after the Soviet withdrawal in 1960.[87] The Chinese Academy also had two research departments—615A and 615B—that worked on uranium hexafluoride and gaseous diffusion research and production.[88] It houses a

81. Fieldhouse 1991, p. 57.

82. Fieldhouse 1991a, p. 38.

83. Norris, Burrows and Fieldhouse 1994, p. 348.

84. Fieldhouse 1991, p. 53.

85. Lewis and Xue 1988, pp. 140–141.

86. *Dangdai Zhongguo* Series Editorial Committee 1988, p. 2.

87. Lewis and Xue 1988, pp. 99–100.

88. Lewis and Xue 1988, p. 100.

Soviet-supplied reactor and cyclotron that went into operation in 1958.[89] The Physics and Chemistry Engineering Academy in Tianjin has also conducted research on gaseous diffusion technology.[90] According to Fieldhouse, the Shanghai Atomic and Nuclear Institute was combined in 1962 with the Chinese Academy's Isotope Applications Research Office.[91]

The Uranium Mining and Metallurgy Design and Research Academy was established in 1958 in Hengyang, Hunan Province, to design plants for mass producing uranium oxides. The Uranium Mining and Metallurgical Processing Institute in Tongxian, a few miles east of Beijing, was created in 1958 by the Second Ministry. Also known as the Sixth Institute and later renamed the Fifth Institute, it has conducted research on uranium ores and their processing.[92]

Research also occurs at production sites. For example, the Jiuquan Integrated Atomic Energy Enterprise is one of China's main nuclear weapons sites. A Jiuquan research unit cooperated with the Beijing Institute of Nuclear Weapons "to organize joint attacks in Beijing on key problems concerning the purification, casting, and machine processing of enriched uranium."[93]

The Northwest Nuclear Technology Institute, near Malan, about 200 kilometers northwest of ground zero at the Lop Nur Test Site, was established in 1963. Lewis and Xue report that it "maintains a large archive on nuclear explosions, anti-nuclear warfare, nuclear weapons designs and so on."[94] According to *Contemporary China's Nuclear Industry*, "Almost one thousand pieces of technical materials concerning the scientific results of nuclear test explosions are classified top secret and preserved in the archives."[95]

The Second Ministry created a research unit on Underground Nuclear Testing Phenomena, write Lewis and Xue, after Premier Zhou Enlai "ordered the development of underground testing speeded up" in the 1960s.[96] Its location is unknown. Some of China's nuclear weapons engineers are based at the Harbin Military Engineering University

89. Lewis and Xue 1988, p. 99.

90. Lewis and Xue 1988, p. 120.

91. Fieldhouse 1991, p. 57.

92. Lewis and Xue 1988, pp. 90–91, 95.

93. *Dangdai Zhongguo* Series Editorial Committee 1988, p. 36.

94. Lewis and Xue 1988, p. 180.

95. As quoted in Lewis and Xue 1988, p. 180 (original in Chinese). Lewis and Xue call *Contemporary China's Nuclear Industry* "the most complete Chinese history to date on the nuclear weapons program" (Lewis and Xue 1988, pp. 253–254).

96. Lewis and Xue 1988, p. 285.

in Harbin. The school trains military research and design personnel for work on guided missiles and atomic weapons.[97]

Uranium Conversion Facilities

China initially converted uranium ore to uranium oxide and uranium tetrafluoride at Tongxian, Hengyang, and Baotou. The Baotou facility, known as the Uranium Nuclear Fuel Component Plant (No. 202), was based on Soviet plans and designed and built with Soviet assistance until 1960. In 1965, this plant produced uranium fuel for a light-water–cooled, graphite-moderated plutonium production reactor.[98]

The Uranium Oxide Production Plant (Plant Two) was established on 12 August 1960, at Tongxian, a few kilometers east of Beijing. In this small one-story building, Chinese technicians processed the uranium for China's first atomic bomb. Plant Four, built later at the same site, also manufactured uranium tetrafluoride.[99]

In July 1960, the Second Ministry assigned the code name Project 161 to research and development on making uranium hexafluoride from uranium tetrafluoride. In three years, the Institute of Atomic Energy's Gaseous Diffusion Laboratory in Tuoli, also known as Research Department 615B, produced more than 10 metric tons of uranium hexafluoride for China's first atomic bomb.[100] Uranium hexafluoride was also produced in the uranium hexafluoride plant at the Jiuquan Complex.[101] Later uranium hexafluoride production was moved to China's main enrichment plant at Lanzhou.

Uranium Enrichment Plants

According to Fieldhouse, China chose the gaseous diffusion method for enriching uranium because of Soviet technical assistance and the U.S. example.[102] China has enriched uranium in at least five plants.[103] It developed its first gaseous diffusion uranium enrichment process on a lab scale at the Beijing Institute of Atomic Energy. The Lanzhou Gaseous Diffusion Plant is China's main facility for enriching uranium.

97. Lewis and Xue 1988, p. 264, note 84.

98. Fieldhouse 1991, p. 67.

99. Fieldhouse 1991, p. 58.

100. Fieldhouse 1991, p. 58.

101. Fieldhouse 1991, p. 66.

102. Fieldhouse 1991a, p. 40.

103. Hahn 1987, p. 12.

Based on Soviet plans, it is 25 kilometers northeast of Lanzhou on the Yellow River.[104] It began operation in 1964.[105] A second uranium enrichment plant, built at Lanzhou in 1980, has an annual capacity of 200,000 separative work units.[106]

China has built two uranium enrichment plants in Ningxia. Helan Shan I uses centrifuge technology and Helan Shan II, gaseous diffusion.[107] Arkin and Fieldhouse report other uranium enrichment plants in Hong Yuan and Urümqi.[108]

Uranium Fuel Fabrication

Historically, chemical processing of enriched uranium hexafluoride and fabrication of enriched uranium metal and oxide fuel took place at the Uranium Nuclear Fuel Component Plant in Baotou. This plant produces uranium fuel components. It also made the uranium cores for China's first nuclear weapon explosion in 1964.[109]

Plant 812 at Yibin produces uranium fuel rods for light-water plutonium production reactors and for submarine propulsion reactors.[110] It began operating in 1987 and will have an estimated capacity of 150 metric tons per year by 1995.[111] The Nuclear Fuel Processing Plant at the Jiuquan Atomic Energy Complex converts enriched uranium hexafluoride into weapons-grade uranium metal, and it also machines uranium cores.[112]

Tritium Production

China built a domestically designed tritium plant at Baotou during the 1960s, which went into production in May 1968. Four years later, China completed a new deuterium/tritium/lithium-6 production line.[113] China has dealt with tritium-contaminated waste water either by con-

104. Lewis and Xue 1988, p. 107; Fieldhouse 1991, p. 55.

105. Arkin and Fieldhouse 1985, p. 290; Fieldhouse 1991, p. 66.

106. *Nuclear Engineering International* 1991, p. 157.

107. Hahn 1987, p. 12.

108. Arkin and Fieldhouse 1985, pp. 290–291.

109. Fieldhouse 1991, pp. 53, 66.

110. Fieldhouse 1991, p. 60.

111. *Nuclear Engineering International* 1991, p. 157.

112. Fieldhouse 1991, p. 55, 66.

113. *Dangdai Zhongguo* Series Editorial Committee 1988, p. 31.

verting it into steam and discharging it into the atmosphere or by dumping it into rivers, lakes, and seas.[114]

Production of Plutonium from Military Reactors

Construction of China's first plutonium production reactor began in 1958, with Soviet assistance, at the Jiuquan Atomic Energy Complex. This complex, known by the code name Plant 404, is not in Jiuquan City but in the desert part of the Jiuquan Prefecture near Subei Mongolian Autonomous County.[115] The Jiuquan project was suspended almost immediately after it began, however, due to the Soviet pullout and China's subsequent decision to base the country's first atomic weapon on enriched uranium instead of plutonium. The Jiuquan reactor, which follows the Soviet graphite-moderated design, was completed in 1966.[116] A second plutonium production reactor, China's largest, was built later at Plant 821 in Guangyuan, Sichuan.[117]

Lewis and Xue note the lack of reliable evidence on the development of China's plutonium facilities: "Several sources assert that the Jiuquan nuclear reactor was one of 'at least seven known nuclear reactors in China' in 1976." They add that a 1960 U.S. government publication referred to at least four atomic reactors.[118] Lewis and Xue also note that some reports have stated that

> in the 1960s the Chinese succeeded in extracting sufficient amounts of weapons-grade plutonium to warrant building a number of chemical separation plants throughout the country in addition to the one near Subei in Jiuquan prefecture. However, we know of no Chinese sources that confirm the location or date of construction of any military plutonium facility other than the one in the Jiuquan complex and one in Sichuan Province.[119]

Plutonium Production from Dual-Use Civilian Power Reactors

China's first reactors were research-scale plants built jointly by academies of sciences of the Soviet Union and China and located at Xi'an, Shenyang, and Chongqing.[120] China's first large-scale nuclear power plant, a 300-megawatt (electrical) reactor in Qinshan, near Shanghai

114. Xu 1990.

115. Fieldhouse 1991, p. 55.

116. Fieldhouse 1991, p. 68.

117. Fieldhouse 1991, p. 54.

118. Lewis and Xue 1988, p. 113.

119. Lewis and Xue 1988, p. 113.

120. Fieldhouse 1991, pp. 53, 57, and 59.

in Zhejiang Province, began operating in 1991.[121] Two 900-megawatt power reactors at Daya Bay in Guangdong Province were expected to come on line in the early 1990s.[122] China plans to significantly expand its nuclear power capacity over the coming years, and the total capacity may reach 6,000 megawatts by the year 2000.[123] No definitive information is available on whether the spent fuel from any of these reactors has been or will be reprocessed into plutonium for nuclear weapons. Ke Youzhi and Wang Rengtao estimate that China will have accumulated over 6,000 kilograms of plutonium in its reactor spent fuel.[124]

Chemical Separation of Plutonium

A plant for reprocessing irradiated fuel was built at the Jiuquan Atomic Energy Complex and put into operation in September 1968.[125] Called the "intermediate testing plant," it operates for 250 days per year and has a capacity for processing 400 kilograms of uranium per day (100 metric tons per year).[126] This would give China the capacity for producing about one-half to one metric ton of plutonium per year from this plant.

The country's largest plutonium separation plant, known as Plant 821, was built at Guangyuan, Sichuan Province, in the late 1960s.[127] It is believed that the plutonium plant, along with other facilities at the Guangyuan site, was built entirely underground to protect it from attack.[128] China uses the PUREX process for reprocessing.[129]

Production of Plutonium Components

China makes plutonium components for nuclear weapons at the Jiuquan Atomic Energy Complex in the Nuclear Component Manufacturing Plant, also known as the Plutonium Processing Plant.[130]

121. Yuan 1989.

122. Travis 1991, p. 8.

123. Xu 1992.

124. Ke and Wang 1987, table 1.

125. *Dangdai Zhongguo* Series Editorial Committee 1988, p. 27.

126. *Dangdai Zhongguo* Series Editorial Committee 1988, p. 25.

127. Fieldhouse 1991, p. 54.

128. Fieldhouse 1991, p. 54.

129. Ke and Wang 1987, p. 528.

130. Fieldhouse 1991, p. 55; Lewis and Xue 1988, p. 111.

Production of Other Components

• *Beryllium:* China produces beryllium for its nuclear weapons and other uses and also exports it to the former Soviet Union.[131]

• *Chemical high explosives:* China produced and assembled conventional explosives for its nuclear weapons in Donghuayuan, in Huailai County, northwest of Beijing.[132] Construction of this plant began in 1960. The Lanzhou Chemical Physics Institute at Lanzhou, in Gansu Province, conducted explosives experiments for the first atomic bomb.[133]

No information is available regarding the production of depleted uranium metal used in warhead casings or the production of electronic components and neutron generators. The processing facilities required would be similar to the ones used for the enriched uranium components made in the Baotou plant.

• *Lithium-6 and lithium deuteride:* Lithium-6 was first produced at the Uranium Nuclear Fuel Component Plant at Baotou in the Lithium-6 Deuteride Workshop, which began operation in 1964 based on Soviet plans.[134] Heavy water has been provided by the Ministry of Chemical Industry from domestic plants.[135]

Plant 812 at Yibin, in Sichuan Province, is thought to be a key production site for lithium-6 deuteride. Plant 812 is a larger, more modern version of the Baotou plant.[136]

Assembly and Disassembly of Nuclear Weapons

The Jiuquan Atomic Energy Complex built and assembled China's first nuclear bomb in April 1964.[137] Public reports indicate that China "currently operates at least three fission and fusion warhead manufacturing installations."[138] The Northeast Nuclear Weapons Research and Design Academy in Haiyan, Qinghai Province, and Plant 821 in Guangyuan, Sichuan Province, are both capable of making thermonuclear weapons. The third plant is supposedly near Harbin in Heilongjiang,[139] but according to Fieldhouse, the information on this

131. Lewis and Xue 1988, pp. 155–156.

132. Fieldhouse 1991, p. 54.

133. Lewis and Xue 1988, pp. 154, 280.

134. Fieldhouse 1991, pp. 53, 67.

135. Lewis and Xue 1988, p. 200.

136. Fieldhouse 1991, p. 60.

137. Fieldhouse 1991, p. 66.

138. Hahn 1987, p. 12.

139. Hahn 1987, p. 12; Fieldhouse 1991, p. 54.

facility is questionable.[140] China also may have a nuclear materials production facility at Yumen in Gansu, although Fieldhouse believes this is probably a mistaken name for the Jiuquan Complex.[141]

Testing of Nuclear Weapons

From 1964 to 1993, China conducted 39 nuclear weapons tests at its only test site, the Lop Nur Nuclear Weapons Test Base.[142] Also known as the Northwest Nuclear Weapons Research Base Area, the 100,000-square-kilometer facility is in the Xinjiang desert, 150 kilometers northwest of the Lop Nur Marshes and 1,440 kilometers southeast of Semipalatinsk, the Soviet test site in Kazakhstan. Malan ("Atom City") is headquarters for the test center.[143]

Fieldhouse quotes a former base commander's description of the site, as of 1984:

> More than 2,000 kilometers of highways have been built on the base. At each test site, there is a command center, a communication hub, a control center, and a permanent survey station. At the air testing grounds, there also are some simple houses, airports, and underground water pipes. In the distance, there is an airport and a factory to assemble test items.[144]

Nuclear weapons tests have been conducted with aircraft-dropped bombs and guided missiles, from towers, and underground in horizontal and vertical shafts.[145] All tests since 1980 have been underground.[146] A test conducted on 29 September 1988 was believed to be a small-scale device based on a neutron bomb design.[147] China conducted its largest underground nuclear test in 1992. It was estimated to be a 660-kiloton test.[148]

140. Fieldhouse 1991, p. 54.

141. Fieldhouse 1991, p. 60.

142. Norris and Arkin 1993a. According to Norris and Arkin, China conducted 38. nuclear tests until the end of 1992. The October 1993 test brings the total to 39.

143. Shen 1990, p. 22.

144. As quoted in Fieldhouse 1991, pp. 55–56.

145. Hahn 1987, p. 12.

146. Shen 1990, p. 22.

147. Shen 1990, p. 23.

148. Norris, Burrows, and Fieldhouse 1994.

Radioactive Waste Disposal Sites

China's National Nuclear Safety Administration is responsible for disposing of nuclear wastes.[149] The administration was established in 1960 under the name of the Nuclear Security and Protection Bureau.

Assuming total plutonium production of 2.5 metric tons and strontium-90 and cesium-137 at 3.3 curies (about 122 gigabecquerels) and 3.6 curies (about 133 gigabecquerels) per gram of plutonium, we estimate that China has generated about 8.25 million curies (about 305,000 terabecquerels) of strontium-90 and 9 million curies (333,000 terabecquerels) of cesium-137. When the radioactive daughters of these fission products are included, the non-decay-corrected high-level waste in China would amount to 34.5 million curies (about 1.28 million terabecquerels).

We have no data on high-level waste storage or disposal practices in China. A paper by Ke Youzhi and Wang Rengtao of the Beijing Institute of Nuclear Engineering states that "vitrification is resorted to" for treating high-level reprocessing wastes, but the status of the program is unclear. Ke and Wang further state that "in order to guarantee the safety of the public, and the protection of the environment, the vitrified product must be isolated in a deep geologic formation. For this purpose, an R and D program has been established and implemented in China."[150] From this statement, it would appear that the program was in its early stages in the late 1980s. We have no data on waste disposal practices in early periods.

Ke and Wang state that currently the Chinese have "gained lots of experience in treating the low and medium level wastes" and that these wastes are "now stored in repositories."[151] We have no data on the nature of these repositories, or whether shallow land burial sites are referred to here as "repositories." Power plant wastes are stored on-site. According to journalist Seth Faison, China plans to build a centralized storage site in the coastal province of Zhejiang for waste from the Qinshan and the Daya Bay nuclear power plants. China is also reportedly investigating sites in the Gobi desert for a deep geologic repository for high-level waste.[152]

In 1987, China and a West German company discussed the possibility of storing West German nuclear wastes in China in exchange for West German power plants. The plan was abandoned partially due to

149. Lewis and Xue 1988, p. 55; Shen 1993.

150. Ke and Wang 1987, p. 530.

151. Ke and Wang 1987, p. 530.

152. Lenssen 1991, p. 24.

transportation problems.[153] Australian Nuclear Science and Technology Organization, an Australian firm, is helping China develop an artificial combination of naturally occurring rocks (called Synroc) designed to contain nuclear wastes.[154] Synroc is being developed with properties that would help prevent the migration of radioactive materials into groundwater and the environment. Radioactive waste would be incorporated in Synroc and put in special containers for disposal in a deep geologic repository.

ENVIRONMENTAL EFFECTS OF THE NUCLEAR WEAPONS COMPLEX

No specific official information is available about the environmental effects of nuclear weapons production. The Chinese government has been highly secretive about related environmental problems. An inquiry to the Chinese Embassy in the United States did not turn up any information about China's problems with radioactive contamination from any source.[155]

Occasionally, a news report makes its way to the West, and it is by gleaning these reports that we have composed a crude, incomplete picture of the current situation. In addition, "Environmental Issues in China," a report by P. S. Travis of the Los Alamos National Labortory, is informative:

> Since China does not have many major civilian reactors running, most of its nuclear waste is generated by the military, laboratories, and hospitals. China is building storage facilities for radioactive waste from nuclear power plants; in fact, facilities have been built in 12 provinces and are planned for eight others. The Chinese reportedly plan to establish four regional disposal facilities, one each in eastern, southern, northwestern and southwestern China.[156]

Lewis and Xue fill in some of the historical background to China's handling of radioactive materials. The Security and Protection Bureau oversaw nuclear waste disposal, environmental protection, and public health and safety. "For each weapons test, it assigned special officers to monitor radiation protection and general safety," they note. "The bureau cooperated with the Ministry of Health in running special hospitals and maintaining personnel health records. When, in 1960, the State Council passed codes for the protection of personnel in the nuclear industry, they fell under its purview."[157]

153. Faison 1989.

154. AFP 1992.

155. Zhan 1993.

156. Travis 1991, p. 9.

157. Lewis and Xue 1988, p. 55.

The Security and Protection Bureau passed some codes in 1960 that defined certain safety precautions, such as protective clothing for workers in the maintenance area of the plutonium separation plant at Jiuquan. According to Lewis and Xue, workers use robots and remote controls to reduce their exposure to reprocessing operations, which are conducted behind a 1.8-meter-thick reinforced concrete wall.[158]

The China Today Series Editorial Committee's book on the nuclear industry cites problems with waste management at the Jiuquan reprocessing plant:

> The large amount of radioactive and acidic waste liquid created during the post-processing process formerly was dealt with using an addition-subtraction neutralization method after "continuous evaporation-formol denitration" that was common in foreign countries. This process required complex operations, efficiency was low, purification was poor and it consumed large amounts of reagent. Moreover, accidents caused by contamination of the system were common. The result was that the waste liquid did not meet discharge standards.[159]

The book describes changes made to rectify these problems, but it does not state where the waste that did not meet specifications was discharged and what environmental consequences might have followed. Nor does it describe the accidents and their health consequences for the workers who were present or for those who had to clean up afterward.

In March 1989, an official with the Environmental Protection Office told the news agency that inadequate waste storage had contributed to some accidental releases in recent years. "In June 1985, 25 metric tons of radioactive material were dumped down a well in Hebei in northeastern China, raising ambient levels of radioactivity several hundred times normal levels."[160] It is not known if any of these sites contain wastes related to the production of nuclear weapons. Wang Yi, of the China Eco-Environmental Research Group of the Chinese Academy of Sciences, has told the press that his group will focus its research on "the handling of radioactive dust and nuclear waste," among other environmental problems, such as acid rain.[161]

There are few details about contamination of soil and water from radionuclides in China. The following information from the China Nuclear Information Center in Beijing describes the pollution of the Yangtze River:

> Large number of samples taken from main stream of Yangtze River and its tributary were analyzed and investigated in the dry season and

158. Lewis and Xue 1988, p. 112.

159. *Dangdai Zhongguo* Series Editorial Committee 1988, p. 30.

160. *Nucleonics Week* 1989, pp. 8–9.

161. Xinhua 1992, p. 13.

rich water season in 1984. The number of sampling sites were 119. The sampling mediums included water, clay, fishes, and soil. The items of analysis included total (alpha), total (beta), hydrogen-3, potassium-40, strontium-90, cesium-137, uranium, thorium, radon-226, polonium-210, and plutonium-239.[162]

Soil contamination from fallout from atmospheric nuclear weapons testing has been measured in China and is in accord with U.S levels.[163] However, these measurements do not shed any light on the possibility of additional contamination of soil from weapons production activities.

In 1983, the State Environmental Protection Bureau carried out a nationwide survey of background radiation. That same year, the bureau built China's first storage facilities for radioactive materials in major urban centers.[164] The Environmental Protection Bureau has also promulgated a series of radiation protection regulations as well as regulations and standards on radioactive environmental control methods. According to a report by Lou Yibo, the bureau is drafting a Radioactive Contamination Prevention Law. "The State Environmental Protection Bureau," Yibo writes, "is devoting close attention to associated work regarding radioactive environmental control policies, laws, and standards for routine administration of radioactive environmental control work nationwide."[165]

Yibo also reports several measures the bureau has already instituted. These include:[166]

- a two-level examination system for evaluating environmental impacts, with a state-level bureau examining nuclear facilities and province-level units examining the use of isotopes and associated radiation matters;
- the storage of waste materials in urban radioactive waste storage facilities in each province, with environmental protection units responsible for storage and control;
- a two-track monitoring system for major nuclear facilities, with both nuclear facilities and the environmental protection system setting up monitoring systems;
- designating nuclear accident emergency committees within local governments for regions possessing nuclear facilities. These bodies are responsible for responding to nuclear accidents;

162. Li et al. 1988.

163. Estimated from Sha et al. 1991, pp. 50–52, and Eisenbud 1987, p. 331. and p. 335.

164. Yibo 1993.

165. Yibo 1993.

166. Yibo 1993.

• the proper handling of several radioactive contamination accidents (note the reference to accidents having occurred);

• the standardization of radioactive waste control.

In 1993, the International Campaign for Tibet, based in Washington, D.C., released a report accusing China of conducting nuclear weapons research on the Tibetan plateau. ICT charged that China had disposed of nuclear waste from the Ninth Academy in Tibet in a "roughshod and haphazard manner" through the 1960s and 1970s.[167] ICT's charges have received indirect corroboration by a report released by the Office of the Dalai Lama at the 1992 Earth Summit in Rio de Janeiro. Entitled *Tibet Environment and Development Issues 1992*, it states that the "disposal of nuclear and other toxic waste in hazard-ridden surface sites with minimal safety measures is contaminating parts of the Tibetan Plateau." The degradation of the Tibetan environment reportedly mirrors that in China, "where few environmental safeguards exist."[168] It must be borne in mind that these reports on Tibet are in the context of the nationalist movement there.

HEALTH IMPACT OF THE ENTIRE COMPLEX

There is an almost complete absence of hard information on the health effects of China's nuclear weapons industry. That the potential exists for serious health effects due to both occupational and environmental exposures cannot be denied. However, the same secrecy that envelops China's environmental problems resulting from nuclear weapons production also obscures the collection, analysis, and dissemination of any data regarding health effects.

Clearly, China has a bureaucracy for collecting health data. As detailed by Lewis and Xue, the same Security and Protection Bureau that has been concerned with environmental protection and safety also has responsibility for health.[169] In cooperation with the Ministry of Health, this bureau runs special hospitals and maintains personnel health records, presumably for workers directly involved in the nuclear weapons industry. These responsibilities are quite distinct from monitoring health effects from nonmilitary uses of radiation, which have come under the purview of health departments at the county level or higher.[170]

It is also clear that China has a skilled group of medical and public health scientists familiar with toxicology, radiation dosimetry, and

167. International Campaign for Tibet 1993, pp. 1, 18–19.

168. Sharma 1992.

169. Lewis and Xue 1988, p. 55.

170. *Renmin Ribao* 1989.

basic environmental epidemiology. For example, the High–Background Radiation Research Group at the Laboratory of Industrial Hygiene in the Ministry of Public Health is quite active in studying health effects among populations in China that have been exposed to high amounts of natural sources of ionizing radiation from thorium and radon. This group has studied site-specific cancer mortality, congenital defects, and thyroid nodularity in exposed populations,[171] doing some of the work in collaboration with the U.S. National Cancer Institute and publishing it in the West.[172]

However, most published research by these and other Chinese medical scientists dealing with radiation has covered laboratory investigations, methodology, or other topics with little connection to the nuclear weapons industry. Most have appeared exclusively in the Chinese medical literature. Perhaps the main exceptions have been scattered articles on cancer and cytogenetic abnormalities among uranium miners,[173] as well as press reports summarizing Chinese conferences (mostly closed to outsiders) on occupational health in mines.[174] However, while these articles demonstrate the ability of Chinese researchers to conduct fairly sophisticated research on mortality and morbidity using clinical data and biological markers that provide sophisticated indications of toxic exposure and effects, none have provided quantitative data that would help others estimate industry-wide or even work site–specific radiation exposures or resulting health effects.

Press and Government Reports and Other Indirect Evidence

This paucity of information in the medical and public health publications means that the only sources of information are the unscientific and speculative reports that have appeared in the press and indirect information on fallout in other countries. In 1981, the *International Herald Tribune* quoted officials in the region of Lop Nur, the major site of Chinese nuclear weapons tests, as saying that cases of cancer of the liver, lung, and skin had increased.[175] According to the officials, "Many years ago, people never died of cancer, but in recent years they have been dying this way. Some people say it is because of the testing." Four years later, the *New York Times* reported that Muslim students in Ürümqui, 500 miles northwest of Lop Nur, alleged that

171. High-Background Radiation Research Group 1980.

172. Luxin et al. 1990, p. 131.

173. *Health Physics* 1981.

174. *Zhongguo Xinwen* 1981.

175. *International Herald Tribune*, 23 August 1981.

nuclear weapons tests had spread radiation sickness and death among a relatively large percentage of the population.[176] An Associated Press article by Elaine Kurtenbach in 1989 cited an "official report" as estimating that "nuclear accidents caused mainly by careless handling of radioactive materials killed 20 people and injured 1,200 people in China from 1980 to 1985." But the official of the State Environmental Protection Bureau quoted in the article, Luo Guozhen, did not state how many accidents occurred or give details of those that did occur.[177]

Kurtenbach also cites an article on mishaps by an official of the Ministry of Nuclear Industry's Bureau of Safety, Zhou Zhumou. According to Kurtenbach, a "serious accident took place at a production reactor in northwestern China" during the 1966–1976 Cultural Revolution. "About 10 people were exposed to radiation in that accident and ... another man got uranium poisoning in 1973, but there were no cases of acute radiation sickness."[178]

In 1993, a Russian periodical attributed to a Kazakh nuclear physicist, I. Chastnikov, the observation that infant mortality in Alma-Ata and Alma-Ata Oblast doubled after every nuclear test at the Lop Nur area.[179] This observation certainly overstates the case because such huge radiation effects appearing so soon after every test would probably give rise to many other effects, such as radiation sickness. It is illustrative of the kind of statements that are in the public domain in the absence of official published data on fallout.

Also in 1993, the report by the International Campaign for Tibet cited cases of cancer with features suggestive of leukemia among children living near the Ninth Academy in Qinghai Province and quoted a local resident as saying that poor Tibetans sometimes ate local meat that was banned by the Chinese authorities because of possible radioactive contamination.[180] Uranium mining in the northwestern province of Gansu and the release of poisonous waste into water apparently caused 50 Tibetan deaths, the report said, naming 24 of those who died. However, the symptoms cited, such as high temperatures, are not directly associated with radiation exposure, and chemical pollution of drinking water appears to have been involved.[181]

China itself has only occasionally admitted to adverse health effects associated with its nuclear weapons industry. Even then, the reports

176. *New York Times*, 31 December 1985.

177. Kurtenbach 1989.

178. Kurtenbach 1989.

179. Kirinitsiyanov 1993, p. 74.

180. International Campaign for Tibet 1993, pp. 21–22.

181. International Campaign for Tibet 1993, pp. 44–45.

have largely been limited to acute accidents. In May 1986, Qian Xuesen, a senior Chinese military official at the time, was quoted as saying: "Facts are facts. A few deaths have occurred, but generally China has paid great attention to possible accidents. No large disasters have happened."[182] As another example, a 1989 article in the *China Daily*, a government-sponsored publication, described an accident the year before that exposed 15 people to radiation, one of whom suffered third-degree burns, after a piece of uranium material was "lost" from a factory in Nanjing.[183] In the same article, Pan Ziqiang, director of the safety department of China National Nuclear Industry Corporation, claimed that over the past 30 years, radiation levels from the country's nuclear industry had stayed far below the state safety standard of about 5 milligray equivalent person per year. He further noted the lowering of the radiation dose limit for the nuclear industry to 1 milligray equivalent person per year.

Travis cites a 1991 report from China that 1,200 people were injured and 20 people died between 1980 and 1985 because of careless handling of radioactive materials. The report, he says, "attribute[d] those accidents to incompetent managers and managers' and workers' ignorance about the dangers of nuclear materials."[184]

Evidence collected in other countries of fallout from Chinese atmospheric tests provides some insight into the extent of contamination likely to have occurred in China itself. After a nuclear weapons test in Lop Nur on 17 September 1977, the U.S. EPA monitored fallout radionuclides by activating its Environmental Radiation Ambient Monitoring System. By extrapolating from fallout radionuclides on airborne particulates, in precipitation, and in cow's milk at several sampling locations, EPA calculated collective doses of 1,502 person-sieverts to the lung, 1,277 person-sieverts to the thyroid, and 1,076 person-sieverts to the bone for all U.S. residents. Japanese scientists reported similar measurements of fission products following Lop Nur weapons test from 1976 to 1981.[185] Doses to Chinese residents were probably considerably higher, although an accurate estimate is impossible without knowing more about the tests and prevailing meteorological conditions.

Raising the Shroud of Secrecy

Despite the current lack of hard data on the health impacts of China's nuclear weapons industry, it is likely that the nature and degree of

182. Qian Xuesen, a consultant for the national defense committee of the scientific and technological industry, as quoted in May 1989, p. 145.

183. Yuan 1989.

184. Travis 1991, p. 9.

185. Momoshima and Takashima 1983; Kojima and Furukawa 1986.

health effects were similar to those in other nuclear weapons–producing countries. Indirect evidence indicates that uranium miners suffered exposure to radon with resultant carcinogenic effects. However, apart from anecdotal reports of acute radiation poisoning, there is no information with which to estimate the health effects from exposure to radiation and toxic chemicals among either the production workers in the nuclear weapons plants themselves or among the residents living in those areas.

It is likely that capable Chinese scientists have conducted investigations that would shed light on these and other matters related to the environmental and health effects of building nuclear weapons. The China Today Series Editorial Committee's book on the nuclear industry indicates the conditions of work in the early years of the Chinese program due to the urgency with which the country pursued its efforts to acquire nuclear weapons. These pressures greatly intensified after the Soviet withdrawal in 1960. This would make the situation in China until about the mid-1960s roughly comparable with that in the early Soviet nuclear program, when exposures and environmental contamination were severe. We have cited one admission that radioactive wastes from reprocessing did not meet specifications but were discharged anyway; however, we have no further information. There is also evidence of accidents in the reprocessing plant at Jiuquan, but their nature and consequences remain secret.

It is vitally important that China abandon this secrecy and give an open and direct accounting of the exposures and health effects sustained by workers and the community. Such information has implications not only for the Chinese people but for residents in neighboring countries as well. Thus, noting that Kazakhstan still receives radioactive contamination from nuclear tests in Lop Nur, Kazakh Ecology Minister Svyatoslav Medvedev added, "The world press hasn't dealt with this at all. Meantime, our death rate keeps on going up."[186] Perhaps most important, Russia, France, and the United States have agreed to a moratorium on nuclear testing. China has not yet stopped its own testing, despite anecdotal reports that are beginning to appear of cancers that may be attributable to past Chinese atmospheric nuclear weapons tests.

186. As quoted in Josephson 1993, p. 586.

11 Near-Nuclear and De Facto Nuclear Weapons Countries

Albert Donnay and Arjun Makhijani

The nuclear weapons powers, notably the United States, have placed great political, military, and economic emphasis on preventing all but a select few other countries from getting even one nuclear weapon, or indeed, even acquiring the ability to make nuclear weapons. At the same time, the superpowers have built up their own arsenals of tens of thousands of nuclear weapons. This behavior derives in large part from the strong interest of the United States, its close ally Britain, and Russia in maintaining their near-monopoly of these weapons. The United States fears that other countries may seek nuclear weapons not just for deterrence but to actually use them.[1]

"The manner in which the West responded to a possible single weapon in the hands of Iraq shows the exaggerated importance attached to the possession of a weapon or two," notes K. Subramanyam, an Indian critic of the global nuclear regime and chair of the UN Study Group on Nuclear Deterrence. "Yet the five nuclear powers continue to maintain their nuclear arsenals, and the fact that they also happen to be the five permanent members of the [UN] Security Council reinforces the association of power and prestige with the possession of nuclear weapons."[2] Subramanyam goes on to point out the failures of such policies: "Preaching to other nations that nuclear weapons have no utility while continuing to maintain thousands of warheads not only fails to evoke credibility, it raises legitimate suspicions and fears about motivations."[3]

Deterrence and the Manhattan Project

The actions of the United States, which leads the world's efforts to stem proliferation, have derived from its own strategy of nuclear

1. U.S. non-proliferation policy has also begun to include reduction of risk of nuclear weapons use in regional conflicts in South Asia and the Middle East. See Spector 1990, p. 12.

2. Subramanyam 1993, p. 39.

3. Subramanyam 1993, p. 39.

weapons use and planning ever since it launched the Manhattan Project to develop such weapons in 1942. Since that time, the concepts of deterrence and nonproliferation have gone hand in hand for, first, U.S. and British leaders, and, later, those of the Soviet Union.

As we have discussed in chapter 6, when the United States entered World War II, one of its greatest fears was that Germany, with its expertise in nuclear physics, would build the world's first nuclear weapon. This fear played a large role in targeting atomic weapons at Japan rather than at Germany. Indeed, although Germany surrendered in May 1945, more than two months before the first atomic bomb was ready for testing, targeting decisions appear to have been initiated long before anyone knew when the war might end or when such weapons would be ready.

Other nonproliferation factors played a role in the Manhattan Project and targeting decisions. The desire on the part of the United States to monopolize experience relating to nuclear weapons meant excluding its allies to the extent possible. This exclusion was total in the case of the Soviet Union and near-total in the case of France. The British were close collaborators in the Manhattan Project, but even so, there were internal tensions in the U.S. government as to what and how much information to share. Having all the potential targets in the Pacific theater meant that logistics and other operational matters related to the use of the bombs would be essentially a U.S. preserve.[4]

After World War II, the strategies of both superpowers included the possibility and threat of being the first to use nuclear weapons, although the Soviet Union renounced this strategy from 1978 to 1993, at least at a formal level. "First use" continues as part of official U.S. policy, which was originally spelled out in 1950 in National Security Council memorandum NSC-68. While arguing for flexible U.S. capabilities, NSC-68 said that "the only deterrent we can present to the Kremlin is the evidence we give that we may make any of the critical points which we cannot hold the occasion for a global war of annihilation."[5]

In U.S. planning, "deterring" the Soviet Union had a broad meaning. It did not just mean preventing a Soviet nuclear attack on the United States. While that was part of the goal, deterrence encompassed the ability to project U.S. power around the world without Soviet interference. Moreover, NSC-68 advocated a policy of readiness for a range of wars, from local conventional ones to limited nuclear wars to a war of global annihilation. The option of limited nuclear war was updated in the late 1970s and early 1980s.[6]

4. Makhijani and Kelly 1985, chapter 5.

5. National Security Memorandum 68, as quoted in Etzold and Gaddis 1979, p. 414.

6. Ellsberg 1986.

Nuclear threats and the potential first use of nuclear weapons have been integral to U.S. policy. Though nuclear weapons have not been exploded in war since Nagasaki, the United States has issued several nuclear threats, against both nuclear and nonnuclear countries, including China in the 1950s and India in 1971. As Daniel Ellsberg has noted, nuclear threats constitute a use of nuclear weapons in the same way "a gun is used when you point it at someone's head in direct confrontation, whether or not the trigger is pulled."[7]

Nonproliferation Policy

The policies of the superpowers, notably the United States, played a role in fueling the desires of many countries to acquire nuclear weapons. In the competition after World War II, both the United States and Soviet Union attempted to use the promise of civilian uses of atomic energy to halt nuclear proliferation and maintain control over nuclear weapons. For instance, this approach was a part of U.S. policy in 1953 when President Eisenhower announced "Atoms for Peace"—countries that forswore nuclear weapons would get U.S. help in developing nuclear power. This was a complex program with many goals, including providing a convenient civilian face, more acceptable internationally, to nuclear technology that had weapons as its main purpose domestically.

As nuclear power technology developed in the 1960s, the United States and the Soviet Union exported civilian nuclear reactors for research and nuclear power to their allies, as well as nuclear materials and expertise to less industrialized countries eager to become part of the nuclear age. These exports became vital to the growth of the superpowers' domestic nuclear industries, which had to compete with exports of other industrialized countries that were equally eager to promote their own nuclear technologies, including France, the United Kingdom, Canada, and West Germany.

The nuclear supplier countries instituted various controls over the use of their technologies to discourage the development of nuclear weapons. Most important, in 1968 the United States, the Soviet Union, and the Disarmament Committee of the United Nations negotiated the Treaty on the Non-Proliferation of Nuclear Weapons (NPT) as a legally binding framework for both deterring the spread of nuclear weapons and promoting the further expansion of nuclear power among nonnuclear countries.[8] The treaty allowed the nuclear weapons powers to keep their weapons and even expand and modernize them. In exchange, these countries made a commitment to negotiate in good faith

7. Ellsberg 1986, p. 36.

8. Spector 1988, pp. 459–461.

to end the nuclear arms race, achieve nuclear disarmament, and sell and transfer civilian nuclear facilities to nonweapons states. The NPT went into effect in 1970. Of the declared nuclear powers, only the United Kingdom joined the United States and Soviet Union in immediately signing the treaty. France and China signed 22 years later.[9] By 1994, over 160 non–nuclear weapons countries had signed the NPT, accepting the application of International Atomic Energy Agency (IAEA) safeguards on all of their nuclear facilities and forswearing the manufacture, acquisition, and testing of nuclear weapons in exchange for technological assistance in developing peaceful uses of nuclear energy, such as nuclear medicine and nuclear power.[10]

One part of the overall dispute between nonsignatories of the NPT and signatory nuclear weapons countries concerns Article V of the NPT, which makes specific provisions for non–nuclear weapons states to obtain any "benefits" from peaceful uses of nuclear explosions on a "non-discriminatory basis."[11] By the mid-1970s, some threshold nuclear weapons states, notably India, Argentina, and Brazil, were "accusing the nuclear weapon states of delaying the establishment of a PNE [peaceful nuclear explosions] regime in order to avoid sharing PNE technology with the rest of the world. . . ."[12] In addition, under the NPT, the IAEA relies on reporting requirements, audits of nuclear materials, and on-site inspections of nuclear facilities to detect and deter the development of nuclear weapons programs.[13] Nuclear suppliers must require purchasers to agree to the "full scope" of IAEA safeguards as a condition of sales to non–nuclear weapons countries, regardless of their NPT status. However, the five declared nuclear weapons countries do not have to observe these safeguards in their own facilities. Thus, the NPT created two distinct classes of states: those allowed to have nuclear weapons and the rest of the signatories, who agree not to have them. The first category of states—the nuclear powers—was frozen at the five (United States, Soviet Union, United Kingdom, France, and China) who already had that status in 1968. Critics often refer to this system as "nuclear apartheid."[14] India, Israel, and Pakistan, which have the technological capability to produce nuclear weapons, have refused to sign a discriminatory treaty.

9. Norris and Arkin 1993b.

10. Arms Control Today 1994.

11. Findlay 1990, chapter 7.

12. Findlay 1990, p. 241.

13. Spector 1990, pp. 422–430. The IAEA was founded in 1957 and its safeguards program established in the early 1960s. By 1990, it was responsible for monitoring about 900 installations in over 50 countries.

14. Subramanyam 1993, p. 38.

In 1976 France joined with the United States, Britain, the Soviet Union, and other nuclear exporters and became part of the Nuclear Suppliers Group. The group's self-directed role is to implement and enforce safeguards on all exports of nuclear materials and equipment.[15] Two other countries with the technological capability to manufacture nuclear weapons, Germany and Japan, are also part of the group. In all, the Nuclear Suppliers Group has 27 members.[16] China has not joined the Nuclear Suppliers Group, even after signing the NPT. China favors nondiscriminatory guidelines for global regulation of nuclear trade.

The Nuclear Suppliers Group maintains a "trigger list" of nuclear weapons–related facilities, equipment, components, material, and technologies—including all reactors and anything related to uranium enrichment or plutonium reprocessing. Nuclear Suppliers Group members agree to export such items to nonnuclear states only with appropriate safeguards.[17] The main requirements are that recipient countries must pledge:

• not to use any imported items to manufacture nuclear explosives;

• to transfer imported items to a third country only under similar safeguards and with the approval of the original supplier;

• to accept IAEA safeguards on all materials and facilities incorporating or made from imported items; and

• to provide adequate security for imported items to protect against theft and sabotage.[18]

These safeguards were strengthened in 1992 by the adoption of a ban on exports to any non-nuclear weapons state unless that country signs a separate safeguards agreement with the IAEA covering all its nuclear activities.[19] This agreement aims at "partly safeguarded" enrichment and processing facilities, which IAEA safeguards cover only when the facilities handle nuclear material obtained under IAEA safeguards.

Despite the broad scope of these safeguards, the Nuclear Suppliers Group's efforts to stop proliferation have not been entirely successful. Many countries have demonstrated they can develop most nuclear

15. Spector 1990, pp. 434–436.

16. The 27 members were Austria, Australia, Belgium, Bulgaria, Canada, the Czech and Slovak Federal Republic, Denmark, Finland, France, Germany, Greece, Hungary, Ireland, Italy, Japan, Luxembourg, the Netherlands, Norway, Poland, Portugal, Romania, Russia, Spain, Sweden, Switzerland, the United Kingdom, and the United States.

17. *Nuclear Fuel* 1992.

18. Spector 1990, p. 435.

19. *Nuclear Fuel* 1992.

weapons technologies and materials domestically, with varying degrees of foreign technology and materials.[20] Some technology supplied for civilian uses has been directed to military purposes, as has been alleged in the case of India.[21] In other cases, countries have attempted to make clandestine purchases from companies willing to violate export controls; this was the case for Iraqi attempts to purchase from European and U.S. firms.[22] And some countries have participated in secret collaborations with countries not party to the Nuclear Suppliers Group or the NPT. According to Spector, China sold some nuclear materials in the late 1970s and early 1980s without requiring IAEA safeguards, but in 1984 told the U.S. that it would require safeguards on transactions from then on.[23]

Civilian nuclear cooperation agreements that encourage the transfer of technology between countries are also potential vehicles for nuclear proliferation. Agreements between states that are not signatories to the NPT, or when one party is not a signatory to the NPT, permit transfers of technology that are not fully within the inspections system of the IAEA. For example, South Africa and Israel used a nuclear cooperation agreement to help each other's nuclear weapons programs. As another example, India has exported nuclear power technology.

The Spread of Nuclear Weapons

International safeguards and sanctions have not stopped the spread of nuclear weapons technologies and materials to several nations that have wanted them, most of whom did not sign the NPT. Since China became the last self-declared nuclear weapons power in 1964, four de facto nuclear weapons states have emerged. Israel developed and began deploying nuclear weapons in the late 1960s. India tested a nuclear device in 1974. South Africa began building nuclear weapons in 1980 or 1981. Pakistan apparently achieved this capability in 1986.[24] Three more states have become possessors of nuclear weapons as a result of the breakup of the Soviet Union. While Russia automatically inherited the Soviet Union's U.N. Security Council seat and nuclear weapons power status under the NPT, questions remain about the eventual status of Ukraine, Kazakhstan, and Belarus, which acquired those parts of the Soviet arsenal remaining on their territories. The United States is pushing for these three countries to accede to the NPT as non–nuclear

20. Albright 1993.

21. De la Court, Pick, and Nordquist 1982, p. 108.

22. Spector 1990, p. 41.

23. Spector 1988, p. 461.

24. Spector 1988, p. 3; Spector 1990, p. 8.

weapons states, and they have agreed to do so. Despite this agreement, an increase in military tensions between Russia and Ukraine could be damaging because Ukraine is believed by some to possess the know-how to reactivate warheads still on its territory, despite the lack of Russian codes. Moreover, the three countries, especially Ukraine, possess a significant portion of the technical infrastructure needed for building nuclear weapons, and they may build dismantling and plutonium storage facilities. Since this work is about production of weapons, we will not consider these states further.

Of the four de facto nuclear weapons powers, only South Africa has signed the NPT, and then only in 1991, after the white minority government of President F. W. de Klerk began moving to dismantle South Africa's nuclear weapons program and arsenal, neither of which he publicly acknowledged until 1993.[25] It is felt that de Klerk's decision to dismantle the program and join the NPT was part of an effort to keep nuclear weapons and materials from falling into the hands of a future black-majority government.[26]

In addition to these de facto nuclear countries, several near-nuclear countries have some of the technology needed to make nuclear weapons. Argentina and Brazil, neither of which has signed the NPT, were engaged in their own nuclear arms race, though their rivalry has cooled under civilian governments that came to office in the 1980s.[27] Both have renounced their nuclear weapons programs and each now allows the other to inspect its nuclear facilities. Both also insist that their programs are entirely peaceful. In 1991, they signed an agreement with the IAEA to allow inspection of all their nuclear facilities.[28]

Several NPT signatories have reportedly sought to acquire nuclear weapons. They include Iraq, Iran, Libya, Taiwan, and North Korea.[29] These countries have been subject to much stronger international pressure—including sanctions and even armed attacks to stop their nuclear weapons programs—than the de facto and near-nuclear states that have not signed the NPT. Iraq in particular suffered the Israeli bombing of its Osiraq reactor facility in 1981 and the almost total destruction of many of its other nuclear facilities in the 1991 Persian Gulf War. IAEA inspectors, acting under special UN mandates, have since found and dismantled most of what the war left standing, including calutrons (electromagnetic enrichment devices) and centrifuges for uranium enrichment and laboratory-scale plutonium reprocessing facili-

25. De Klerk 1993.

26. Albright and Hibbs 1993.

27. Spector 1988, pp. 231–277; Albright, Berkhout, and Walker 1993, pp. 179–185.

28. Albright, Berkhout, and Walker 1993, p. 179.

29. Spector 1990, pp. 7, 60, 175–183.

ties.[30] Iraq is now years away from being able to manufacture the various materials and components needed for nuclear weapons. The extent of Iraq's program has focused considerable discussion on the inadequacy of the NPT framework to detect or prevent active nuclear weapons programs.

Iran lacks key nuclear facilities, and its small research programs probably leave it years from having the capacity to manufacture nuclear weapons.[31] Libya has no domestic production capabilities. Reportedly it sought to buy nuclear weapons directly from China but was rebuffed.[32] With the dissolution of the Soviet Union, interested countries may find it easier and cheaper to purchase nuclear materials and even complete nuclear weapons on the black market rather than to develop them domestically.[33]

Of the Third World NPT signatories suspected of having nuclear ambitions, only Taiwan has a large nuclear power program. According to Spector, in both the mid-1970s and in 1987, Taiwan tried to build a secret facility to separate plutonium for nuclear weapons from the spent fuel of its civilian reactors. The United States detected both efforts and persuaded Taiwan to dismantle the facilities.[34] Taiwan probably no longer has an active nuclear weapons program.

According to some estimates, North Korea, which signed the treaty in 1985 under strong pressure from the Soviet Union, appeared to be, until 1994 agreements, the NPT signatory closest to achieving a nuclear weapons capability.[35] Although it signed a joint declaration with South Korea in 1991 pledging not to reprocess plutonium and calling for bilateral inspections of nuclear facilities, North Korea refused to implement this agreement until 1994. It also delayed implementing safeguards inspections by the IAEA required by the NPT, until 1992.[36]

After initial IAEA inspections revealed suspected diversions of plutonium, North Korea at first refused to allow the IAEA any further access to nuclear sites unless the inspections were linked to other unrelated issues, such as North Korea's demands for nonaggression pledges from South Korea and the United States. When negotiations over these issues broke down in March 1993, North Korea notified the IAEA of its intention to withdraw from the treaty. However, North

30. Albright 1993; Albright, Berkhout, and Walker 1993, pp. 168–173.

31. Spector 1990, p. 6 and chapter 12.

32. Spector 1990, p. 175.

33. Sharp 1993; Goldanskii 1993; Potter 1993, appendix one; Proliferation Watch 1993, p. 13; Makhijani and Hoenig 1991.

34. Spector 1990, p. 6.

35. Albright 1993.

36. Albright and Hibbs 1992.

Korea put its withdrawal from the NPT on hold in the wake of extraordinary pressure from the United States. In July 1993, North Korea said it would allow IAEA inspections of its nuclear reactors as long as they were impartial, but it still refused to allow inspections of two nuclear waste sites that could reveal information about plutonium production.[37] In March 1994, North Korea again allowed only a partial IAEA inspection. In late 1994, the United States and North Korea arrived at an agreement in which North Korea suspended its existing reprocessing program in exchange for aid in the acquisition of large civilian nuclear reactors.

It is important to consider North Korea's actions in context. During the Korean War, the U.S. government worked out specific plans to use nuclear weapons against Korea, as well as against China and the Soviet Union. The United States seriously considered several targets in Korea for nuclear attack, including an area that had been designated through armistice negotiations as a sanctuary.[38] Until 1992, the United States maintained hundreds of nuclear weapons aimed at North Korea on the South Korean side of the demilitarized zone, and it still denies North Korea the right to verify for itself whether these weapons have actually been removed. At the same time, the United States and other western powers fear North Korea would contribute to a regional arms race in East Asia and to nuclear proliferation. For instance, one issue of concern is that North Korea may sell nuclear weapons or weapons-usable fissile materials to Iran, to which North Korea is thought to have supplied other weaponry.[39]

Two other NPT signatories, Japan and Germany, have significant nuclear weapons capabilities based on their nuclear facilities and stockpiled nuclear materials. However, nonproliferation debates usually exclude these countries, which are part of, and theoretically protected by, U.S.-led military alliances. Japan's post–World War II constitution, written by the United States, bars it from deploying military forces. The Japanese government interprets this as meaning that "defensive" forces are allowed, but "offensive" forces are not. In practice, however, it is difficult to distinguish these two categories. Japan's Atomic Power Basic Law also explicitly prohibits the development of nuclear weapons. What this means in practice is open to interpretation.

The legacy of the bombing of Hiroshima and Nagasaki has left most Japanese with a strong aversion to nuclear weapons, commonly known as Japan's "nuclear allergy." Nevertheless, the country has a large nuclear power program, including plans for plutonium breeder reactors. It already operates a plutonium-reprocessing plant at Tokai-

37. *New York Times*, 20 July 1993.

38. Ege and Makhijani 1982, p. 11.

39. Spector 1990, pp. 133–134.

mura and a second, large facility is under construction at Rokkashomura.[40] Through reprocessing contracts with France and the United Kingdom, Japan owns large stockpiles of plutonium in those countries, all of which are to be returned eventually to Japan. Some of this plutonium has already been separated from spent fuel by reprocessing, while the rest awaits separation. This plutonium supply far exceeds the amount required for Japan's breeder program.[41]

A right-wing political faction in Japan's Liberal Democratic Party (until 1994 the ruling party for decades) has long advocated keeping the nuclear weapons option open for Japan. This idea has recently gained some strength in response to the potential of a nuclear-armed North Korea. There has also been some sentiment in this direction deriving from Japan's proximity to a nuclear-armed China. This right-wing pressure, combined with the country's growing stockpile of plutonium and its eager pursuit of domestic plutonium-reprocessing and breeder reactor technologies, is of great concern to Japan's potential adversaries in the Pacific. For example, the North Korean Foreign Ministry has accused the Japanese government of seeking to acquire nuclear weapons "at any cost."[42] However, this accusation was in the context of North Korea's own reconsideration of NPT membership.

Japan already has the aircraft, missiles, and other delivery systems it would need to deploy nuclear weapons, backed by a large military-industrial complex ready to make more. While none of this means Japan will become a nuclear weapons state, its intentions can change quickly. Thus, the criteria for including countries in these discussions as potential or near–nuclear weapons states are that they have the technical capability to make nuclear weapons and possess the necessary materials and technology. (For a discussion of the criteria used in selecting these countries, see chapter 2.) By these criteria, Japan is a technological near–nuclear weapons state. Germany also belongs in this category in possessing both the technology and the materials (separated plutonium) to make nuclear weapons, but not the current intent.

One capability often left out of nonproliferation discussions is the ability to fashion bombs to disperse radioactive materials, such as plutonium, cesium-137, or strontium-90. Groups or countries unable to manufacture nuclear weapons could wield considerable power and do great damage by releasing large quantities of highly radioactive material without a nuclear explosion.[43] During World War II, the Allies

40. IPPNW and IEER 1992, pp. 46–47.

41. Walker and Berkhout 1992.

42. Anonymous official, as quoted by the Korean Press Agency and cited in the *New York Times*, 13 July 1993.

43. IEER and IPPNW 1992, p. 141.

feared Germany might develop such capability, and the United States considered developing them as well.[44] Several reports published in 1992 and 1993 suggest that Armenia may be working on developing such a radioactive punishment weapon, although whether for deterrence or actual use is not known.[45] The ultranationalist Russian politician Vladimir Zhirinovsky has threatened the use of radioactive weapons against the Baltic states. Ideas of using radioactive materials in warfare go back to the very beginning of deliberations on nuclear weapons. As early as May 1941, a U.S. National Academy of Sciences committee suggested that nuclear energy could be used to create materials for radioactive warfare.[46]

There are also several countries that possess plutonium due to reprocessing contracts they hold with France or Britain. Belgium, Germany, Italy, Switzerland, and the Netherlands belong in this category. Belgium, Germany, and Italy also have operated reprocessing plants in the past. These countries could choose to become nuclear weapons powers or contribute their nuclear materials production capabilities and expertise to a common European nuclear weapons force should political and military structures to enable that arise in the future. However, because a considerable amount of weapons-grade plutonium already exists in France and Britain, it is not likely that reactor-grade plutonium would be used for a European nuclear force. In addition, Spain has operated a small pilot reprocessing plant, and there was a laboratory-scale plant in Yugoslavia.[47] The amounts of plutonium produced by Britain and France for other countries are discussed in chapters 8 and 9 respectively.

A NOTE ON SOURCES

Due to the tight security under which all the countries in this chapter have built and operated their nuclear facilities, there are very few data available on the health and environmental impacts of the facilities. In general, military nuclear facilities do not appear to be regulated by any independent health or environmental agencies. As in the nuclear weapons production complexes of the United States and the former Soviet Union, the priority in these states seems to be on production and not public health, occupational safety, or the environment. And, as is also true in the declared nuclear weapons states, the nuclear weapons establishments in these states seem to be secret governments within governments.

44. IEER and IPPNW 1992, p. 142.

45. Proliferation Watch 1993a, p. 7.

46. Lanouette and Silard 1993, p. 226.

47. Chayes and Lewis 1977, pp. 19–21.

Although the quantities of spent fuel, highly enriched uranium, and separated plutonium produced in these countries are orders of magnitude less than that produced by Russia and the United States, it does not necessarily follow that their production has had less impact on the local environment and public health. Many of these countries' smaller facilities are based on outmoded technologies. These small research reactors and pilot-scale processing plants may result in much greater occupational exposures and levels of environmental pollution per unit of production than full-scale, automated facilities.

Regardless of size, their production facilities may cause more damage to the environment and human health if economic and other pressures lead to less stringent observation of health and safety standards. This is illustrated by the far worse pollution and accident record of the Soviet nuclear complex relative to that of the United States, even during the 1950s and 1960s, when production and technology levels were comparable. Without any published data on environmental or occupational monitoring, it is impossible for us to assess the true impact of these facilities.

In compiling the information presented here, we rely on the few published sources available: *Nuclear Ambitions*, by Leonard Spector; *World Nuclear Industry Handbook 1991*, by Nuclear Engineering International Publications; *World Inventory of Plutonium and Highly Enriched Uranium*, by David Albright, Frans Berkhout, and William Walker; and *The Nuclear Fix*, by Thijs de la Court, Deborah Pick, and Daniel Nordquist of the World Information Service on Energy. Table 11.1 lists the facilities that are being used or could be used for nuclear weapons production in the de facto and near–nuclear weapons states. Countries are discussed below in alphabetical order.

ARGENTINA

Argentina's large and comprehensive nuclear program dates back to 1950, built partly with the assistance of expatriate Nazi scientists.[48] Argentina's program also has been assisted by purchases of heavy water, uranium hexafluoride, and 20 percent–enriched uranium from China in the 1980s.[49] Although Argentina has renounced nuclear weapons plans, it has facilities and capabilities to produce nuclear weapons. As of July 1994, Argentina had not signed the NPT, but its parliament was considering it.[50]

Argentina has three heavy-water power reactors that fall under IAEA safeguards and seven research reactors, six of which are safe-

48. de la Court, Pick, and Nordquist 1982, pp. 21–22.

49. Spector 1990, pp. 38–39.

50. Arms Control Today 1994.

Table 11.1 Location of Nuclear Weapons–Related Activities in Near-Nuclear and Undeclared Nuclear Weapons Countries

Facility Name and Location	Built or Supplied By[1]	Operating Status[2]	Safeguards Status[3]	Comments	Source[4]
Nuclear Research Centers					
India					
Saha Institute of Nuclear Physics, Calcutta					f
Bhabha Atomic Research Center [BARC], Trombay					f
Tata Institute of Fundamental Research, Bombay					f
Indira Gandhi Center for Atomic Research [IGCAR], Kalpakkam					a, b
Israel					
Nuclear Research Center, Dimona, Negev Desert					a
Pakistan					
PINSTECH, Rawalpindi					c
South Africa					
Advena Laboratory [Kentron Circle], 15 km from Pelindaba					i
National Nuclear Research Center, Pelindaba					a

Table 11.1 (continued)

Facility Name and Location	Built or Supplied By[1]	Operating Status[2]	Safeguards Status[3]	Comments	Source[4]
North Korea					
Yongbyon [also Nyongbyon]					d
Argentina					
Ezieza Research Complex, Buenos Aires Brazil					d
Instituto de Pesquicas de Energia Nuclear [IPEN], Sao Paulo					c
Belo Horizonte					c
Army Technological Center [CETEX], Rio de Janeiro					d
Navy's Aramar Experimental Center, Ipero					a
Uranium Mining and Milling (production of yellowcake)[5]					
Argentina[6]					
La Estella		S = 1986		Mine & mill, capacity = 20 t U/yr	a, b
Los Gigantes		S = 1982		Mine & mill, capacity = 20 t U/yr	a, b, c
San Rafael		S = 1970		Mine & mill, capacity = 120 t U/yr	a, b

Sierra Pintada	S = 1992	Mine & mill, capacity = 500 t U/yr, 50% of total reserves	a, b, c
Los Adobes	NO, 1985	Mine & mill, capacity = 55 t U/yr	b, c
Malargue	NO, 1987	Mine & mill, capacity = 70 t U/yr	a, b, c
Don Otto	1962–80	Mine & mill, capacity unknown	c
Brazil[7]			
Pocos de Caldas	S = 1981	Mine & mill, capacity = 420 t U/yr, est. reserves 17,000 metric tons U	b a, b, c
Figueira	NA	Mine, capacity unknown, est. reserves 6,000 t U	c
Itataia	NA	Mine, capacity unknown est. reserves 70,000 t U	c
Lagoa Real	NA	Mine, capacity unknown, est. reserves 15,000 t U	c
Esphinharas	NA	Mine, capacity unknown, est. reserves 4,000 t U	c
India[8]			
Domiasiat, Khasi and Jaintia Hills, Meghalaya	S = 1992		e
Jaduguda region, west of Calcutta	S = 1968, OP	Mine & mill, capacity = 200 t U/yr	a, b

Table 11.1 (continued)

Facility Name and Location	Built or Supplied By[1]	Operating Status[2]	Safeguards Status[3]	Comments	Source[4]
Mosaboni		NA		Mine & mill, capacity = 15 t U/yr	b
Rakha		NA		Mine & mill, capacity = 15 t U/yr	b
Surda		NA		Mine & mill, capacity = 15 t U/yr	b
Israel[9]					
Near Beersheba, Negev Desert		S = pre-1972		Phosphate mine, capacity unknown	a
Two phosphoric acid plants in Haifa		NA		Total uranium processing capacity = 100 t/year	a
One phosphoric acid plant in southern Israel		NA			a
North Korea					
Pyongsan		NA		Mine, capacity unknown	a
Hamhung ?		NA		Mine, capacity unknown	a
Unggi ?		NA		Mine, capacity unknown	a
Hae Kumgang ?		NA		Mine, capacity unknown	a
Kusong ?		NA		Mill, capacity unknown	a

Pakistan[10]			
Dera Ghazi Khan	S = 1985	30 t U/yr	a, b
South Africa[11]			
Buffelsfontein	S = 1957	Mine & mill, capacity = 400 t U/yr	a, b
Chemwes	1979–88	Mine & mill, capacity = 500 t U/yr	a, b
West Rand Cons.	1952–81	Mine & mill, capacity = 500 t U/yr	b
Blyvooruitzicht	1967–84	Mine & mill, capacity = 500 t U/yr	b
Harmony	1955–1989	Mine & mill, capacity = 150 t U/yr	a, b
Hartebeestfontein	S = 1956	Mine & mill, capacity = 350 t U/yr	a, b
St. Helena-Beisa	1982–84		a
Palabora	S = 1971	Mine & mill, capacity = 150 t U/yr	a, b
East Rand Gold & Uranium	S = 1978	Mine & mill, capacity = 250 t U/yr	a, b
Vaal Reefs	S = 1953	Mine & mill, capacity = 2,000 t U/yr	a, b
Western Deep Levels	NO	Mine & mill, capacity = 300 t U/yr	b

Table 11.1 (continued)

Facility Name and Location	Built or Supplied By[1]	Operating Status[2]	Safeguards Status[3]	Comments	Source[4]
Joint Metallurgical Scheme		S = 1977		Mine & mill, capacity = 500 t U/yr	a, b
Metallurgical Scheme		S = 1977		Mill serving serveral mines, capacity = 450 t U/yr	b
Afrikander Lease		NO		Mine & mill, capacity = 300 t U/yr	b
Randfontein		1954–88		Mine & mill, capacity = 400 t U/yr	a, b
Western Areas		S = 1982		Mine mill, capacity = 200 t U/yr	a, b
Driefontein		1958–88		Mine & mill, capacity = 500 t U/yr	a, b
Conversion of Yellowcake (U_3O_8) to Uranium Oxide (UO_2) (for use as reactor fuel or for further conversion to uranium hexafluoride (UF_6))					
Argentina					
Cordoba Phase I	Argentina	S = 1980	Safeguarded	Capacity = 15 t U/yr	a
				Capacity = 55 t U/yr	b
Cordoba Phase II	Argentina	S = 1982	Safeguarded	Capacity = 150 t U/yr	a
Ezeiza, Buenos Aires	Argentina	Planned	Safeguarded	Capacity = 150 t U/yr	b
Brazil					
IPEN, São Paulo	Brazil and West Germany	S = 1981	Safeguarded	Capacity = 10 t U/yr	a

India					
Hyderabad	India	S = 1984	Unsafeguarded	Capacity = 50 t U/yr	a, b
North Korea					
No information available; at least one facility is likely and probably built before 1987.					
Conversion of Uranium Oxide (UO_8) to Uranium Hexafluoride (UF_6) (in preparation for uranium enrichment)					
Argentina					
Pilcaniyeu Pilot Plant	Argentina	S = 1983	Safeguarded	Capacity unknown	a
Brazil					
IPEN, São Paulo	Brazil	S = 1981–82	Safeguarded	Capacity = 15 t UF_6/year	a
IPEN, São Paulo	Brazil	S = 1984	Safeguarded	Capacity = 90 t UF_6/year	a, b
India					
BARC, Trombay	India	S = 1984?	Unsafeguarded		a, b
Israel					
Dimona, Negev Desert	Israel	S = mid-1970s?	Unsafeguarded	Capacity unknown, conversion from U_3O_8 to UF_6	a
North Korea					
No information available; no facilities are likely					

Table 11.1 (continued)

Facility Name and Location	Built or Supplied By[1]	Operating Status[2]	Safeguards Status[3]	Comments	Source[4]
Pakistan					
Dera Ghazi Khan	Pakistan and West Germany	S = 1980	Unsafeguarded	Capacity = 198 t UF_6/year, conversion from U_3O_8 to UF_6	a
South Africa					
Valindaba (pilot plant)	United Kingdom?	S = pre-1975	Unsafeguarded	Both facilities convert from U_3O_8 to UF_6	a
Valindaba	South Africa	S = 1986	Unsafeguarded	Capacity = 700 t UF_6/year	a
Uranium Enrichment[12]					
Argentina					
Pilcaniyeu Phase I	Argentina	S = 1983	Safeguarded	Gaseous diffusion, capacity = 20,00 SWU	b
Pilcaniyeu Phase II	Argentina	S = 1988	Safeguarded	Gaseous diffusion, capacity = 100,00 SWU	b
Brazil					
Belo Horizonte	West Germany	S = 1980	Safeguarded	Jet nozzle, laboratory scale	a
IPEN, São Paulo	Brazil	S = 1986–87	Safeguarded	Ultracentrifuges, laboratory scale 20% enrichment	a
Aramar Research Center, Ipero	Brazil	S = 1988	Safeguarded	Ultracentrifuges, 400 kg to 5% in 1989 plus some 20%	a

Resende	West Germany	S = 1990	Safeguarded	Jet nozzle, 5 metric tons/year to 0.85% projected	a, b
India					
BARC, Trombay	India	S = 1984?	Unsafeguarded	Ultracentrifuges, pilot scale	a
Rare Metals Plant, Ratanhalli	India	S = late 1980s?	Unsafeguarded	Ultracentrifuges, capacity unknown	d
Israel					
Nuclear Research Center, Dimona	Israel	S = 1979–80	Unsafeguarded	Centrifuges, combined capacity = 2–3 kg HEU/year	a
Nuclear Research Center, Dimona	Israel	S = 1981	Unsafeguarded	Laser enrichment	a
North Korea					
No enrichment program; reactors all use natural uranium					a
Pakistan					
Sihala	Pakistan[13]	S = 1979	Unsafeguarded	Ultracentrifuges, pilot scale	a
Kahuta (underground facility)	Pakistan[13]	S = 1984	Unsafeguarded	Ultracentrifuges, capacity = 50 kg HEU/year	a, b
Golra	Pakistan[13]	UC in 1987	Unsafeguarded	Ultracentrifuges, capacity unknown	a
South Africa					
Y Pilot Plant, Valindaba	South Africa and others	1978–90	Unsafeguarded	Stationary-wall centrifuge, capacity = 50–100 kg HEU/year	a

Table 11.1 (continued)

Facility Name and Location	Built or Supplied By[1]	Operating Status[2]	Safeguards Status[3]	Comments	Source[4]
Valindaba (semi-commercial plant)	South Africa	S = 1988	Unsafeguarded	Stationary-wall centrifuge with helicon cascade, capacity = 50 metric tons LEU/year	a
Uranium Processing and Fuel Fabrication[14]					
Argentina					
Constituyentes Pilot Plant, Buenos Aires	Argentina	S = 1976	Partly safeguarded	Research reactor fuel, pilot plant, capacity unknown	a
Ezeiza, Buenos Aires	Argentina and West Germany	S = 1982	Partly safeguarded	HWR fuel, capacity = 300 metric tons/year	a, b
Brazil					
Resende	West Germany	S = 1982	Safeguarded	LWR fuel, planned capacity = 100 metric tons/year	a, b
IPEN Pilot Plant, São Paulo		NA	Safeguarded	Pilot plant, capacity unknown	a
India					
BARC, Trombay	India	S = late 1960s	Unsafeguarded	Research reactor and breeder reactor fuel, capacity = 135 metric tons/year	a, b

Nuclear Fuel Complex, Hyderabad	India	S = 1971	Partly safeguarded	HWR fuel, capacity = 250 metric tons/year	a, b
Nuclear Fuel Complex, Hyderabad	India	OP	Partly safeguarded	LWR fuel, capacity = 50 metric tons/year	b
Israel					
Nuclear Research Center, Dimona	France	S = 1963	Unsafeguarded	Natural uranium fuel for Dimona HWR, capacity unknown	a
North Korea					
No information available; at least one facility is likely, probably built before 1987.					a
Pakistan					
Chashma	Pakistan and Canada	S = 1980	Unsafeguarded	Natural uranium fuel for KANUPP HWR, capacity unknow	a, b
South Africa					
Elprod, Pelindaba	South Africa	S = 1981	Unafeguarded	LEU and HEU for LWRs, capacity unknown	a, i
Production of Heavy Water					
Argentina					
Arroyito	Argentina and Switzerland	S = 1991	Safeguarded	Capacity = 250 metric tons/year after 2 upgrades	a, b

Table 11.1 (continued)

Facility Name and Location	Built or Supplied By[1]	Operating Status[2]	Safeguards Status[3]	Comments	Source[4]
Atucha	Argentina	NA	Unsafeguarded	Capacity = 2–4 metric tons/year (pilot plant)	a, b
Brazil					
No heavy-water reactors.					
India					
Nangal	West Germany	S = 1962, OP	Unsafeguarded	Capacity = 14 metric tons/year	a, b
Baroda	Switzerland and France	S = 1977, OP	Unsafeguarded	Capacity = 45–67 metric tons/year	a, b
Tuticorin	Switzerland and France	S = 1978, OP	Unsafeguarded	Capacity = 49–62 metric tons/year	a, b
Talcher	West Germany	S = 1979, OP	Unsafeguarded	Capacity = 62 metric tons/year	a
Kota	India and Canada	S = 1984, OP	Unsafeguarded	Capacity = 85–100 metric tons/year	a, b
Thal-Vaishet	India	S = 1987, OP	Unsafeguarded	Capacity = 110 metric tons/year	a, b
Manuguru	India	S = 1990	Unsafeguarded	Capacity = 185 metric tons/year	a, b
Hazira	India	S = 1990	Unsafeguarded	Capacity = 110 metric tons/year	a

Israel					
Rehovot	Israel	S = 1954?		Capacity unknown	a
North Korea					
No heavy-water reactors.					
Pakistan					
Karachi	Canada?	S = 1976	?	Capacity unknown	a
Multan	Belgium	S = 1980?	Unsafeguarded	Capacity = 13 metric tons/ year	a
South Africa					
No heavy-water reactors.					
Nuclear Research Reactors (potential sources of weapons-grade plutonium)[15]					
Argentina					
RA-0, Cordoba	Argentina	S = 1965	Unsafeguarded	MEU fuel, 1 Wth	a, c
RA-1, Constituyentes, Buenos Aires	Argentina	S = 1958	Safeguarded	MEU fuel, 70 KWth	a, c
RA-2, Constituyentes, Buenos Aires	Argentina	S =1966	Safeguarded	HEU fuel, <1 MWth	a, c
RA-3, Ezeiza, Buenos Aires	Argentina	S = 1968	Safeguarded	MEU fuel, 3.5 MWth, upgraded to 8MWth	a, c
RA-4, Rosario	West Germany	S = 1972	Safeguarded	MEU fuel, <1 MWth	a, c
RA-6, San Rafael de Bariloche	Argentina	S = 1982	Partly safeguarded	HEU fuel, 500 KWth	a, c

Table 11.1 (continued)

Facility Name and Location	Built or Supplied By[1]	Operating Status[2]	Safeguards Status[3]	Comments	Source[4]
Brazil					
IEAR-1, São Paulo	USA	S = 1957	Safeguarded	HEU fuel, 5 MWth	a
RIEN-1, Rio de Janeiro	Brazil	S = 1965	Safeguarded	MEU fuel, 10 KWth	a
Triga-UMG, Belo Horizonte	USA	S = 1960	Safeguarded	MEU fuel, 100 KWth	a
IPEN Zero Power, São Paulo	Brazil	S = 1988	Unsafeguarded	LEU fuel, 100Wth	a
India					
Apsara Reactor, BARC, Trombay	India and UK	S = 1956	Unsafeguarded	LWR, MEU fuel, 1 MWth	a
Cirus Reactor, BARC, Trombay	India and Canada	S = 1960	Unsafeguarded	HWR, natural uranium fuel, 40 MWth	a, d
Zerlina Reactor, BARC, Trombay	India	1961–83	Unsafeguarded	HWR, variable fuel, 400 Wth	a
Purmima II, BARC, Trombay	India	1984–87	Unsafeguarded	U-233 fuel	a
Purmima III, BARC, Trombay	India	S = 1989	Unsafeguarded	U-233 fuel	a
Dhruva [R-5] Reactor, BARC, Trombay	India	S = 1985	Unsafeguarded	HWR, natural uranium fuel, 100 MWth	a
Kamini Reactor, IGCAR, Kalpakkam	India	S = 1988	Unsafeguarded	U-233 fuel, 30 KWth	a
Fast Breeder Test Reactor, IGCAR, Kalpakkam	India and France	S = 1985	Unsafeguarded	Breeder reactor, 42 MWth	a
Israel					
IRR1, Nahal Soreq	USA	S = 1960	Safeguarded	LWR, HEU fuel, 5 MWth	a

IRR2, NRC Institute No. 1, Dimona	France[16]	S = 1963	Unsafeguarded	HWR, natural uranium fuel, 24 MWth, upgraded to 70 MWth	a
North Korea					
Yongbyon [also Nyongbyon] IRT	USSR	S = 1965	Safeguarded	HEU fuel, 4 MWth	a
Yongbyon [also Nyongbyon]	North Korea	S = 1987?	Unsafeguarded	Gas-graphite reactor, natural uranium fuel, 30 MWth	a
Pakistan					
PARR, Rawalpindi	USA	S = 1965	Safeguarded	LWR, HEU fuel, 5 MWth upgraded to 10 MWth	a
PARR-II, Rawalpindi	Pakistan and China	S = 1989	Unsafeguarded	LWR, HEU fuel, 27 KWth	a
No information available	Pakistan[13]	UC in 1980s	Safeguarded	HEU fuel, 50 MWth	a
South Africa					
Safari I, Pelindaba	USA	S = 1965	Safeguarded	LWR, HEU fuel, 20 MWth	a
Pelinduna Zero Power Research Reactor	South Africa	Dismantled	Safeguarded	HWR, LEU fuel, no MWth	a
Nuclear Power Reactors (only unsafeguarded reactors are detailed)[17]					
Argentina					
Three safeguarded reactors; unlikely sources of weapos-grade plutonium					a
Brazil					
Three safeguarded reactors; unlikely sources of weapons-grade plutonium					a

Table 11.1 (continued)

Facility Name and Location	Built or Supplied By[1]	Operating Status[2]	Safeguards Status[3]	Comments	Source[4]
India					
Four safeguarded reactors not listed					
Madras I, Kalpakkam	India	S = 1983, OP	Unsafeguarded	HWR, natural uranium fuel, 220 to 235 MWe, CANDU type	a
Madras II, Kalpakkam	India	S = 1985, OP	Unsafeguarded	HWR, natural uranium fuel, 220 (net) MWe, CANDU type	a
Narora I, Narora	India	S = 1989, OP	Unsafeguarded	HWR, natural uranium fuel, 220 (net) MWe, CANDU type	a
Narora II, Narora	India		Unsafeguarded	HWR, natural uranium fuel, 220 (net) MWe, CANDU type	a
Kakrapar I, Kakrapar	India		Unsafeguarded	HWR, natural uranium fuel, 220 (net) MWe, CANDU type	a
Kakrapar II, Kakrapar	India		Unsafeguarded	HWR, natural uranium fuel, 220 (net) MWe, CANDU type	a
Kaiga I, Karnataka	India	UC	Unsafeguarded	HWR, natural uranium fuel, 220 (net) MWe, CANDU type	a
Kaiga II, Karnataka	India	UC	Unsafeguarded	HWR, natural uranium fuel, 220 (net) MWe, CANDU type	a
Rajasthan III, Kota	India	UC	Unsafeguarded	HWR, natural uranium fuel, 220 (net) MWe, CANDU type	a

Rajasthan IV, Kota	India	UC	Unsafeguarded	HWR, natural uranium fuel, 220 (net) MWe, CANDU type	a
Israel					
No nuclear power reactors.					
North Korea					
Taechoon	North Korea	UC	Unsafeguarded	Gas-graphite reactor, natural uranium fuel, 100 MWth	d
Pakistan					
One safeguarded reactor; unlikely source of weapons-grade plutonium, although Canada cut off all supplies in 1976 (a)					a
South Africa					
Two safeguarded reactors; unlikely sources of weapons-grade plutonium.					a
Chemical Separation of Plutonium (reprocessing)					
Argentina					
Ezeiza, Buenos Aires	Argentina	1969–73	Unsafeguarded	Purex process, laboratory scale	a, c
Ezeiza, Buenos Aires	Argentina and Italy	NC 1990	Partly safeguarded	Purex process, capacity = 5 metric tons spent fuel/year	a, b
Brazil					
Resende	West Germany	NC	Safeguarded	Purex process, capacity = 10 kg spent fuel/day	a, b
IPEN, São Paulo	Brazil and West Germany	NO	Partly safeguarded	Laboratory scale	a

Table 11.1 (continued)

Facility Name and Location	Built or Supplied By[1]	Operating Status[2]	Safeguards Status[3]	Comments	Source[4]
India					
BARC, Trombay	India and USA	S = 1966[18]	Unsafeguarded	Purex process, capacity = 30 metric tons spent fuel/year from Cirus and Dhruva	a, b
PREFRE, Tarapur	India	S = 1979	Partly safeguarded	Purex process, capacity = 100–150 metric tons spent fuel/year from CANDU	a, b
IGCAR, Kalpakkam	India	S = 1985	Unsafeguarded	Purex process	a, b
IGCAR, Kalpakkam Phase I	India	S = 1991	Unsafeguarded	Purex process, capacity = 100–200 metric tons spent fuel/year	a, b
IGCAR, Kalpakkam Phase II	India	UC	Unsafeguarded	Purex process, capacity = 1000 metric tons spent fuel/year	b
Israel					
Nahal Soreq	United Kingdom	S = 1960?	Unsafeguarded	Laboratory scale	a
NRC Institute No. 2, Dimona	Israel and France	S = 1966	Unsafeguarded	Capacity = 15–50 kg Pu/yr	a
North Korea					
Yongbyon [also Nyongbyon]	North Korea and China	S = 1993?	Unsafeguarded	Capacity unknown	a

Pakistan					
PINSTECH, Rawalpindi	Pakistan and UK	S = 1973	Unsafeguarded	Experimental scale	a, h
Chasma	France[19]	UC since 1978	Safeguards apply but not in force	Capacity = 100–200 kg Pu/yr	a
New Labs, Pinstech, Rawalpindi	Pakistan[20]	NO since 1982	Unsafeguarded	Capacity = 10–20 kg Pu/yr	a
South Africa					
Pelindaba	South Africa?	S = 1987	Partly safeguarded	Hot cell only	c
Testing of Nuclear Devices					
Argentina					
No testing or test preparations suspected					
Brazil					
Test preparations at Cachimbo in Amazon reported by press Brazilian press in 1986 but denied by government					a
India					
One "peaceful" nuclear explosion conducted underground near Pokharan in Thar [Rajashtan] Desert (near Pakistan border) in 1974					h
Israel					
One test suspected in 1979 in collaboration with South Africa in Indian Ocean; see South Africa, below					a

Table 11.1 (continued)

Facility Name and Location	Built or Supplied By[1]	Operating Status[2]	Safeguards Status[3]	Comments	Source[4]
North Korea					
No testing or test preparations suspected					
Pakistan					
Test preparations suspected by USA satellites in 1979 in Cholistan Desert near Baluchistan					g
South Africa					
Testing shafts dug but abandoned in 1977 after detection (and objection) by USA					c
One atmospheric test suspected in 1979 in collaboration with Israel about 1,500 miles southest of South Africa in Indian Ocean					a

1. Unless otherwise stated, countries built their own facilities or supplied their own materials.

2. Abbreviations:
NA = no information available
NC = not completed (construction suspended or abandoned)
NO = not operating and either certain or likely not to resume, with year stopped, if known
OP = operating at some capacity
P = projected startup year
S = startup year (implies still operating)
US = under construction

3. Safeguards status refers to safeguards of the Nuclear Non-Proliferation Treaty as enforced by the International Atomic Energy Agency (IAEA). Safeguarded means all operations of facility are convered by IAEA safeguards. Unsafeguarded means facility is not subject to IAEA safeguards. Partly safeguarded means facility is subject to IAEA safeguards only when handling safeguarded fuel.

4. Sources:
a = Spector 1990
b = Nuclear Engineering International Publications 1991
c = de la Court, Pick, and Nordquist 1982
d = Albright, Berkhout, and Walker 1993
e = Khasi-Jaintia Environment Protection Council, n.d.
f = Sharma 1983
g = IPPNW & IEER 1991
h = Breier 1992
i = Smith 1993
j = Coll and Taylor 1993
k = The Uranium Institute, 1993

5. Capacities and estimated reserves are given in metric tons of uranium.
6. U_3O_8 produced via heap leaching and ion exchange. All mines are open pit. Estimated reserves in 1981 were 31,000 metric tons of uranium in 31 deposits.
7. U_3O_8 produced via acid leaching and solvent extraction. Uranium reserves as estimated in 1982.
8. U_3O_8 produced via acid leaching and ion exchange.
9. Uranium is obtained as a by-product of phosphate ore mining. Total estimated uranium reserves from phosphate ore mining: 30,000–60,000 metric tons.
10. U_3O_8 produced via acid leaching and solvent extraction.
11. U_3O_8 produced via acid leaching and solvent extraction. Total estimated reserves: 317,000 metric tons; 1992 production was 1,769 metric tons (ninth largest in the world). Uranium is obtained mostly as a by-product of gold mining.
12. SWU = separative work units (a measure of enrichment capacity); HEU = highly enriched uranium; LEU = low-enriched uranium.
13. Facility built by Pakistan with extensive use of clandestinely obtained technology from Western countries.
14. HWR = heavy-water reactor; LWR = light-water reactor; HEU = highly enriched uranium; LEU = low-enriched uranium.
15. Wth = watts themal (a measure of reactor power); KWth = kilowatts thermal; MWth = megawatts thermal; MEU = medium-enriched uranium. Other abbreviations as in previous note.
16. Fuel for IRR2 supplied by serveral sources, including Argentina and possibly South Africa, Belgium, France, Niger, Gabon, and the Central African Republic.
17. MWe = megawatts electric; MWth = megawatts thermal; HWR = heavy-water reactor.
18. This facility was closed for decontamination and expansion from 1974 to 1983 or 1984 (Albright, Berkhout, and Walker 1993).
19. France withdrew assistance for this facility in 1978 but construction may have continued.
20. Built by Pakistan using French blueprints from uncompleted Chasma facility (Spector 1990).

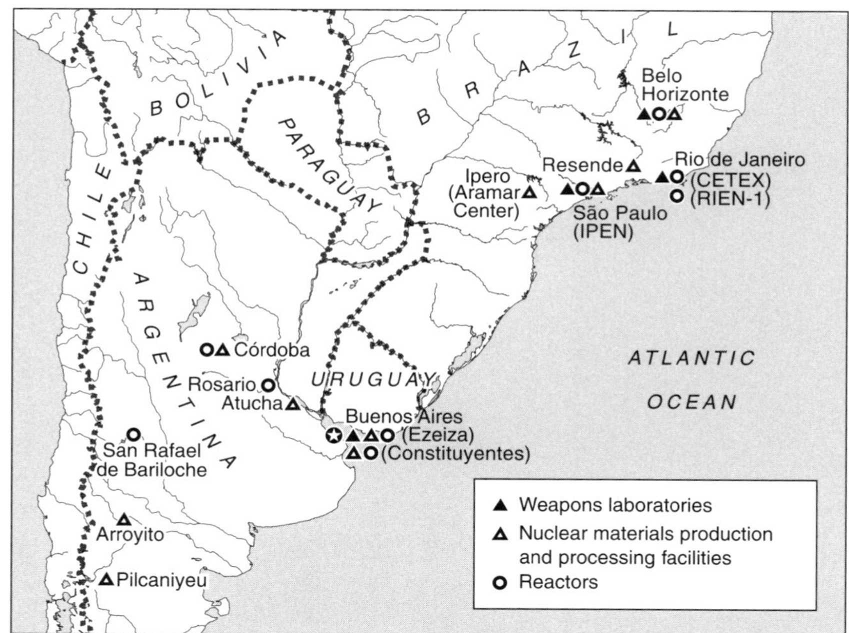

Figure 11.1 Potential nuclear weapons production sites in Argentina and Brazil.

guarded. In addition, it has facilities for heavy-water production, plutonium reprocessing, and uranium mining, milling, conversion, and enrichment, none of which are safeguarded. The navy-controlled Comisisón Nacional de Energia Atsmica (CNEA) operates the country's nuclear facilities.

Argentina has signed nuclear cooperation agreements with most South American countries and with India, Iran, Iraq, Israel, South Korea, Mexico, Nigeria, and Turkey.[51] Although Argentina has not signed the NPT, it requires that IAEA safeguards cover its wide-ranging nuclear exports as a contribution to nonproliferation efforts.[52] It has the capability of exporting all the technologies noted above, in addition to materials such as uranium and fuel cycle services.[53] Among the recipients have been Algeria, Brazil, Egypt, India, Iran, Peru, and Romania.[54]

Of greatest concern from a proliferation perspective is Argentina's facility at Pilcaniyeu for enriching uranium. Construction of this gaseous diffusion plant began in secret in 1978, and it has operated since

51. De la Court, Pick, and Nordquist 1982, p. 26; Spector 1990, p. 47.

52. Spector 1990, pp. 44–45.

53. Spector 1990, pp. 44–45; Spector 1988, pp. 243–245.

54. Spector 1990, p. 47.

1983.[55] The CNEA claims the plant only enriches uranium to 20 percent for use as fuel in research reactors, but it could be used to make highly enriched uranium for nuclear weapons.[56] No estimates of the country's actual production or inventory are available, but the Pilcaniyeu plant has a capacity of 20,000 separative work units and is theoretically capable of producing up to 500 kilograms per year of 20 percent–enriched uranium or, at greatly reduced efficiency, 10 kilograms per year of weapons-usable, 80 percent–enriched uranium.[57] Before building this facility, Argentina bought enriched uranium from China and the Soviet Union.[58]

Argentina's plutonium-reprocessing plant at the Ezeiza Research Complex in Buenos Aires was designed to separate 15 kilograms of plutonium per year from the spent fuel of CNEA power reactors for eventual reuse in either the same reactors or new breeder reactors that it may develop jointly with Brazil.[59] In 1990, Atomic Energy Commission chairman Manuel Mondino indefinitely suspended work on this project, citing Argentina's economic crisis.[60] As part of their respective decisions to renounce nuclear weapons programs, Argentina and Brazil signed an agreement in 1991 to allow the IAEA to inspect all their nuclear facilities.[61]

BRAZIL

Following the lead of Argentina, Brazil established a nuclear industry in the late 1950s, with mostly West German and some U.S. assistance.[62] The industry includes a mix of safeguarded and unsafeguarded facilities for plutonium reprocessing and uranium mining, milling, conversion, and enrichment, as well as three safeguarded light-water power reactors and four research reactors, all but one of which is safeguarded. Brazil has renounced plans to construct nuclear weapons and has signed an agreement with Argentina to verify this as noted above. As of July 1994, Brazil was not a signatory to the NPT.

Since the mid-1970s, Brazil's civilian and military nuclear programs have been developed separately.[63] The open civilian activities, based

55. Spector 1990, pp. 224, 239.

56. Spector 1990, pp. 228–229.

57. Albright, Berkhout, and Walker 1993, pp. 180–182.

58. Albright, Berkhout, and Walker 1993, p. 180.

59. Albright, Berkhout, and Walker 1993, p. 182.

60. Spector 1990, p. 232.

61. Albright, Berkhout, and Walker 1993, p. 179.

62. De la Court, Pick, and Nordquist 1982, pp. 29–30; Spector 1990, pp. 39, 260–262.

63. Spector 1990, p. 243.

mainly on technologies imported from West Germany under a 1975 development agreement, are all under IAEA safeguards.[64] A parallel program focusing on uranium enrichment, led by the Brazilian Navy, used clandestinely developed, unsafeguarded facilities.[65] Meanwhile, the army has developed a small reactor, but it may not be built. Brazil has a laboratory-scale reprocessing plant.[66] While much of the technology for the parallel program was developed indigenously, Brazil has allegedly transferred some components from the safeguarded civilian program and imported others secretly.[67]

Like Argentina, Brazil has demonstrated the capacity to produce uranium enriched up to the 20 percent level, although it needs only low-enriched uranium for its research and power reactors. It theoretically could produce 90 percent–enriched uranium for weapons from the same ultracentrifuge technology. The capacities of Brazil's enrichment plants are not known, but it would apparently take several years to make enough highly enriched uranium for a nuclear weapon, and the country's plutonium-reprocessing capacity is even smaller.[68] A constitution adopted in 1989 limits nuclear activities to peaceful purposes. In September 1990 then-President Fernando Collor de Mello cancelled the country's secret unclear weapons program, code-named "Solimes," after a river in the Amazon region, and also closed Brazil's potential nuclear weapons test site.[69]

Like Argentina, Brazil is striving to become an international supplier of nuclear reactors and the full range of fuel cycle technologies and services. It has nuclear cooperation agreements with China, Egypt, India, Iraq, Israel, Spain, and several South American countries, and has exported to Iraq, Libya, and Somalia, among others. Brazil has no announced safeguards policy with regard to these exports but has imposed some restrictions in specific bilateral agreements.[70]

INDIA

India's ambitious nuclear program began in the late 1940s for peaceful purposes, so far as the public literature allows one to determine. India's first Prime Minister, Jawaharlal Nehru, was convinced of the potential of nuclear energy. If India was "to remain abreast in

64. Spector 1990, pp. 39–40.

65. Spector 1990, p. 39.

66. Albright, Berkhout, and Walker 1993, p. 185.

67. Spector 1990, pp. 39–40.

68. Albright, Berkhout, and Walker 1993, pp. 182–185.

69. Albright, Berkhout, and Walker 1993, p. 179.

70. Spector 1990, pp. 44–45, 47.

the world as a nation which keeps ahead of things," he declared, "we must develop this atomic energy quite apart from war—indeed, I think we must develop it for the purpose of using it for peaceful purposes."[71]

India's program as it has been planned relies mainly on heavy-water reactors fueled by natural uranium and reprocessing the plutonium from these reactors for use in plutonium breeder reactors. India also plans eventually to build breeder reactors to run on uranium-233 bred from thorium-232, of which the country has very large reserves.[72] It also may buy pressurized-water reactors from France.[73]

India has among the largest nuclear programs in the Third World, with seven power reactors in place and seven under construction, plus eight research reactors, eight heavy-water production plants, two minor uranium enrichment facilities, and five plutonium-reprocessing plants (see figure 11.2).[74] Except for the two General Electric power reactors built in the late 1960s and early 1970s at Tarapur near Bombay and the two heavy-water reactors built with Canadian assistance at the Kota site in Rajasthan, none of these facilities are safeguarded. India likewise has no formal safeguards policy for its nuclear exports. It has the capability to provide the full range of nuclear fuel cycle services as well as research and production reactors. The government has nuclear cooperation agreements with Afghanistan, Algeria, Argentina, Brazil, Egypt, Iraq, Libya, Syria, and the former Yugoslavia.[75]

India has obtained most of the reactor and reprocessing technology it needed to produce nuclear weapons from its own nuclear power and research programs. This has resulted from substantial investments in research, which escalated rapidly in the 1950s.[76] The decision to develop nuclear explosives was made in 1964, shortly after China's first atomic test and also after Prime Minster Nehru's death. The new Prime Minister Lal Bahadur Shastri stated in November 1964 that he still opposed nuclear weapons but supported the development of nuclear explosives for such peaceful purposes as digging tunnels.[77] His successor, Indira Gandhi, reasserted this view in 1969 after refusing to sign the NPT, which many in India believe is discriminatory.[78] Even after India's first and only nuclear test in 1974, Prime Minister Indira

71. As quoted in Parthasarathi and Singh 1992.

72. De la Court, Pick, and Nordquist 1982, p. 107.

73. *Economic and Political Weekly* 1992.

74. Spector 1990, pp. 83–84; Albright, Berkhout, and Walker 1993, p. 157.

75. Spector 1990, p. 43.

76. Parthasarathi and Singh 1992.

77. Spector 1990, p. 64.

78. Spector 1990, p. 64; Perkovich 1992, p. 26.

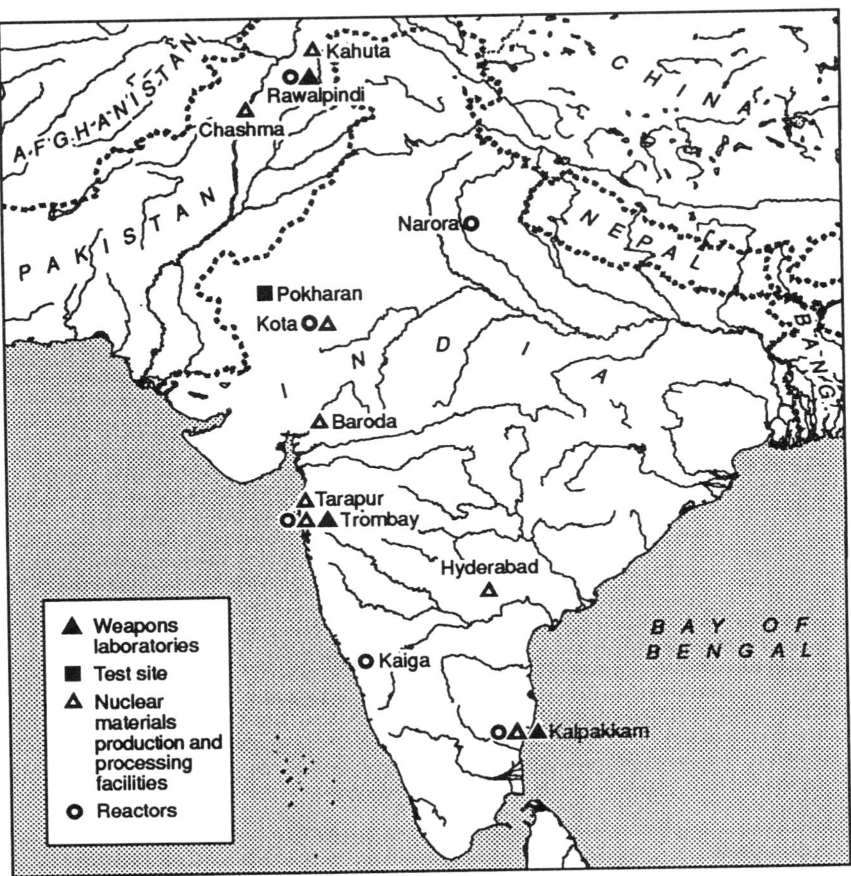

Figure 11.2 Actual and potential nuclear weapons production and testing sites in Pakistan and India.

Gandhi insisted that the country was interested in nuclear technology exclusively for peaceful purposes.[79]

The U.S. "tilt" toward Pakistan during the 1971 war between India, Pakistan, and Bangladesh also played a role in India's development of nuclear weapons technology. During that war, the United States sent a nuclear-capable aircraft carrier to the Bay of Bengal. Many observers felt that this event contributed to India's determination to acquire nuclear weapons capability.[80] In this case, the U.S. nuclear arsenal had been implicitly extended to cover Pakistan's interests during a war.

79. Findlay 1990, pp. 210–211. The controversy over peaceful nuclear explosions is a long one. The Soviet Union had an extensive program; the U.S. program was called the "Plowshare" program. There are two kinds of references to a paradigm of swords-into-plowshares in the Bible. Isaiah chapter 2 is unambiguous about the conversion of swords into plowshares. In the book of Joel, plowshares can also be turned back into swords; so, too, can "peaceful" nuclear explosives be used as weapons.

80. Findlay 1990, p. 210.

India claims that the nuclear device it tested in 1974 was peaceful, but this is disingenuous. The underground test was conducted at Pokharan near the border with Pakistan, in the Thar Desert in Rajasthan. The circumstances of India's development of a capability to detonate a nuclear explosive clearly point to military motives; its strategic significance has been widely discussed in India and elsewhere. It alarmed Pakistani officials, who accelerated their nascent nuclear weapons program in response.

That India did not go on to develop a large nuclear arsenal or a PNE program suggests that the 1974 test was intended to serve mostly as a symbolic demonstration of "great power" status, in part to counter China's growing arsenal while enhancing India's prestige as a regional superpower vis-à-vis Pakistan.[81] Pakistan has also developed the capability to make nuclear weapons without building a large arsenal. Indeed it appears that neither country has actually assembled any nuclear weapons at all, while both seem to maintain an unknown degree of readiness to do so. George Perkovich of the W. Alton Jones Foundation has called the confrontation between India and Pakistan "non-weaponized deterrence."[82] In a similar fashion, Jasjit Singh, an Indian analyst, has dubbed it "recessed deterrence," a lower-risk form of nuclear deterrence.[83] Besides this implicit and informal balancing act, India and Pakistan also have a formal agreement not to attack each other's nuclear facilities.

India's code words for its 1974 nuclear test were "the Buddha is smiling." The country's nuclear suppliers, mainly the United States, Great Britain, and Canada, were not pleased.[84] Canada immediately cut off all nuclear exports and assistance including heavy water for the Rajasthan reactors. Because India had no indigenous enriched uranium, the United States agreed to continue to supply some enriched uranium for the Tarapur reactor after 1974; this source of supply ended in 1982. To reconcile the need to fuel these reactors with a 1978 U.S. nonproliferation law, India and the United States arrived at an agreement in 1982 that France (not then an NPT signatory) would supply the needed low-enriched uranium.[85] India manufactures its own heavy water. It has also obtained some from the Soviet Union.[86]

The Department of Atomic Energy controls India's nuclear facilities. The department makes no formal distinction between military and

81. Spector 1990, pp. 64–65.

82. Subramanyam 1993, p. 39.

83. Subramanyam 1993, p. 39.

84. De la Court, Pick, and Nordquist 1982, p. 109.

85. Manchanda 1992.

86. Spector 1990, pp. 36–38.

Figure 11.3 Bhabha Atomic Research Centre, Trombay, India.

civilian activities but has made some in practice.[87] Although India has the capability to obtain plutonium from the spent fuel of all its reactors, it appears to use only two research reactors for this purpose. These are the Cirus and Dhruva reactors at the Bhabha Atomic Research Center (BARC) in Trombay, which can produce up to 10 kilograms and 30 kilograms of plutonium per year, respectively.[88] Under normal operation, India's CANDU-type power reactors produce non-weapons-grade plutonium. Only about 5 kilograms of weapons-grade plutonium is recoverable from the initial fuel loading of these reactors. If dedicated to producing weapons-grade plutonium, however, each of India's 235-megawatt electric power reactors could produce about 150 kilograms of plutonium per year.[89]

India's first PUREX-type plutonium-reprocessing plant, also located at BARC, began operating in 1966. After India's nuclear test in 1974, the plant was shut down for nine years for decontamination and expansion. Its capacity was estimated at 30 metric tons of spent fuel per year, enough to handle the output of the Cirus and Dhruva reactors. India's total inventory of weapons-grade plutonium was an estimated

87. Perkovich 1992, p. 28; Albright, Berkhout, and Walker 1993, p. 158.

88. Albright, Berkhout, and Walker 1993, p. 158.

89. Albright, Berkhout, and Walker 1993, p. 159. Production estimates assume operation at 60 percent of capacity. The net capacity of the plants is 220 megawatts electrical.

290 kilograms in 1991, or enough for almost 60 nuclear weapons using 5 kilograms each. This supply is projected to grow to 425 kilograms by the end of 1995.[90]

A second PUREX facility at Tarapur was started up in 1979 to reprocess the lower-grade plutonium created in India's power reactors, although it also reprocessed some weapons-grade plutonium from the Cirus reactor. This plant can handle 100 metric tons of spent fuel per year, with an estimated separation capacity of 100 to 150 kilograms per year of reactor-grade plutonium.[91] India also has a reprocessing plant at Kalpakkam near Madras. India claims that this (unsafeguarded) stockpile of plutonium is needed for its planned network of breeder reactors. India's breeder program has suffered numerous delays, like that of every other country attempting to develop this technology.

India wants to use the Tarapur reprocessing plant to recover plutonium from the spent fuel that has accumulated from reactors at the site. It may fuel these reactors with a mixture of natural uranium oxide and plutonium oxide because France has informed India that it will discontinue supplying enriched uranium unless the latter opens all nuclear facilities to international inspections. India refuses to do that, fueling a controversy between India and the United States over whether India has the right to reprocess the Tarapur fuel, which was originally provided by the United States, without specific and prior U.S. consent.[92]

India has established a nuclear waste treatment plant called the Solid Storage Surveillance Facility near the Tarapur reactors.[93] Liquid wastes are solidified in glass and stored in steel canisters pending development of an underground repository. However, no repository site has been selected. India is conducting waste disposal tests in a closed gold mine in the Kolar gold fields in southern India.[94]

Poor efficiency and numerous shutdowns have plagued India's existing reactors and heavy-water production plants. The Baroda heavy-water plant has suffered two explosions, and the Nuclear Fuel Complex in Hyderabad had a major fire.[95] India's test reactor for plutonium breeding has allegedly suffered serious problems and, according to critics, "has never really been functional for more than a few minutes at a time."[96]

90. Albright, Berkhout, and Walker 1993, p. 161.

91. Albright, Berkhout, and Walker 1993, p. 160.

92. Jayaraman 1993.

93. Kumar 1993.

94. Sarkar 1990, p. 84.

95. Sharma 1983, p. 120.

96. *Economic and Political Weekly* 1990.

Serious environmental contamination and radiation-related public health problems are alleged at many facilities. Independent investigators have had difficulty obtaining data since India's Atomic Energy Act prohibits the release of information related to India's nuclear facilities.[97] A 1978 investigation by *Business India* magazine alleged that at Tarapur, the atomic plants in Rajasthan, and even at the Bhabha Atomic Research Center in Bombay, "those responsible for personnel safety show little regard for those very safety standards that they have set themselves." According to this report, at the Tarapur and Rajasthan power plants, "the 5 rem limit for radiation workers' exposure means very little in practice: it has been breached so frequently as to make one wonder why it exists at all."[98]

The Department of Atomic Energy, which has sole authority over nuclear safety and environmental protection, releases little data on accidents or environmental contamination.[99] Dr. Raja Ramanna, one of the principal scientific leaders of India's nuclear establishment, questioned whether India should "follow international standards blindly" and advised that India "should have the courage to look at these standards especially where they are leading to runaway costs."[100] According to an editorial in the Bombay-based *Economic and Political Weekly*, the Indian nuclear establishment has not investigated cancer mortality and morbidity near nuclear installations.[101] However, Sanghimitra Gadekar and other observers report evidence that people living near India's nuclear facilities receive exposures far in excess of allowable limits, leading to abnormally high rates of cancer and birth defects.[102] According to Mynore Singh, a former employee of a nuclear power plant, "The people who work [at the plant] are poor ... [so they] ... go into the reactor without their dosimeter.... The longer they work, the more they get paid.... The management don't care about it." Similar unconcern has been reported at other sites.[103]

At Tarapur, workers' unions have taken up the problem of contract workers being exposed to excessive doses of radiation; they have also

97. *Economic and Political Weekly* 1992a; Sharma 1983, pp. 58–59, describes working conditions at the Tarapur Atomic Power Station, as does *Nuclear India*, a 50-minute documentary produced in 1992. by Chameleon Television for Britain's Channel Four, which provides a critical view of health and environmental problems throughout India's nuclear complex.

98. *Business India* 1978.

99. Sharma 1983, pp. 121–122.

100. As quoted in Sharma 1983, p. 110.

101. *Economic and Political Weekly* 1992a.

102. *Nuclear India* 1992.

103. *Nuclear India* 1992.

alleged that health records are not well maintained. Doctors near the test site at Pokharan have begun reporting cases of radiogenic cancers, such as lymphomas and leukemias.[104] Such claims cannot be scientifically assessed until the Department of Atomic Energy agrees to release radiation dosimetry and other relevant records (with appropriate protection for workers' privacy). But the main thrust of the official response to these complaints continues to be one of secrecy in regard to the records and a denial that serious problems exist. However, according to an editorial in the *Economic and Political Weekly*, there are some signs that a more responsive attitude to public complaints and allegations is beginning to emerge.[105]

ISRAEL

Israel is believed to have a stockpile of 60 to 100 nuclear weapons, by far the largest arsenal of the undeclared nuclear powers.[106] Even after Mordechai Vanunu's detailed revelations about Israeli nuclear weapons production,[107] Israel continues in effect to deny that it possesses nuclear weapons by stating that it will not be "the first country to introduce nuclear weapons into the Middle East."[108]

Israel launched its nuclear weapons program after the 1956 Suez crisis left the country feeling vulnerable to growing Arab nationalism.[109] The program was made possible only by the support of France, which provided an unsafeguarded gas-cooled, heavy-water-moderated reactor for producing plutonium and an underground reprocessing plant for plutonium separation for Israel's Nuclear Research Center at Dimona.[110] France may even have given Israel information on the design, manufacture, and testing of nuclear weapons until it was cut off in 1960.[111]

104. *Economic and Political Weekly* 1994.

105. *Economic and Political Weekly* 1994.

106. Spector 1990, p. 162.

107. Spector 1990, pp. 150, 158. Vanunu worked as a technician in Israel's nuclear weapons complex from 1977. to 1985. After the London *Sunday Times* published his story and photographs on 5. October 1986, Israeli agents abducted Vanunu. A secret tribunal tried and convicted him of espionage and treason and sentenced him to 18. years in prison.

108. As quoted in Cohen 1993, p. 40.

109. Spector 1990, p. 151; see also Cohen 1993. for more on Israel's decision to acquire nuclear weapons.

110. Spector 1990, p. 151.

111. Spector 1990, p. 152.

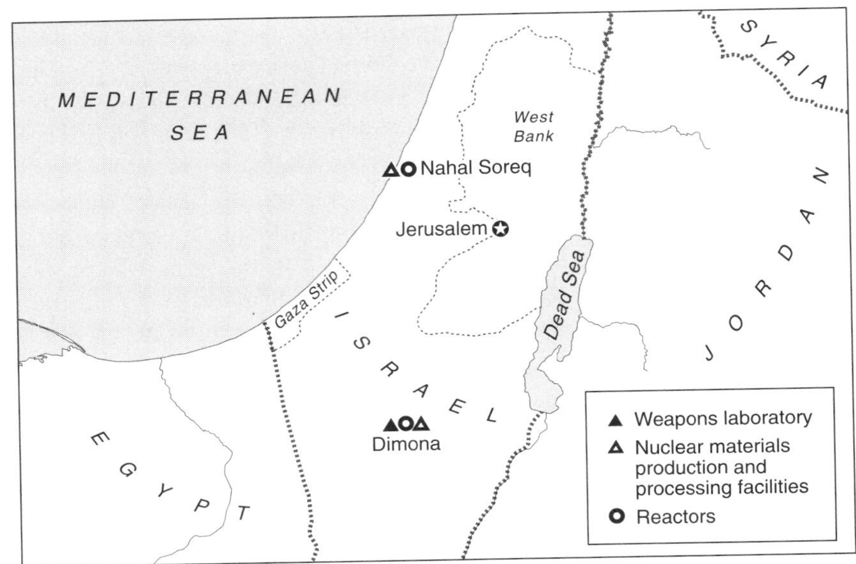

Figure 11.4 Nuclear weapons production sites in Israel.

Figure 11.5 Inside the Dimona Reactor, Israel. London Sunday Times photo.

This top-secret Dimona facility is located in the Negev Desert, about 65 kilometers from Beersheba.[112] The reactor is housed in NRC Institute No. 1 and the reprocessing plant in Institute No. 2.[113] The unsafeguarded Dimona site also contains two research-scale uranium enrichment plants (a centrifuge facility built in 1979–1980 and a laser built in 1981) and facilities for producing lithium-6, processing tritium, and treating and storing nuclear waste.[114] These NRC facilities can produce 40 kilograms of plutonium per year, and Israel's stockpile as of 1991 was an estimated 240 to 415 kilograms.[115] Based on Vanunu's revelations, Albright, Berkhout, and Walker estimate that Israel's cumulative plutonium production through the end of 1991 could be fashioned into 48 to 83 warheads, based on an assumption that about 5 kilograms of plutonium are used in each Israeli nuclear warhead.[116]

The waste treatment facility, known as Institute No. 4, handles all the NRC's high-level and low-level radioactive waste, as well as the mostly low-level waste generated by about 300 medical and industrial users of nuclear materials in Israel.[117] The high-level liquid waste generated in reprocessing is stored in underground vats, while gases are routinely vented and carried by prevailing winds toward Jordan, 40 kilometers east of Dimona.[118] Low-level waste is packed in barrels, sealed with tar, and buried at a site about 1 kilometer from the reactor complex.[119]

Uranium for Dimona has come from several foreign sources, first from French-controlled mines in Africa and then from Argentina and South Africa. According to press reports, South Africa provided Israel with 50 metric tons of uranium yellowcake in the late 1970s in exchange for a little less than 30 grams of tritium.[120] Israel also developed a domestic capability to extract uranium as a by-product of phosphate mining and processing. Heavy water for Dimona was first imported from Norway, which required it be used only for peaceful purposes, and then produced domestically at an unsafeguarded plant in Rehovot.[121]

112. Spector 1990, p. 151.

113. Rabinowitz 1993.

114. Spector 1990, pp. 172–173; Rabinowitz 1993.

115. Albright, Berkhout, and Walker 1993, pp. 155–157.

116. Albright, Berkhout, and Walker 1993, table 9.2, p. 157.

117. Rabinowitz 1993.

118. As cited in Rabinowitz 1993.

119. Rabinowitz 1993.

120. Albright 1993a.

121. Spector 1990, pp. 153, 172.

Israel operates a smaller and safeguarded research reactor at Nahal Soreq, which the United States provided in 1955. The United States supplied the highly enriched uranium fuel for this reactor until 1977, when a bilateral agreement with Israel expired.[122]

Although Israel has the capability to export a wide range of nuclear fuel cycle technologies, it has had limited nuclear cooperation with other countries, mainly with South Africa and Brazil.[123] The former has played a very important role in Israel's development of nuclear weapons, and vice versa. In addition to supplying Israel with uranium, South Africa may have collaborated with Israel in the alleged 1979 test of a nuclear device. There is considerable controversy as to whether this event, which U.S. satellites detected in the Indian Ocean 2,400 kilometers southeast of South Africa, was actually a nuclear test. Various U.S. evaluations have not come to an agreed, definitive conclusion.[124]

Very little information is available about the health and environmental effects of Israel's nuclear weapons program. Unconfirmed reports suggest that several scientists have died in plutonium-related accidents at the Dimona reprocessing plant.[125] Yosi Sarid, Israel's Minister of Environmental Protection, denied news reports in 1993 that radioactive waste from Dimona was seeping into a tourist site in the Negev Desert near the Egyptian border, although he confirmed that a leak had occurred on 2 August 1992.[126] The Israeli Defense Forces ordered the area around the leak closed, but they have banned independent examinations of the site.[127] After this incident was reported, Prime Minister Yitzak Rabin ordered the Nuclear Research Center to allow the Environmental Protection Ministry to review its confidential data on soil, water, and air monitoring. However, this civilian oversight is limited.[128] In the Negev Desert south of Dimona, reports of higher than normal radiation have led to an agreement between Israel and Egypt to study radioactivity levels along their border.[129] It is not known when and if data from this study will be made public.

122. Spector 1990, p. 47.

123. Spector 1990, pp. 44–45, 164–165.

124. Spector 1990, p. 157.

125. De la Court, Pick, and Nordquist 1982, p. 94.

126. *Qol Yisra'el* 1993; WISE 1993a.

127. WISE 1993.

128. Rabinowitz 1993.

129. WISE 1993a.

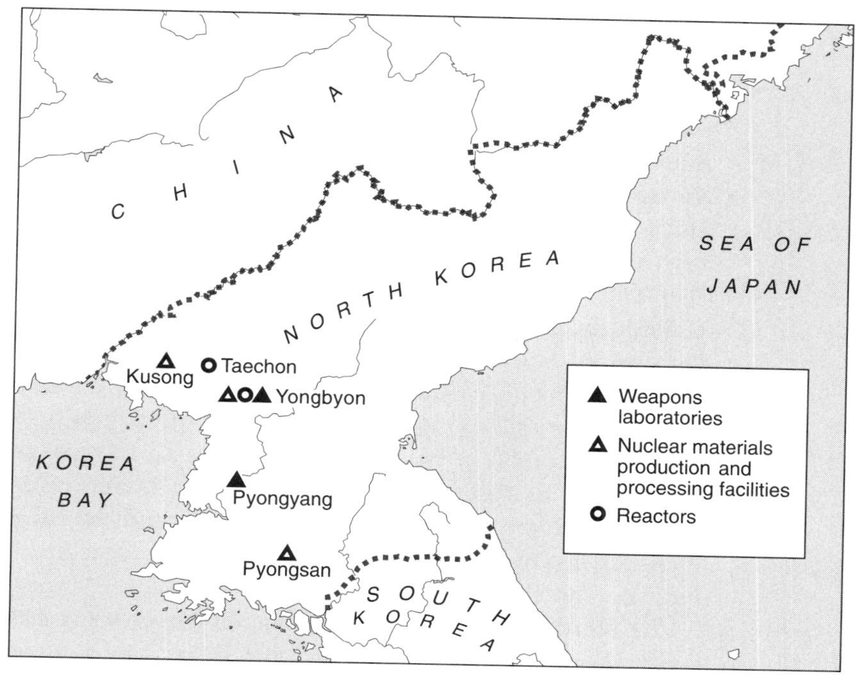

Figure 11.6 Nuclear weapons production sites in North Korea.

NORTH KOREA (DEMOCRATIC PEOPLE'S REPUBLIC OF KOREA)

The Soviet Union began exploring uranium in North Korea in the late 1940s.[130] The country established a nuclear research center at Yongbyon, about 100 kilometers north of Pyongyang, in 1964, and in 1965 a small Soviet safeguarded reactor was installed there.[131] In the 1980s, North Korea began building unsafeguarded nuclear weapons–related facilities. These include a 30-megawatt thermal gas-graphite reactor, well suited to producing plutonium without heavy water or enriched uranium, and a reprocessing plant for separating plutonium.[132] Both are located at North Korea's main nuclear research center in Yongbyon. The reactor's plutonium production capacity is estimated at 4 to 6 kilograms per year, with an estimated 5 to 10 kilograms produced through the end of 1991.[133] North Korea is believed to have built the currently operating Yongbyon production reactor itself, while China may have assisted with the plutonium-reprocessing plant.[134] North

130. Spector 1990, p. 121.

131. Spector 1990, p. 121.

132. Spector 1990, pp. 123–124, 139.

133. Albright, Berkhout, and Walker 1993, pp. 173–175.

134. Spector 1990, p. 139.

Korea also is assumed to have facilities for uranium purification and fuel fabrication at unknown locations.[135]

The Soviet Union signed an agreement with North Korea in 1985 to provide four power reactors. The two countries also strengthened military ties in the late 1980s.[136] Two gas-graphite reactors were under construction in 1993: a 200- to 300-megawatt thermal unit in Yongbyon and a 1,000-megawatt thermal unit (estimated size) in Taechon. These reactors could produce an estimated 250 kilograms of plutonium per year.[137]

Aside from China and the Soviet Union, Iran is the only country with which North Korea may have signed a bilateral nuclear cooperation agreement.[138] Iran is reportedly helping fund North Korea's missile program, which could improve North Korea's ability to launch nuclear weapons, in exchange for manufactured missiles and help in building its own missile system. This collaboration might eventually include nuclear weapons technologies, although North Korea's export capabilities are currently limited to research reactors, uranium conversion, and fuel fabrication.[139]

The North Korean regime is believed to want nuclear weapons for prestige, international legitimacy, and as an economic bargaining chip —these weapons would provide security by both threatening and deterring the South Korean and U.S. forces deployed against it. Until recently, these included U.S. tactical nuclear weapons. It is possible that North Korea does not want to rely on Russia or China to support its nuclear programs as in the past.[140]

North Korea signed the NPT in 1985. According to Spector, this was apparently to avoid international pressure and not with any intent to abide by its terms.[141] The government then put off signing an inspection agreement with the IAEA until 1992, after completing its production reactor and starting its reprocessing facility.[142] IAEA inspections of North Korean facilities in 1992 revealed outmoded equipment and technical difficulties with the safe storage of spent fuel.[143]

135. Spector 1990, p. 139.

136. Spector 1990, p. 126.

137. Albright, Berkhout, and Walker 1993, pp. 174–175.

138. Spector 1990, p. 47.

139. Spector 1990, pp. 44–45; see also sources quoted in Proliferation Watch 1993a, p. 5.

140. Spector 1990, pp. 124–125.

141. Spector 1990, p. 127.

142. Proliferation Watch 1993, p. 6.

143. Imai 1993.

IAEA officials described North Korea's radiation shielding systems as "inferior to those of the advanced countries" and, more generally, declared the design of North Korean nuclear facilities to be "clearly substandard."[144] They reported the absence of protective devices on large cranes moving nuclear waste and the lack of any independent supervisory organizations or control committees to monitor the safety of nuclear facilities.[145] The IAEA also found discrepancies in North Korea's plutonium accounting and tried to gain access to two suspected nuclear waste sites to verify the government's claim that only one batch of plutonium had been reprocessed.[146] North Korea denied that either site contained nuclear waste and refused to allow further inspections.

Faced with continuing IAEA pressure, North Korea announced its intention to withdraw from the NPT in March 1993.[147] It suspended this withdrawal, however, pending intense negotiations with the United States. A partial agreement, announced in July 1993, was based on U.S. assurances that it would not use nuclear weapons or force of any kind against North Korea. Even though U.S. President Bill Clinton threatened that same month to retaliate if North Korea developed nuclear weapons,[148] North Korea said it would allow IAEA inspections as long as they were conducted impartially.

In its negotiations with North Korea, the United States also offered to replace the country's dangerously outmoded graphite reactors with light-water reactors less suited to producing weapons-grade plutonium. North Korea agreed that this would be desirable, even though the slightly enriched uranium fuel these reactors require would make the country dependent on foreign supplies. Still, the United States insists that no assistance will be forthcoming until North Korea unambiguously complies with its obligations under the NPT.[149] A partial IAEA inspection was held in March 1994, but this was not sufficient to determine the extent of North Korea's plutonium production.

Rising military tensions were defused when former U.S. President Carter visited North Korea in June 1994. This paved the way for a resumption of negotiations; a summit meeting between the heads of state of North and South Korea was also planned as a result of the Carter

144. Maeng-ho 1992.

145. Anonymous IAEA officials as quoted in Maeng-ho 1992.

146. Albright 1993; Proliferation Watch 1993, p. 6.

147. Albright 1993; Proliferation Watch 1993, p. 6.

148. Associated Press 1993.

149. Stevenson 1993.

initiative. The death of North Korea's president, Kim Il Sung, in early July 1994 led to a postponement of the meeting and to renewed uncertainty on the issue of IAEA verification of North Korean plutonium production. Shortly thereafter, North Korea agreed to shelve its reprocessing program and shut existing reactors in exchange for $4 billion in foreign assistance to provide light-water nuclear power reactors.

PAKISTAN

Prime Minister Ali Bhutto formally launched Pakistan's program to acquire nuclear weapons in 1972, after the 1971 war between India, Pakistan, and Bangladesh.[150] Driving the program was a desire to counter India's military superiority, reinforced by India's 1974 nuclear test, as well as a desire to boost Pakistan's prestige and influence in the Islamic world.[151] As early as 1965, then-Foreign Minister Bhutto had said that "if India builds the bomb, we will eat grass or leaves, even go hungry. But we will get one of our own. We have no alternative."[152]

Since Pakistan was indeed too poor to build nuclear weapons on its own, Bhutto framed Pakistan's nuclear ambitions in religious tones. The project was expected to have backing in the Muslim world.[153] "There's a Hindu bomb, a Jewish bomb, and a Christian bomb," he declared. "There must be an Islamic bomb."[154] Libya apparently was the first of many oil-rich Arab countries to sign on, agreeing in 1974 at the first Conference of Islamic Nations to help finance the Pakistani program in exchange for nuclear material and information.[155]

Pakistan received the greatest assistance from China, which had its own interest in countering India's nuclear capability. China allegedly provided Pakistan with a working design for nuclear weapons (alleviating the need for nuclear testing) and there are unconfirmed reports that it also provided it with enough highly enriched uranium for two warheads.[156] Pakistan, in exchange, allegedly supplied China with more modern enrichment technology.[157]

150. Spector 1990, p. 90.

151. Spector 1990, pp. 89–90, 95.

152. As quoted in de la Court, Pick, and Nordquist 1982, p. 117.

153. Hoodbhoy 1993.

154. "CBS Evening News" broadcast on 11. June 1979, as quoted in de la Court, Pick, and Nordquist 1982, p. 117.

155. De la Court, Pick, and Nordquist 1982, pp. 117–118.

156. Albright, Berkhout, and Walker 1993, p. 166.

157. Gillespie 1992, p. 141.

Pakistan relied heavily on foreign sources for most of its nuclear technology. It bought some facilities openly, including the country's first research reactor, supplied by the United States in 1965.[158] Pakistan also acquired a CANDU power reactor (the KANUPP reactor near Karachi) from Canada in 1972, which suspended the supply of uranium and spare parts after Pakistan refused to accept upgraded safeguards.[159] In addition, Pakistan has a research reactor built with Chinese assistance (started up in 1989), a uranium conversion facility (obtained clandestinely from West Germany),[160] two heavy-water production plants (built with Canadian and Belgian assistance), two indigenously developed uranium enrichment plants (a third is under construction), and two small plutonium-reprocessing plants (built with British, French, and Belgian assistance; the first enrichment plant probably no longer operates, and a third is under construction).[161] Pakistan obtained much of the equipment for these facilities—all unsafeguarded—by clandestine means from Western suppliers, not all of whom were unwittingly involved.[162]

Pakistan has the capability to export practically all these technologies.[163] In addition to its links with China, Pakistan has nuclear cooperation agreements with France, Malaysia, Niger, Romania, and Turkey.[164] It has reportedly trained Iranian nuclear specialists and may have positioned itself to share nuclear know-how with Iraq before the 1991 Persian Guff war.[165]

Pakistan's main enrichment plant is a secret underground facility at Kahuta near Islamabad. Its capacity is estimated at 45 to 75 kilograms of highly enriched uranium per year, although since coming on line in 1984 it has rarely produced even half this amount. Pakistan's total inventory of weapons-grade uranium through 1991 is estimated to be roughly 130 to 220 kilograms, enough for about 6 to 10 nuclear warheads based on the solid-core design provided by China.[166] The capacity of its only operable reprocessing plant is 10 to 20 kilograms of

158. Spector 1990, p. 115.

159. De la Court, Pick, and Nordquist 1982, pp. 117–118.

160. Gillespie 1992, p. 141; Spector 1990, p. 116.

161. Spector 1990, pp. 114–116; Albright, Berkhout, and Walker 1993, pp. 163–165.

162. De la Court, Pick, and Nordquist 1982, p. 119; Spector 1990, p. 91.

163. Spector 1990, pp. 44–45.

164. Spector 1990, p. 47; Proliferation Watch 1993, p. 15.

165. Spector 1990, pp. 110–111.

166. Albright, Berkhout, and Walker 1993, pp. 166–167.

plutonium per year, but it is not known if this plant ever actually operated beyond cold testing.[167] Figure 11.2 shows the location of Pakistan's nuclear weapons facilities.

Little is known about the health and environmental conditions at these facilities. When General Hamid Gul, former director-general of Pakistan's Inter-Services Intelligence, was asked about this in 1992, he said, "We are not that educated or sophisticated about any long-term effects of the nuclear complex."[168]

Pakistan refuses to acknowledge that it has developed the capability to produce nuclear weapons and claims its nuclear program is peaceful.[169] As recently as 1993, then Prime Minister Sharif allegedly stated, "Pakistan does not have the bomb, and we will not build it."[170] But Pakistan's nuclear capability is clearly a source of great national pride. General Gul described it in 1992 as a "symbol of Pakistan's sovereignty. It is the one issue of the nation's honor."[171] Like India, it appears that Pakistan has stopped just short of assembling or at least deploying nuclear weapons.[172] How long this will last is unclear, especially given the two sides' continuing armed standoff over the fate of Kashmir.[173] Both countries did agree in 1988 not to attack each other's nuclear facilities.[174]

SOUTH AFRICA

South African Foreign Minister Pik Botha admitted in 1988 that his country had the capability to build a nuclear bomb, "should we want to."[175] Given the sanctions arrayed against South Africa over apartheid, he apparently felt the country had little to lose by this declaration. Just one year later, his government claims, it secretly began to dismantle all South African nuclear weapons.[176] In 1991, South Africa signed the NPT and opened its facilities to IAEA inspectors. Reportedly, it has begun negotiating with the United States to trade

167. Spector 1990, p. 115.

168. As quoted by Perkovich 1992, p. 23.

169. Spector 1990, pp. 89, 97.

170. As quoted in Proliferation Watch 1993a, p. 7.

171. As quoted by Perkovich 1992, p. 20.

172. Spector 1990, pp. 105, 111.

173. Hersh 1993; Spector 1990, p. 111.

174. Gillespie 1992, p. 142.

175. Pik Botha, as quoted in Gillespie 1992, p. 136.

176. De Klerk 1993.

its stockpile of weapons-grade highly enriched uranium for reactor-grade low-enriched uranium.[177]

As noted above, President de Klerk appears to have taken these steps to avoid any chance that nuclear weapons or materials might fall into the hands of the black majority.[178] However, given the extensive secret security and intelligence apparatus of the apartheid regime, there have been suspicions that pro-apartheid forces may have retained some nuclear weapons. An IAEA inspection in September 1993 indicates that South Africa may have made a complete declaration of its highly enriched uranium inventory, though some uncertainty remains. However, South Africa has so far refused to reveal the names of its foreign nuclear suppliers and collaborators.[179]

Whatever the reason, this declared disarmament represents an extraordinary reversal for a country whose nuclear activities began in the mid-1950s. South Africa is one of the world's largest exporters of uranium, most of which it processes as a by-product of its extensive gold-mining operations. It also had uranium-mining operations in Namibia while it still occupied that country in violation of UN Security Council resolutions. South Africa's first uranium exports were to the United States and the United Kingdom for their weapons programs.

South Africa first acquired extensive expertise in nuclear technology via a program for civilian nuclear power. This program received considerable support from Europe and the United States, notably in the form of training and assistance.[180] In the early 1960s, South Africa established the National Nuclear Research Center at Pelindaba, a Zulu word for "we don't talk about this anymore."[181] South Africa operated a safeguarded light-water research reactor there; it opened in 1965.[182] The United States provided it under its Plowshare program.

A second, indigenously designed heavy-water-moderated, sodium-cooled reactor, known as the "Pelinduna critical facility," was also built at the Pelindaba site; it was started up in 1967. It was safeguarded and was fueled with 2 percent–enriched uranium from the United States; it was shut in 1969. Before 1971, the slightly irradiated fuel from this reactor was reprocessed in Britain.[183]

177. Albright 1993; Proliferation Watch 1993, p. 17.

178. Albright and Hibbs 1993.

179. Albright 1994, box on p. 3.

180. Albright 1994, p. 3.

181. De la Court, Pick, and Nordquist 1982, p. 78.

182. Spector 1990, p. 288.

183. Albright 1994, p. 4.

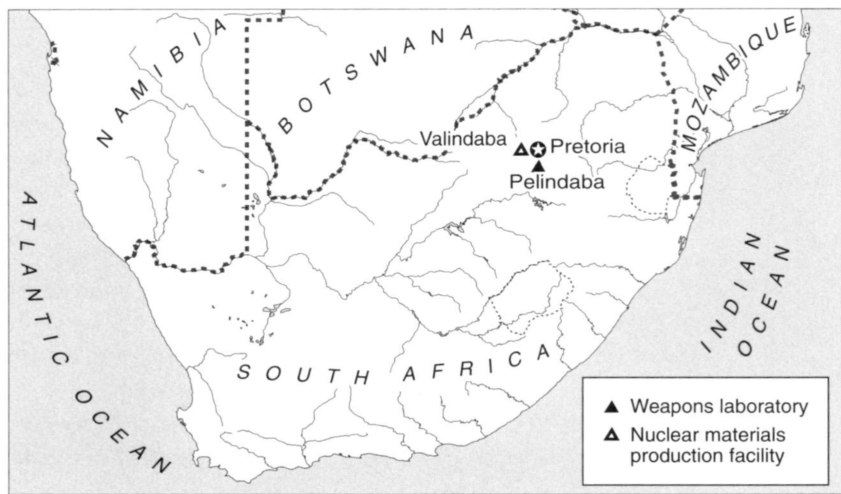

Figure 11.7 Nuclear weapons production sites in South Africa.

Also in the early 1960s, South Africa began secret work on uranium enrichment. In 1970, it started work on a facility next to the National Nuclear Research Center, to house an unsafeguarded uranium enrichment complex, the centerpiece of its nuclear weapons program.[184] It is known as Valindaba, which is Zulu for "we don't talk about this at all."[185] This facility includes two plants for converting uranium yellowcake to uranium hexafluoride (the first built by Great Britain and the second by South Africa). It also has built two plants for uranium enrichment: a pilot plant which commenced producing highly enriched uranium in 1978 (for which some equipment may have been provided by U.S., West German, French, and Swiss companies) and a plant for low-enriched uranium built to provide fuel for the country's power and research reactors.[186] The pilot plant that produced the highly enriched uranium was called the Y-plant.[187]

Both enrichment plants use a centrifuge/jet-nozzle technology developed in West Germany, although the low-enriched uranium plant incorporates a more advanced aerodynamic "helicon" cascade.[188] According to research by David Albright, there are significant differences between the German "Becker process" and the South African enrich-

184. Albright, Berkhout, and Walker 1993, p. 186.

185. De la Court, Pick, and Nordquist 1982, p. 78.

186. Spector 1990, pp. 187, 288; Albright, Berkhout, and Walker 1993, pp. 186–187.

187. Albright 1994, p. 6.

188. Albright, Berkhout, and Walker 1993, p. 187; de la Court, Pick, and Nordquist 1982, p. 78.

ment plant.[189] The HEU plant had an initial annual capacity of 10,000 to 15,000 separative work units, which was increased to 20,000 by design changes. If it averaged 10,000 separative work units of actual annual output, the production of weapons-grade HEU would amount to about 60 kilograms per year.[190] When South Africa's weapons were dismantled nearly 400 kilograms of HEU were recovered from them. South Africa has also accumulated a similar quantity of uranium enriched to between 20 percent and 80 percent uranium-235 content.[191]

South Africa also has researched gas-centrifuge and molecular laser isotope separation enrichment technologies. The former was abandoned in 1991, while the laser work continues, with a pilot plant planned for the late 1990s.[192] It also built a partially safeguarded "hot cell" for handling spent fuel at Pelindaba in 1987, although, as discussed by Albright, Berkhout, and Walker, it has not been determined if this equipment can be used for small-scale plutonium separation.[193]

South Africa has the capability to export all these technologies, and is believed to have cooperated with Israel, Brazil, and Taiwan.[194] Since 1987, it has agreed to comply with the export guidelines of the NPT and the Nuclear Suppliers Group.[195]

South Africa's nuclear program has been controlled by the Atomic Energy Corporation (formerly Atomic Energy Board), which began work on atomic energy research and on materials production in the 1960s. The Atomic Energy Board selected a test site in the Kalahari desert and a test with a dummy core (without the nuclear explosive) was planned for 1977, but later cancelled. Sometime between 1974 and 1978 the nuclear explosives program acquired a specific military mission of threatening to use nuclear weapons—and thereby obtaining outside, Western assistance—"in the case of an overwhelming military threat."[196] Armscor, the government's arms conglomerate, took over the program and assembled South Africa's first nuclear weapon in 1979 at Pelindaba. It then moved its operations to Advena, a secret facility 15 kilometers east of Pelindaba, where it built more weapons.[197]

189. Albright 1994, p. 4.

190. Albright 1994, p. 6.

191. Albright 1994, p. 18.

192. Albright, Berkhout, and Walker 1993, p. 188.

193. Spector 1990, p. 287; Albright, Berkhout, and Walker 1993, p. 189.

194. Spector 1990, pp. 44–47.

195. Spector 1990, p. 44.

196. Albright 1994, p. 8.

197. Albright 1994, p. 9.

South Africa has apparently dismantled its nuclear weapons program. According to Albright, the transition away from weapons production has not been entirely smooth. He reported that in March 1994, 16 disaffected employees attempted to blackmail the government by threatening to sell nuclear secrets unless they were given more than a million dollars in unemployment benefits. They have been barred from doing so.[198]

198. Albright 1994, p. 19.

12 The Global Picture: Summary and Recommendations

Howard Hu and Arjun Makhijani

This century has witnessed the birth of the only industry ever engineered with the potential to destroy humanity itself: the production of vast arsenals of nuclear weapons. The 15-kiloton blast that obliterated the lives of about 100,000 people in Hiroshima 50 years ago also ushered in an era of unprecedented competition to build immense nuclear arsenals. Figure 12.1 shows the world's main actual—and some suspected and potential—nuclear weapons facilities.[1] It excludes uranium mining and milling sites: these are too numerous to show on one map. (Please note that references for facts and figures discussed elsewhere in this book are not repeated in this chapter.)

The production of nuclear weapons requires the mining of large quantities of uranium ore and the transformation of it and other materials over many steps. In the process, this industry creates highly radioactive materials and wastes and uses large quantities of non-radioactive hazardous substances. Some of the radioactive substances fabricated in the course of nuclear weapons production have half-lives of thousands or tens of thousands of years or longer, necessitating consideration of health and environmental protection measures on unprecedented time scales. For instance, plutonium-239, one of two principal fissile materials in nuclear weapons (the other is uranium-235), has a half-life of about 24,000 years, and is a dangerous carcinogen.

This book surveys both the technologies required for making nuclear weapons and the countries that possess these armaments or are close to that point. It delineates the generic health and environmental impacts from the processes required to mine and make the materials and fabricate the warheads and, within the limits of available data and resources, describes these for the countries under study. It analyzes the environmental and health effects of hazards associated with making nuclear weapons by studying production processes, analyzing official and independent environmental data and analyses, and using the existing bodies of knowledge available in radiobiology, toxicology, and

1. Many of the facilities were shut in the late 1980s and early 1990s, while others are operating. Their operating status was discussed earlier in this book.

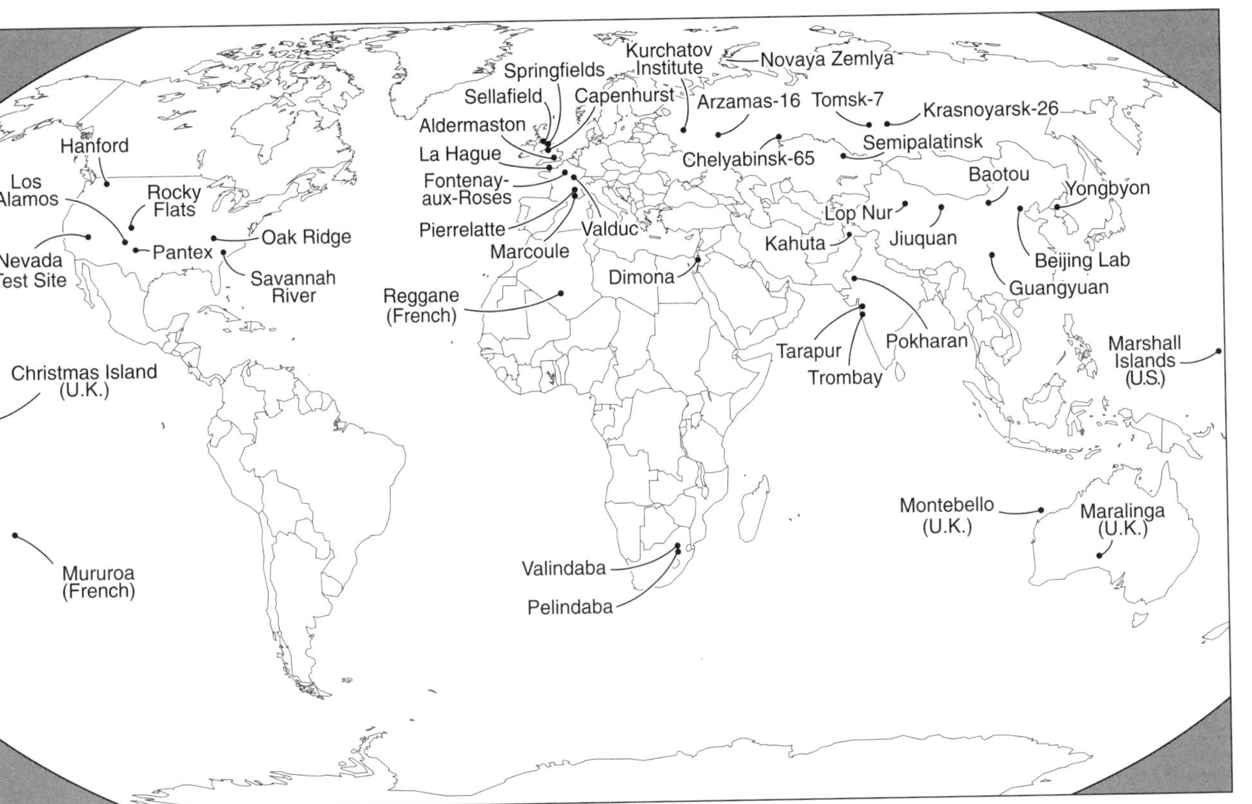

Figure 12.1 Principal nuclear weapons–related production and test sites in the world.

environmental epidemiology. However, accurate estimates of the actual toll on human health of nuclear weapons production worldwide are at present impossible, given the uncertainties regarding the numbers of people exposed through industrial and community exposures, the type, level, and duration of chemical and radioactive exposures sustained by these populations, and the health effects of chronic exposures to radiation and chemical toxins. These problems are compounded by others: the suppression and distortion of information, the poor state of much of the official data, and even fabrication of data by governments responsible for nuclear weapons production.

The extension of governmental and corporate secrecy to include questions affecting health and environment has posed some of our most difficult methodological problems. Even in the United States, by far the most open of the nuclear weapons states, much of the relevant data is still secret. This obstacle is compounded by the incompleteness and poor quality of much of the official data. In the absence of sound, verified official information about releases of pollutants, dumping of wastes, and exposures to radioactive and nonradioactive toxic materials, independent assessments, including this one, must necessarily rely on all publicly available data. Thus, with a critical eye, we make use of a wide range of sources of information, pointing out uncertainties and inconsistencies.

A great deal of uncertainty exists regarding the long-term hazards posed by the introduction of various radioisotopes into the environment. This partly reflects uncertainty regarding the pathways of exposure and hence the amounts of exposure to people in the future. It is also due in part to uncertainties about the extent of the risk of cancer from chronic exposure to low-level radiation.

Serious health damage is well known to occur at high radiation doses (acute doses on the order of 1 sievert or more). There is also firm evidence about long-term damage from moderate doses below 1 sievert, such as those received by survivors of the atomic bombings of Hiroshima and Nagasaki (on the order of 0.2 sievert). At lower levels of exposure, particularly chronic low-dose exposure, there is evidence for an increased risk of cancer, although this is not universally accepted. In general, with the accumulation of data, official scientific assessments, notably those made by the Committee on the Biological Effects of Ionizing Radiation (BEIR) of the U.S. National Academy of Sciences, have increased the magnitude of risk estimates per unit dose of radiation with each new study.

Large amounts of nonradioactive hazardous materials have been used in nuclear weapons production, but exposures to them are poorly understood. At the same time, concern is rising over the health effects of the wide variety of ordinary chemical pollutants emitted by nuclear weapons facilities, including metals, solvents, and PCBs. Epidemiology studies on many of these contaminants indicate that low-level exposures may significantly raise the risk for such chronic illnesses as cancer, hypertension, and kidney disease.

SUMMARY OF FINDINGS

Production and Waste Estimates

The production of nuclear weapons begins with research, design, and development. After that, the chief steps required to produce weapons are mining and milling of uranium, uranium conversion and enrichment, production of plutonium in nuclear reactors, extraction of plutonium from irradiated reactor fuel ("reprocessing"), processing of plutonium and highly enriched uranium (and other materials) into forms suitable for nuclear weapons parts, the fabrication of these parts, warhead assembly, disassembly of obsolete warheads, testing, and storage and disposal of radioactive and nonradioactive hazardous wastes. Some types of weapons can be made with either plutonium or highly enriched uranium and do not require both.

Albright, Berkhout, and Walker estimate that the total worldwide inventory of military plutonium production as of the end of 1990 was about 257 metric tons.[2] The total inventory of separated plutonium, civilian and military—all of which is usable for military purposes—was about 400 metric tons. The cumulative production of highly enriched uranium is about 2,300 metric tons, though there is considerable uncertainty due to the lack of reliable knowledge about Soviet production. Table 12.1 shows the estimated arsenals and weapons-usable fissile materials inventories for the five nuclear weapons states and many other countries.

Using U.S. data as a guide, table 12.2 shows order-of-magnitude estimates for the amount of ore needed per kilogram of highly enriched uranium production, as well as the mining and millings wastes generated in the process. It also shows some estimates of other radioactive wastes, and uranium processing–related air emissions. Table 12.3 shows similar figures per nuclear weapon, assuming that each warhead contains four kilograms of plutonium and 20 kilograms of highly enriched uranium (93 percent U-235). These figures do not include uranium needed for plutonium production. Military uranium (including that required for plutonium production and that which has been used as naval reactor fuel) has accounted for about 360,000 metric tons of natural uranium. This is over 20 percent of the total worldwide cumulative natural uranium production of about 1.7 million metric tons.

Tables 12.2 and 12.3 can provide a starting point for evaluating specific situations in the absence of official information, although some

2. Albright, Berkhout, and Walker 1993, p. 197.

Table 12.1 Nuclear Weapons and Weapons-Usable Materials Inventories, 1990

State	Warhead Inventory	Military Plutomium (metric tons)	Separated Civilian Plutonium (metric tons)	Highly Enriched Uranium (metric tons)[6]
United States	16,750	103	1.5	994
Former USSR[1]	32,000	125	25	1,250
United Kingdom[2]	200	11	45.8	10
France	525	6	19.3	15
China	435	2.5	0	15
India[3]	?	0.29		0
Israel[4]	60–100	0.33	0	0
Pakistan	?	0	0	0.1–0.2
North Korea[5]	?	?		
South Africa	0?	0	0	0.2–0.5
Belgium	none	none	1.17	0
Germany	none	none	15.75	0
Italy	none	none	2.6	0
Japan	none	none	2.77	0
Netherlands	none	none	0.67	0
Switzerland	none	none	1.11	0

Sources: Norris and Arkin 1993e; Albright, Berkhout, and Walker 1993; DOE 1994b; National Academy of Sciences 1994.

1. Figures include about 3,000 warheads in Ukraine, Kazakhstan, and Belarus.
2. For U.K. weapons data, see chapter 8.
3. India's reprocessing program includes civilian and potentially military plutonium.
4. For Israel weapons data, see chapter 11.
5. North Korea's plutonium stocks are unknown, but could be sufficient for one or more nuclear weapons.
6. Figures for highly enriched uranium are production rather than inventory figures.

aspects of waste generation and emissions differ greatly from one country to the next, particularly the volume of wastes and quantities of emissions to the environment. Some of the figures in the tables will yield more precise (or less uncertain) estimates than others. For instance, total metric tons of mill tailings will vary greatly with the quality of the ore, but the amounts of radium-226 and thorium-230 in the tailings per unit of uranium production will be relatively uniform. The quantities of fission products per unit of plutonium production will vary somewhat more than radium-226 varies per unit uranium. This is because radium-226 is generally in equilibrium with uranium in ore bodies, while the amounts of fission products per gram of plutonium depend on the design of the nuclear reactors and the burn-up of the fuel. The amount of land and water contaminated and the effects

Table 12.2 Radioactive Waste Production per Kilogram of Highly Enriched Uranium (order of magnitutde U.S. estimates)[1]

Uranium mining waste[2]	On the order of 100 metric tons, with 1 to 10 or more becquerels of uranium per gram of soil
Uranium mill tailings[3]	About 100 metric tons, total
	0.06 curies (about 2.2 gigabecquerels) thorium-230
	0.06 curies (2.2) gigabecquerels radium-226
	Heavy metals such as copper, arsenic, molybdenum, vanadium
	0.02 to 2 kg of uranium emissions to the air (0.01 to 1 percent of production)
Uranium processing[4]	200 kilograms of depleted uranium
	Air emissions of uranium 0.02 to 0.2 kg (0.01 to 0.1 percent of production)
	Solid waste uranium content on the order of 2 kilograms (1 percent of production)

1. Uranium requirements for plutonium production are not included. All figures except unit conversions are estimated to one significant figure.
2. Overburden assumed to be the same order of magnitude as the ore in weight (Dehemel and Rogers 1993, table B.1–2).
3. Uranium ore grade 0.2%. Uranium emissions from mills and from processing are order of magnitude estimates based on limited U.S. data.
4. Uranium conversion losses to UF_6 alone are about 0.5% (Lamarsh 1983, p. 150).

of such contamination on the local ecology and the health of populations tend to be the most variable and uncertain factors.

Nonradioactive hazardous materials are often mixed within radioactive wastes, but there is little data on the quantities involved. Tables 12.2 and 12.3 do not show estimates of nonradioactive pollutants. This is an area where a great deal of research remains to be done. Finally, tables 12.2 and 12.3 do not include contamination due to accidents, such as that resulting from the 1957 explosion of a high-level waste tank in the Chelyabinsk-65 complex in the former Soviet Union.

Environmental Contamination: Some Global Estimates

Roughly 70,000 nuclear warheads have been fabricated worldwide. This does not include reworking of materials and components of obsolete weapons into new ones. As we have discussed, there are many aspects of environmental contamination resulting from nuclear weapons production that we cannot estimate due to lack of data. But we can make some order-of-magnitude estimates of waste generation and environmental contamination from some of the principal processes based on the information in tables 12.2 and 12.3 (all waste and discharge estimates, except krypton-85, rounded to one significant figure):

Table 12.3 Radioactive Waste Production per Nuclear Weapon (order of magnitude U.S. estimates)[1]

Uranium mining waste[2]	2,000 metric tons, with a total of 2 to 20 gigabecquerels of uranium
Uranium mills	2,000 metric tons
	1.2 curies (about 44 gigabecquerels) thorium-230
	1.2 curies (about 44 gigabecquerels) radium-226
	Heavy metals such as copper, arsenic, molybdenum, vanadium
	0.4 to 40 kg uranium emissions to the air
Uranium processing	4 metric tons depleted uranium
	Air emissions 0.4 to 4 kilograms
	Solid waste uranium content on the order of 40 kilograms
Reprocessing, high-level waste[3]	12,000 curies (440 terabecquerels) each of strontium-90 and cesium-137, and equal amounts of yttrium-90 and barium-137 (non-decay-corrected)
"Low-level" waste[4]	50 cubic meters containing 10 terabecquerels of radioactivity
Transuranic waste	7 cubic meters containing 700 gigabecquerels of alpha radioactivity

1. Each nuclear weapon is assumed to contain 4 kg Pu-239 and 20 kg 93% uranium-235. Figures are rounded to one or two significant places, as indicated.
2. Uranium-related data were taken from table 12.2 and applied to 20 kilograms of highly enriched uranium.
3. Strontium-90 and cesium-137 figures assume that roughly 100 to 150 gigabecquerels of each are produced per gram of plutonium production.
4. Low-level waste and transuranic waste numbers are derived from U.S. DOE 1992 and assumed to be evenly spread over the 60,000 weapons produced in the United States (including partially disassembled and reassembled warheads).

• one hundred to two hundred million metric tons of uranium mill tailings containing 100,000 curies (about 4,000 terabecquerels) each of radium-226 and thorium-230 from the estimated 400,000 metric tons of natural uranium used for military purposes;

• over 400,000 metric tons of depleted uranium;

• about 3 billion curies (100 million terabecquerels) of high-level radioactive waste from plutonium production, including only strontium-90, cesium-137, and their daughter radionuclides yttrium-90 and barium-137m (this estimate is not corrected for radioactive decay; such a correction would reduce it by about one-half);

• twenty million curies (about 700,000 terabecquerels) of other radioactive wastes;

• emissions of about 145 million curies (5.5 million terabecquerels) of krypton-85 (non-decay-corrected) into the atmosphere due to reprocessing;[3]

• thousands of square kilometers of land highly contaminated from production processes and accidents;

• global contamination from fallout due to atmospheric nuclear weapons tests amounting to 30 million curies (one million terabecquerels) combined of strontium-90 and cesium-137 (decay-corrected) and 10 million curies (0.4 million terabecquerels) of carbon-14. Additional inventories of fission products and unfissioned plutonium have been left underground due to underground testing.

These summary estimates provide a starting point for the work ahead in making estimates of the contamination in specific areas and countries. They are to be regarded as indicative rather than definitive. Moreover, they do not convey the real extent of the damage. Some of the worst damage has been in the former Soviet Union. Entire river systems have been contaminated in some cases, as for instance with the river system near the Chelyabinsk-65 plant. Lake Karachay at Chelyabinsk-65 is perhaps the most contaminated body of water on Earth. The dose rate near the pipe that discharges radioactive wastes into it is 6 grays per hour, which would yield a lethal (LD50) dose in about 45 minutes.

Highly radioactive liquid wastes that result from reprocessing have been responsible for the worst accident resulting from nuclear weapons production, the explosion of a tank at Chelyabinsk-65 containing highly radioactive waste in September 1957. It resulted in the contamination of about 15,000 square kilometers of land and the evacuation of over 10,000 people. Dozens of tanks in the United States and elsewhere are at risk of explosions.

Uranium mining has been responsible for contamination not only in the nuclear weapons states but also in many other countries. Some of the most polluted areas from nuclear weapons production are in East Germany, which supplied the Soviet nuclear weapons program, and in the Third World, which supplied the programs of the United States, United Kingdom, and France. Even within the nuclear weapons states, uranium mining and resultant contamination have disproportionately affected tribal peoples.

The nuclear weapons industry has contaminated groundwater, surface waters, seas, and oceans. For instance, the sea off Sellafield in England and the seas off Russia have been the dumping grounds for large amounts of radioactivity. In the United States and elsewhere, groundwater at many of the sites where weapons factories are located has become highly contaminated. While this water is not now being

3. Albright, Berkhout, and Walker 1993, p. 37.

used for domestic consumption, it is not evident how its use can be regulated once institutional control of the sites is lost or once they have been designated for other uses. Decommissioning and cleaning up nuclear weapons plants will produce additional large quantities of waste, the magnitude of which will become clear only over the next decade or so as decommissioning proceeds in the United States and possibly in other countries. Dismantlement of unwanted nuclear weapons and disposition of the fissile materials they countain present further formidable security and environmental challenges.[4]

Exposed Populations

Broadly speaking, the making of nuclear weapons has exposed five groups of people to environmental and health dangers:

(1) workers at nuclear weapons facilities;

(2) armed forces personnel who participated in atmospheric weapons testing;

(3) people living near nuclear weapons sites;

(4) people who were subjects of experiments;

(5) the world's inhabitants for centuries to come.

These categories include only those affected by the production and testing of nuclear weapons. As we have noted, the transportation, deployment, and possible use of nuclear weapons are not within the scope of this book.

Generally, the most intensely exposed people have been workers in nuclear weapons plants and testing facilities and members of the armed forces. Within these two populations, the extent of exposure varies according to the specific nature of their duties and length of service. The third set of victims, often called "downwinders," are people who live near nuclear weapons facilities. The definition of "near" extends in some cases to hundreds of kilometers downwind, especially in the case of atmospheric nuclear weapons testing and large intentional or accidental releases, such as those that occurred at Chelyabinsk-65 or at Hanford in the United States. Some downwinders have been as highly exposed as workers and armed forces personnel. This is certainly the case for some affected by the explosion at Chelyabinsk-65, for iodine-131 exposures from the first two decades of operation of the Hanford plant, and for nuclear testing downwinders among the people of Rongelap in the Marshall Islands and people living near the Soviet test site near Semipalatinsk in Kazakhstan. Recent revelations in the United States have brought to light human experiments involving thousands of people. Finally, there have been and

4. Makhijani and Makhijani 1995; NAS 1994.

will continue to be exposures to the entire global population, mainly due to atmospheric nuclear weapons testing[5] but also to releases of krypton-85 and other gaseous radionuclides from plutonium production. Given the long-lived nature of some of the radionuclides involved, these exposures will persist for thousands of years.

It is possible to make rough, order-of-magnitude estimates of the number of exposed armed forces and worker populations in some instances. The figure for exposed "downwinders" is considerably more fluid, mainly because of the interlinked problem of defining the boundary of the "downwind" area and uncertainties about doses to people off-site.

About 250,000 members of the U.S. armed forces participated in the atmospheric nuclear weapons testing program. The number of workers in the U.S. nuclear weapons complex at any time has been on the order of 100,000 since the mid to late 1950s, excluding workers in uranium mining and milling. (Current employment during the decommissioning phase is actually higher.) Considering some turnover of workers and recent increases in employment for cleanup operations, several hundred thousand people have at one time or another worked in the U.S. nuclear weapons complex.

In the Soviet Union, the number of workers involved in the nuclear weapons complex has been reported to be on the order of 1 million, including people engaged in uranium mining and milling. No reliable estimate is available for armed forces personnel involved. Large numbers of people were involved in uranium mining and milling in other countries. Perhaps the largest number in a single place was the 450,000 uranium mine and mill workers in East Germany, which supplied much of the uranium for the Soviet nuclear arsenal. Tens of thousands of people, at the very least, have been involved in uranium mining in China, including the period of particularly labor-intensive mining during the Great Leap Forward in the late 1950s and early 1960s. While this book has not attempted to gather comprehensive data on the number of workers involved in this global industry, it would appear that at least two million people have been involved in various aspects of nuclear weapons production worldwide; the true figure is probably considerably higher.

The levels of exposure to radiation of the four population groups vary widely. Exposures due to global fallout are on the order of a few tens of microsieverts (a few millirems) per year. However, the dispersed nature of fallout has resulted in exposure of billions of people to such levels of radiation.

Researchers have made various estimates of levels of exposures to downwinders. The most highly exposed groups that we know about

5. See IPPNW and IEER 1991.

are those living downwind and downriver of the Chelyabinsk-65 and downwind of the Semipalatinsk sites in the former Soviet Union, the people of Rongelap, and, in the case of thyroid doses, children living downwind of Hanford in the early years of production. The downwind exposures near Oak Ridge, Tennessee, may also be high, but this remains the subject of study and controversy.

The most highly exposed groups have tended to be workers. The most severe exposures of workers for whom some data is available were in the Chelyabinsk-65 gas-graphite reactor and reprocessing plant. Worker doses in the early years averaged about 1 sievert, according to data published so far. Under many circumstances, notably in facilities that processed uranium, internal exposures may have been high among certain groups of workers. For instance, at the uranium-processing plant near Fernald, Ohio, data on employees indicate cumulative lung doses of several sieverts for many production workers. Yet neither the plant's corporate contractors nor the Department of Energy calculated internal doses from urine and lung-counting data that were collected at the plant.

Even greater uncertainties exist in regard to internal exposures for armed forces personnel, notably to alpha-emitting radionuclides.[6] Thus, the overall exposures to workers, armed forces personnel, and downwind populations will remain the subject of considerable uncertainty, and controversy, for some time. Because most official data on these subjects in most countries is still secret, it is impossible to know whether reliable quantitative estimates can be produced at least for an appreciable fraction of the exposed population.

The Burden of Disease

As noted above, estimating the total toll on human health of nuclear weapons production worldwide is almost impossible given the types of uncertainties discussed. Aside from the global fallout effects of nuclear weapons testing, estimated to produce hundreds of thousands of excess cancer fatalities over the centuries,[7] uranium mining has been responsible for the largest collective exposures to workers. While precise global estimates are at present impossible, we note that one estimate puts the number of workers who have died of lung cancer and silicosis due to mining and milling in East Germany alone at twenty thousand people. But such estimates are often questionable or preliminary as yet.

Unfortunately, we cannot make similar estimates on a global level of the disease burden that may have resulted from occupational ex-

6. IPPNW and IEER 1991, pp. 17–20.

7. IPPNW and IEER 1991, chapter 3.

posures in uranium mining, milling, and the industries related to plutonium reprocessing and nuclear weapons manufacturing. It is instructive to note that many of the occupational mortality studies of uranium miners in the United States and Canada have estimated lung cancer risks two to six times higher than expected. To the extent that this reflects generic risks shared by all uranium miners, and that working conditions have been similar or worse in other uranium-mining countries, this would mean that the mining of uranium for nuclear weapons has led to thousands of excess lung cancers. It is also apparent that a disproportionate share of that burden fell on indigenous or colonized peoples who lived in the areas of and were employed in the mines.

Findings of excess cancers in workers and off-site populations have been noted in many epidemiological studies discussed in this book, while in others they have not been detected. In general, it is difficult to determine the validity of these studies in the face of serious problems with the quality and completeness of the data. For instance, in 1994, U.S. officials admitted that even external dose data for workers have some serious deficiencies. In fact, portions of the data were fabricated in that zeros were entered into the radiation dosimetry records of workers when the badges were not turned in.

Russian data on health are clearly suspect. Even for groups of workers and off-site populations living near Chelyabinsk-65 with high exposures, health outcome data show far fewer than expected leukemia or other cancer fatalities. This result is at considerable variance with well-established risk factors from medical radiation exposure studies and follow-up of Hiroshima-Nagasaki survivors. It is reported that doctors were forbidden to make radiation-related diagnoses, on pain of punishment. Thus, while some dose data indicate that one should find relatively high levels of fatal cancers, the health findings do not correspond to the dose estimates. New diagnoses such as "weakened vegetative syndrome" and even "ABC disease," unknown elsewhere, were created in Russia, possibly to fill the void for radiation-related diagnoses banned by nuclear authorities.

In Britain there appears to be the opposite problem. The existence of a comprehensive public health system and public health data has enabled researchers to confirm clusters of childhood leukemias near the Aldermaston and Burghfield sites (which are in the same general area) and near Sellafield. However, official data on worker doses and environmental releases data indicate expected excess cancers based on standard coefficients far lower than those found in the clusters. For this reason, the suggestion that these clusters are linked to nuclear weapons as well as civilian reprocessing has been controversial. The cause of the excess leukemias has not been established. It may be that releases and exposures were higher than has been estimated, or that routes of exposure created vulnerabilities among populations in ways that are not yet clear.

It is also impossible at present to estimate the disease burden due to community exposures to nonradioactive chemical pollution emitted by industries associated with nuclear weapons. The database is so inadequate that it does not permit even qualitative discussion of the health impact for individual countries, to say nothing of worldwide estimates. There are anecdotal reports of damage that are inconsistent with radiation damage. Such damage may be linked to chemical discharges. However, such emissions have not been monitored carefully, or indeed at all for most of the period of nuclear weapons production, so far as publicly available data indicate.

In sum, from the data that are available on environmental releases, discharges, accidents, and radiation doses and the current state of knowledge regarding the risks posed by exposure to radiation, it appears likely that health effects have been experienced on a significant scale. Continuing health risks will persist for decades.

Cleanup and Waste Management

There is a vast amount of polluted soil and water at hundreds of sites associated with nuclear weapons production. Remediation of these sites and decommissioning of highly contaminated factories and equipment will be necessary if a substantial portion of the thousands of square kilometers of land occupied by these sites is to become available for general use or, in some cases, even for restricted non-weapons uses. These cleanup efforts will be very expensive. In the United States, remedial measures are expected to cost hundreds of billions of dollars over three decades or more.

Environmental remediation means that waste now widely dispersed throughout these sites will be gathered together in some form for processing and disposal. Thus, the cleanup process itself will generate volumes of radioactive and mixed radioactive and nonradioactive hazardous wastes. The disposal of long-lived waste in ways that reduce risk of exposure and consequent disease is a major unresolved issue throughout the world. For instance, not a single repository has been opened for high-level radioactive wastes, such as those that arise from plutonium production. As another example, remedial measures for uranium-mill tailings have not taken adequate account of the longevity of radium-226 and thorium-230.

It is also noteworthy that many countries that are not nuclear weapons states, including former colonies of some of the nuclear powers, are suffering adverse environmental and health consequences from nuclear weapons production. This is especially true of uranium mining and milling locations and nuclear weapons test sites.

RECOMMENDATIONS

1. End secrecy surrounding all nuclear weapons issues related to health and environmental protection.

At the end of the Cold War, there is no justification for keeping any information on nuclear weapons production secret except design data. All records relating to health and environmental issues, including materials production and accounting data, must be made public. This is not only essential for reconstructing the damage that has been done in the past, it is crucial for designing effective cleanup programs and for verification and materials accounting that are central to nonproliferation. This information should not only be declassified but also made public in a manner that is accessible to all interested parties, including independent researchers and activists.

2. Stop production of all weapons-usable fissile materials.

Whatever the post–Cold War level of nuclear arsenals, the world has an immense surplus of plutonium and highly enriched uranium, both civilian and military. The health, environmental, and waste disposal costs associated with their production are large and stopping their production would be justified on these grounds alone. Nonproliferation and disarmament goals reinforce this recommendation.

3. Identify, monitor, and assist exposed populations.

Greater efforts are needed to identify exposed populations, assess the extent to their exposures, and provide them with health monitoring and related assistance. In such health assessments, it is essential to account carefully for both external and internal radiation exposures and also to take proper consideration of exposure to nonradioactive hazardous materials.

4. Separate all nuclear warheads from their delivery vehicles and store the warheads in criticality-safe containers.

In view of the serious proliferation threats and the end of the Cold War, all nuclear weapons should be taken off alert, removed from their delivery systems, and put into secure storage in continuously monitored facilities.

5. Treat plutonium as a hazardous waste material rather than as a resource.

Plutonium should be treated as a waste material never again to be used. Disposal techniques compatible with this policy should be used. Transportation of plutonium should be limited to that needed to store warheads, dismantle them, or process plutonium as a waste.[8]

8. This recommendation is discussed in more detail in IPPNW and IEER 1992, chapters 7 and 8.

6. Pay special attention and compensation to non–nuclear weapons states or territories that are suffering adverse consequences.

The adverse consequences of nuclear weapons production and testing on nonnuclear states have fallen especially hard on uranium mining and milling locations and nuclear test sites. Many of these sites are in areas that belong traditionally to tribal people or oppressed minorities. This is almost uniformly the case with major nuclear weapons test sites.[9] Many production sites within the weapons states are also located on or near tribal lands. The restoration of these sites and adequate health and environmental monitoring for the affected populations is essential to redress the manifest injustice of the pollution.

7. Fully involve the public in cleanup and waste management decisions.

The past history of neglect regarding environmental and health issues is tied to a history of secrecy and public disenfranchisement. Decisions regarding site remediation and waste management must be made with full disclosure of documentation and full participation of the public.

8. Establish radiation standards that protect public health even in the face of uncertainty.

Given current large uncertainties and emerging new data, further revisions of radiation risk estimates are likely. But the continuing uncertainties over risk of low-level radiation exposures should not obscure the fundamental principle that standards to limit public and worker exposure should protect health in the face of uncertainty. This is especially important in view of the long half-lives of plutonium-239, thorium-230, and other highly carcinogenic materials.

9. Reexamine the concept of nuclear deterrence in view of its role in the health and environmental damage that has been caused in the name of national security and in view of its effect of promoting nuclear ambitions and arsenals.

As an ancillary part of our work we examined the history of how the nuclear arms race started and spread. The concept of deterrence has played a strong role in the history of nuclear weapons development. The first instance was during World War II, when the United States began a program to develop nuclear weapons in response to the fear that Germany would do it first. At the same time, discussions that took place as early as 1943 indicate that Germany was not targeted partly out of fear of its nuclear capability.

The Manhattan Project had one eye on the Soviet Union, despite the fact that the Soviet Union was a wartime ally of the United States. As project head General Leslie Groves put it, "There was never ... any illusion on my part but that Russia was our enemy and the Project was

9. IPPNW and IEER 1991, p. 170.

conducted on that basis."[10] From a Soviet perspective, the end of World War II and the start of the Cold War added considerable urgency to the development of a Soviet nuclear arsenal.

Proliferation was fueled by a similar logic as China developed its bomb, first with Soviet aid and then independently, following U.S. nuclear threats. China continued its nuclear weapons program partly in response to Soviet hostility after the Sino-Soviet split. France developed its arsenal partly due to a Soviet threat during the Suez crisis of 1956. India's decision to go nuclear was heavily influenced by its defeat in the 1962 border war with China, China's first nuclear test in 1964, and the U.S. decision to throw symbolic military weight behind Pakistan during the 1971 Pakistan-Bangladesh-India war. Pakistan followed with its own program.

Historically, the five nuclear weapons states have had high status in the global hierarchy and commanded attention. All five permanent members of the United Nations Security Council are nuclear weapons states. The prestige and power this confers and the exclusion of others from power that it provides with official sanction are implicit invitations to countries to pursue nuclear weapons production.

At the end of the Cold War, the existence of large amounts of plutonium and highly enriched uranium presents a severe proliferation problem. It is for historians to do more definitive research and analysis into the exact role of the doctrine of deterrence in the spread of nuclear weapons. It has created a division of the world into militarily powerful nuclear haves and weak have-nots. We find the evidence compelling that this dynamic has provided a strong impetus to proliferation. The world needs to come to grips with the counterproductive aspects of this doctrine both from the point of view of security and from the point of view of the enduring damage to health and the environment that it has engendered.

10. As quoted in Sherwin 1987, p. 62.

Glossary

Alpha radiation: Radiation consisting of a helium ion that is emitted upon radioactive disintegration of the nuclei of certain heavy elements like uranium-238 and radium-226.

Atomic number: The number of protons that an element has in its nucleus. This number determines the chemical properties of the element. The periodic table of elements is organized in ascending order of atomic number.

Becquerels: A measure of radioactivity of a substance equalling one disintegration per second. One becquerel equals about 27 picocuries.

BEIR Committee: The committee of the U.S. National Academy of Sciences that studies the Biological Effects of Ionizing Radiation (BEIR).

Beta radiation: Radiation consisting of high-speed electrons or positrons.

CEA: The Atomic Energy Commission of France.

Critical mass: The minimum amount of a substance that will result in a self-sustaining chain reaction. This amount depends on the properties of the material, the geometry into which it is shaped, and other factors.

Curies: A measure of radioactivity of a substance equaling 37 billion disintegrations per second. This is the traditional measure of radioactivity, and is based on the number of disintegrations per second undergone by 1 gram of pure radium-226. One curie equals 37 billion becquerels.

Decay correction: The amount by which the radioactivity of a substance must be reduced after a period of time to account for its radioactive decay during that time.

DOE: U.S. Department of Energy.

Electron: An elementary particle with a negative electrical charge that has about 1/1836 of the mass of a proton.

Electron capture: The transmutation of a radionuclide that occurs when the nucleus captures an electron, thereby decreasing its atomic number by one.

Electron volt (eV): The energy acquired by an electron after being accelerated through a one-volt electrical potential difference.

External radiation dose: Radiation dose from sources of radioactivity located outside the body.

Fission: The splitting of the nucleus of an element into fragments. Heavy elements such as uranium release energy when fissioned.

Fission product: An isotope of an element created by the fission of a heavy element.

Free radical: An atom or group of atoms that is chemically reactive because it contains one or more unpaired electrons. It may be electrically charged or neutral.

Fusion: The fusion of the nuclei of two elements. Fusion of certain light elements such as heavy hydrogen isotopes gives a net energy release.

Gamma radiation: Electromagnetic radiation with very energetic photons (that is, high-frequency electromagnetic radiation). The frequencies are far higher than the visible range and high enough to cause ionization of elements. Gamma radiation is identical to X rays of high energy. This is the most penetrating form of radiation.

Giga-: Prefix used to indicate one billion.

Gray: A unit of dose equal to one joule of energy absorbed by one kilogram of material. It is equal to 100 rads.

Half-life: The time in which half of a radioactive substance decays away.

IAEA: Acronym for International Atomic Energy Agency. This is the U.N. agency responsible for overseeing both promotion of civilian nuclear power and enforcing nuclear nonproliferation safeguards.

Induced radioactivity: Radioactivity produced in certain materials as a result of nuclear reactions, especially by the absorption of neutrons.

Internal radiation dose: Radiation dose to internal organs due to radioactivity inside the body; may consist of any combination of alpha, beta, and gamma radioactivity. Neutron doses may also be involved.

Isotope: One of two or more variant forms of an element. Isotopes have the same number of protons in the nucleus (and hence essentially the same chemical properties), but a different number of neutrons, and therefore different weights. Some isotopes are radioactive, while others are not. Many isotopes occur naturally, while a large number have been created artificially due to nuclear reactions. The various radioactive isotopes of an element have different half-lives.

Ionization: The knocking out of one or more electrons from their orbit around the nucleus of an atom or from a group of atoms.

Ionizing radiation: Radiation energetic enough to ionize atoms.

Kilo-: Prefix used to indicate one thousand.

Kiloton: One thousand tons. Used in combination with the concept of TNT equivalent as a measure of yield of nuclear explosives.

Linear energy transfer (LET): The average amount of energy transferred by radiation into the surrounding medium over a unit length of the medium.

Mass number: The nominal atomic weight of an isotope equal to the sum of the number of protons and neutrons in the nucleus.

Mega-: Prefix used to indicate one million.

Megaton: One million tons. Used in combination with the concept of TNT equivalent as a measure of yield of nuclear explosives.

Micro-: Prefix used with rads, rems, grays, sieverts, and other units to indicate one-millionth of the unit.

Milli-: Prefix used with rads, rems, grays, sieverts, and other units to indicate one-thousandth of the unit.

MINATOM: The Russian Ministry of Atomic Energy.

MoD: The British Ministry of Defence.

Nano-: Prefix used with rads, rems, grays, sieverts, and other units to indicate one-billionth of the unit.

Neutron: An elementary particle that is electrically neutral. Together with protons, it forms the nucleus of an element (except the normal hydrogen nucleus, which consists only of a single proton). Neutrons are stable in the nucleus but unstable in free air, disintegrating into a proton and an electron, with a half-life of about 12 minutes. Energetic neutrons, such as those produced by fission, constitute indirectly ionizing radiation.

Pico-: Prefix used with rads, rems, grays, sieverts, and other units to indicate one-trillionth of the unit.

Positron: An elementary particle with a positive electrical charge that is in other ways identical to an electron.

Proton: An elementary particle with a positive electrical charge, with a mass slightly less than that of a neutron. Protons and neutrons make up the nuclei of elements.

Quality factor: An empirical ratio used to convert radiation dose measured in terms of deposited energy from different kinds of radiation into radiation dose measured in terms of biological damage. This ratio is a simplification of a more complex concept called relative biological effectiveness of different kinds of radiation.

Rad: A unit of dose equal to the deposition of 100 ergs of energy per gram of material being irradiated.

Radioactivity: The spontaneous release of energy from the nucleus of an atom, generally in the form of gamma, beta, and/or alpha radiation. Beta and alpha particle emissions from a nucleus result in the transformation of an atom into a different element (known as transmutation). Transmutation of some radionuclides also occurs by spontaneous fission and by electron capture in the nucleus.

Radionuclide: The radioactive isotope of an element.

Rem: A unit of dose that takes into account the relative biological damage due to various kinds of radiation energy absorbed by tissue. In general, the larger the amount of energy deposited per unit length of tissue, the greater the radiation damage per unit of absorbed radiation energy, that is, the greater the ratio of rems to rads. Linear energy transfer (LET) is a measure of the ratio of the relative damage that a unit of deposited radiation energy can do. For gamma radiation, where the energy transfer per unit length is low (low-LET radiation), rems and rads are essentially equivalent units of radiation dose. For beta radiation, rads and rems are also usually considered to be equivalent, though the energy transfer per unit length is greater than for gamma radiation. For radiation due to heavy particles—that is neutrons, protons, and alpha particles—where the linear energy transfer is high (high-LET radiation), the ratio of rems to rads ranges from 2 to 40. An empirical ratio, called the "quality factor" of the radiation, is used to convert rads (or grays) into rems (or sieverts).

Roentgen: A unit measuring gamma radiation dose. It is the quantity of gamma radiation that will produce electrons (in ion pairs) with a total electrical charge of 0.258 millicoulombs in a kilogram of dry air. A roentgen is equal to 0.93 rads. Given the uncertainty in doses that prevails in the circumstances discussed in this study, a roentgen may be taken as essentially equivalent to one rad.

Sievert: A unit of dose measuring the biological effectiveness of 1 gray of gamma radiation. It is equal to 100 rems.

Tera-: Prefix used to indicate one trillion.

Thermonuclear weapon: A nuclear weapon that gets a substantial part of its explosive energy from fusion reactions.

TNT equivalent: The unit most commonly used to measure the energy released in nuclear explosions. One short ton of TNT (2,000 pounds or 907 kilograms) is assumed to be equivalent to one billion calories of energy. The energy released by nuclear explosions is generally measured in units of kilotons and megatons of TNT equivalent.

Working level: Any combination of the radioactive decay products of radon gas resulting in the emission of 130,000 megaelectron-volts (MeV) of energy in one liter of air. This unit is used to measure the extent of radioactivity in uranium mines.

Working level month (WLM): Exposure to air with an average of one working level of radioactivity over a period amounting to 170 hours (the nominal hours of work in a month).

Yield: The energy released by a nuclear explosion.

References

Acquavella, J. F., L. D. Wiggs, R. J. Waxweiler, D. G. Macdonell, G. L. Tietjen, and G. S. Wilkinson. 1985. Mortality among workers at the Pantex weapons facility. *Health Physics*, vol. 48, pp. 735–746.

Acquavella, J. F., G. S. Wilkinson, G. L. Tietjen, C. R. Key, J. H. Stebbings, Jr., and G. L. Voelz. 1982. Malignant melanoma incidence at the Los Alamos National Laboratory. *Lancet*, 17 April 1982, pp. 883–884.

Acquavella, J. F., G. S. Wilkinson, G. L. Tietjen, C. R. Key, J. H. Stebbings, Jr., and G. L. Voelz. 1983. A melanoma case-control study at the Los Alamos National Laboratory. *Health Physics*, vol. 45, pp. 587–592.

ACRO. 1992. La pollution de la rivière Ste-Hélène: Un essai inter-laboratoire le confirme, l'ACRO avait raison! *Gazette Nucléaire*, no. 113/114, March 1992, pp. 27–31.

AFP. 1992. [Nuclear waste research agreement with China reported.] Transcript of report broadcast on Hong Kong AFP, 0406 GMT, 2 April 1992. In *Foreign Broadcast Information Service*, JPRS-TEN-92-009, 22 May 1992, p. 17.

Ahearne, J. (Chairman, Advisory Committee on Nuclear Facility Safety) 1989. Letter-report on DOE's Pantex Plant to James D. Watkins, U.S. Secretary of Energy. 15 December 1989.

Ahearne, J. (Chairman, Advisory Committee on Nuclear Facility Safety) 1990. Letter-report on Y-12 Plant, Rocky Flats, and Savannah River to James D. Watkins, U.S. Secretary of Energy. 28 March 1990.

Akleyev, A. V. 1993. Health effects of radiation incidents in the southern Urals. Presentation, Prague, April 1993.

Akleyev A., P. Golohchanov, M. Degteva, M. Kossenko, B. Kostiychenko, P. Malkin, R. Pogodin, G. Romanov, and B. Shvedov. 1989. Radioactive pollution of the surrounding environment in the southern Ural region and its effect on the health of the population (in Russian). Information Bulletin no. 3. Center for Public Information on Atomic Energy, Moscow.

Akleyev, A., P. Golohchanov, M. Degteva, M. Kossenko, V. Kostiychenko, P. Malkin, R. Pogodin, G. Romanov, and V. Shvedov. 1992. Radiation, ecology and human health: Radioactive pollution in the environmental region of the southern Urals and its effects on the health of the population (in Russian). Center for Public Information on Atomic Energy, Moscow. Information Bulletin no. 3, 1992.

Akleyev, A. V., and S. A. Shalaginov. 1991. Possibilities of biological dosimetry in reconstruction of population radiation doses in southern Urals. Presented at USSR-Japan joint meeting on radiation dosimetry in epidemiological research in Southern Urals. Chelyabinsk, June 1991.

Albright, D. 1993. A proliferation primer. *Bulletin of the Atomic Scientists,* June 1993, pp. 14–23.

Albright, D. 1993a. Slow but steady. *Bulletin of the Atomic Scientists,* July/August 1993, pp. 5–6.

Albright, D. 1994. South Africa's secret nuclear weapons. *ISIS Report.* (Institute for Science and International Security, Washington, D.C.) vol. I, no. 4, May 1994.

Albright, D., F. Berkhout, and W. Walker. 1993. *World Inventory of Plutonium and Highly Enriched Uranium, 1992.* Oxford: Oxford University Press.

Albright, D., J. Cabasso, T. Carpenter, B. Costner, H. J. Geiger, P. Gray, D. Hancock, M. Kelley, D. G. Kimball, A. Makhijani, G. Pollet, D. W. Reicher, J. Stroud, J. D. Werner, and T. Zamora. 1992. *Facing reality: The future of the U.S. nuclear weapons complex.* Ed. Peter Gray. San Francisco: Tides Foundation, May 1992.

Albright, D., and M. Hibbs. 1992. North Korea's plutonium puzzle. *Bulletin of the Atomic Scientists,* November 1992, pp. 36–40.

Albright, D., and M. Hibbs. 1993. South Africa: The ANC and the atomic bomb. *Bulletin of the Atomic Scientists,* April 1993, pp. 32–37.

Albright, D., and N. Schonbeck. 1993. Report of the Incident Investigation Subcommittee—Incident: 1957 fire in Building 771. Submitted to the State of Colorado, 26 May 1993.

Aleksrashin, L. I., O. Prokofiev, I. Andrev, V. Sharalapov, A. Shramchenka, L. Anicimova, V. Chirkov, U. Karelin, and A. Petyhov. 1990. Report of the Interdepartmental Commission for the Evaluation of Radiation Situation in the Tomsk Region (in Russian). N.p., July 1990.

Alexander, V. 1991. Brain tumor risk among United States nuclear workers. *Occupational Medicine: State of the Art Reviews,* vol. 6, pp. 695–714.

Alperovitz, G. 1985. *Atomic Diplomacy.* 2d edition. New York: Penguin.

Alvarez, R., and A. Makhijani. 1988. Radioactive waste: Hidden legacy of the arms race. *Technology Review,* vol. 91, no. 6, pp. 42–51.

ANDRA (l'Agence Nationale pour la Gestion des Déchets Radioactifs). 1988. L'ANDRA, Un service public pour une gestion sure des dchets radioactifs. Paris.

ANDRA (l'Agence Nationale pour la Gestion des Dchets Radioactifs). 1993. Inventaire national des dchets radioactifs de l'ANDRA. Paris.

Archer, V. E. 1980. Epidemiologic studies of lung disease among miners exposed to increased levels of radon daughters. In *Health Implications of New Energy Technologies* [presented at Park City Environmental Conference (1979)], ed. W. N. Rom and V. E. Archer, pp. 13–35. Ann Arbor, Michigan: Ann Arbor Science Publishers.

Archer, V. E., J. D. Gillan, and J. K. Wagoner. 1976. Respiratory disease mortality among uranium miners. *Annals New York Academy of* Sciences, vol. 271, pp. 280–293.

Archer, V. E., and J. K. Wagoner. 1973. Lung cancer among uranium miners in the United States. *Health Physics,* vol. 25, pp. 351–371.

Archer, V. E., J. K. Wagoner, and F. E. Lundin. 1973. Cancer mortality among uranium mill workers. *Journal of Occupational Medicine,* vol. 15, pp. 11–14.

Arkin, W., and R. Fieldhouse. 1985. *Nuclear Battlefields: Global Links in the Arms Race.* Cambridge, Massachusetts: Ballinger Publishing Company.

Arms Control Today. Factfile: Non-signatories to the Nuclear Non-Proliferation Treaty. 1994. *Arms Control Today,* July/August 1994.

Arnold, L. 1987. *A Very Special Relationship.* London: Her Majesty's Stationery Office.

Arnold, L. 1992. *Windscale 1957: Anatomy of a Nuclear Accident.* Basingstoke, England: Macmillan.

Athas, W.F., and C. R. Key. 1993. *Los Alamos Cancer Rate Study: Phase I.* Final Report. Santa Fe, New Mexico: The State of New Mexico, March, 1993.

ATSDR (Agency for Toxic Substances and Disease Registry). 1988. *The Nature and Extent of Lead Poisoning in Children in the United States: A Report to Congress.* Atlanta, Georgia: U.S. Department of Health and Human Services, Public Health Service.

Austin, D. F., P. J. Reynolds, M. A. Snyder, M. W. Biggs, and H. A. Stubbs. 1981. Malignant melanoma among employees of Lawrence Livermore National Laboratory. *Lancet,* 3 October 1981, pp. 712–716.

Australia, Department of Primary Industries and Energy. 1991. *Annual Report of the Director of Safeguards 1990–1991.* Canberra: Australian Government Publishing Service.

Australian Department of Trade and Resources. 1981. *Australia's Mineral Resources: Uranium.* Canberra: Australian Government Publishing Service.

Aycoberry, C. (Director of Reprocessing Branch, Cogéma). 1990. Letter to l'Ingénieur Général des Mines, Chef du Service Central de Sûreté des Installations Nucléaires, 12 July 1990.

Bailar, J. C., and J. L. Young. 1966. Oregon malignancy pattern and radioisotope storage. *Public Health Reports,* vol. 81, no. 4, pp. 311–317.

Bair, W., and R. Thompson. 1974. Plutonium: biomedical research. *Science,* vol. 183, pp. 715–722.

Baisogolov, G. D., M. G. Bolotnikova, I. A. Galstyan, A. K. Guskova, H. A. Koshirnokova, A. F. Lezov, B. V. Nikipelov, B. C. Pesternikova, and N. S. Shilnikova. 1991. Malignant neoplasms of hematapoietic and lymphoid tissues in the personnel of the first plant of the atomic industry (in Russian). *Voprosy Onkologii,* vol. 37, pp. 553–9.

Barrillot, B. 1991. The manufacture of nuclear weapons in France. Lyons, France: Documentation and Research Center for Peace and Conflicts, September 1991.

Barrillot, B., and M. Davis. 1994. *Les Déchets Nucléaires Militaires Français* (in French). Lyon, France: Centre de Documentation et de Recherche sur la Paix et les Conflits.

Barron, B. 1977. Effect of misclassification on estimates of relative risk. *Biometrics,* vol. 2, pp. 414–418.

Bastable, J. 1993. Secret scientists threaten to strike. *New Scientist,* 3 July 1993, p. 9.

Bebbington, A. 1990. History of Du Pont at the Savannah River Plant. Wilmington, Delaware: E. I. Du Pont de Nemours and Company.

Beebe, G. W., H. Kato, and C. Land. 1977. Life span study, Report F, mortality experience of atomic bomb survivors, 1950–1974. RERF TR 1–77. Radiation Effects Research Foundation.

Behar, A. 1992. Déchets radioactifs. *Médecine et Guerre Nucléaire,* vol. 7, no. 4, October/November/December 1992, pp. 11–17.

Belboech, R. 1991. Les risques de cancer chez les mineurs d'uranium. *La Gazette Nucléaire*, no. 111/112, pp. 6–9.

Beleites, M. 1988. Pechblende-der Uranbergbau in der DDR und seine Folgen (Pitchblende-uranium mining in the GDR and its consequences). *Dokumentation* (Frankfurt, Germany), no. 40/88.

Belyaninov, K. 1993. Uranium caravans on the country's roads. There is a chance that Russia will become an international dumping ground. *Literaturnaya Gazeta*, 3 March 1993, p. 13. Translated from the Russian in *Foreign Broadcast Information Service* JPRS-TEN-93-08, 31 March 1993, pp. 22–23.

Benedict, M., T. Pigford, and H. W. Levi. 1981. *Nuclear Chemical Engineering*. 2d ed. New York: McGraw-Hill.

Benes, J., ed. 1991. *Environment of the Czech Republic*. Prague: Prace.

Beral, V., P. Fraser, L. Carpenter, M. Booth, and A. Brown. 1988. Mortality of employees of the Atomic Weapons Establishment, 1951–82. *British Medical Journal*, vol. 297, pp. 757–770.

Beral, V., H. Inskip, P. Fraser, M. Booth, D. Coleman, and G. Rose. 1985. Mortality of employees of the United Kingdom Atomic Energy Authority, 1946–1979. *British Medical Journal*, vol. 291, pp. 440–447.

Berkhout, F. 1991. *Radioactive Waste: Politics and Technology*. London: Routledge.

Berkhout, F., A. Diakov, H. Feiveson, M. Miller, and F. von Hippel. 1992. Plutonium: true separation anxiety. *Bulletin of the Atomic Scientists*, November 1992, pp. 28–34.

Biychaninova, A. 1989. Information on the situation developing in Tomsk in connection with the functioning of the atomic industrial plant. Translated from the Russian. N.P., December 1989.

BNFL (British Nuclear Fuels Limited). 1991. *BNFL's submission to Her Majesty's Inspectorate of Pollution for the review of Springfields Liquid Discharge Authorisation*. Risley, England, January 1991.

BNFL (British Nuclear Fuels Limited). 1992. *BNFL Newsletter* (Risley, England). 14 May 1992.

Bobrova, M. 1990. Big secret for a small company (in Russian). *Pravda Ilycha*, 30 March 1990.

Bolshakov, V., R. Aleksakhin, L. Bolshov, L. Kochetkov, A. Tsyb, V. Chukanov, and V. Petukhov. 1991. Conclusions of the Commission for Estimating the Ecological Situation in the Neighborhood of the Mayak Production Complex of the USSR Ministry of Atomic Power and Industry (in Russian). Report of commission appointed by the Presidium of the USSR Academy of Sciences, Resolution 1140–501, N.p., 12 June 1990.

Bolsunovsky, A. 1992. Russian nuclear materials production and environmental pollution. Draft paper presented at CIS Nonproliferation Project Conference on The Proliferation Predicament in the Former Soviet Union. Monterey, California, 6–9 April 1992.

Bolsunovksy, A. 1993. Telephone conversation with A. Donnay (IEER). 3 February 1993.

Boltyanskaya, P. 1991. Who made the bomb? *Literaturnaya Gazeta*, 10 April 1991, p. 5. Translated from the Russian in *JV Dialogue: Soviet Press Digest*, 10 April 1991.

Bradley, D. 1991. Travel to Russia to conduct technology exchange workshops as part of the DOE U.S./U.S.S.R. Joint Coordinating Committee on Environmental Restoration and

Waste Management. Foreign travel report for the U.S. Department of Energy. N.p., 11 November 1991. Photocopy.

Brandom, W. F., L. McGavran, R. W. Bistline, and A. D. Bloom. 1990. Sister chromatid exchanges and chromosome aberration frequencies in plutonium workers. *International Journal of Radiation Biology*, vol. 58, no. 1, pp. 195–207.

Breier, D. 1992. Country faced with disposal of weapons-grade uranium. *Sunday Star* (Johannesburg, South Africa), 27 December 1992, p. 9. Reprinted in *Foreign Broadcast Information Service* JPRS-TEN-93-006, 25 March 1993, p. 2.

Brilliant, M. D., et al. 1990. Determining the accumulated dose of gamma radiation in the tooth enamel (in Russian). *Gematologiia i Transfuziologiia*, vol. 35, pp. 11–16.

Bromley, D., and P. Perrole, eds. 1980. *Nuclear Science in China*. Washington, D.C.: National Academy Press.

Brown, E. G. 1980. Sociological problems encountered in western boom towns. In *Health Implications of New Energy Technologies*, Rom, W. N., and V. E. Archer, eds. pp. 687–695. Ann Arbor, MI: Ann Arbor Science.

Brown, P. A. 1991. The risk business. *SCRAM, The Safe Energy Journal*, No. 83, June/July 1991.

Brown, P. A., ed. 1993. *Ribble Roundup* (Paul A. Brown, Southport, Merseyside), issue 1, May 1993.

Bukharin, O. 1992. The threat of nuclear terrorism and the physical security of nuclear installations and materials in the former Soviet Union. Occasional Paper no. 2. Monterey, California: Monterey Institute of International Studies, Center for Russian and Eurasian Studies.

Bukharin, O. 1993. The structure and the production capabilities of the nuclear fuel cycle in the countries of the former Soviet Union. PU/CEES no. 274, Center for Energy and Environmental Studies, Princeton University, January 1993.

Buldakov, L. A. 1989. Interview by Arjun Makhijani. Interpreting provided by the Soviet Committee of Physicians for the Prevention of Nuclear War. Moscow, 7 December 1989.

Buldakov, L. A. 1991. Radiobiology and radiation hygiene. *Radiobiologia*, vol. 31, pp. 527–36.

Buldakov, L. A., C. N. Demin, V. A. Kostyuchenko, N.A. Koshurnikova, L. Krestinina, M. M. Saurov, I. A. Ternovsky, Z. B. Toskorskaya, and V. L. Shvedov. 1989. Medical consequences of the radiation accident in the southern Urals in 1957. IAEA International symposium on recovery operations in the event of a nuclear accident or radiological emergency. Vienna, 6 November 1989.

Buldakov, L. A., C. N. Demin, V. A. Kostyuchenko, et al. 1990. Medical consequences of the radiation accident in the southern Urals 1957 (in Russian). *Meditsinskaia Radiologiia*, vol. 36, pp. 38–43.

Buldakov, L. A., A. M. Vorobiev, V. V. Kopaev, A. Koshurnikova, A. F. Lyzlov, A. V. Simakov, and V. M. Chistokhin. 1991. The irradiation of the personnel of industrial and power-generating atomic reactors (in Russian). *Meditsinskaia Radiologiia*, 27 July 1990, pp. 38–43.

Bunyan, T. 1977. *The History and Practice of the Political Police in Britain*. London: Quartet Books.

Burke, P., ed. 1988. *The Nuclear Weapons World: Who, How & Where*. London: Oxford Research Group.

Burnazyan, A., ed. 1990. A case of accidental regional contamination by uranium fission products: Study results and cleanup experience. Report translated from the Russian by Frank C. Farnham Company, Inc., for Lawrence Livermore National Laboratory, Livermore, California. UCRL Report TT-106911, July 1991.

Business India. 1978. Nuclear power in India: A white elephant? *Business India*, 4–17 September 1978, pp. 20–35.

California Engineer. 1990. Editorial. Vol. 68, no. 3, March 1990, p. 23.

Campaign against the Namibian uranium contracts. 1986. Namibia: A contract to kill. London: Namibia Support Committee.

Canada, National Film Board. 1991. Uranium: A discussion guide. Montreal.

Carnegie Endowment for International Peace and Monterey Institute of International Studies. 1994. Nuclear successor states of the Soviet Union. Nuclear Weapons and Sensitive Export Report, no. 2, December 1994.

Carpenter, A. V., W. D. Flanders, P. Cole, and S. A. Fry. 1987a. Brain cancer and nonoccupational risk factors: A case-control study among workers at two nuclear facilities. *American Journal of Public Health*, vol. 77, no. 9, pp. 1180–1182.

Carpenter, A. V., W. D. Flanders, E. L. Frome, D. J. Crawford-Brown, and S. A. Fry. 1987. CNS cancers and radiation exposure: A case-control study among workers at two nuclear facilities. *Journal of Occupational Medicine*, vol. 29, no. 7, pp. 601–604.

Carter, L. 1987. *Nuclear Imperatives and Public Trust: Dealing with Radioactive Waste*. Washington, D.C.: Resources for the Future.

Cate, S., A. J. Ruttenber, and A. W. Conklin. 1990. The feasibility of an epidemiologic study of thyroid disease in persons exposed to environmental releases of radioiodine from the Hanford nuclear facility. *Health Physics*, vol. 59, pp. 169–178.

CDI (Center for Defense Information). 1993. Nuclear weapons after the Cold War: Too many, too costly, too dangerous. *Defense Monitor* (CDI, Washington, D.C.), vol. 22, no. 1, p. 2.

CFDT (Confédération Française Démocratique du Travail). 1984. Marcoule et son avenir. *Gazette Nucléaire*, no. 59/60, January/February 1984, pp. 2–11.

CGT (Confédération Général du Travail). 1989. Déclaration de la CGT à la Commission locale d'information de la centrale nucléaire de Gravelines. *Gazette Nucléaire*, no. 98/99, December 1989, pp. 19–22.

Chambley, R. 1993. Interview with Rebecca Johnson. 26 February 1993.

Chang, H. 1989. Nuclear energy program makes further progress: Power plant imports discussed. *China Daily*, 14 December 1989, p. 1. Reprinted in *Foreign Broadcast Information Service* JPRS-TND-90-001, 4 January 1990, p. 3.

Charles, D. 1992. Will Russia open its secret cities? *New Scientist*, 15 August 1992, p. 4.

Charles, D., M. Jones, and J. R. Cooper. 1990. Memorandum: Radiological impact on EC member states of routine discharges into north European waters—Report of Working Group IV of CEC Project MARINA. Didcot, Oxon, England: National Radiological Protection Board.

Chayes, A., and W. B. Lewis, eds. 1977. *International Arrangements for Nuclear Fuel Reprocessing*. Cambridge, Massachussetts: Ballinger Publishing Company.

Checkoway, H., R. M, Mathew, C. M. Shy, J. E. Watson, W. G. Tankersley, S. H. Wolf, J. C. Smith, and S. A. Fry. 1985. Radiation, work experience and cause-specific mortality among workers at an energy research laboratory. *British Journal of Industrial Medicine*, vol. 42, pp. 525–533.

Checkoway, H., N. E. Pearce, D. J. Crawford-Brown, and D. L. Cragle. 1988. Radiation doses and cause-specific mortality among workers at a nuclear materials fabrication plant. *American Journal of Epidemiology*, vol. 127, no. 2, pp. 255–266.

Chelnokov, A. 1992. Tomsk-7: What the film "The Resident's Mistake" did not tell about. *Komsomolskaya Pravda*, 14 July 1992. Translated from the Russian in *Foreign Broadcast Information Service* JPRS-TEN-92-016, 3 September 1992, pp. 61–64.

ChemRisk. 1993. Identification of important environmental pathways for materials released from the Oak Ridge Reservation. Draft Report of Project Tasks 3 and 4 prepared for the Tennessee Department of Health Division of Environmental Epidemiology. McLaren/Hart Environmental Engineering, Alameda, CA, June 1993.

ChemRisk. 1994. Estimating historical emissions from Rocky Flats. Project Task 5 Report prepared for the Colorado Department of Health. McLaren/Hart Environmental Engineering, Alameda, CA, March 1994.

Cherepanov, A. 1993. Confession on an atomic subject. *Rossiya* (Moscow), no. 14, 31 March–6 April 1993, p. 9. Translated from the Russian in *Foreign Broadcast Information Service* JPRS-TEN-93-009, 9 April 1993, pp. 25–27.

Chernoff, A. (Project Manager, U.S. DOE Uranium Mill Tailings Remediation Action Program). 1993. DOE presentation on UMTRA nationwide. Briefing for Federal Republic of Germany Bundestag Delegation. Albuquerque, New Mexico, 9 June 1993.

Chernykh, A. 1992. Where death should be buried. *Rossiyskaya Gazeta* (Moscow), 3 November 1992, p. 3. Translated from the Russian in *Foreign Broadcast Information Service* JPRS-TEN-92-022, 17 December 1992, pp. 63–65.

Churchill, W. 1953. *Triumph and Tragedy*. Boston: Houghton Mifflin Co.

Chykanov, V., Y. Drozhko, A. Kuligin, G. Mesyats, A. Penyagin, A. Trapeznikkov, and P. Bolbuev. 1991 Ecological conditions for the creation of atomic weapons at the atomic industrial complex near the city of Kyshtym. Paper presented at the Conference on the Environmental Consequences of Nuclear Weapons Development, 11–14 April 1991. University of California, Irvine.

Clapp, R. (Director of the Center for Environmental Health Studies, John Snow, Inc., Boston). 1993. Telephone communication with Katherine Yih, 24 September 1993.

Coates, K., ed. 1986. *China and the Bomb*. Nottingham, England: Spokesman.

Cochran, T. B., W. M. Arkin, and M. M. Hoenig. 1984. *Nuclear Weapons Databook*. Vol. 1, *U.S. Nuclear Forces and Capabilities*. Cambridge, Massachusetts: Ballinger.

Cochran, T. B., W. M. Arkin, R. S. Norris, and M. M. Hoenig. 1987a. *Nuclear Weapons Databook*. Vol. 2, *U.S. Nuclear Warhead Production*. Cambridge, Massachusetts: Ballinger.

Cochran, T. B., W. M. Arkin, R. S. Norris, and M. M. Hoenig. 1987b. *Nuclear Weapons Databook*. Vol. 3, *U.S. Nuclear Warhead Facilities Profiles*. Cambridge, Massachusetts: Ballinger Publishing Company.

Cochran, T. B., W. M. Arkin, R. S. Norris, and J. I. Sands. 1989. *Nuclear Weapons Databook*. Vol. 4, *Soviet Nuclear Weapons*. New York: Harper & Row.

Cochran, T. B., and R. S. Norris. 1990. Soviet nuclear warhead production. Nuclear Weapons Databook Working Papers, NWD 90–3 (2d rev.). Washington, D.C.: Natural Resources Defense Council.

Cochran, T. B., and R. S. Norris. 1992. Russian/Soviet nuclear warhead production. Nuclear Weapons Databook Working Papers, NWD 92-4. Washington, D.C.: Natural Resources Defense Council.

Cochran, T. B., and R. S. Norris. 1993. Russian/Soviet nuclear warhead production. Nuclear Weapons Databook Working Papers, NWD 93-1. Washington, D.C.: Natural Resources Defense Council.

Cogéma. 1988. *Retraitement.* Velizy-Villacoublay, France: Cogéma.

Cogéma. 1989. Rejects d'UP2. *Gazette Nucléaire,* No. 98/99, December 1989, pp. 21–22.

Cognetti, G. 1990. Technetium in the sea. *Marine Pollution Bulletin,* vol. 21, p. 409.

Cohen, A. 1993. A sacred matter: Israel's nuclear project was not for national glory or prestige. It was perceived as a sacred matter of national survival. *Bulletin of the Atomic Scientists,* June 1993, pp. 39–41.

Coker, C. 1988. *Less important than opulence: The conservatives and defence.* Institute for European Defence and Strategic Studies.

Coll, S. and P. Taylor. 1993. Tracking S. Africa's elusive A-program. *Washington Post,* 18 March 1993, p. A1.

Commission for Investigation of the Ecological Situation in the Chelyabinsk Region. 1991. Proceedings of the Commission for Investigation of the Ecological Situation in the Chelyabinsk Region (in Russian). Vol. 1–2. By decree #P 1283 of the President of the USSR, 3 January 1991.

Cook-Mozaffari, P., et al. 1989. Geographical variation in mortality from leukaemia and other cancers in England and Wales in relation to proximity to nuclear installations 1969–78. *British Journal of Cancer,* vol. 59, pp. 476–485.

Coyle, D., L. Finaldi, E. Greenfield, M. Hamilton, E. Hedemann, W. McDonnell, M. Resnikoff, J. Scarlott, and J. Tichenor. 1988. Deadly defense: Military radioactive landfills. Radioactive Waste Campaign, New York, N.Y.

Cragle, D. L., D. R. Hollis, J. R. Qualters, W. G. Tankersley, and S. A. Fry. 1984. A mortality study of men exposed to elemental mercury. *Journal of Occupational Medicine,* vol. 26, no. 11, pp. 817–821.

Cragle, D. L., D. R. Hollis, C. M. Shy, and T. H. Newport. 1984a. A retrospective cohort mortality study among workers occupationally exposed to metallic nickel powder at the Oak Ridge Gaseous Diffusion Plant. In *Nickel in the Human Environment,* F. W. Sunderman, et al., eds, pp. 57–63. IARC Scientific Publications, no. 53. Lyon, France: International Agency for Research on Cancers.

Cragle, D. L., R. W. McLain, J. R. Qualters, J. L. S. Hickey, G. S. Wilkinson, W. G. Tankersley, and C. C. Lushbaugh. 1988. Mortality among workers at a nuclear fuels production facility. *American Journal of Industrial Medicine,* vol. 14, no. 4, pp. 379–401.

Crigger, B. 1994. Yes, they knew better. *Bulletin of the Atomic Scientists,* vol. 50, no. 2., March/April 1994, p. 29.

CRII-Rad (Commission de recherche et d'information independantes sur la radioactivité). 1987. Stockage de matières radioactives a istres: la population se mobilise. *Le Cri du Rad,* no. 6, December 1987, p. 42.

CRII-Rad (Commission de recherche et d'information indépendantes sur la radioactivité). 1989. Dossier rejets. *Le Cri du Rad,* 1st Trimester 1989, pp. 36–48.

CRII-Rad (Commission de recherche et d'information indépendantes sur la radioactivité). 1990. Pollutions radioactives a Saint-Aubin. *Le Cri du Rad,* Autumn 1990, pp. 4–16.

CRII-Rad (Commission de recherche et d'information indépendantes sur la radioactivité). 1992. Centre nucléaire de Marcoule: Une pollution a très long terme. *Le Rem, Bulletin d'information des adhérents de la CRII-RAD*, June 1992, pp. 1–4.

CRII-Rad (Commission de recherche et d'information indépendantes sur la radioactivité). 1992a. Que faire des déchets radioactifs? *Le Rem, Bulletin d'information des adhérents de la CRII-RAD*, October 1992, pp. 1–4.

Cross, F. T., R. F. Palmer, R. H. Busch, et al. 1981. Development of lesions in Syrian golden hamsters following exposures to radon daughters and uranium ore dust. *Health Physics*, vol. 41, pp. 135–153.

Crump, K. S., T-H. Ng, and R. G. Cuddihy. 1987. Cancer incidence patterns in the Denver metropolitan area in relation to the Rocky Flats Plant. *American Journal of Epidemiology*, vol. 126, no. 1, pp. 127–135.

Cutter Information Corp. 1992. Local disposal of foreign nuclear waste favored: Add it up yourself. *Environment Watch: East Europe, Russia and Eurasia*, November 1992, p. 15.

Dabrowski, T. 1979. Memo from T. E. Dabrowski (UNI Nuclear Industries) to J. W. Riches. 24 May 1979.

Dahlburg, J. T. 1992. The atom sows crop of sadness. *Los Angeles Times*. Washington Edition. 2 September 1992, pp. A1, A3.

Dangdai Zhongguo Series Editorial Committee, ed. 1987. *Modern China's Nuclear Industry*. Beijing: Dangdai Zhongguo. Excerpts translated from the Chinese in *Foreign Broadcast Information Service* JPRS-CST-88-002, 15 January 1988, pp. 1–46.

Darby, S. 1986. Epidemiological evaluation of radiation risk using populations exposed at high doses. *Health Physics*, vol. 51, pp. 269–81.

Davis, M. D. 1988. *The Military-Civilian Nuclear Link: A Guide to the French Nuclear Industry*. Boulder, Colorado: Westview Press.

Davis, M. 1992. French radioactive waste management: A preliminary overview. The Foundation for Global Sustainability, Knoxville, Tennessee.

Dayton, L. 1991. Ways to make the old mining territory safe. *New Scientist*, 22 June 1991, p. 45.

de Klerk, F. W. 1993. The Nuclear Non-Proliferation Treaty and related matters. Text of speech to a joint session of the South African Parliament, 24 March 1993.

de la Court, T., D. Pick, and D. Nordquist. 1982. *The Nuclear Fix: A Guide To Nuclear Activities in The Third World*. Amsterdam: WISE Publications.

Del Tredici, R. 1987. *At Work in the Fields of the Bomb*. New York: Harper & Row.

Deutsches Atomforum e.V., Bonn (Germany). 1991. The radiological situation in the uranium mining of the SDAG Wismut—taking stock. *Inforum Verl.* (Bonn), no. 27.

Dickman, S. 1991. Uranum mining's legacy. *Nature*, vol. 352, p. 8.

Dircks, W. (Executive Director for Operations, U.S. Nuclear Regulatory Commission). 1981. Memorandum to the Commissioners. Disposal or onsite storage of residual thorium or uranium (either as natural ores or without daughters present) from past operations. SECY 81–576. Washington, D.C., 5 October 1981.

Directorate of Fisheries Research, MAFF (Ministry of Agriculture, Fisheries and Food). 1992. Aquatic environment monitoring reports (years 1977 to 1992). Lowestoft, England: MAFF.

Dirschl, H. J., N. S. Novakowski, and L. C. N. Burgess. 1992. An overview of the biophysical environmental impact of existing uranium mining operations in Northern Saskatchewan. ESAS, Inc., Ottawa, Canada.

Dogudjiev, V. H. (USSR Council of Ministers). 1990. Letter to A. Biychaninova (in Russian), 12 February 1990.

Doll, R., H. J. Evans, and S. C. Darby. 1994. Paternal exposure not to blame. *Nature*, vol. 367, 24 February 1994.

Doschenko, V. 1991. *The Truth about Radiation* (in Russian). Moscow: Transport Publishing House.

Dousset, M. 1989. Cancer mortality around La Hague nuclear facilities. *Health Physics*, vol. 56, no. 6, June 1989, pp. 875–884.

Dousset M., and H. Jammet. 1983. La mortalité par affections malignes dans le département de la Manche (1960–1982) *Radioprotection*, GEDIM, 1983, vol. 18, no. 4, pp. 223–232.

Draper, G. J., C. A. Stiller, R. A. Cartwright, A. W. Craft, and T. J. Vincent. 1993. Cancer in Cumbria and in the vicinity of the Sellafield nuclear installation, 1963–90. *British Medical Journal*, vol. 306, pp. 89–94.

Dreckmann, H. 1993. [High radiation recorded at Zagorsk nuclear disposal site.] Transcript of report broadcast by ARD Television Network (Hamburg, Germany), 2130 GMT, 3 February 1993. Translated from the German in *Foreign Broadcast Information Service* JPRS-TEN-93-004, 8 March 1993, p. 54.

Dropkin, G., and D. Clark. 1992. Past exposure: Revealing health and environmental risks of Rossing uranium. Partizans, London.

Dubois, C. 1988. Démantèlement réussi à Marcoule. *Cogémagazine*, December 1988, pp. 17–19.

Dunayeva, L. 1992. [Government approves program to investigate radiation levels.] Transcript of report broadcast by ITAR-TASS World Service, 1234 GMT, 10 June 1992. Translated from the Russian in *Foreign Broadcast Information Service* JPRS-TEN-92-014, 9 July 1992, p. 65.

Dupree, E. A., D. L. Cragle, R. W. McLain, D. G. Crawford-Brown, and M. J. Teta. Mortality among workers at a uranium processing facility: The Linde Air Products Company ceramics plant: 1943–1949. *Scandinavian Journal of Work, Environment and Health*, vol. 13, pp. 100–107.

Durbin, P. 1994. Testimony before the House Subcommittee on Energy Conservation and Power, and House Committee on Energy and Commerce, U.S. House of Representatives, 18 January 1994.

Dzeylton, Tom. 1984. Wollaston Lake people speak out against mining. *Akwesasne Notes*, Fall, pp. 8–9.

Economic and Political Weekly. 1990. Clinging to a forgotten dream. *Economic and Political Weekly*, 15 December 1990, p. 2692.

Economic and Political Weekly. 1992. Limited options. *Economic and Political Weekly*, 3 October 1992, p. 2139.

Economic and Political Weekly. 1992a. Nuclear Power: Unhealthy? *Economic and Political Weekly*, 4–11 January 1992, p. 6.

Economic and Political Weekly. 1994. Nuclear safety: Secrecy is poor defence. *Economic and Political Weekly*, 12 March 1994.

ECOTASS. 1991. Kazakhstan closes uranium mines. *ECOTASS*, 9 December 1991, p. 6. Translated from the Russian in *JV Nova*, 9 December 1991.

EDF (Eléctricité de France). 1989. Bilan économique du nucléaire en France. Brochure of Eléctricité de France, no. JA 5195. Paris.

Edwards, G. (founder of the Canadian Coalition for Nuclear Responsibility). 1985. Fuelling the arms race: Canada's nuclear trade. *Ploughshares Monitor*, vol. 6, no. 2. Project Ploughshares, Waterloo.

Edwards, G. 1993. Unpublished writings on uranium and Canada's nuclear industry sent to K. Yih, August 1993.

Ege, K., and A. Makhijani. 1982. U.S. nuclear threats: A documentary history. *CounterSpy*, vol. 6 no. 4, July-August 1982, pp. 8–23.

Eisenbud, M. 1987. *Environmental Radioactivity*. 3d ed. Orlando, Florida: Academic Press.

Ellsberg, D. 1986. Call to mutiny. In *The Deadly Connection: Nuclear War and U.S. Intervention*, Joseph Gerson, ed., pp. 36–59. Philadelphia: New Society Publishers.

Energizdat. 1981. Radiation safety norms RSN-76 and basic sanitary regulations BSR-72/80 (in Russian). Moscow.

Energy Research Foundation. 1993. The future of production activities at SRS. Fact sheet. Columbia, South Carolina, July 1993.

Energy Research Foundation. 1993a. Radioactive and hazardous waste at the Savannah River Site. Fact sheet. Columbia, South Carolina, February 1993.

Energy Research Foundation. 1993b. SRS reactor restart cancelled. Fact sheet. Columbia, South Carolina, July 1993.

Erickson, D. M. 1981. *Uranium Development in Less Developed Countries: A Handbook of Concerns*. Lincoln Institute Monograph 81-4.

Erickson, K. 1994. Out of site out of mind. *New York Times Magazine*, 6 March 1994, p. 36.

Etzold, T., and J. Gaddis. 1979. *Containment: Documents on American Foreign Policy and Strategy 1945–50*. New York: Columbia University Press.

Evans, D. 1993. Big uranium deal in works. *Chicago Tribune*, 7 February 1993, p. A3.

Eyster, G. 1944. Memorandum to Chief, C.W.S. ETOUSA, Subject: Operation "PEPPERMINT", 1 May 1944 with attachments. Correspondence ("Top Secret") of the Manhattan Engineer District 1942–1946. Roll 1109, no. 1, folder 7D, National Archives, Washington, D.C, pp. 0803–0850.

Fadeley, R. C. 1965. Oregon malignancy pattern physiographically related to Hanford, Washington radioisotope storage. *Journal of Environmental Health*, vol. 27, pp. 883–897.

Fairlie, I. 1992. Tritium: The overlooked nuclear hazard. *Ecologist*, vol. 22, pp. 228–232.

Faison, S. 1989. First nuclear waste sites to be built. *South China Morning Post*, 20 February 1989, pp. 1–2. Reprinted in *Foreign Broadcast Information Service* JPRS-TND-89-005, 20 March 1989, p. 1.

Feshbach, M., and A. Friendly, Jr. 1992. *Ecocide in the USSR*. New York: Basic Books.

Fieldhouse, R. 1991. Chinese nuclear weapons: A current and historical overview. Nuclear Weapons Databook Working Papers, NWD-91-1. Natural Resources Defense Council, Washington, D.C.

Fieldhouse, R. 1991a. China's mixed signals on nuclear weapons. *Bulletin of the Atomic Scientists*, May 1991, pp. 37–42.

Filin, V. 1992. Nuclear bomb assembly technology: Yardman Minayev speaks (in Russian). *Moscow Komsomoloskaya Pravda*, 6 February 1992.

Findlay, T. 1990. *Nuclear Dynamite: The Peaceful Nuclear Explosions Fiasco.* Rushcutters Bay, New South Wales: Brassey's Australia.

Fleming, R. M., and New, C. B. (National Institute for Occupational Safety and Health). 1981. A plan to reevaluate risks to miners from radiation exposure. In *Radiation Hazards in Mining*, Manuel Gómez, ed. New York: Society of Mining Engineers of American Institute of Mining, Metallurgical, and Petroleum Engineers, Inc.

Fonotov, M. 1993. Who is closer to the atom? The nuclear shield is heavy and dangerous. *Chelyabinskiy Rabochiy* (Chelyabinsk), 6 March 1993, p. 2. Translated from the Russian in *Foreign Broadcast Information Service* FBIS-TEN-93-013, 14 May 1993, p. 15.

Fox, A. J., P. Goldblatt, and L. J. Kinlen. 1981. A study of the mortality of Cornish tin miners. *British Journal of Industrial Medicine*, vol. 38, pp. 378–380.

Fradkin, P. 1989. *Fallout: An American Nuclear Tragedy*. Tuscon: University of Arizona Press.

Franke, B., and K. R. Gurney. 1994. Estimates of lung burdens for workers at the Feed Materials Production Center, Fernald, Ohio. Takoma Park, Maryland: Institute for Energy and Environmental Research, June 1994.

Franke, B., K. R. Gurney, A. Makhijani, and M. M. Hoenig. 1992. Uranium doses to workers at the Feed Materials Production Center—Six case studies. Takoma Park, Maryland: Institute for Energy and Environmental Research, 23 December 1992.

Friedlander, G., J. Kennedy, E. Macias, and J. Miller. 1981. *Nuclear and Radiochemistry*. 3d ed. New York: John Wiley and Sons.

Friends of the Earth. 1991. Profits before safety: Radioactive discharges from the British Nuclear Fuels Springfields Works. Friends of the Earth, London, November 1991.

Frumpkin, H. 1995. Carcinogens. In *Occupational Health: Recognizing and Preventing Work-Related Disease*, ed. B. Levy and D. Wegman, chapter 14. New York: Little Brown.

Fry, R., and M. Carter. 1992. A career dose limit for underground uranium miners. Paper I1–1. In Proceedings of the International Conference on Radiation Safety in Uranium Mining, Saskatoon, Canada, 1992.

Fry, R. M., and I. W. Morison. 1982. Regulation of the management of waste from uranium mining and milling in Australia. In *Management of Waste from Uranium Mining and Milling*, pp. 42–54. Vienna: International Atomic Energy Agency.

Galinka, A. 1990. Radiation threatens to waste Chelyabinsk (in Russian). *Glasnost*, 12 November 1990, p. 38.

Gardner, M. J., M. P. Snee, A. J. Hall, C. A. Powell, S. Downes, and J. D. Terrell. 1990. Results of case-control study of leukaemia and lymphoma among young people near Sellafield nuclear plant in West Cumbria. *British Medical Journal*, vol. 300, pp. 423–429.

Gattis, B. 1993. Personal written communication with Ellen Kennedy (IEER), 4 August 1993.

Gazette Nucléaire. 1983. Incident à Marcoule le 22 mars 1983. *Gazette Nucléaire*, no. 53, June/July 1983, pp. 13–14.

Gazette Nucléaire. 1988. Nouvelles en provenance de la commission spéciale et permanente d'information près l'établissment de la Hague. *Gazette Nucléaire*, no. 90/91, November 1988, pp. 8–14.

Gazette Nucléaire. 1991. Les Stériles des mines. *Gazette Nucléaire*, no. 111/112, November 1991, pp. 14–19.

Gazette Nucléaire. 1991a. Les problèmes posés par la fermeture d'une mine d'uranium. *Gazette Nucléaire*, no. 111/112, November 1991, pp. 2–5.

Gazette Nucléaire. 1991b. Anciens sites ou installations contenant des substances radioactives. *Gazette Nucléaire*, no. 107/108, April 1991, pp. 13–14.

Gazette Nucléaire. 1991c. Saint Aubin. *Gazette Nucléaire*, no. 105/106, January 1991, pp. 11–16.

Gazette Nucléaire. 1991d. Centre du Bouchet. *Gazette Nucléaire*, no. 105/106, January 1991, pp. 4–10.

Gazette Nucléaire. 1991e. Etude de faisabilité de la saisine sur les sites de stockage du CEA. *Gazette Nucléaire*, no. 107/108, April 1991, pp. 10–12.

Gazette Nucléaire. 1991f. Site "Hague." *Gazette Nucléaire*, no. 107/108, April 1991, pp. 29–30.

Gazette Nucléaire. 1991g. Annexe No. 1: Reprise et conditionnement des coques et embouts stockés dans l'usine UP2 400. *Gazette Nucléaire*, no. 107/108, April 1991, pp. 27–28.

Gazette Nucléaire. 1991h. La radioprotection dans les mines d'uranium, la législation. *Gazette Nucléaire*, no. 111/112, November 1991, pp. 10–11.

Gazette Nucléaire. 1991i. La reglementation des mines d'uranium pour la protection de l'environment. *Gazette Nucléaire*, no. 111/112, November 1991, p. 13.

Gazette Nucléaire. 1992. Petit Glossaire. *Gazette Nucléaire*, no. 121/122, December, p. 2.

Gazette Nucléaire. 1992a. Rapport Desgraupes. *Gazette Nucléaire*, no. 113/114, March 1992, pp. 18–26.

Gazette Nucléaire. 1992b. Dérnière minute, bis, 27-1-92. *Gazette Nucléaire*, no. 113/114, March 1992, p. 31.

Geiger, H. J., D. Rush, D. Michaels, D. B. Baker, J. Cobb, E. Fischer, A. Goldstein, H. S. Kahn, J. L. Kirsch, P. J. Landrigan, E. Mauss, and D. E. McLean. 1992. *Dead reckoning: A critical review of the Department of Energy's epidemiologic research*. Physicians for Social Responsibility, Washington, D.C.

Gilbert, E. S. 1991. Studies of workers exposed to low doses of external radiation. In *Occupational Medicine: State of the Art Reviews*, vol. 6, no. 4, October–December 1991, pp. 665–680.

Gilbert, E. S., D. L. Cragle, and L. D. Wiggs. 1993. Updated analyses of combined mortality data for workers at the Hanford Site, Oak Ridge National Laboratory, and Rocky Flats Weapons Plant. *Radiation Research*, vol. 136, pp. 408–421.

Gilbert, E. S., S. A. Fry, L. D. Wiggs, G. L. Voelz, D. L. Cragle, and G. R. Peterson. 1989. Analyses of combined mortality data on workers at the Hanford Site, Oak Ridge National Laboratory, and Rocky Flats Nuclear Weapons Plant. *Radiation Research*, vol. 120, pp. 19–35.

Gilbert, E. S., and S. Marks. 1979. An analysis of the mortality of workers in a nuclear facility. *Rad Research*, vol. 79, pp. 122–148.

Gilbert, E. S., and S. Marks. 1980. An updated analysis of mortality of workers in a nuclear facility. *Radiation Research*, vol. 83, pp. 740–741.

Gilbert, E. S., G. R. Petersen, and J. A. Buchanan. 1989. Mortality of workers at the Hanford site: 1945–1981. *Health Physics*, vol. 56, pp. 11–25.

Gilles, C., M. Reed, and J. Seronde. 1990. Our uranium legacy. (Available from Southwest Research and Information Center, Albuquerque, New Mexico.)

Gillespie, A. 1992. Current Third World Nuclear Programmes. In *Plutonium and Security: The Military Aspects of the Plutonium Economy*. Ed. F. Barnaby, pp. 133–153. New York: St. Martin's Press.

Gofman, J. W. 1979. The question of radiation causation of cancer in Hanford workers. *Health Physics*, vol. 37, pp. 617–39.

Goldanskii, V. 1993. Russia's "red-brown hawks." *Bulletin of the Atomic Scientists*, June 1993, pp. 24–27.

Goldman, M. I. 1972. *Environmental Pollution in the Soviet Union: The Spoils of Progress*. Cambridge, Massachusetts: MIT Press.

Goldschmidt, B. 1990. *Atomic Rivals*. New Brunswick, New Jersey: Rutgers University Press.

Goldsmith, J. R. 1989a. Childhood leukemia mortality before 1970 among populations near two US nuclear installations. *Lancet*, vol. i, pp. 793.

Goldsmith, J. R. 1989b. Childhood leukemia mortality before 1970 among populations near two US nuclear installations. *Lancet*, vol. ii, pp. 1443–1444.

González, A. J., B. G. Bennett, and G. A. M. Webb. 1993. Mission report: Radiological accident at Tomsk-7. Vienna: IAEA International Atomic Energy Agency, 6 April 1993.

Goskomstat. 1990. Public health in the USSR: Statistical yearbook (in Russian). Moscow.

Gotlieb, L. S., and L. A. Husen. 1982. Lung cancer among Navajo uranium miners. *Chest*, vol. 81, April 1982, pp. 449–452.

Government Commission on the Questions Related to the Dumping of Radioactive Waste at Sea. 1993. Facts and problems related to the dumping of radioactive waste in the seas surrounding the territory of the Russian Federation. (Translated from the Russian by Greenpeace Russia). Greenpeace, Moscow.

Government Statistical Service. 1991. *Digest of Environmental Protection and Water Statistics*. No. 13. London: Her Majesty's Stationery Office.

Gowing, M. (with L. Arnold) 1974. *Independence and Deterrence*. 2 volumes. Basingstoke, Hampshire, England: Macmillan.

Goyer, R. A. 1991. Lead toxicity: Current concerns. *Environmental Health Perspectives*, vol. 100, pp. 177–187.

Green, F. H. Y., R. Althouse, J. L. Frost, et al. 1986. Forensic investigation of coal mine fatalities. *Annals of the American Conference of Governmental Industrial Hygienists*, vol. 14, p. 117.

Greenpeace. 1993. Aldermaston: Inside the Citadel. London, January 1993.

Groves, L. 1943. Actions reported by General Groves. Summary of the 5 May 1943 meeting of the Military Policy Committee of the Manhattan Project, Record Group 77, Records of the Office of the Chief of Engineers, Records of the Manhattan Engineer District, 1942–48, National Archives, Washington, D.C.

Groves, L. 1962. *Now It Can Be Told: The Story of the Manhattan Project*. New York: Harper and Row.

Gubarev, V. 1989. The nuclear trace (in Russian). *Pravda*, 25 August 1989, p. 4.

Guskova, A. K., and G. D. Baisogolov. 1985. Human radiation disease (in Russian). Moscow.

Gustavson, T. (Senior Research Scientist, University of Texas at Austin) 1993. Letter to Roger Mulder (Office of the Governor, the State of Texas), 28 July 1993.

Gydesen, C. H. 1981. Preliminary engineering study report: Dry storage of nuclear fuel. UNC Nuclear Industries, 30 January 1981.

Hahn, B. 1987. Beijing's growing global missile reach. *Pacific Defence Reporter* (Melbourne, Australia), February 1987.

Handy, T. 1945. Memorandum from General Handy to General Hull, with attachments, 1 June 1945. Manhattan Project, National Archives, Washington, D.C.

Harrison, J. E., M. S, Frommer, E. A. Ruck, and F. M. Blyth. 1989. Deaths as a result of work-related injury in Australia, 1982–1984. *Medical Journal of Australia*, vol. 150, pp. 118–125.

Health and Safety Executive. 1990. Statement of nuclear incidents. News release. London, 28 December 1990.

Health and Safety Executive. 1993. *HSE Investigation of Leukaemia and other Cancers in the Children of Male Workers at Sellafield*. London: HSE Books.

Health Physics. 1981. Health physics in the People's Republic of China. Partial list of titles published in the People's Republic of China by Members of the Chinese Radiation Protection Society. *Health Physics*, vol. 41, pp. 585–588.

Henshaw, S.L., et al. 1990. Radon as a causative factor in the induction of myeloid leukaemia and other cancers. *Lancet*, vol. 335, pp. 1008–1012.

Hersh, S. 1993. A reporter at large: On the nuclear edge. *New Yorker*, 29 March 1993, pp. 56–73.

Hertsgaard, M. 1992. From here to Chelyabinsk. *Mother Jones*, January/February 1992, pp. 51–55, 70–72.

Hewlett, R., and O. Anderson, Jr. 1990. *The New World: A History of the United States Atomic Energy Commission: Vol. I, 1939–1946*. Berkeley: University of California Press.

Hewlett, R., and F. Duncan. 1990. *Atomic Shield: A History of the United States Atomic Energy Commission: Volume II 1947–1952*. 2d ed. Berkeley, California: University of California Press.

Hibbs, M. 1993. Minatom may dispose spent RBMK fuel in Arctic permafrost at Novaya Zemlya. *Nuclear Fuel*, 18 January 1993, p. 4.

Hibbs, M. 1993a. Pu vitrification economics: technology must be improved, Russian expert says. *Nuclear Fuel*, 18 January 1993, pp. 6–7.

High-Background Radiation Research Group, China. 1980. Health survey in high background radiation areas in China. *Science*, vol. 209, pp. 877–880.

Hill, C., and A. Laplanche. 1990. Overall mortality and cancer mortality around French nuclear sites. *Nature*, vol. 347, 25 Oct. 1990, pp. 755–757.

HMIP (Her Majesty's Inspectorate of Pollution) and MAFF (Ministry of Agriculture, Fisheries and Food). 1993. Report on the public consultation by Her Majesty's Inspectorate of Pollution and the Ministry of Agriculture, Fisheries and Food (The Authorising Departments) on British Nuclear Fuels plc's applications for revision of the certificate of authorization to discharge liquid and gaseous low level radioactive wastes and for a new authorisation to dispose of combustible waste oil from the Sellafield Works at Seascale in Cumbria. August 1993.

Holaday, D. A. 1969. History of the exposure of miners to radon. *Health Physics*, vol. 16, pp. 547–552.

Holloway, D. 1984. *The Soviet Union and the Arms Race*. 2d ed. New Haven: Yale University.

Hoodbhoy, P. 1993. Myth-building: The "Islamic" bomb. *Bulletin of the Atomic Scientists*, June 1993, pp. 42–49.

Hornung, R. W., and T. J. Meinhardt. 1987. Quantitative risk assessment of lung cancer in U. S. uranium miners. *Health Physics*, vol. 52, pp. 417–430.

Howe, G. R., R. C. Nair, H. B. Newcombe, A. B. Miller, and J. D. Abbatt. 1986. Lung cancer mortality (1950–80) in relation to radon daughter exposure in a cohort of workers at the Eldorado Beaverlodge uranium mine. *Journal of the National Cancer Institute*, vol. 77, pp. 357–362.

Howe, G. R., R. C. Nair, H. B. Newcombe, A. B. Miller, J. D. Burch, and J. D. Abbatt. 1987. Lung cancer mortality (1950–80) in relation to radon daughter exposure in a cohort of workers at the Eldorado Port Radium Uranium Mine: Possible modification of risk by exposure rate. *Journal of the National Cancer Institute*, vol. 79, pp. 1255–1260.

Hughes, J. S., and G. C. Roberts. 1984. *Radiation exposure of the UK population—1984 review*. NRPB-R173. Didcot: National Radiation Protection Board, November 1984.

Hughes, J., K. Shaw, and M. O'Riordan. 1989. *Radiation exposure of the UK population—1988 review*. NRPB-R227. Didcot, Oxon, England: National Radiation Protection Board, March 1989.

Hughes, L., ed. 1992. Nuclear notes. *NACE News* (Native Americans for a Clean Environment, Tahlequah, Oklahoma), vol. 2, no. 4, July-August 1992, p. 10.

Huisman, Ruurd. 1990. The African uranium mining countries. Unpublished report of Hermes R & C, Zeist, Netherlands.

Hull, J. 1945. Memorandum from General Hull to General Eaker, 13 September 1945. Manhattan Project, National Archives, Washington, D.C.

IAEA (International Atomic Energy Commission). 1992. *Nuclear Power Reactors in the World*. Reference Data Series no. 2, IAEA-RDS-2/12. Vienna: International Atomic Energy Agency.

ICRP (International Commission on Radiological Protection). 1977. Recommendations of the International Commission on Radiological Protection. ICRP Publication 26. *Annals of the ICRP*, vol. 1, no. 3. Oxford, New York: Pergamon Press.

ICRP (International Commission on Radiological Protection). 1985. Quantitative bases for developing a unified index of harm. ICRP Publication 45. *Annals of the ICRP*, vol. 15, no. 3. Oxford, New York: Pergamon Press.

ICRP (International Commission on Radiological Protection). 1989. Age dependent doses to members of the public from intake of radionuclides: Part 1. ICRP Publication 56. *Annals of the ICRP*, vol. 20, no. 2. Oxford, New York: Pergamon Press.

ICRP (International Commission on Radiological Protection). 1991. 1990 Recommendations of the International Commission on Radiological Protection. ICRP Publication 60. *Annals of the ICRP*, vol. 21, no. 1–3. Oxford, New York: Pergamon Press.

I. F. Stone's Weekly. 1958. Why the AEC retracted that falsehood on nuclear testing. *I. F. Stone's Weekly*, vol. 6, 17 March 1958, p. 1.

Illesh, A., and V. Kostyukovskiy. 1993. Nuclear monster has spoken in Siberia: Danger has been sensed all over the world. *Izvestiya* (Moscow). Translated from the Russian in *Foreign Broadcast Information Service* JPRS-TEN-93-011, 27 April 1993, pp. 19–20.

Imai, R. 1993. Asian ambitions, rising tensions. *Bulletin of the Atomic Scientists*, June 1993, pp. 33–36.

International Campaign for Tibet. 1993. Nuclear tibet: Nuclear weapons and waste on the Tibetan Plateau. Washington, D.C., April 1993.

IPPNW (International Physicians for the Prevention of Nuclear War). 1992. Nuclear test tally 1992. *Vital Signs* (IPPNW, Cambridge, Massachusetts), vol. 5, no. 3, September 1992, p. 4.

IPPNW (International Physicians for the Prevention of Nuclear War). 1992a. Map of storage sites of high-level radioactive waste from plutonium production. Press release. Washington, D.C., 18 November 1992.

IPPNW (International Physicians for the Prevention of Nuclear War). 1992b. Nuclear weapons in the Commonwealth of Independent States: A report of the International Physicians for the Prevention of Nuclear War. IPPNW, Cambridge, Massachusetts, 24 April 1992. Photocopy.

IPPNW and IEER (International Physicians for the Prevention of Nuclear War and Institute for Energy and Environmental Research). 1991. *Radioactive Heaven and Earth: The Health and Environmental Effects of Nuclear Weapons Testing In, On, and Above the Earth.* New York: The Apex Press.

IPPNW and IEER (International Physicians for the Prevention of Nuclear War and Institute for Energy and Environmental Research). 1992. *Plutonium: Deadly Gold of the Nuclear Age.* Cambridge, Massachusetts: International Physicians Press.

Isherwood, J. 1992. Uranium fear for Baltic. *Daily Telegraph,* 3 September 1992, p. 7.

Isle of Man Local Government Board. 1977. Report to Tynwald, British Nuclear Fuels Ltd., Sellafield Operations.

ITAR-TASS. 1993. [Tomsk-7: Varying opinions on gravity of leakage.] Transcript of report broadcast by ITAR-TASS World Service, 0829 GMT, 9 April 1993. Translated from the Russian in *Foreign Broadcast Information Service* JPRS-TEN-93-011, 27 April 1993, p. 22.

Ivanov, A., M. Ashanin, and A. Nosov. 1991. Report on the radioactive situation of the Yenisey River (in Russian). Institute of Applied Geophysics, State Committee for Hydrometeorology, N.p.

Izvestiya. 1991. Untitled article. *Izvestiya* (Moscow), 13 December 1991, p. 1. Translated from the Russian in SOVECON, n.d.

Izvestiya. 1993. Burial ground for nuclear waste will be created in Maritime Kray. *Izvestiya* (Moscow), 23 March 1993, p. 1. Translated from the Russian in *Foreign Broadcast Information Service* JPRS-TEN-93-008, 31 March 1993, p. 28.

Izvestiya. 1993a. [Tomsk-7: Latest data on contamination.] Untitled report. Translated from the Russian in *Foreign Broadcast Information Service* JPRS-TEN-93-012, 3 May 1993, pp. 24–25.

Jablon, S., Z. Hrubec, and J. D. Boice. 1991. Cancer in populations living near nuclear facilities: A survey of mortality nationwide and incidence in two states. *Journal of the American Medical Association,* vol. 265, pp. 1403–1408.

Jaimes, M. A., ed. 1992. *The State of Native America: Genocide, Colonization and Resistance.* Boston: South End Press, 1992, pp. 172–175.

Jayaraman, K. S. Indo-US plutonium dispute flares up. *Nature,* vol. 363, 17 June 1993, p. 575.

Johnson, C. J. 1981. Cancer incidence in an area contaminated with radionuclides near a nuclear installation. *Ambio,* vol. 10, no. 4, pp. 176–182.

Johnson, C. J. 1984. Cancer incidence in an area of radioactive fallout downwind from the Nevada Test Site. *Journal of the American Medical Association,* vol. 251, pp. 230–236.

Joint Chiefs of Staff (U.S.). 1945. Joint War Plans Committee Details of the Campaign Against Japan. J.W.P.C. 369/1. 15 June 1945. National Archives, Washington, D.C.

Joint Chiefs of Staff (U.S.) Evaluation Board for Operation Crossroads. 1947. The evaluation of the atomic bomb as a military weapon. Final report, enclosure to Joint Chiefs of Staff Document Number JCS 1691/7, 1947. Modern Military Branch, National Archives, Record Group 218, Washington, D.C. pp. 86–87.

Joint Federal-Provincial Panel (on Uranium Mining Developments in Northern Saskatchewan). 1993. *Uranium Mining Developments in Northern Saskatchewan: Dominique-Janine Extension, McClean Lake Project, and Midwest Joint Venture*. Ottawa: Minister of Supply and Services Canada, October 1993.

Joint Federal-Provincial Panel (on Uranium Mining Developments in Northern Saskatchewan). 1993a. Transcripts of meeting. Ottawa, 7 May 1993.

Jones, K. P., and A. W. Wheater. 1989. Obstetric outcomes in West Cumberland Hospital: Is there a risk from Sellafield? *Journal of the Royal Society of Medicine*, vol. 82, pp. 524–527.

Jones, R. R., and R. Southwood, eds. 1987. *Radiation and Health*. Chichester, England: John Wiley & Sons.

Jones Williams, W. 1988. Beryllium disease. *Postgraduate Medical Journal*, vol. 64, no. 753, pp. 511–516.

Jones Williams, W., J. H. Lawrie, and H. J. Davies. 1967. Skin granulomata due to beryllium oxide. *Journal of Surgery*, vol. 54, pp. 292–297.

Jordan, J. M. (National Conference of State Legislatures). 1984. Low-level radioactive waste management: An update. October 1984.

Josephson, J. 1993. Cleanup in the Arctic. *Environmental Science and Technology*, vol. 277, no. 4, April 1993, pp. 585–586.

Kaazik, F. 1990. Emeralds instead of uranium. *Soyuz*, no. 49, December 1990, p. 10. Translated from the Russian in *JV Dialogue: Soviet Press Digest*, December 1990.

Kadhim, M. A., D. A. Macdonald, D. T. Goodhead, S. A. Lorimore, S. J. Marsden, and E. G. Wright. 1992. Transmission of chromosomal instability after plutonium alpha-particle irradiation, *Nature*, vol. 355, pp. 738–740.

Kahn, P. 1993. A grisly archive of cancer data. *Science*, vol. 259, pp. 448–451.

Kaplan, S. 1981. *Diplomacy of Power: Soviet Armed Forces as Political Instrument*. Washington, D.C.: The Brookings Institution.

Ke, Y., and R. Wang. 1987. Prospect of spent fuel reprocessing and back-end cycling in China in 1990s. In Editorial Committee of the Sixth PBNC Proceedings. In Proceedings of the Sixth Pacific Basin Nuclear Conference, Beijing, China, 7-11 September 1987, L. Rongguang and Z. Renkai, eds. Reprinted in *Transactions* (American Nuclear Society, La Grange Park, Illinois), vol. 56, supplement 1, 1988, pp. 527–530.

Kendall, G. M., et al. 1992. Mortality and occupational exposure to radiation: First analysis of the National Registry for Radiation Workers. *British Medical Journal*, vol. 304, 25 January 1992, pp. 220–225.

Kendall, G. M., C. R. Muirhead, B. H. MacGibbon, J. A. O'Hagan, A. J. Conquest, A. A. Goodhill, B. K. Butland, T. P. Fell, D. A. Jackson, M. A. Webb, R. G. E. Haylock, J. M. Thomas, and T. J. Silk. 1992a. *First Analysis of the National Registry for Radiation Workers: Occupational Exposure to Ionising Radiation and Mortality*. NRPB-R251. Didcot, Oxon, England: National Radiation Protection Board, January 1992.

Kerber, R. A., J. E. Till, S. L. Simon, J. L. Lyon, D. C. Thomas, S. Preston-Martin, M. L. Rallison, R. D. Lloyd, and W. Stevens. 1993. A cohort study of thyroid disease in relation to fallout from nuclear weapons testing. *Journal of the American Medical Association*, vol. 270, pp. 2076–2082.

Kershner, D. 1994. Some examples of radiation experiments on humans. Table compiled from various sources. *Science for Democratic Action* (IEER, Takoma Park, MD), vol. 3, no. 1, Winter 1994.

Khariton, Y., and Y. Smirnov. 1993. The Khariton version. *Bulletin of the Atomic Scientists*, May 1993, pp. 20–31.

Khasi-Jaintia Environment Protection Council. Official statement submitted by Khasi-Jaintia Environment Protection Council (Shillong, India) to World Uranium Hearing. Salzburg, Austria, n.d.

Khokhlov, A. 1992. 100 thousand people make our bomb (in Russian). *Komsomolskaya Pravda*, 31 January 1992.

Khots, Y. 1992. [Plutonium-producing reactor shut down in Krasnoyarsk.] Transcript of report broadcast on *ITAR-TASS*, 0518 GMT, 29 September 1992. Translated from the Russian in *Foreign Broadcast Information Service* JPRS-TEN-92–020, 28 October 1992, p. 58.

Kimball, D., L. Siegel, and P. Tyler. 1993. Covering the map: A survey of military pollution sites in the United States. Physicians for Social Responsibility, Washington, D.C. and Military Toxics Project, Litchfield, Maine, May 1993.

King, L., and W. McCarley. 1961. *Plutonium Release Incident of November 20, 1959*. ORNL-2989. Oak Ridge, Tennessee: Oak Ridge National Laboratory, 1 February 1961.

Kinlen, L. J. 1990. Evidence from population mixing in British New Towns 1946–85 of an infective basis for childhood leukeamia. *Lancet*, vol. 336, pp. 577–582.

Kinlen, L. J. 1993. Can paternal preconceptional radiation account for the increase of leukaemia and non-Hodgkin's lymphoma in Seascale? *British Medical Journal*, vol. 306, pp. 1718–1721.

Kirinitsiyanov, Y. 1993. Where does the radioactive wind come from? From China? *Rabochaya Tribuna* (Moscow), 8 December 1992, p. 1. Translated from the Russian in *Foreign Broadcast Information Service* JPRS-TEN-93-001, 8 January 1993, p. 74.

Klimov, G., and V. Shtengelov. 1992. [Ukraine's 'Atomgrad' seeks government environmental aid.] Transcript of *Novosti* newscast report broadcast on Teleradiokompaniya Ostankino Television First Program Network, 0200 GMT, 4 August 1992. Translated from the Russian in *Foreign Broadcast Information Service* JPRS-TEN-92-017, 21 September 1992, p. 68.

Kneale, G. W., T. F. Mancuso, and A. M. Stewart. 1978. Reanalysis of data relating to the Hanford study of cancer risks of radiation workers, Hanford 11a. In: *Late Biological Effects of Ionizing Radiation, IAEA*. Vienna: IAEA, pp. 387–410.

Kneale, G. W., T. F. Mancuso, and A. M. Stewart. 1981. Hanford radiation study III: a cohort study of the cancer risks from radiation to workers at Hanford (1944–77 deaths) by the method of regression models in life-tables. *British Journal of Industrial Medicine*, vol. 38, pp. 156–166.

Kneale, G. W., and A. M. Stewart. 1993. Reanalysis of Hanford data: 1944–1986 deaths. *American Journal of Industrial Medicine*, vol. 23, pp. 371–389.

Kojima, S., and M. Furukawa. 1986. The measurement of neutron-induced radionuclides from Chinese nuclear weapons tests. *Journal of Radioanalytic Nuclear Chemistry*, vol. 100, pp. 231–240.

Kolesnikov, A. 1993. Warning: MR reactor is hazardous. *Moscow News*, 20–27 December 1992, p. 2. Reprinted in *Foreign Broadcast Information Service* JPRS-TEN-93-008, p. 18.

Kolesnikov, P. 1992. We are turning the Urals into a model nuclear storage facility. *Megapolis Ekspress*, 9 January 1992, p. 22. Translated from the Russian in *Foreign Broadcast Information Service* FBIS-SOV-92-034, 20 February 1992, p. 9.

Korogodin, B., B. Idimechev, C. Lopatko, A. Masliok, G. Matrocov, A. Pakulo, A. Samarkin, B. Hizhnyak, and O. Shamov. 1990. *Report of the Interministerial Commission on the Assessment of the Radioactive Situation in the Region of the city of Kransoyarsk* (in Russian). Prepared by decree of the Ministry of Natural Resources and the Soviet Council of Ministers, Order No. #PP23511D, N.p.

Koshurnikova, N. A., L. A. Buldakov, G. D. Bysogolov, M. G. Bolotnikova, N. S. Shilnikova, and V. S. Pesternikova. 1991. Mortality from malignant tumors in hematopoietic and lymphatic tissues among personnel of the first atomic power plant in the USSR. Soviet-Japanese seminar on radiation epidemiology. Tokyo.

Koshurnikova, N. A., L. A. Buldakov, G. D. Bysogolov, M. G. Bolotnikova, N. S. Shilnikova, and V. S. Pesternikova. 1992. Tables presented at conference in Chelyabinsk, May 1992.

Kossenko, M. M. 1992. Radiation incidents in the southern Urals. Unpublished paper. Chelyabinsk, Russia, May 1992.

Kossenko, M. M. 1992a. Effects of radiation exposure in the personnel of the nuclear facilities in the Urals and in the population of Chelyabinsk-40. Unpublished literature review for IPPNW. June 1992.

Kossenko, M. M. 1992b. Medical effects of population irradiation on the river Techa: Radiation risk assessment. Prepared for IPPNW and IEER, Chelyabinsk, June 1992.

Kossenko, M. M., M. O. Degteva, and N. A. Petrushova. 1991a. Assessment of radiation risk of leukemia induction based on the analysis of the consequences of population irradiation in southern Urals (in Russian). *Vestnik Akademii Metitsinskikh Nauk SSSR*. vol. 8, pp. 23–28.

Kossenko, M. M., M. O. Degteva, and N. A. Petrushova. 1991b. Leukemia risk estimate on the basis of nuclear incidents in the southern Urals. Chelyabinsk Branch Office of Institute of Biophysics of the USSR Ministry of Health.

Kossenko, M. M., M. O. Degteva, and N. A. Petrushova. 1992. Estimate of the risk of leukemia to residents exposed to radiation as a result of a nuclear accident in the Southern Urals. *The PSR Quarterly*, December 1992, pp. 187–197.

Kostyukovskiy, V., Y. Perepletkin, V. Van Der Naald, and M. Hoffman. 1991. Secrets of a closed city. *Izvestiya* (Moscow), 2 August 1991. Translated from the Russian in *Foreign Broadcast Information Service* JPRS-TEN-91–018, 11 October 1991, pp. 71–73.

Kovalenko, Y. 1992a. Chetek: Explosions for sale, but at what price? *Megapolis Continent*, 22 May 1992, 10. Translated from the Russian in *Foreign Broadcast Information Service* JPRS-TEN-92–015, pp. 58–60.

Krause, J. 1991. Results of safeguarding the WISMUT health data (in German). Paper presented at Conference of Citizen Groups Against Uranium Mining in Europe, Zwickau (Germany), 1–3 August 1991.

Krenzelok, E. P. 1992. Hydrofluoric acid. In *Hazardous Materials Toxicology*, J. B. Sullivan, Jr., and G. R. Krieger, eds., pp. 785–790. Philadelphia: Williams & Wilkins.

Krestinina, L. Yu., M. M. Kossenko, and V. A. Kostiuchenk. 1991. Lethal developmental defects in the offspring of a population living in a territory with traces of radioactivity (in Russian). *Meditsinskaia Radiologiia*, vol. 36, pp. 30–32.

Kruger, J. 1980. Surveying and assessing the hazards associated with the processing of uranium. In *Fifth International Congress of the International Radiation Protection Association*, Jerusalem, Israel, March 8, 1980, pp. 454–459. Oxford: Pergamon Press.

Kumar, P. 1993. Digging a grave for N-waste. *Financial Express* (Bombay), 6 May 1993, p. 8.

Kurtenbach, E. 1989. China reports 20 deaths, 1,200 injured in nuclear accidents. Associated Press. 5 August 1989.

Kusiak, R. A., A. C. Ritchie, J. Muller, and J. Springer. 1993. Mortality from lung cancer in Ontario uranium miners. *British Journal of Industrial Medicine*, vol. 50, no. 10, pp. 920–928.

Kusiak, R. A., J. Springer, A. C. Ritchie, and J. Muller. 1991. Carcinoma of the lung in Ontario gold miners: Possible aetiological factors. *British Journal of Industrial Medicine*, vol. 48, pp. 808–817.

Lamarsh, J. 1983. *Introduction to Nuclear Engineering*. 2d ed. Reading, Massachusetts: Addison-Wesley Publishing Company.

Lambrecht, Bill. 1991. Poisoned land: Cold War brought mining jobs to the Indians, but uranium mines also paid off in misery. *St. Louis Post-Dispatch*, November 19, 1991, pp. 1A, 6A, 7A.

Lancet. 1994. Comments. *Lancet*, vol. 343, 8 January 1994, p. 106.

Land, C. E., F. W. McKay, and S. G. Machado. 1984. Childhood leukemia and fallout from the Nevada nuclear tests. *Science*, vol. 223, pp. 139–144.

Lanouette, W., with B. Silard. 1992. *Genius in the Shadows: A Biography of Leo Szilard, the Man Behind the Bomb*. New York: Macmillan.

Lapham, M., and J. Samet. 1986. Radionuclide levels in cattle raised near uranium mines and mills in northwest New Mexico. New Mexico Environmental Improvement Division, Santa Fe, June 1986.

Lapp, R. 1958. *The Voyage of the Lucky Dragon*. New York: Harper & Brothers.

Lariviere, M., and J. Denis-Lampereur. 1991. URSS: le plutonium par la racine. *Science et Vie*, February 1991, pp. 102–103.

Lawson, D. 1993. Health assessment of Po-210 in caribou from the NWT. Memo submitted to the public hearings on uranium mineral development, Saskatoon, Saskatchewan, 4 May 1993.

LeFaure, C., and J. Lochard. 1990. La dosimétrie des travailleurs des enterprises extérieures dans les centrales nucléaires. *Risque et Prévention*. Bulletin du CEPN. no. 9, November 1990.

Leggett, R. W. 1989. The behavior and chemical toxicity of uranium in the kidney: A reassessment. *Health Physics*, vol. 57, pp. 365–383.

Lehman, L. 1990. Briefing to the ACNW (U.S. Nuclear Regulatory Commission Advisory Committee on Nuclear Waste) on July 30th. L. Lehman & Associates, Burnsville, Minnesota, August 1990. Photocopy.

Lehman, L. 1992. International technology exchange program. Trip report of Linda Lehman (Russia, June 1992). L. Lehman & Associates, Burnsville, Minnesota, 15 July 1992. Photocopy.

Lehman, L. 1993. Telephone communication with A. Brooks (IEER). 9 March 1993.

Leigh, D. 1991. Ways to make the old mining territories safe. *New Scientist*, 22 June, p. 45.

Lemert, A. A. 1979. *First You Take a Pick and Shovel: A History of the Mason Companies*. Lexington, Kentucky: Bradford Press.

Lenssen, N. 1991. Nuclear waste: The problem that won't go away. World Watch Paper 106. World Watch Institute, Washington, D.C., December 1991.

Leonov, L. 1991. A measure of secrecy: The today and tomorrow of a closed city. *Sovetskaya Rossiya*, 20 April 1991. Translated from the Russian in *Foreign Broadcast Information Service* JPRS-TEN-91-012, 26 June 1991, pp. 64–67.

Leskov, S. 1992. Russia intends to significantly boost uranium exports. *Izvestiya* (Moscow), 17 January 1992, p. 2. Translated from the Russian in *JV Dialogue: Soviet Press Digest*, 17 January 1992.

Leskov, S. 1993. Nuclear dumping: Lies and incompetence. *Bulletin of the Atomic Scientists*, June 1993, p. 13.

Levine, R. 1991. Soviet Union. In *Mining Annual Review 1991*. London, England: Mining Journal Ltd. Reprinted by U.S. Department of the Interior, Bureau of Mines, Washington, D.C.

Lewis, J., and L. Xue. 1988. *China Builds the Bomb*. Stanford: Stanford University Press.

Lewis, K. 1986. China's nuclear industry comes of age. *Nuclear Engineering International*, vol. 31, no. 383, June 1986, pp. 23–25.

Li, Z., Z. Wang, W. Sun, M. Chen, and Z. Zhu. 1988. Survey and assessment of radioactivity level in Yangtze River water system (in Chinese), No. CNIC-00223; BSPHN-0001. Report of the China Nuclear Information Center, Beijing, August 1988. Abstracted in English in National Technical Information Service Database, Accession Number DE90628023/XAB.

Lindberg, K. (NUEXCO International Corp.). 1992. Telephone communication with A. Brooks (IEER). 4 December 1992.

Linden, C. H. 1992. Inorganic acids and bases. In *Hazardous Materials Toxicology*, J. B. Sullivan, Jr., and G. R. Krieger, eds., pp. 762–775. Philadelphia: Williams & Wilkins.

Lippard, L. M., and L. W. Davis. 1991. Depleted uranium hexafluoride management study. Louisiana Energy Services, Claiborne, Louisiana, 1 October 1991.

Lipschutz, R. D. 1980. *Radioactive Waste: Politics, Technology and Risk*. Cambridge, Massachusetts: Ballinger.

Litvinov, Y. 1991. De-classified city. *Trud*, 25 January 1991, p. 4. Translated from the Russian in *JV Dialogue: Soviet Press Digest*, 25 January 1991.

Lloyd, P. J. D. 1980. Ninety years experience in the preservation of uranium ore dumps. In C. O. Brawner, ed., *First International Conference on Uranium Mine Waste Disposal*, May 19–21, 1980, Vancouver, British Columbia, Canada. New York: Society of Mining Engineers of the American Institute of Mining, Metallurgical, and Petroleum Engineers, Inc.

Lockwood, D. 1993. Dribbling aid to Russia. *Bulletin of the Atomic Scientists*, July/August 1993, pp. 39–42.

Lundin, F. E., J. W. Lloyd, E. M. Smith, V. E. Archer, and D. A. Holaday. 1969. Mortality of uranium miners in relation to radiation exposure, hard-rock mining and cigarette smoking—1950 through September 1967. *Health Physics*, vol. 16, pp. 571–578.

Luxin, W., Z. Yongru, T. Zufan, H. Weihui, C. Deqing, and Y. Yongling. 1990. Epidemiological investigation of radiological effects in high background radiation areas of Yangjiang, China. *Journal of Radiation Research* (Japan), vol. 31, pp. 119–136.

Lyon, J. L., J. W. Gardner, D. W. West, and L. Schussman. 1980. Further information on the association of childhood leukemias with atomic fallout. In *Cancer Incidence in Defined Populations,* Cairns, J., J. L. Lyon, and M. Skolnick, eds. Long Island: Cold Spring Harbor Laboratory; Banbury Report 4, vol. 145, p. 60.

Lyon, J. L., M. Klauber, J. W. Gardner, K. S. Udall. 1979. Childhood leukemias associated with fallout from nuclear tests. *New England Journal of Medicine*, vol. 300, pp. 397–402.

Machado, S. G., C. E. Land, and F. W. McKay. 1987. Cancer mortality and radioactive fallout in southwestern Utah. *American Journal of Epidemiology*, vol. 125, pp. 44–61.

Mackenzie, D., and J. Bastable. 1993. Plutonium missing after Tomsk blast. *New Scientist*, 24 April 1993, p. 5.

MacLachlan, A. 1990. Chelyabinsk deputies agree that FBR project can save environment. *Nucleonics Week*, 20 December 1990, p. 3.

MacLachlan, A. 1990a. Nuclear phobia haunts the Southern Urals power plant. *Nucleonics Week*, 26 July 1990, p. 13.

Maclean, I., D. Breen, and J. Chalmers. 1992. The third annual report of the Chief Administrative Medical Officer and Director of Public Health. Dumbries, Scotland: Department of Public Health Medicine and Galloway Health Board, December, p. 73.

Madel, J. 1990. Wismut AG: Past, present and future of the largest uranium producer in Europe. Paper presented at the 17th Annual Meeting of the World Nuclear Fuel Market, Toulouse, France, October 15–16, 1990.

Maeng-ho, C. IAEA finds DPRK nuclear safety 'defective.' *Tong-A Ilbo*, 15 June 1992, p. 1. Tranlated from the Korean in *Foreign Broadcast Information Service* JPRS-TEN-92-014, 9 July 1992.

Makhijani, A. 1988. Release estimates of radioactive and non-radioactive materials to the environment by the Feed Materials Production Center 1951–85. Institute for Energy and Environmental Research, Takoma Park, Maryland, 7 July 1989.

Makhijani, A. 1994. Energy enters guilty plea. *Bulletin of the Atomic Scientists*, March/April 1994, pp. 18–28.

Makhijani, A., and D. Albright. 1983. Irradiation of personnel during Operation Crossroads: An evaluation based on official documents. International Radiation Research Training Institute, Washington, D.C., May 1983.

Makhijani, A., R. Alvarez, and B. Blackwelder. 1986. Deadly crop at the tank farm: An assessment of the management of high-level radioactive wastes in the Savannah River Plant tank farm, based on official documents. Photocopy (available from: Institute for Energy and Environmental Research, Takoma Park, Maryland).

Makhijani, A., and B. Franke. 1989. Addendum to the report: Release estimates of radioactive and non-radioactive materials to the environment by the Feed Materials Production Center 1951–85. Institute for Energy and Environmental Research, Takoma Park, Maryland, May 1989.

Makhijani, A., and M. M. Hoenig. 1991. A black market in red nukes? Behind Bush's plan are concerns about arms chaos in the Soviet Union. *Washington Post*, 29 September, 1991, pp. C1–C2.

Makhijani, A., and J. Kelly. 1985. Target Japan: The decision to bomb Hiroshima and Nagasaki (manuscript). [Published in Japanese as *Why Japan?* Tokyo, Japan: Kyoikusha.]

Makhijani, A. [Arjun], and A. [Annie] Makhijani. 1995. Fissile materials in a glass, darkly: Technical and policy aspects of the disposition of plutonium and highly enriched uranium. Institute for Energy and Environmental Research, Takoma Park, Maryland, January 1995.

Makhijani, A., and S. Saleska. 1992. *High-Level Dollars, Low-Level Sense*. New York: The Apex Press.

Manchanda, R. Annual nuclear hiccup in US-Pakistan Relations. *Economic and Polititical Weekly*, 3 November 1990, pp. 2414–2415.

Mancuso, T. F., A. M. Stewart, G. W. Kneale. 1977. Radiation exposures of Hanford workers dying from cancer and other causes. *Health Physics*, vol. 33, pp. 369–384.

Mann, C. 1994. Radiation: Balancing the record. *Science*, 28 January 1994, p. 470.

MAPI (Ministry of Atomic Power and Industry). 1991. Organizational Chart. Photocopy. Moscow.

Marshall, G. 1945. Memorandum for the Secretary of War. National Archives, Washington, D.C., 7 June 1945.

Martin, J., and A. Thomas. 1990. Origins, concentrations and distributions of artificial radionuclides dicharged by the Rhône River to the Mediterranean Sea. *Journal of Environmental Radioactivity*, vol. 2, pp. 105–139.

Martin, T. 1993. Waste by any other name ... Deciding what goes into Hanford Grout. *Perspective*, (Hanford Education Action League, Hanford, Washington), Winter 1993, pp. 8–9.

Martin Marietta (Martin Marietta Energy Systems, Inc). 1990. *The Ultimate Disposition of Depleted Uranium*, DE91–006414, K/ETO-44. Oak Ridge, Tennessee: U.S. DOE, December 1990.

May, J. 1989. *The Greenpeace Book of the Nuclear Age: The Hidden History, The Human Cost*. New York: Pantheon Books.

McInroy, J. F., and R. L. Kathren. 1990. Plutonium content in marrow and mineralised bone in an occupationally exposed person. *Radiation Protection Dosimetry*, vol. 32, pp. 245–252.

McSorley, J. 1990. *Living in the Shadow*. London: Pan Books.

Médecine et Guerre Nucléaire. 1989. Agenda. *Médecine et Guerre Nucléaire*, vol. 4, no. 3, July/August/September 1989, p. 38.

Médecine et Guerre Nucléaire. 1991. Radioactivité et santé: La guerre des labos? *Médecine et Guerre Nucléaire*, vol. 6, no. 3, July/August/September 1991, p. 28.

Médecine et Guerre Nucléaire. 1991a. Armes et guerre nucléaire: Le plan francais de désarmement. *Médecine et Guerre Nucléaire*, vol. 6, no. 3, July/August/September 1991, p. 27.

Médecine et Guerre Nucléaire. 1992a. La France va signer le Traité de non-prolifération nucléaire TNP. *Médecine et Guerre Nucléaire*, vol. 7, no. 1, January/February/March 1992, p. 12.

Médecine et Guerre Nucléaire. 1992b. Arme nucléaire, armements, en France. *Médecine et Guerre Nucléaire*, vol. 7, no. 1, January/February/March 1992, pp. 13–14.

Médecine et Guerre Nucléaire. 1992c. Agenda. *Médecine et Guerre Nucléaire*, vol. 7, no. 4, October/November/December 1992, pp. 4–10.

Médecine et Guerre Nucléaire. 1993. Surarmament ou reconversion? *Médecine et Guerre Nucléaire*, vol. 8, no. 2, April/May/June 1993, pp. 20–21.

Médecine et Guerre Nucléaire. 1993a. Le Traité de Tlatelolco ratifié par la France. *Médecine et Guerre Nucléaire*, vol. 8, no. 1, January/February/March 1993, p. 2.

Médecine et Guerre Nucléaire. 1993b. Armes nuclaires, armements, écologie, et santé. *Médecine et Guerre Nucléaire*, vol. 8, no. 1, January/February/March 1993, pp. 4–7.

Medvedev, Z. 1976. Nuclear disaster in the Urals. *New Scientist*, 4 November 1976, p. 264.

Medvedev, Z. 1979. *Nuclear Disaster in the Urals*. New York: Norton.

Medvedev, Z. 1990. Bringing the skeleton out of the closet. *Nuclear Engineering International*, November 1990, p. 26.

Melnik, E. (Press Secretary, Krasnoyarsk Region Presidium Committee of People's Deputies). 1992. Have your dollars, with spent fuel in the bargain (in Russian). *Krasnoyarksii Rabochi* (Krasnoyarsk), 22 February 1992.

Mervis, J. 1992. Yeltsin's ecology adviser chips away at a sorry past. *Nature*, vol. 360, 19 November 1992, p. 198.

Meyer, K. 1994. Beryllium and lung disease. *Chest*, vol. 106, pp. 942–946.

Miheev, V. 1991. The nuclear trail of the Yenisey (in Russian). *Ecologicheski Vestnik*, no. 13, 1991.

Miheev, V. (editor, *Ecologicheski Vestnik*, co-founder, Krasnoyarsk Regional Green Movement, and Krasnoyarsk Regional People's Deputy). 1992. Letter to Alex Brooks, IEER, 17 July 1992.

Miheev, V. 1992a. Letter to Alex Brooks, IEER, 17 August 1992.

Miheev, V. 1992b. South Korea is ready to pay (in Russian). *Ecological Bulletin* (Krasnoyarsk), 9 February 1992.

Miheev, V. 1992c. Radionuclides in the Kara Sea (in Russian). *Ecological Bulletin* (Krasnoyarsk), 9 February 1992.

Miheev, V. 1993. Letter to Alex Brooks, IEER, 2 February 1993.

Miller, M. S., ed. 1993. *State of the Peoples*. Boston: Beacon.

Milliken, R. 1986. *No Conceivable Injury*. Harmondsworth, Middlesex, England: Penguin.

Ministry of Defence (U.K.). 1981. List of Assessed Contractors. Interim Issues no. 4, London.

Ministry of Defence (U.K.). 1989. *Radioactive and toxic waste disposals by AWE: 1989 from MoD (PE)[Procurement Executive Atomic Weapons Establishments] nuclear premises under control of Director AWE*. Report No. D/AWE/SFS/A23. London.

Mironova, N. (Spokesperson, Movement for Nuclear Safety). 1991. Soviet non-governmental movements in a situation of radiation ecological catastrophe caused by a radiochemical military complex operating in the Chelyabinsk region. Paper presented at CIS Nonproliferation Project Conference on The Proliferation Predicament in the Former Soviet Union, 6–9 April 1992, Monterey, California.

Momoshima, N., and Y. Takshima. 1983. Variations in radionuclide concentrations and size distribution of radioactive particles from the Chinese nuclear weapon test of October 16, 1980. *Journal of Radioanalytic Chemistry*, Vol. 76, pp. 7–18.

Mongin, D. 1990. Il y a trente ans: La première bombe A française. *Le Monde*. 12 February 1990.

Monroe, S. 1991. Chelyabinsk: the evolution of disaster. *Soviet Environmental Watch* (Monterey Institute of International Studies, Monterey, California), Fall 1991, pp. 18–29.

Monterey Institute of International Studies. 1992. Hungarian nuclear fuel on the rails. *CIS Environmental Watch*, no. 3, Fall 1992, p. 72.

Moody, Roger. 1992. *The Gulliver File*. London: Minewatch.

Moody, Roger (Minewatch). 1993. Telephone communication with K. Yih (IPPNW), 7 May 1993.

Morgachev, S. 1992. Plutonium is a friend, but truth is more costly. *Megapolis Express*, no. 2, 9 January 1992, p. 22. Translated from the Russian in *Foreign Broadcast Information Service* FBIS-SOV-92-034, 20 February 1992, pp. 8–9.

Morozova, V. 1993. [Tomsk-7: Situation reported not dangerous.] Transcript of report broadcast on Novosti newscast, Moscow Ostankino Television First Channel Network, 2000 GMT, 13 April 1993. Translated from the Russian in *Foreign Broadcast Information Service* JPRS-TEN-93-012, 3 May 1993, p. 21.

Morris, J. A., R. Butler, R. Flowerdew, and A. C. Gatrell. 1993. Retinoblastoma in children of former residents of Seascale [letter]. *British Medical Journal*, vol. 306, p. 650.

Moshman, J., and A. H. Holland. 1949. On the incidence of cancer in Oak Ridge Tennessee. *Cancer*, vol. 2, no. 4, pp. 567–75.

Movement against Uranium Mining, NSW. 1991. *Uranium Mining in Australia*, 2d ed. Haymarket, NSW, Australia.

Muller, J., R. Kusiak, G. Suranyi, and A. Ritchie. 1986. *Study of Ontario Miners, 1955–1977*. Ontario: Ontario Ministry of Labour, Ontario Worker's Compensation Board, Atomic Energy Control Board.

Muller, J., W. Wheeler, J. Gentleman, G. Suranyi, and R. Kusiak. 1983. *Study of Ontario Miners, 1955–1977*, Part 1. Ontario: Ontario Ministry of Labour, Ontario Worker's Compensation Board, Atomic Energy Control Board.

Mullican, W., A. Fryar, and N. Johns. 1993. Milestone report: The aerial extent and hydraulic continuity of perched ground water in the vicinity of the Pantex Plant. Bureau of Economic Geology, University of Texas at Austin, Austin, Texas.

Murashev, A. 1993. [Discharge of uranium spray: journalists allowed access.] Transcript of report broadcast on Moscow Radio Rossii Network, 0700 GMT, 7 April 1993. Translated from the Russian in *Foreign Broadcast Information Service* JPRS-TEN-93-009, 9 April 1993.

Mushak, P. 1992. Defining lead as the premiere environmental health issue for children in America: Criteria and their quantitative application. *Environmental Research*, vol. 59, pp. 281–309.

National Academy of Sciences. Committee on International Security and Arms Control. 1994. *Management and Disposition of Excess Weapons Plutonium*. Washington, D. C.: National Academy Press.

National Research Council. 1987. *Safety Issues at Defense Production Reactors*. Washington, D.C.: National Academy of Sciences.

National Research Council. Board of Radioactive Waste Management. 1986. *Scientific Basis for Risk Assessment and Management of Uranium Mill Tailings*. Washington, D.C.: National Academy Press.

National Research Council. Commission on Physical Sciences, Mathematics, and Resources. Committee to Provide Oversight of the DOE Nuclear Weapons Complex. 1989. *The Nuclear Weapons Complex: Management for Health, Safety, and the Environment*. Washington, D.C.: National Academy Press.

National Research Council. Committee on the Biological Effects of Ionizing Radiations. 1988. *Health Risks of Radon and other Internally Deposited Alpha Particle Emitters, BEIR IV*. Washington, DC: National Academy Press.

National Research Council. Committee on the Biological Effects of Ionizing Radiations. 1990. *Health Effects of Exposures to Low Levels of Ionizing Radiation, BEIR V*. Washington, D.C.: National Academy Press.

Nazarov, A. G. 1991. Rezonans, Chelyabinsk (in Russian). Yuzhno-Uralskaya knizhnoye izdatelstvo.

Nazarov, A. G. 1992. Rezonans, Chelyabinsk (in Russian). Yuzhno-Uralskaya knizhnoye izdatelstvo.

Neff, T. 1984. *The International Uranium Market*. Cambridge, Massachusetts: Ballinger.

Neff, T. (Massachusetts Institute of Technology). 1993. Telephone conversation with K. Yih (IPPNW), 1 April 1993.

Nelyubin, V. 1991. If the Genie breaks free ... *Sobesednik*, no. 22, p. 4. Translated from the Russian in *JV Dialogue: Soviet Press Digest*, 1–21 June 1991.

Nelyubin, V. 1992. A radioactive stand: A nuclear ultimatum. *Komsomolskaya Pravda*, 30 January 1992, p. 2. Translated from the Russian in *JV Dialogue: Soviet Press Digest*, 20 January 1992.

New Scientist. 1993. Traces of Tomsk. *New Scientist*, 29 May 1993, p. 11.

Nikipelov, B. V. 1989. Experience in managing the radiological and radioecological consequences of the accidental release of radioactivity which occured in the Southern Urals in 1957. Paper presented to the IAEA international symposium on recovery operations in the event of a nuclear accident or radiological emergency, Vienna, 6 November 1989. Translated from the Russian by the U.S. Department of Energy. 89-12472 (5626e/586e). Washington, D.C.

Nikipelov, B. V., and Eu. G. Drozhko. 1990. Vzruv na Yuzhnom Urale (in Russian). *Priroda*, vol. 5, pp. 48–49.

Nikipelov, B. V., Eu. G. Drozhko, G. N. Romanov, A. S. Voronov, D. A. Spirin, and R. M. Alexakhin, et al. 1990. The Kyshtym disaster: A close-up (in Russian). *Priroda*, May 1990, pp. 47–75.

Nikipelov, B. V., A. F. Lyzlov, and N. A. Koshurnikova. 1990. Experience with the first Soviet nuclear installation, irradiation doses and personnel health. *Priroda*, vol. 2, pp. 30–38. Translated from the Russian by V. A. Kokhryakov.

Nikipelov, B. V., E. I. Mikerin, G. N. Romanov, D. A. Spirin, Yu. B. Kholina, and L. A. Buldakov. 1990a. Radiation accident in the southern Urals in 1957 and mitigation of its consequences. In *Recovery Operations in the Event of a Nuclear Accident or Radiological Emergency*. Vienna: IAEA, pp. 373–404.

Nikipelov, B. V., G. N. Romanov, L. A. Buldakov, N. S. Babaev, Y. B. Kholina, and E. I. Mikerin. 1989. Radiation accident in the Southern Urals on 29 September 1957 (in Russian). *Atomic Energy*, no. 67, August 1989.

NIOSH (National Institute of Occupational Safety and Health). 1981. Health hazard evaluation report, Cotter Corporation, Canon City, Colorado. HETA 81–055–954. Atlanta, Georgia: Center for Disease Control.

Norris, [R.] S. 1992. The Soviet nuclear archipelago. *Arms Control Today*. January/February 1992, pp. 24–31.

Norris, R. S., and W. M. Arkin. 1990. Nuclear Notebook. *Bulletin of the Atomic Scientists*, July/August 1990, p. 48.

Norris, R. S., and W. M. Arkin. 1990a. Lesser nuclear powers: France. *Bulletin of the Atomic Scientists*, December 1990, p. 57.

Norris, R. S., and W. M. Arkin. 1991. Nuclear pursuits. *Bulletin of the Atomic Scientists*, May 1991, p. 49.

Norris, R. S., and W. M. Arkin. 1993. Nuclear pursuits. *Bulletin of the Atomic Scientists*, May 1993, pp. 48–49.

Norris, R. S., and W. M. Arkin. 1993a. Known nuclear tests worldwide, 1945 to December 31, 1992. *Bulletin of the Atomic Scientists*, April 1993, p. 49.

Norris, R. S., and W. M. Arkin. 1993b. NPT–onward and upward. *Bulletin of the Atomic Scientists*, July/August 1993, p. 56.

Norris, R. S., and W. M. Arkin. 1993c. Estimated Russian (CIS) nuclear stockpile (July 1993). *Bulletin of the Atomic Scientists*, July/August 1993, p. 57.

Norris, R. S., and W. M. Arkin. 1993d. Russia's Pantexes. *Bulletin of the Atomic Scientists*, January/February 1993, p. 56.

Norris, R. S., and W. M. Arkin. 1993e. Estimated nuclear stockpiles 1945–1993. *Bulletin of the Atomic Scientists*, December 1993, p. 57.

Norris, R. S., and W. M. Arkin. 1994. Known nuclear tests worldwide, 1945–1993. *Bulletin of the Atomic Scientists*, May/June 1994, pp. 62–63.

Norris, R. S., A. S. Burrows, and R. W. Fieldhouse. 1994. *Nuclear Weapons Data Book*. Vol. 5, *British, French, and Chinese Nuclear Weapons*. Boulder, San Francisco, Oxford: Westview Press.

NRPB (National Radiation Protection Board). 1990. Follow-up to Professor Gardner's case-control study of leukaemia and lymphoma among young people near Sellafield Nuclear Plant in West Cumbria. Report no. R242. Workshop held at NRPB, Chilton, 9 May 1990.

NRPB (National Radiation Protection Board). 1991. Consultative document—Board advice following publication of the 1990 recommendations of ICRP. Report no. NRPB-M321. London: Her Majesty's Stationery Office.

NRPB (National Radiation Protection Board). 1993. Board statement on the 1990 recommendations of ICRP. *Documents of the NRPB* (Chilton, Didcot, England), vol. 4, no. 1.

NSC (National Security Council). 1951. Report to the National Security Council by the Chairman, National Security Resources Board on the recommended policies and actions in light of the grave world situation. NSC 100, 11 January 1951. [Available from Civil Reference Division, National Archives, Washington, D.C..]

Nuclear Energy Agency. Committee on Radiation Protection and Public Health. 1982. Applicability of the ICRP principle of justification of a practice to radiological protection standards. *Journal of the Society for Radiological Protection*, vol. 2, no. 4, pp. 15–16.

Nuclear Engineering International. 1991. *World Nuclear Industry Handbook 1991*. Sutton, Surrey, England: Business Press International.

Nuclear Fuel. 1990. Canada: Cameco fined for mine water spill. *Nuclear Fuel*, 5 February 1990, p. 15.

Nuclear Fuel. 1991. Wismut uranium cleanup will cost $10 billion, U.S. consultant says. *Nuclear Fuel*, 21 January 1991, pp. 5–6.

Nuclear Fuel. 1992. Suppliers agree on dual-use controls, full-scope safeguards export policy. *Nuclear Fuel*, 13 April 1992, p. 4.

Nuclear India. 1992. Report produced by Chameleon Television for Channel 4 in the United Kingdom. London.

Nuclear Installations Inspectorate. 1980. Inquiry into Sellafield discharges. London.

Nucleonics Week. 1989. China nuclear waste sites being readied. *Nucleonics Week*, 30 March 1989, pp. 8–9.

NUEXCO. 1991. *NUEXCO Monthly Report* (NUEXCO, Denver, Colorado), no. 274, June 1991, pp. 17–23.

NUEXCO. 1991a. The present and future of nuclear energy and industry in the USSR. *NUEXCO Monthly Report* (NUEXCO, Denver, Colorado), no. 275, July 1991, pp. 25–29.

NUEXCO. 1992. *International Directory of Uranium Producers*, 4th ed. Denver, Colorado.

Odell, M. (Public Affairs, International Atomic Energy Agency). 1992. Telephone conversation with A. Brooks (IEER), 17 October 1992.

O'Donnell, F. R. 1992. Additional documentation for the 1988 letter report. Martin Marietta internal correspondence, 14 October 1992.

OECD (Organization for Economic Cooperation and Development) Nuclear Energy Agency and the International Atomic Energy Agency. 1990. *Uranium Resources, Production and Demand 1989*. Paris: OECD.

O'Toole, T. (Assistant Secretary for Environment, Safety and Health, U.S. Department of Energy). 1994. Testimony before the Subcommittee on Oversight and Investigations, U.S. House of Representatives, Washington, D.C., 17 March 1994.

Owen, A. 1985. *The Economics of Uranium*. New York: Praeger Special Studies and Praeger Scientific.

Parker, Honorable J. 1978. *The Windscale Inquiry*. Report presented to the Secretary of State for the Environment. Vol. 1. Her Majesty's Stationery Office: London, 26 January 1978.

Parthasarathi, A., and B. Singh. 1992. Science in India: The first ten years. *Economic and Political Weekly*, 29 August 1992, pp. 1852–1858.

Patrick, C. H. 1977. Trends in public health in the population near nuclear facilities: A critical assessment. *Nuclear Safety*, vol. 18, pp. 647–662.

Patterson, R. 1945. Memorandum for the Secretary of War, 25 February 1945. Manhattan Project, Record Group 77, National Archives, Washington, D.C.

Pelt, A., and G. Goryunov. 1993. [Tomsk-7: Failure to protect people questioned.] Transcript of video report broadcast on Vesti newscast, Moscow Russian Television Network, 1900 GMT, 14 April 1993. Translated from the Russian in *Foreign Broadcast Information Service* JPRS-TEN-93-013, 14 May 1993, p. 15.

Penyagin, A. (former Chairman, Subcommittee on Atomic Power and Nuclear Ecology, USSR Supreme Soviet). 1991. Interview by K. Yih (IPPNW) and A. Brooks (IEER). Moscow, 30 October and 6 November 1991.

Penyagin, A. (former Chairman, Subcommittee on Atomic Power and Nuclear Ecology, USSR Supreme Soviet). 1991a. Interview by K. Mbulawa and A. Brooks (IEER). Moscow, 21 December 1991.

Perkovich, G. 1992. Trip report: Pakistan and India, 10–22 September 1992. Secure Society Program, W. Alton Jones Foundation, Charlottesville, Virginia. Photocopy.

Polednak, A. P. 1980. Mortality among men occupationally exposed to phosgene in 1943–1945. *Environmental Research*, vol. 22, pp. 357–367.

Polednak, A. P., and D. R. Hollis. 1985. Mortality and causes of death among workers exposed to phosgene in 1943–45. In *Toxicology and Industrial Health*, vol. 1, pp. 137–147.

Polsgrove, Carol. 1980. In hot water: uranium mining and water pollution. *Sierra Club Bulletin*, November/December, pp. 29–31.

Popova, L. (Socio-Ecological Union, Moscow). 1993. Letter to Alex Brooks (IEER). 9 March 1993.

Portanskiy, A. 1993. Weapons-grade uranium from Russia will go to fuel U.S. nuclear electric power stations. *Izvestiya* (Moscow), 26 February 1993, p. 3. Translated from the Russian in *Foreign Broadcast Information Service* JPRS-TEN-92-007, 29 March 1993, p. 23.

Potter, W. 1993. Nuclear profiles of the Soviet successor states. Monterey, California: Program for Nonproliferation Studies, Monterey Institute of International Studies.

Potter, W., and E. Cohen. 1992. *Nuclear assets of the former Soviet Union*. Monterey, California: CIS Nonproliferation Project, Center for Russian and Eurasian Studies, Monterey Institute of International Studies.

Proliferation Watch. 1993. Country developments. *Proliferation Watch*, vol. 4, no. 1, January-February 1993, pp. 11–18.

Proliferation Watch. 1993a. Country developments. *Proliferation Watch*, vol. 4, no. 2, March-April 1993, pp. 3–10.

PSR (Physicians for Social Responsibility). 1988. Physicians group calls for independent review of health risks at nation's nuclear weapons facilities. Photocopy. Washington, D.C., 16 October 1988.

Qol Yisra'el. 1993. [Sarid rejects Egyptian concern about radioactive waste.] Transcript of report broadcast by Qol Yisra'el (Jerusalem), 0700 GMT, 16 April 1993. Translated from the Hebrew in *Foreign Broadcast Information Service* JPRS-TEN-93-014, 28 May 1993, p. 29.

QUEST Radiation Data Base. Vol. 2.6, 1992. [Produced and distributed by Radiation Technology, Inc., P.O. Box 10457, Silver Spring, MD 20914, USA.]

Rabinowitz, D. 1993. Need to open Dimona reactor after leakage. *Ha'aretz* (Tel Aviv), 16 April 1993, p. B5. Translated from the Hebrew in *Foreign Broadcast Information Service* JPRS-TEN-93-014, 28 May 1993, pp. 29–31.

Rapoport, R. 1971. *The Great American Bomb Machine*. New York: Ballantine Books.

Ray-Press Information Agency. 1991. *Register Ray-Press: Product and Ore Manufacturers, USSR 1991*. Vol.1, *Product and Ore Manufacturers, A to O*. Moscow: Russian American University Press.

Reicher, D., and K. Suokko. 1993. Radioactive waste and contamination in the former Soviet Union. *Environmental Science and Technology*, vol. 27, pp. 602–4.

Renmin Ribao. 1989. Regulations governing the protection against radiation from radioisotopes and beam installations. Text of decree by Li Peng. *Renmin Ribao* (Beijing), 23 November 1989, p. 6. Translated from the Chinese in *Foreign Broadcast Information Service* JPRS-TND-90-001, 4 January 1990, pp. 4–7.

Reynolds P., and D. F. Austin. 1985. Cancer incidence among employees of Lawrence Livermore National Laboratory (1969–1980). *Western Journal of Medicine*, vol. 142, pp. 214–218.

Rhodes, R. 1988. *The Making of the Atomic Bomb*. New York: Simon and Schuster.

Rich, V. 1990. 'Years of neglect' led to beryllium blast. *New Scientist*, 17 November 1990, p. 15.

Ringholz, R. 1989. *Uranium Frenzy: Boom and Bust on the Colorado Plateau*. New York: W. W. Norton & Co.

Ripple, S. R. 1992. Looking back at nuclear weapons facilities: The use of retrospective health risk assessments. *Environmental Science and Technology*, vol. 26, pp. 1270–1277.

Robertson, A. M. 1982. Site selection and design options of mining and metallurgical institutions. In *Twelfth Congress of the Council of Mining and Metallurgical Institutions*, Johannesburg, South Africa, May 3, 1982, pp. 861–865.

Robinette, C. D., S. Jablon, and T. L. Preston. 1985. Studies of participants in nuclear tests. DOE/EV/01577. Washington, D.C.: National Research Council.

Robinson, P. 1980. Responsible uranium mining and milling: An overview. In Proceedings of the First International Conference on Uranium Mine Waste Disposal, Vancouver, British Columbia, 19–21 May 1980. C. O. Brawner, ed. Society of Mining Engineers of the American Institute of Mining, Metallurgical, and Petroleum Engineers, Inc., New York.

Robinson, P. (Research Director, Southwest Research and Information Center, Albuquerque, New Mexico). 1991. An introduction to the giant uranium facilities of Eastern Germany: A report of observations of the uranium mines and mills operated by Wismut AG in Saxony and Thuringia in the former German Democratic Republic. Prepared For IPPNW—German chapter. June 1991.

Robinson, P. 1994. Personal written communication with Arjun Makhijani (IEER), 25 February 1994.

Robinson, P. (Research Director, Southwest Research and Information Center, Albuquerque, New Mexico). 1994a. Personal written communication with Arjun Makhijani (IEER), 17 March 1994.

Rocky Mountain Peace Center. 1992. Citizen's guide to Rocky Flats: Colorado's nuclear bomb factory. Boulder, Colorado, May 1992.

Rogers, W. (manager of energy and environmental programs, Council of Resource Tribes). 1992. Telephone interview with Lenore Azaroff (Harvard School of Public Health), 2 March 1992.

Roman, E., V. Beral, L. Carpenter, A. Watson, C. Barton, H. Ryder, and D.L. Aston. 1987. Childhood leukaemia in the West Berkshire and Basingstoke and North Hampshire District Health Authorities in relation to nuclear establishments in the vicinity. *British Medical Journal*, vol. 294, pp. 597–602.

Roman, E., A. Watson, V. Beral, S. Buckle, D. Bull, K. Baker, H. Ryder, and C. Barton. 1993. Case-control study of leukaemia and non-Hodgkin's lymphoma among children aged 0–4 years living in west Berkshire and north Hampshire health districts. *British Medical Journal*, vol. 306, pp. 615–621.

Romanov, G. N., L. A. Buldakov, and V. L. Shvedov. 1990. Irradiation of the population and medical consequences of the accident (in Russian). *Priroda*, vol. 5, pp. 67–72.

Romanov, G. N., L. A. Buldakov, V. L. Shvedov, I. K. Dibobes, P. V. Goloshchapov, E. M. Kravtsova, D. A. Spirin, and E. G. Drozhko. 1991. Rationale for and effectiveness of civil radiation protection measures after the Kyshtym accident. Paper presented to the U.S. DOE delegation, Chelyabinsk, Russia, 21 October 1991.

Romanov, G. N., I. G. Teplyakov, and V. P. Shilov. 1990. Restoration of economic activity (in Russian). *Priroda*, vol. 5, pp. 63–67.

Romanov, G., I. Teplyakov, and V. Shilov. 1990a. Return to economic activities (in Russian). *Priroda*, no. 5, pp. 67–72.

Rooney, C., V. Beral, N. Maconochie, P. Fraser, and G. Davies. 1993. Case-control study of prostatic cancer in employees of the United Kingdom Atomic Energy Authority. *British Medical Journal*, vol. 307, pp. 1391–1397.

Roscoe, R. J., K. Steenland, W. E. Halperin, J. J. Beaumont, and R. J. Waxweiler. 1989. Lung cancer mortality among nonsmoking uranium miners exposed to radon daughters. *Journal of the American Medical Association*, vol. 262, pp. 629–633.

Rosenberg, D. 1982. 'A smoking radiating ruin at the end of two hours': Documents on American nuclear plans for nuclear war with the Soviet Union, 1954–55. *International Security*, Winter 1981–82, pp. 1–38.

Rosenman, K. D., J. A. Valciukus, I. Glickman, B. R. Meyers, and A. Cinotti. 1986. Sensitive indicators of inorganic mercury toxicity. *Archives of Environmental Health*, vol. 41, pp. 2008–2015.

Rossiyskaya Gazeta. 1992. Based on the data of Gosatomnadzor. *Rossiyskaya Gazeta* (Moscow), 15 September 1992, p. 2. Translated from the Russian in *Foreign Broadcast Information Service* JPRS-TEN-92-019, 7 October 1992, pp. 50–52.

Rowley, D., P. Turri, and D. Paschal. 1986. An assessment of human exposure to mercury in soil, Oak Ridge, Tennessee. Atlanta, Georgia: Centers for Disease Control, Division of Environmental Hazards and Health Effects.

Royal Commission (on the Health and Safety of Workers in Mines). 1976. *Report of the Royal Commission on the Health and Safety of Workers in Mines*. Toronto: Government of Ontario.

Ruggles, R. G., et al. 1978. A study of water pollution in the vicinity of two abandoned uranium mills in northern Saskatchewan. Ottawa: Environment Canada.

Rush, D. 1992. Response to the paper of Kossenko et al. *The PSR Quarterly*, vol. 2, no. 4, December 1992, pp. 221–222.

Russell, W. 1977. Mutation frequencies in female mice and the estimation of generic hazards of radiation in women. *Proceedings of the National Academy of Science*, USA, vol. 74, pp. 3523–7.

Ruttenber, A. J. 1994. Evaluating health risks in communities near nuclear facilities. In *Radiation and Society*, J. P. Young and R. S. Yalow, eds. Washington, D.C.: American Chemical Society.

Ruttenber, A. J., K. Kriess, R. L. Douglas, T. E. Buhl, and J. Millard. 1984. The assessment of human exposure to radionuclides from a uranium mill tailings release and mine dewatering effluent. *Health Physics*, vol. 47, pp. 21–35.

Saccamano, G. S., G. C. Huth, O. Auerbach, and M. Kuschner. 1988. Relationship of radioactive daughters and cigarette smoking in the genesis of lung cancer in uranium miners. *Cancer*, vol. 62, pp. 1402–1408.

Saleska, S. 1992. New evidence on low-dose radiation exposure. *Science for Democratic Action* (IEER, Takoma Park, Maryland), Spring 1992, pp. 1, 5–7.

Saleska, S. 1992a. Ecological consequences of nuclear weapons development in the Southern Urals: A conference and working trip in Chelyabinsk, Russia, 15–30 May 1992. Trip report for Institute for Energy and Environmental Research, Takoma Park, Maryland. 15 July 1992. Photocopy.

Saleska, S., with K. Bolley, D. Borson, K. Bossong, G. Davis, and N. Rader. 1989. *Nuclear Legacy: An Overview of the Places, Politics, and Problems of Radioactive Waste in the United States*. Washington, D.C.: Public Citizen.

Saleska, S., and A. Makhijani. 1990. To process or not to process: The PUREX question: The alternatives for the management of N-Reactor irradiated fuel at the U.S. Department of Energy's Hanford Nuclear Reservation. Report prepared for the Hanford Education Action League. Institute for Energy and Environmental Research, Takoma Park, Maryland, July 1990.

Saleska, S., and A. Makhijani. 1992. Environmental Issues at Sequoyah Fuels Corporation's uranium conversion plant near Gore, Oklahama. Report prepared for Native Americans for a Clean Environment. Institute for Energy and Environmental Research, Takoma Park, Maryland, July 1992.

Samet, J. M. 1986. Radiation and disease in underground miners. *Annals of the American Conference of Governmental Industrial Hygienists*. vol. 14, pp. 27–35.

Samet, J. M., D. M. Kutwirt, R. J. Waxweiler, and C. R. Key. 1984. Uranium mining and lung cancer in Navajo men. *New England Journal of Medicine*, vol. 310, pp. 1481–1484.

Samet, J. M., D. R. Pathak, M. V. Morgan, M. C. Marbury, C. R. Key, and A. A. Valdivia. 1989. Radon progeny exposure and lung cancer risk in New Mexico U Miners: A case control study. *Health Physics*, vol. 56, pp. 415–421.

Sanatin, V. 1991. Uran-gate (in Russian). *Komsomolskaya Pravda*, 27 February 1991.

Sandback, M. 1992. Uranium waste in Estonia poses threat to Gulf of Finland. *Eesti Paevaleht*, 7 February 1992, p. 1. Translated from the Estonian in *Foreign Broadcast Information Service* JPRS-TEN-92-008, 5 May 1992, p. 107.

Sanger, S. L., with R. W. Mull. 1989. *Hanford and the Bomb: An Oral History of World War II*. Seattle: Washington: Living History Press.

Sarkar, S. 1990. Nuclear waste: The white man's burden? *Economic and Political Weekly*, 13 January 1990, pp. 84–85.

SC&A and Rogers (& Associates Engineering Corp.). 1993. Diffuse norm wastes: Waste characterization and preliminary risk assessment. Report prepared for U.S. Environmental Protection Agency, Washington, D.C. RAE-9232/1-2, vol. 1. April 1993.

Schüttman, W. 1993. Schneeberg lung disease and uranium mining in the Saxon Ore Mountains (Erzgebirge). *American Journal of Industrial Medicine*, vol. 23, pp. 355–368.

Schwartz, S. 1993. Department of Energy nuclear weapons complex facilities—Principal managing and operating contractors. Military Production Network, Washington, D.C. 22 June 1993. Photocopy.

S. Cohen & Associates, Inc. 1989. Radiological monitoring at inactive surface uranium mines. Report prepared for the U.S. Environmental Protection Agency, February 1989.

Sevc, J., L. Tomasek, E. Kunz, V. Placek, D. Chmelevsky, D. Barclay, and A. M. Keller. 1993. A survey of the Czechoslovak follow-up of lung cancer mortality in uranium miners. *Health Physics*, vol. 64, pp. 355–369.

Sevcova, M., J. Sevc, and J. Thomas. 1978. Alpha irradiation of the skin and the possibility of late effects. *Health Physics*, vol. 35, pp. 803–806.

Sever, L. E., E. S. Gilbert, N. A. Hessol, and J. M. McIntyre. 1988. A case-control study of congenital malformation and occupational exposure to low-level ionizing radiation. *American Journal of Epidemiology*, vol. 127, pp. 226–241.

Sever, L. E., N. A. Hessol, E. S. Gilbert, and J. M. McIntyre. 1988a. The prevalence of birth of congenital malformations in communities near the Hanford Site. *American Journal of Epidemiology*, vol. 127, pp. 243–254.

Sha, L., M. Yamamoto, K. Komura, and K. Ueno. 1991. Pu-239/240, Am-241 and Cs-137 in soils from several areas in China. *Journal of Radioanalytical and Nuclear Chemistry: Letters*, vol. 155, no. 1, pp. 45–53.

Shabad, T. 1969. *Basic Industrial Resources of the U.S.S.R.* New York: Columbia University Press.

Shanghai City Service 1989. [Shanghai's Zhu Rongji attends nuclear conference.] Transcript of a report broadcast on "Morning News" program, 2200 GMT 16 April 1989. Translated from the Chinese in *Foreign Broadcast Information Service* JPRS-TND-89-010, 23 May 1989, p. 2.

Sharma, D. 1983. *India's Nuclear Estate.* New Delhi: Lancers Publishers.

Sharma, Y. 1992. Tibet's ecological degradation seen as 'Asian Issue'. *South China Morning Post*, 9 June 1992, p. 21. Reprinted in *Foreign Broadcast Information Service* JPRS-TEN-92-014, 9 July 1992, pp. 19–20.

Sharp, J. 1993. Europe's nuclear dominoes: U.S. leadership in NATO is the best protection against nuclear proliferation in Europe. *Bulletin of the Atomic Scientists*, June 1993, pp. 29–33.

Sheleketov, V. 1992. Northern lights over Kyshtym. *Russian Gazette*, 25 June 1992.

Shen, D. 1990. The current status of Chinese nuclear forces and nuclear policies. PU CEES Report no. 247. The Center for Energy and Environmental Studies, Princeton University, Princeton, New Jersey.

Shen, D. 1993. Letter to A. Makhijani (IEER), 17 April 1993.

Sherwin, M. 1987. *A World Destroyed: Hiroshima and the Origins of the Arms Race.* New York: Vintage Books.

Shields, L. M., W. H. Wiese, B. J. Skipper, B. Charley, and L. Benally. 1992. Navajo birth outcomes in the Shiprock uranium mining area. *Health Physics*, vol. 63, pp. 542–551.

Shifrin, A. 1982. *The First Guidebook to Prisons and Concentration Camps of the Soviet Union.* Translation from the Russian. New York: Bantam Books.

Shimizu, Y. H., H. Kato, W. J. Schull, D. L. Preston, S. Fujita, and D. A. Pierce. 1987. Life span study report aa, Part 1, comparison of risk coefficients for site specific cancer mortality based on the DW86 and T65 Dr sheilded kerma and organ doses. RERF TR 12–87. Hiroshima. Radiation Effects Research Foundation.

Shleien, B., A. J. Ruttenber, and M. Sage. 1991. Epidemiologic studies of cancer in populations near nuclear facilities. *Health Physics*, 1991, vol. 61, pp. 699–713.

Shulman, S. 1990. Hanford nuclear radiation doses assessed. *Nature*, vol. 346, 19 July 1990, p. 205.

Shusterman, D. 1993. The Limited Test Ban Treaty: A twenty-year follow-up. *PSR Newsletter*, vol. 4, p. 1.

Shy, C. M. 1985. Chemical contamination of water supplies. *Environmental Health Perspectives*, vol. 62, pp. 399–406.

Simpson, J. 1986. *The Independent Nuclear State*. 2d ed. Basingstoke, Hampshire, England: Macmillan.

Singham, A. W. 1980. The illegal exploitation of Namibia. *Nation*, October 18, pp. 371–373.

Smetana, J., and J. Jech. 1992. Effect of uranium ore mining and dressing on the environment in the surroundings of Czechoslovak uranium industry facilities. *Energetika* (Prague), vol. 42, p. 1.

Smit, M. T. R., and C. P. Brent. 1991. Water management at Rössing uranium mine, Namibia. *African Mining '91*, pp. 191–201.

Smith, D. 1991. The impact of agriculture on the High Plains Trade Area. Panhandle Area Neighbors and Landowners, Amarillo (PANAL), Texas, June 1991.

Smith, M. T. 1993. Molybdenum toxicity in people living near a uranium mill in southern Colorado, USA. Paper presented at the International Society for Environmental Epidemiology Annual Conference, Stockholm, Sweden, 16 August 1993.

Smith, P. G., and A. J. Douglas. 1986. Mortality of workers at the Sellafield plant of British Nuclear Fuels. *British Medical Journal*, vol. 293, pp. 845–854.

Smith, R. 1993. South Africa's 16–year secret: The nuclear bomb. *Washington Post*, 12 May 1993, p. A1.

SocEco Agency. (Socio Ecological Union Centre for Coordination and Information). 1992. Russian nuclear power plants and their safety (in Russian). *SocEco Agency*, vol. 31a, 7 September 1992.

Soler-Sala, P. 1992. *Letters to Ecologia: Fourteen months as an ecological traveller during the collapse of the Soviet Union*. Ecologia, Harford, Pennsylvania.

Solntsev, R. 1992. A Munchhausen Windlass. *Moscow Kultura*, No. 5, 1 February 1992, pp. 2–3. Translated from the Russian in *Foreign Broadcast Information Service* JPRS-TEN-92-009, 22 May 1992, pp. 62–65.

South China Morning Post. 1989. Soviet technology to be used. Editorial in *South China Morning Post*, 4 April 1989, p. 20. Reprinted in *Foreign Broadcast Information Service* JPRS-TND-89-010, 23 May 1989, p. 4.

Sovetskaya Kultura. 1991. A life hazard. *Sovetskaya Kultura*, 13 April 1991, p. 2. Translated from the Russian in *JV Dialogue: Soviet Press Digest*, 13 April 1992.

Soyfer, V. N., M. O. Degteva, M. M. Kossenko, A. A. Akleev, V. P. Kozheurov, and G. N. Romanov. 1992. Radiation accidents in the Southern Urals (1949–1967). Laboratory of Molecular Genetics, Department of Biology, George Mason University, Virginia. Photocopy.

Spector, L. 1988. *The Undeclared Bomb*. Cambridge, Massachusetts: Ballinger.

Spector, L. 1990. *Nuclear Ambitions*. Boulder, Colorado: Westview Press.

Spector, L., and J. Smith. 1990. *Nuclear Ambitions: The Spread of Nuclear Weapons 1989–1990*. Boulder, Colorado: Westview Press.

Spiridonov, B. 1992. Underground chamber open. *Rossiyskaya Gazeta* (Moscow), 6 May 1992, p. 3. Translated from the Russian in *Foreign Broadcast Information Service* JPRS-TEN-92-011, 23 June 1992, pp. 93–94.

Stannard, J. N. 1988. *Radioactivity and Health: A History*. Prepared for the U.S. Department of Energy, Office of Health and Environmental Research. Oak Ridge, Tennessee: Office of Scientific and Technical Information, U.S. DOE. October 1988.

Stather, J. W., A. D. Wrixon, and J. R. Simmonds. 1984. *The Risks of Leukaemia and Other Cancers in Seascale from radiation exposure*. NRPB-R171. London: Her Majesty's Stationery Office.

Stebbings J. H., and G. L. Voelz. 1981. Morbidity and mortality in Los Alamos, New Mexico, l. Methodological issues and preliminary results. *Environmental Research*, vol. 25, pp. 86–105.

Stefashin, O. 1991. Direct connection (in Russian). *Izvestiya* (Moscow), 19 December 1991.

Stefashin, O. 1992. Information on Semipalatinsk Test Range declassified. *Izvestiya* (Russia), 20 May 1992. Translated from the Russian in *Foreign Broadcast Information Service* JPRS-TEN-92-013, 7 July 1992, p. 64.

Stevens, W., D. C. Thomas, J. L. Lyons, J. E. Till, R. A. Kerber, S. L. Simon, R. D. Lloyd, N. A. Elghany, and S. Preston-Martin. 1990. Leukemia in Utah and radioactive fallout from the Nevada Test Site. *Journal of the American Medical Association*, vol. 264, no. 5, pp. 585–591.

Stevenson, R. W. U.S.-North Korea meeting yields some gains on arms. *New York Times*, 20 July 1993, p. A2.

Stewart, A. M., and G. W. Kneale. 1991. An overview of the Hanford controversy. *Occupational Medicine: State of the Art Reviews*, vol. 6, pp. 641–663.

Stewart, A. M., T. F. Mancuso, and G. W. Kneale. 1980. The Hanford data—a reply to recent criticisms. *Ambio*, vol. 9, pp. 66–73.

Stimson, H. 1944. Letter to Harry S. Truman, March 13, 1944. Manhattan Project Record Group 77, National Archives, Washington, D.C.

Stimson, H. 1945. Memorandum discussed with the President, April 25, 1945. Harrison-Bundy File 62, National Archives, Washington, D.C.

Subramanyam, K. 1993. An equal-opportunity NPT. *Bulletin of the Atomic Scientists*, June 1993, pp. 37–39.

Sullivan, J. B., Jr., and G. R. Krieger, eds. 1992. *Hazardous Materials Toxicology*. Philadelphia: Williams & Wilkins.

Supervising Scientist for the Alligator Rivers Region. 1991. *Annual Report 1990–1991*. Canberra: Australian Government Publishing Service.

Supervising Scientist for the Alligator Rivers Region. 1992. *Annual Report 1991–1992*. Canberra: Australian Government Publishing Service.

Swanson, J. 1988. Recent studies related to Head-end Fuel Reprocessing at PUREX. Pacific Northwest Laboratory, Richland, Washington, PNL-6609, August 1988.

Szymanski, W. 1992. The uranium industry of the Commonwealth of Independent States. In *Uranium Industry Annual 1991*. Washington, D.C.: Energy Information Administration, U.S. Department of Energy, DOE/EIA-0478(91).

Tarasov, A. 1992. Siberians refuse to accept waste from Ukranian nuclear power stations. *Izvestiya* (Moscow), 11 January 1992, p. 2. Translated from the Russian in *JV Dialogue: Soviet Press Digest*, 11 January 1992.

Tarasov A. 1992a. People of Krasnoyarsk demand 6.730 billion [roubles] from atomic specialists. *Izvestiya*, 6 August 1992, p. 2. Translated from the Russian in *Foreign Broadcast Information Service* JPRS-TEN-92-017, 21 September 1992, pp. 61–62.

Taylor, P. J. 1987. The interpretation of monitoring results. In *Radiation and Health*, R. R. Jones and R. Southwood, eds., pp. 19–45. Chichester, England: John Wiley & Sons.

Teeple, D. 1955. *Atomic Energy*. New York: Little, Brown, and Company.

Texas Department of Agriculture. 1988. *Agriculture and the Uranium Industry*. Austin, Texas, September 1988.

Thomas, P., et al. 1993. *Uranium series radionuclides, polonium-210 and lead-210 in lichen-caribou-wolf food chain of the Northwest Territories* (draft). Prepared for Environment Canada, Atomic Energy Control Board, Department of Indian Affairs and Northern Development, and Department of Renewable Resources (NWT). Ottawa: Environment Canada, March 1993.

Thompson, R. C. 1989. Life-span effects of ionizing radiation in the beagle dog. PNL-6822. Battelle Pacific Northwest Laboratory, Richland, Washington.

Tkachenko, Y. 1992. 'Mayak' develops safe method of storing nuclear waste. Transcript of report broadcast by *ITAR-TASS* in English, 1339 GMT, 10 April 1992. Translated from the Russian in *Foreign Broadcast Information Service* JPRS-TEN-92-009, 23 May 1992, p. 65.

Tolley, H. D., S. Marks, J. A. Buchanan, E. S. Gilbert. 1983. A further update of the analysis of mortality of workers in a nuclear facility. *Radiation Research*, vol. 95, pp. 211–213.

Tona, F. 1986. Uranium exploration and mining operations in central and western Africa. Uranium and Nuclear Energy: Proceedings of the 11th International Symposium held by the Uranium Institute, London, Sept. 2–4, 1986.

Toro, T. 1991. Uranium mines leave heaps of trouble for Germany. *New Scientist*, 22 June 1991, p. 29.

Toro, T. 1991a. How to close the uranium mine. *New Scientist*, 22 June 1991, pp. 42–45.

Travis, P. 1991. Environmental issues in China. Los Alamos National Laboratory, Los Alamos, New Mexico, LA-12153-MS; DE92 003736.

Truman, H. 1944. Letter to Henry Stimson, March 10, 1944. Manhattan Project, Record Group 77, National Archives, Washington, D.C.

Truman, H. 1955. *Year of Decision*. Garden City, New York: Doubleday.

Turner, R., G. Kamp, M. Bogle, J. Switek, and R. McElhaney. 1985. *Sources and Discharges of Mercury in Drainage Waters at the Oak Ridge Y-12 Plant*. Y/TS-90. Oak Ridge, Tennessee: Oak Ridge National Laboratory, June 1985.

UKAEA (U.K. Atomic Energy Authority). 1987. *Report on Radiological Protection and Occupational Health for the Year 1986*. Report no. AHRM R6, London.

U.K. Department of the Environment. *Statement of Government Policy on Reprocessing and Operation of the Thermal Oxide Reprocessing Plant at Sellafield*. London, July 1993.

U.K. Nirex, Ltd. and U.K. Department of the Environment. 1991. *1991 United Kingdom Radioactive Waste Inventory*. UK Nirex Report No. 284, DOE/RAS/92.010, Waste Stream No. 7A18, Harwell, Didcot, Oxon, England.

Underhill, D., and E. Muler-Kahle. 1993. *IAEA Bulletin*, vol. 35, no. 3, p. 8.

Unger, W. 1951. *Design considerations in RaLa Processes: Experimental unit for product purification by ion exchange process*. ORNL-622. Oak Ridge, Tennessee: Oak Ridge National Laboratory, 8 November 1951.

UNSCEAR (United Nations Scientific Committee on the Effects of Atomic Radiation). 1988. *Sources, Effects and Risks of Ionizing Radiation*. New York: United Nations.

UNSCEAR (United Nations Scientific Committee on the Effects of Atomic Radiation). 1993. *Sources, Effects and Risks of Ionizing Radiation*. New York: United Nations.

Uranium Institute. 1993. Uranium: From mine to mill. Fact Sheet. London, April 1993.

Uranium Institute. Committee on Supply, Demand and Trade. 1991. *Uranium in the New World Market: Supply and Demand 1990–2010*. London: The Uranium Institute.

Uranium Institute. Committee on Supply, Demand and Trade. 1992. *Uranium in the New World Market: A Statistical Update of Supply and Demand 1991–2010*. London: The Uranium Institute.

U.S. AEC (Atomic Energy Commission). 1971. *In the Matter of J. Robert Oppenheimer: Transcript of Hearing Before Personal Security Board and Texts of Principal Documents and Letters*. Cambridge, Massachusetts: MIT Press.

U.S. Army Corps of Engineers (Tulsa, Oklahoma District Office). 1991. U.S. Department of Energy Pantex Plant, Final Work Plan, RCRA Facility Investigations for Ditches and Playas: Volume II (Site Specific). Amarillo, Texas: U.S. Department of Energy, May 1991.

U.S. Bureau of Mines. Various years, 1945–1990. *Minerals Yearbook*. Washington, D.C.: Government Printing Office. [Note: Years given in the citations (footnotes) are from *Minerals Yearbook* titles; actual publication date is the following year.]

U.S. Centers for Disease Control. 1980. *Biological Assessment after Uranium Mill Tailings Spill, Church Rock, New Mexico*. EPI-79-94-2. Atlanta: U.S. Public Health Service, December 24.

U.S. CIA (Central Intelligence Agency). 1959. *Accident at the Kasli Atomic Plant*. Report no. CS-3/389, 785. 4 March 1959.

U.S. CIA (Central Intelligence Agency). 1985. *USSR Energy Atlas*. Washington D.C.: Central Intelligence Agency.

U.S. Congress. 1971. Subcommittee on Raw Materials of the Joint Committee on Atomic Energy. *Hearings the Use of Uranium Mill Tailings for Construction Purposes*. 92nd Congress, First Session, October 28–29, 1971. Washington, D.C.: U.S. Government Printing Office.

U.S. Congress, House. 1983. Subcommittee on Investigations and Oversight and Subcommittee on Energy Research and Production. 1983. *Hearings on the Impact of Mercury Releases at the Oak Ridge Complex*. 97th Congress, 2d Session, July 11, 1983.

U.S. Congress, House. 1986. Committee on Energy and Commerce, Subcommittee on Energy, Conservation and Power. *American Nuclear Guinea Pigs: Three Decades of Radiation Experiments on U.S. Citizens*. 99th Congress, 2d session. Washington, D.C.: U.S. Government Printing Office.

U.S. Congress, House. 1987. Committee on Government Operations. *NRC's Regulation of Fuel Cycle Facilities: A Paper Tiger*. 100th Congress, 1st session. Washington, D.C.: U.S. Government Printing Office.

U.S. Congress, OTA (Office of Technology Assessment). 1991. *Complex Cleanup: The Environmental Legacy of Nuclear Weapons Production*. OTA-O-484. Washington, D.C.: U.S. Government Printing Office, February 1991.

U.S. Congress, OTA (Office of Technology Assessment). 1993. *Dismantling the Bomb and Managing Nuclear Materials*. OTA-O-572. Washington, D.C.: U.S. Government Printing Office, September 1993.

U.S. DOE (Deparment of Energy). 1987. *Environmental Survey Preliminary Report, Y-12 Plant, Oak Ridge, Tennessee*. Washington, D.C., November 1987.

U.S. DOE (Department of Energy). 1988. *Environmental Survey: Preliminary Summary Report of the Defense Production Facilities*. DOE/EH-0072. Washington, D.C.: U.S. DOE, September 1988.

U.S. DOE (Department of Energy). 1988a. *Integrated Database for 1988: Spent Fuel and Radioactive Waste Inventories, Projections and Characteristics*. Revision 4. DOE/RW-0006. Washington, D.C.: U.S. DOE, November 1988.

U.S. DOE (Department of Energy). 1989. *Announced United States Nuclear Tests, July 1945 through December 1988*. NVO-209. Washington, DC: U.S. DOE, September 1989.

U.S. DOE (Department of Energy). 1989a. *Integrated Database for 1989: Spent Fuel and Radioactive Waste Inventories, Projections and Characteristics*. Revision 5. DOE/RW-0006. Washington, D.C.: U.S. DOE, November 1989.

U.S. DOE (Department of Energy). 1990. *1989 Population, Economic and Land Use Survey for Rocky Flats Plant*. Golden, CO: U.S. DOE, August 1990.

U.S. DOE (Department of Energy). 1990a. *Integrated Database for 1990: U.S. Spent Fuel and Radioactive Waste Inventories, Projections and Characteristics*. Revision 6. DOE/RW-0006. Oak Ridge, Tennessee: U.S. DOE, October 1990.

U.S. DOE (Department of Energy). 1991. *Closure Plan for the Feed Materials Production Center: Fernald's Main Priority Is Cleanup*. Washington, D.C.: U.S. DOE, February 1991.

U.S. DOE (Department of Energy). 1991a. *Final Report on DOE Nuclear Facilities*. Washington, D.C.: Advisory Committee on Nuclear Facility Safety to the U.S. Department of Energy.

U.S. DOE (Department of Energy). 1991b. *Annual Status Report on the Uranium Mill Tail Remedial Action Program*. Washington D.C.: Office of Environmental Restoration and Waste Management. DOE/EM-0001. January 1991.

U.S. DOE (Department of Energy). 1992. *Integrated Database for 1992: U.S. Spent Fuel and Radioactive Waste Inventories, Projections and Characteristics*. Revision 8. DOE/RW-0006. Oak Ridge, Tennessee: U.S. DOE, October 1992.

U.S. DOE (Department of Energy). 1992a. Predecisional Environmental Assessment for Interim Storage of Plutonium Components at Pantex. Amarillo, Texas: U.S. DOE, 17 December 1992.

U.S. DOE (Department of Energy). 1992b. *Oak Ridge Reservation Environmental Report for 1991: Volume 1: Narrative, Summary, and Conclusions*. ES/ESH-22/V1. Oak Ridge, Tennessee: U.S. DOE, October 1992.

U.S. DOE (Department of Energy). 1993. *Environmental Restoration and Waste Management Five-Year Plan*. Vol. 1. DOE/S-00097P. Washington, D.C.: U.S. Government Printing Office, January 1993.

U.S. DOE (Department of Energy). 1993a. *Environmental Restoration and Waste Management Five-Year Plan*. Vol. 2, *Installation Summaries*. DOE/S-00097P. Washington, D.C.: U.S. Government Printing Office, January 1993.

U.S. DOE (Department of Energy). 1993b. *Pantex Plant*. GA93–0036/HB. Amarillo, Texas: U.S. DOE Amarillo Area Office. Booklet.

U.S. DOE (Department of Energy). 1993c. *Pantex Plant Environmental Restoration Program Zone 12 Groundwater Assessment*. GA93–0528/HB. Amarillo, Texas: U.S. DOE Amarillo Area Office. Fact sheet.

U.S. DOE (Department of Energy). 1993d. Openness Press Conference fact sheets. Washington, D.C., 7 December 1993. Photocopy.

U.S. DOE (Department of Energy). 1994. *Deficiencies in Reporting of Worker Exposure to Radiation and Toxic Material*. Submitted to the Subcommittee on Oversight and Investigations, U.S. House of Representatives, Washington, D.C., 17 March 1994.

U.S. DOE (Department of Energy). 1994a. *Integrated Database for 1993: U.S. Spent Fuel and Radioactive Waste Inventories, Projections and Characteristics*. Revision 9. DOE/RW-0006. Oak Ridge, Tennessee: U.S. DOE, March 1994.

U.S. DOE (Department of Energy). 1994b. Openness press conference fact sheets. Washington, D.C., 27 June 1994.

U.S. EPA (Environmental Protection Agency). 1983. Environmental standards for uranium and thorium mill tailings at licensed commercial processing sites: Final rule. 40 CFR, Part 192, *Federal Register*, vol. 48, no. 196, 7 October 1983, pp. 45, 926–7.

U.S. EPA (Environmental Protection Agency). 1985. *Ambient Water Quality Criteria for Mercury—1984*. EPA 440/5-84-227452. Washington, D.C.: U.S. EPA, January 1985.

U.S. EPA (Environmental Protection Agency). 1986. National emission standard for radon-222 emissions from licensed uranium mill tailings. *Code of Federal Regulations*, 40 CFR Part 61, Subpart W, September 1986.

U.S. EPA (Environmental Protection Agency). 1989. Background information document proposed NESHAPS for radionuclides. Draft report prepared by SC&A, Inc. for the U.S. EPA, 520/1-89-006, February 1989.

U.S. EPA (Environmental Protection Agency). 1991. *Radiation and Mixed Waste Incineration: Background Information Document*. Vol. 1, *Technology*. EPA 520/1-91-010-1. Washington, D.C.: U.S. EPA, Office of Radiation Programs, May 1991.

U.S. EPA (Environmental Protection Agency). 1993. *Issues Paper on Radiation Site Cleanup Regulations*. EPA 402-R-93-084. Washington, D.C.: Office of Radiation and Indoor Air, September 1993.

U.S. GAO (General Accounting Office). 1985. *Environment, Safety and Health: Information on Three Ohio Defense Facilities*. GAO/RCED-86-51FS. Washington, D.C.: U.S. General Accounting Office. November 1985.

U.S. GAO (General Accounting Office). 1989. *Nuclear Waste: DOE's Management of Single-Shell Tanks at Hanford, Washington*. GAO/RCED-89-157. Washington, D.C.: U.S. General Accounting Office, July 1989.

U.S. NRC (Nuclear Regulatory Commission). 1980. *Final Generic Environmental Impact Statement on Uranium Milling*. Project M-25. NUREG-0706. Vols. I and II. September 1980.

U.S. NRC (Nuclear Regulatory Commission). 1985. Supplemental statement of the Environmental Defense Fund, Sierra Club, Environmental Policy Institute, and Southwest Research and Information Center, in the matter of proposed rules to conform uranium mill tailings regulations to standards adopted by the EPA, 10 September 1985. Docket no. PR-40.

U.S. NRC (Nuclear Regulatory Commission). 1985a. Uranium Recovery Field Office, Uranium Recovery Facility Detection Monitoring Programs, 29 May 1985.

Usoltsev, A. 1992. Thinking aloud: How I spoiled a nice tea party for journalists. *Rossiyskaya Gazeta* (Moscow), 28 July 1992, p. 4. Translated from the Russian in *Foreign Broadcast Information Service* JPRS-TEN-92-016, 3 September 1992, p. 65.

van As, D. 1980. Uranium mill tailings management. South African Association of Physicists in Medicine and Biology Summer School on Uranium Health Physics, Pretoria, April 14–15, 1980, lecture 11.

Vancl, V. 1985. Procedures for personnel protection from ionizing radiation in uranium mines. *Rudy* (Prague), vol. 33, p. 10.

Van der Linde, A. 1992. Radiation monitoring in South African mines: some basic considerations. *Journal of the Mine Ventilation Society of South Africa*, December, pp. 190–196.

Vaughn, J. 1976. Plutonium—a possible leukemic risk. In *Health Effects of Plutonium and Radium*. Ed. W. S. S. Jee, pp. 691–705. Salt Lake City, Utah: J. W. Press.

Vaughn, J., B. Bleaney, and D. M. Taylor. 1973. Distribution, excretion and effects of plutonium as a bone-seeker. In *Uranium-Plutonium-Transplutonic Elements*, H. C. Hodge, J. N. Stannard, and J. B. Hursh, eds. pp. 349–502. New York: Springer-Verlag.

Velichko, O. and I. Ivantsov. 1993. [Tomsk-7: Committee, scientist report on radiation levels, cloud.] Transcript of report broadcast on ITAR-TASS World Service, 1031 GMT, 8 April 1993. Translated from the Russian in *Foreign Broadcast Information Service* JPRS-TEN-93-011, 27 April 1993, p. 18.

Viel, J. F., and S. T. Richardson. 1990. Childhood leukemia around the La Hague nuclear waste reprocessing plant. *British Medical Journal*, vol. 300, pp. 580–581.

Viel, J. F., S. Richardson, P. Danel, P. Boutard, M. Malet, P. Barrelier, O. Reman, and A. Carre. 1993. Childhood leukemia incidence in the vicinity of La Hague nuclear-waste reprocessing facility (France). *Cancer Causes and Control*, vol. 4, no. 4, pp. 341–343.

Voelz, G. L., R. S., Grier, and L. H. Hempelmann. 1985. A 37-year medical follow-up of Manhattan Project plutonium workers. *Health Physics*, vol. 48, pp. 249–259.

Voelz, G. L., L. H. Hempelmann, J. N. P. Lawrence, and W. D. Moss. 1979. A 32-year medical follow-up of Manhattan Project plutonium workers. *Health Physics*, vol. 37, pp. 445–485.

Voelz, G. L., and J. N. P. Lawrence. 1991. A 42-year medical follow-up of Manhattan Project plutonium workers. *Health Physics*, vol. 61, pp. 181–190.

Voelz, G. L., G. S. Wilkinson, J. F. Acquavella, et al. 1983. An update of epidemiologic studies of plutonium workers. *Health Physics*, vol. 44 (suppl. l), pp. 493–503.

Voillequé, P., K. Meyer, D. Schmidt, S. Rope, G. Killough, M. Case, R. Moore, B. Shleien, and J. Till. 1993. The Fernald dosimetry reconstruction project: Task 2 and 3: Radionuclide source terms and uncertainties (draft report for comment). Radiological Assessments Corporation, Neeses, South Carolina, no. CDC-5, November 1993.

von Hippel, F., D. Albright, and B. Levi. 1986. Quantities of fissile materials in U.S. and Soviet nuclear weapons arsenals. Center for Energy and Environmental Studies, Princeton University, PU/CEES no. 168, July 1986.

von Hippel, F., T. Cochran, and C. Paine. 1993. Report of an international workshop on the future of reprocessing and arrangements for the storage and disposition of already-separated plutonium (Moscow, 14–16 December 1992), and an international workshop on nuclear security problems (Kiev, 17 December 1992). Natural Resources Defense Council, Washington, D.C., 10 January 1993.

Wagoner, J. K., V. E. Archer, B. E. Carroll, D. A. Holaday, and P. A. Lawrence. 1964. Cancer mortality patterns among U.S. uranium miners and millers, 1950 through 1962. *Journal of the National Cancer Institute*, vol. 32, pp. 787–801.

Waite, D. W. et al. 1988. The effect of uranium mill tailings on radionuclide concentrations in Langley Bay, Saskatchewan, Canada. *Archives of Environmental Contamination and Toxicology*, vol. 17, pp. 373–380.

Walker, W., and F. Berkhout. 1992. Japan's plutonium problem—and Europe's. *Arms Control Today*, September 1992.

Walton, T. (Public Affairs Officer, U.S. Department of Energy). 1993. Fascimile letter to Jim Werner (at the Natural Resources Defense Council), 19 January 1993.

Wang, D. 1992. Lanzhou accelerates implementation of 'Blue Sky Plan' to bring back beautiful scenery of bright mountains, clear waters and a blue sky. *Renmin Ribao Overseas*

Edition 10 June 1992, p. 1. Translated from the Chinese in *Foreign Broadcast Information Service* JPRS-TEN-92-019, 7 October 1992, p. 10.

Warren, S. L. (Colonel). 1945. Report on Test II at Trinity, 16 July 1945. Memorandum to Major General Groves. Modern Military Branch, National Archives, 21 July 1945.

Waxweiler, R. J., R. J. Roscoe, V. E. Archer, M. J. Thun, J. K. Wagoner, and F. E. Lundin. 1981. Mortality follow-up through 1877 of the white underground uranium miners cohort examined by the United States Public Health Service. pp. 823–830. In *Radiation Hazards in Mining: Control, Measurement, and Medical Aspects*. M. Gomez, ed. New York: Society of Mining Engineers, American Institute of Mining, Metallurgical, and Petroleum Engineers.

Weart, S. 1979. *Scientists in Power*. Cambridge: Harvard University Press.

Weast, R. C., ed. 1988. *CRC Handbook of Chemistry and Physics: 69th Edition 1988–1989*. Boca Raton, FL: CRC Press, Inc.

Wells, J. 1994. Protecting Department of Energy Workers Health and Safety. Testimony before the Subcommittee on Oversight and Investigations, U.S. House of Representatives. GAO/T-RCED-94-143. Washington, D.C.: U.S. General Accounting Office.

West, G. 1991. United States nuclear warhead assembly facilities (1945–1990). Mason & Hanger-Silas Mason Co., Amarillo, Texas, March 1991.

Whittemore, A. S., and McMillan, A. (1983). Lung cancer mortality among U.S. uranium miners: A reappraisal. *Journal of the National Cancer Institute*, vol. 71, pp. 489–499.

WHO (World Health Organization). 1980. *Manual of the international statistical classification of diseases, injuries, and causes of death*. Geneva: World Health Organization.

Wiese, W. H., and B. J. Skipper. 1986. Survey of reproductive outcomes in uranium and potash mine workers: A first analysis. *Annals of the American Conference of Governmental Industrial Hygienists*, vol. 14, pp. 187–192.

Wiggs, L. D., C. A. Cox-DeVore, and G. L. Voelz. 1991. Mortality among a cohort of workers monitored for Po-210 Exposure: 1944–1972. *Health Physics*, vol. 61, pp. 71–76.

Wiggs L. D., C. A. Cox-DeVore, G. S. Wilkinson, and M. S. Reyes. 1991. Mortality among workers exposed to external ionizing radiation at a nuclear facility in Ohio. *Journal of Occupational Medicine*, vol. 33, pp. 634–637.

Wilkinson, G. S. 1991. Epidemiologic studies of nuclear and radiation workers: An overview of what is known about health risks posed by the nuclear industry. *Occupational Medicine: State of the Art Reviews*, vol. 6, pp. 715–724.

Wilkinson, G. S., and N. A. Dreyer. 1991. Leukemia among nuclear workers with protracted exposure to low-dose ionizing radiation. *Epidemiology*, vol. 2, pp. 305–309.

Wilkinson, G. S., G. L. Tietjen, L. D. Wiggns, et al. 1987. Mortality among plutonium and other radiation workers at a plutonium weapons facility. *American Journal of Epidemiology*, vol. 125, pp. 231–250.

Wilson, Ellen. 1985. *Environmental Action*. November/December 1985, pp. 28–32.

Wing, S., C. M. Shy, J. L. Wood, S. Wolf, D. L. Cragle, and E. L. Frome. 1991. Mortality among workers at Oak Ridge National Laboratory: Evidence of radiation effects in follow-up through 1984. *Journal of the American Medical Association*, vol. 265, pp. 1397–1402.

Wing, S., C. M. Shy, J. L. Wood, S. Wolf, D. L. Cragle, W. Tankersley, and E. L. Frome. 1993. Job factors, radiation and cancer mortality at Oak Ridge National Laboratory: Follow-up through 1984. *American Journal of Industrial Medicine*, vol. 23, pp. 265–279.

WISE. 1991. Soviet/French reactor deal. *WISE News Communique*, no. 355.3522, 22 June 1991.

WISE. 1993. Israel covered up leak at Dimona. *WISE News Communique*, no. 392.3825, 11 June 1993.

WISE. 1993a. Tomsk follow-up. *WISE News Communique*, no. 391.3809, 21 May 1993.

WISE. 1993b. Cogéma to acquire U-monopoly in France. *WISE News Communiqué*, no. 392, 11 June 1993, p. 8.

Woodward, A., et al. 1991. Radon daughter exposure at the Radium Hill uranium mine and lung cancer rates among former workers, 1952–87. *Cancer Causes and Control*, vol. 2, pp. 213–220.

Xinhua. 1989. [Physicist recalls nuclear weapons development.] Transcript of report broadcast by Xinhua, 1117 GMT, 14 September 1989. In *Foreign Broadcast Information Service* JPRS-TND-89-019, 6 October 1989, p. 1.

Xinhua. 1989a. [Nuclear industry makes military, civilian goods.] Transcript of report broadcast by Xinhua, 1042 GMT, 6 December 1989. In *Foreign Broadcast Information Service* JPRS-TND-90-001, 4 January 1990, p. 3.

Xinhua. 1992. [Academy research group warns of environmental deterioration.] Transcript of report broadcast by Xinhua, 1345 GMT, 11 May 1992. In *Foreign Broadcast Information Service* JPRS-TEN-92-014, 9 July 1992, p. 13.

Xu, L. 1990. Treatment and disposal of waste water containing tritium. *He Dongli Gongcheng*, vol. 2, no. 3, 10 June 1990, pp. 86–89. Translated from the Chinese in *Foreign Broadcast Information Service* JPRS-CEN-90-016, 21 December 1990, p. 23.

Xu, Y. 1992. [Feature: China battles for clean environment.] Transcript of report broadcast by Xinhua, 1337 GMT 2 October 1992. In *Foreign Broadcast Information Service* JPRS-TEN—92-020, 28 October 1992, p. 14.

Yablokov, A. V., A. E. Vorobiev, and V. I. Pakrovski. 1992. On the conditions of the health of the population of the Russian Federation in 1991 (in Russian). State lecture. Moscow, 28 August 1992.

Yakushev, V. 1993. [Tomsk complex to restore site at own expense.] Transcript of report broadcast by Moscow ITAR-TASS, 1227 GMT, 24 April 1993. Translated from the Russian in *Foreign Broadcast Information Service* JPRS-TEN-93-016, 16 June 1993, p. 41.

Yibo, L. 1993. Full control of the country's radiation environment. Urban radiation waste storage operation regular. *Zhongguo Hanjing Bao*, 6 October 1992, p. 1. Translated from the Chinese in *Foreign Broadcast Information Service* JPRS-TEN-93-006, 25 March 1993, p. 4.

Yuan, Z. 1989. Permanent nuclear waste dumps planned. *China Daily*, 20 February 1989, p. 1. Reprinted in *Foreign Broadcast Information Service* JPRS-TND-89-005, 20 March 1989. p. 2.

Zaloga, S. 1991. The Soviet nuclear bomb programme: The first decade. *Jane's Soviet Intelligence Review*, April 1991, pp. 174–181.

Zelenyy Mir. 1992. Yenisey radiation levels detailed. *Zelenyy Mir* (Moscow), May 1992, pp. 6–7. Unattributed report. Translated from the Russian in *Foreign Broadcast Information Service* JPRS-TEN-92-017, 21 September 1992, pp. 62–65.

Zerbib, J. C. 1979. Les recommandations de la CIPR et les travailleurs IAEA. IAEA SR 36/32.

Zerbib, J. C. 1983. Evaluations comparées du plutonium dispersé lors des essais nucléaires aériens et des déchets plutonifères evacués, stockés ou en attente de conditionnement dans le cycle du combustible. Société Française de Radioprotection; Journées plutonium et radioprotection, 14–16 June 1983.

Zerbib, J. C. 1985. Les rayonnements ionisants: Les risques du travail. Editions la Découverte, March 1985.

Zerbib, J. C. 1988. Pour un réexamen des données qui fondent la protection des travailleurs contre les rayonnements ionisants. CFDT-Press, Montauban, 21–23 January 1988.

Zerbib, J. C., and H. Forest. 1991. Le suivi dosimétrique des "entreprises extérieures" intervenant dans les installations nucléaires. CFDT, October 1991.

Zhan, Y. (Culture Office, Embassy of the Peoples Republic of China). 1993. Telephone conversation with D. Kershner (IEER), 16 July 1993.

Zhao, X. 1993. China's first environmental radiation monitoring station passes examination and acceptance. *Zhongguo Hunanjing Bao,* 15 May 1993, p. 1. Translated from the Chinese in *Foreign Broadcast Information Service* JPRS-TEN-93-027, 15 December 1993, p. 8.

Zhongguo Xinwen (Hong Kong). 1981. [China's research into radioactive environments in mines progresses], p. 3. Translated from the Chinese in Foreign Broadcast Information Service JPRS 79453, 16 November 1981.

Zubov, Y. 1992. The radioactive contamination of the water of the Yenisey River. Committee of Hydrometeorological and Ecological Monitoring of the Ministry of Natural Resources of the Russian Federation, Moscow, 7 April 1992.

Contributors

Alexandra Brooks is a chemical engineering student whose studies are aimed at cleaning up pollution, especially from radiation caused by nuclear power plants and nuclear weapons facilities. Formerly she was a researcher at IEER, where she contributed to both *Plutonium: Deadly Gold of the Nuclear Age* and *Radioactive Heaven and Earth*.

Martin Cherniack, M.D., M.P.H., is an associate professor of occupational and environmental medicine and of epidemiology and public health at the Yale School of Medicine. He is the project director for a long-term National Cancer Institute study of thyroid cancer in Belarus resulting from Chernobyl radiation.

Richard Clapp is an epidemiologist and assistant professor of environmental health at the Boston University School of Public Health. He has conducted studies of cancer incidence around nuclear facilities and has visited several U.S. Department of Energy sites to assess health effects in workers and surounding communities.

Albert Donnay earned his master's degree in environmental health engineering from the Johns Hopkins School of Hygiene and Public Health in 1982. He founded and from 1982 to 1990 was executive director of Nuclear Free America, an international resource center for Nuclear Free Zones. He contributed to the present book while working as a staff scientist for IEER in 1993 and is now executive director of MCS Referral & Resources in Baltimore, a project of the Chemical Injury Information Network.

Amy Hopkins, M.D., M.P.H., Ph.D., is an instructor of occupational and environmental medicine at the Yale School of Medicine. She is currently conducting research on leukemia and other health effects of radiation in Chelyabinsk, Russia.

Howard Hu, M.D., M.P.H., Sc.D., is a physician specializing in internal medicine and occupational environmental medicine. He also holds a doctoral degree in epidemiology. He is associate professor of occupational medicine and assistant professor of medicine at the Harvard schools of Medicine and Public Health. His research concentrates on environmental epidemiology. He has been director of the IPPNW Commission to Investigate the Health and Environmental Effects of Nuclear Weapons Production since 1991.

Rebecca Johnson is a consultant based in London and Geneva, specializing in security and nuclear issues. She has conducted research into U.K. nuclear weapons facilities over many years.

Ellen Kennedy is education coordinator at IEER, where she organizes technical training workshops for citizen activists and edits the quarterly newsletter *Science for Democratic Action*. She has worked extensively on immigration issues and holds a master's degree in Latin American studies from the University of California at Berkeley.

Martin Kuster, M.D., M.P.H., is an occupational medicine resident in the Occupational

Health Program, Department of Environmental Health of the Harvard School of Public Health. He is also an active member of the IPPNW affiliate in Switzerland.

Arjun Makhijani, president of the Institute for Energy and Environmental Research, holds a Ph.D. in engineering from the University of California at Berkeley. He has a Bachelor of Engineering (Electrical) degree from the University of Bombay in India. He has produced studies on nuclear fuel cycle–related issues, including weapons production, testing, and nuclear waste, for well over a decade. He has served on the Radiation Advisory Committee of the Science Advisory Board of the U.S. Environmental Protection Agency, and he is a member of the EPA's advisory subcommittee on Radiation Cleanup Standards of the National Advisory Committee on Environmental Policy and Technology. He has written extensively on ozone layer protection and is the coauthor of three other books on nuclear weapons–related and nuclear waste issues.

William Peden for the last 10 years has been a consultant to Greenpeace, the British American Security Information Council (BASIC), and local authorities and emergency planning offices on a wide range of nuclear issues. He is the author or coauthor of many weapons-related reports, including BASIC reports on nuclear weapons safety and Greenpeace's "Inside the Citadel."

A. James Ruttenber is an associate professor of preventive medicine and biometrics at the University of Colorado School of Medicine. He holds doctorates in ecology and medicine and specializes in the field of environmental and occupational health. He has worked as a medical epidemiologist at the Centers for Disease Control and Prevention, where he investigated public exposures to radiation and chemicals. He is the principal investigator for a multi-city case-control study of the relation between childhood leukemia and parental exposure to ionizing radiation, and for a cancer-incidence study of workers at the Rocky Flats nuclear weapons facility.

Scott Saleska is author or coauthor of numerous studies dealing with radioactive waste and environmental problems at nuclear weapons facilities. He has a bachelor's degree from MIT and is currently in the graduate program of the Energy and Resources Group at the University of California at Berkeley. He is a member of the Sierra Club committee on military impacts on the environment.

David Sumner obtained a B.Sc. in physics at Imperial College, London, and a D.Phil. in high energy nuclear physics at the University of Oxford. After a period as lecturer in physics at Makerere University in Uganda, he took up an appointment with the Department of Clinical Physics and Bio-Engineering, West of Scotland Health Boards. Until 1990 he was based at the Department of Nuclear Medicine, Stobhill General Hospital, Glasgow. He is now a senior research fellow in the Department of Medicine and Therapeutics, University of Glasgow, and a tutor with the Open University.

Alistair Woodward is a physician who has specialized in public health and epidemiology. He led a research team investigating the health effects of uranium mining in South Australia and is now professor of public health at the Wellington School of Medicine, New Zealand.

Annalee Yassi is a physician specializing in occupational and environmental medicine. She is an associate professor and director of the Occupational and Environmental Health Unit at the University of Manitoba and has worked on various environmental impact assessments. She is currently a member of the panel assessing uranium mine developments in northern Saskatchewan.

Katherine Yih is an ecologist and received her Ph.D. in biology from the University of Michigan in 1982. She has done research and writing for several research institutes and other nongovernmental organizations in the United States and abroad. Since 1990, she has coordinated IPPNW's Commission to Investigate the Health and Environmental Effects of Nuclear Weapons Production.

About IPPNW and IEER

IPPNW

The International Physicians for the Prevention of Nuclear War (IPPNW) is a federation of national physicians' groups in 78 countries committed to the abolition of nuclear weapons. IPPNW received the Nobel Peace Prize in 1985 for public education on the medical effects of nuclear warfare.

IPPNW, 128 Rogers St., Cambridge, MA 02142
Telephone: (617) 868-5050, Fax: (617) 868-2560

IEER

The Institute for Energy and Environmental Research (IEER) provides the public and policy-makers with thoughtful, clear, and sound scientific studies on a wide range of issues. IEER's aim is to bring scientific excellence to policy issues to promote the democratization of science and a safer, healthier environment.

IEER, 6935 Laurel Ave., Takoma Park, MD 20912
Telephone: (301) 270-5500, Fax: (301) 270-3029

Index

FRANKLIN PIERCE COLLEGE LIBRARY
00085998